METHODS IN MOLECULAR BIOLOGY

For further volumes:
http://www.springer.com/series/7651

For over 35 years, biological scientists have come to rely on the research protocols and methodologies in the critically acclaimed *Methods in Molecular Biology* series. The series was the first to introduce the step-by-step protocols approach that has become the standard in all biomedical protocol publishing. Each protocol is provided in readily-reproducible step-by-step fashion, opening with an introductory overview, a list of the materials and reagents needed to complete the experiment, and followed by a detailed procedure that is supported with a helpful notes section offering tips and tricks of the trade as well as troubleshooting advice. These hallmark features were introduced by series editor Dr. John Walker and constitute the key ingredient in each and every volume of the *Methods in Molecular Biology* series. Tested and trusted, comprehensive and reliable, all protocols from the series are indexed in PubMed.

Drosophila Oogenesis

Methods and Protocols

Edited by

Michelle S. Giedt and Tina L. Tootle

Department of Anatomy & Cell Biology, Carver College of Medicine, University of Iowa, Iowa City, IA, USA

Humana Press

Editors
Michelle S. Giedt
Department of Anatomy &
Cell Biology, Carver College
of Medicine
University of Iowa
Iowa City, IA, USA

Tina L. Tootle
Department of Anatomy &
Cell Biology, Carver College
of Medicine
University of Iowa
Iowa City, IA, USA

ISSN 1064-3745 ISSN 1940-6029 (electronic)
Methods in Molecular Biology
ISBN 978-1-0716-2969-7 ISBN 978-1-0716-2970-3 (eBook)
https://doi.org/10.1007/978-1-0716-2970-3

This Humana imprint is published by the registered company Springer Science+Business Media, LLC, part of Springer Nature.
The registered company address is: 1 New York Plaza, New York, NY 10004, U.S.A.

Preface

Drosophila melanogaster is a robust model organism for genetic studies of conserved mechanisms controlling cell biology, development, tissue homeostasis, and cellular stress. One tissue that allows all of these processes to be studied is the ovary. In this book, we have compiled a series of methods to explore fly oogenesis. We aimed to address the needs of both those new to and experts in the field. We start with a review discussing the broad utility of *Drosophila* oogenesis as a model. Then, we provide updated protocols from isolating and preparing the ovary to numerous imaging techniques and genetic protocols for cell-specific assessment and CRISPR-mediated mutagenesis. We then cover approaches and/or tools to assess cytoskeletal structures, characterize cell cycle transitions, analyze cell migration, live image morphogenic events, visualize lipid droplets, and assess ovulation. We continue with chapters addressing methods for detecting *Wolbachia* infection, measuring transposon activity, single cell RNA sequencing, chromatin immunoprecipitation, characterizing nuclear actin levels, and uncovering protein-protein interactions. We end with chapters on how *Drosophila* oogenesis can be used in the classroom and in outreach programs to increase interest in biomedical research, STEM education, and STEM careers.

Iowa City, IA, USA

Michelle S. Giedt
Tina L. Tootle

Contents

Contributors

ELIZABETH T. ABLES • *Department of Biology, East Carolina University, Greenville, NC, USA*
CSILLA ABONYI • *Eötvös Loránd Research Network (ELKH), Biological Research Centre, Szeged, Hungary*
CHRISTOPHER P. AICHELE • *Immersion Science Program, Fox Chase Cancer Center, Philadelphia, PA, USA; eCLOSE Institute, Huntingdon Valley, PA, USA*
JENNIFER I. ALEXANDER • *Immersion Science Program, Fox Chase Cancer Center, Philadelphia, PA, USA; eCLOSE Institute, Huntingdon Valley, PA, USA*
MITCHELL T. ANDERSON • *Department of Molecular Genetics and Cell Biology, The University of Chicago, Chicago, IL, USA; Committee on Development, Regeneration, and Stem Cell Biology, The University of Chicago, Chicago, IL, USA*
NICHOLAS F. ARCHER • *Immersion Science Program, Fox Chase Cancer Center, Philadelphia, PA, USA*
Y. JAMES AYKIT • *Immersion Science Program, Fox Chase Cancer Center, Philadelphia, PA, USA*
ANDREW BEARD • *Department of Physiology & Neurobiology, University of Connecticut, Storrs, CT, USA*
ERNEST BOAMAH • *Immersion Science Program, Fox Chase Cancer Center, Philadelphia, PA, USA*
ADAM J. BOLOTSKY • *Immersion Science Program, Fox Chase Cancer Center, Philadelphia, PA, USA*
PÉTER BORKÚTI • *Eötvös Loránd Research Network (ELKH), Biological Research Centre, Szeged, Hungary; Doctoral School of Multidisciplinary Medical Science, University of Szeged, Szeged, Hungary*
MEGAN BOYLE • *Immersion Science Program, Fox Chase Cancer Center, Philadelphia, PA, USA*
AMANDA J. BROWNE • *Immersion Science Program, Fox Chase Cancer Center, Philadelphia, PA, USA*
ALEJANDRO CAMPOY-LOPEZ • *Centro Andaluz de Biología del Desarrollo, CSIC/Universidad Pablo de Olavide/JA, Sevilla, Spain*
YUJUN CHEN • *Division of Biology, Kansas State University, Manhattan, KS, USA*
ILIANA CORREA • *Immersion Science Program, Fox Chase Cancer Center, Philadelphia, PA, USA; eCLOSE Institute, Huntingdon Valley, PA, USA*
JAYSON A. CORTES • *Biology Department, Arcadia University, Glenside, PA, USA*
JENNIFER CRIOLLO • *Immersion Science Program, Fox Chase Cancer Center, Philadelphia, PA, USA*
GABRIELA CRUZ • *Immersion Science Program, Fox Chase Cancer Center, Philadelphia, PA, USA*
TINGTING DUAN • *Department of Biochemistry, Carver College of Medicine, University of Iowa, Iowa City, IA, USA*
EBONY R. DYSON • *Immersion Science Program, Fox Chase Cancer Center, Philadelphia, PA, USA; eCLOSE Institute, Huntingdon Valley, PA, USA*

Oren Eldor • *Immersion Science Program, Fox Chase Cancer Center, Philadelphia, PA, USA*
Valerie N. Fortuona • *Immersion Science Program, Fox Chase Cancer Center, Philadelphia, PA, USA; eCLOSE Institute, Huntingdon Valley, PA, USA*
Maya J. Foster • *Immersion Science Program, Fox Chase Cancer Center, Philadelphia, PA, USA*
Isabela M. Fuentes • *Immersion Science Program, Fox Chase Cancer Center, Philadelphia, PA, USA*
Katherine C. Gallen • *Immersion Science Program, Fox Chase Cancer Center, Philadelphia, PA, USA; eCLOSE Institute, Huntingdon Valley, PA, USA*
Pamela K. Geyer • *Department of Biochemistry, Carver College of Medicine, University of Iowa, Iowa City, IA, USA*
Michelle S. Giedt • *Department of Anatomy and Cell Biology, Carver College of Medicine, University of Iowa, Iowa City, IA, USA*
Acaimo González-Reyes • *Centro Andaluz de Biología del Desarrollo, CSIC/Universidad Pablo de Olavide/JA, Sevilla, Spain*
Shreeya D. Gounder • *Immersion Science Program, Fox Chase Cancer Center, Philadelphia, PA, USA*
Nyim Greenwood • *Immersion Science Program, Fox Chase Cancer Center, Philadelphia, PA, USA*
Raymond Habas • *Department of Biology, Temple University, Philadelphia, PA, USA*
Nicole Harrington • *Immersion Science Program, Fox Chase Cancer Center, Philadelphia, PA, USA*
Ryan A. Henry • *Immersion Science Program, Fox Chase Cancer Center, Philadelphia, PA, USA; Wilkes University, Wilkes-Barre, PA, USA*
Troy J. Holiday • *Immersion Science Program, Fox Chase Cancer Center, Philadelphia, PA, USA*
Sally Horne-Badovinac • *Department of Molecular Genetics and Cell Biology, The University of Chicago, Chicago, IL, USA; Committee on Development, Regeneration, and Stem Cell Biology, The University of Chicago, Chicago, IL, USA*
Jasmin Imran Alsous • *Flatiron Institute, Simons Foundation, New York, NY, USA*
Jonathan A. Jackson • *Department of Biology, Massachusetts Institute of Technology, Cambridge, MA, USA; Graduate Program in Biophysics, Harvard University, Cambridge, MA, USA*
Jennifer C. Jemc • *Department of Biology, Loyola University Chicago, Chicago, IL, USA*
Kayla W. Ji • *Immersion Science Program, Fox Chase Cancer Center, Philadelphia, PA, USA*
Aminah Johnson • *Immersion Science Program, Fox Chase Cancer Center, Philadelphia, PA, USA; eCLOSE Institute, Huntingdon Valley, PA, USA*
Nirupama Kotian • *Division of Biology, Kansas State University, Manhattan, KS, USA*
Zoltán Kovács • *Eötvös Loránd Research Network (ELKH), Biological Research Centre, Szeged, Hungary; Doctoral School of Multidisciplinary Medical Science, University of Szeged, Szeged, Hungary*
Ildikó Kristó • *Eötvös Loránd Research Network (ELKH), Biological Research Centre, Szeged, Hungary*
Sophie Lara • *Immersion Science Program, Fox Chase Cancer Center, Philadelphia, PA, USA*
Eric H. Lee • *Immersion Science Program, Fox Chase Cancer Center, Philadelphia, PA, USA*

JI-LONG LIU • *School of Life Science and Technology, ShanghaiTech University, Shanghai, China; Department of Physiology, Anatomy and Genetics, University of Oxford, Oxford, UK*

SHEILA M. LONGO • *Immersion Science Program, Fox Chase Cancer Center, Philadelphia, PA, USA*

LAUREL D. LORENZ • *Immersion Science Program, Fox Chase Cancer Center, Philadelphia, PA, USA*

MIRIAM MARÍN-MENGUIANO • *Centro Andaluz de Biología del Desarrollo, CSIC/Universidad Pablo de Olavide/JA, Sevilla, Spain*

NYELLIS MARMOL • *Immersion Science Program, Fox Chase Cancer Center, Philadelphia, PA, USA*

ADAM C. MARTIN • *Department of Biology, Massachusetts Institute of Technology, Cambridge, MA, USA*

ESTEBAN MARTINEZ • *Immersion Science Program, Fox Chase Cancer Center, Philadelphia, PA, USA*

JOCELYN A. MCDONALD • *Division of Biology, Kansas State University, Manhattan, KS, USA*

JULIE A. MERKLE • *Department of Biology, University of Evansville, Evansville, IN, USA*

NATHANIEL MEYER • *Department of Anatomy, School of Medicine, University of California, San Francisco, CA, USA*

EMLY MOISE • *eCLOSE Institute, Huntingdon Valley, PA, USA*

BRIANNA MONTANEZ • *eCLOSE Institute, Huntingdon Valley, PA, USA*

BENJAMIN C. NEUBERT • *Immersion Science Program, Fox Chase Cancer Center, Philadelphia, PA, USA*

LINDSAY B. M. NEVALAINEN • *Department of Biology, Indiana University, Bloomington, IN, USA*

IRENE L. G. NEWTON • *Department of Biology, Indiana University, Bloomington, IN, USA*

ANNA NIKONOVA • *Immersion Science Program, Fox Chase Cancer Center, Philadelphia, PA, USA*

TODD NYSTUL • *Department of Anatomy, School of Medicine, University of California, San Francisco, CA, USA*

ALANA M. O'REILLY • *Immersion Science Program, Fox Chase Cancer Center, Philadelphia, PA, USA; eCLOSE Institute, Huntingdon Valley, PA, USA*

REBECCA ORAMAS • *Department of Physiology & Neurobiology, University of Connecticut, Storrs, CT, USA*

JOBELLE PERALTA • *Department of Anatomy, School of Medicine, University of California, San Francisco, CA, USA*

CHRISTOPHER PETIT • *Department of Biology, Loyola University Chicago, Chicago, IL, USA*

DANIEL N. PHIPPS • *Department of Biology, East Carolina University, Greenville, NC, USA; Biomedical Sciences Graduate Program, University of Virginia, Charlottesville, VA, USA*

AMANDA M. POWELL • *Department of Biology, East Carolina University, Greenville, NC, USA*

MARIA RADU • *Immersion Science Program, Fox Chase Cancer Center, Philadelphia, PA, USA*

DEVON J. RIGGS • *Immersion Science Program, Fox Chase Cancer Center, Philadelphia, PA, USA; eCLOSE Institute, Huntingdon Valley, PA, USA*

Felipe Rodriguez-Tirado • *Department of Biochemistry and Department of Molecular Physiology and Biophysics, Carver College of Medicine, University of Iowa, Iowa City, IA, USA*
Alicia E. Rosales-Nieves • *Centro Andaluz de Biología del Desarrollo, CSIC/Universidad Pablo de Olavide/JA, Sevilla, Spain; Instituto de Biomedicina de Sevilla (IBiS), Hospital Universitario Virgen del Rocío/CSIC/Universidad de Sevilla, Sevilla, Spain*
Dara M. Ruiz-Whalen • *Immersion Science Program, Fox Chase Cancer Center, Philadelphia, PA, USA; eCLOSE Institute, Huntingdon Valley, PA, USA*
Savannah Ryan • *Department of Biology, Loyola University Chicago, Chicago, IL, USA*
Jasmine A. Saunders • *Immersion Science Program, Fox Chase Cancer Center, Philadelphia, PA, USA*
Maxwell Saurman • *Immersion Science Program, Fox Chase Cancer Center, Philadelphia, PA, USA*
Ava M. Self • *Immersion Science Program, Fox Chase Cancer Center, Philadelphia, PA, USA*
Kristin Sherrard • *Department of Molecular Genetics and Cell Biology, The University of Chicago, Chicago, IL, USA*
Jacqueline C. Simonet • *Immersion Science Program, Fox Chase Cancer Center, Philadelphia, PA, USA; Arcadia University, Glenside, PA, USA*
Tanu Singh • *Immersion Science Program, Fox Chase Cancer Center, Philadelphia, PA, USA*
Daniel R. Smith • *Immersion Science Program, Fox Chase Cancer Center, Philadelphia, PA, USA*
Shannon C. Smith • *Immersion Science Program, Fox Chase Cancer Center, Philadelphia, PA, USA*
Maria Sokolova • *Institute of Biotechnology, University of Helsinki, Helsinki, Finland*
Adrianna Soriano • *Department of Biology, Loyola University Chicago, Chicago, IL, USA; Houston Baptist University, Houston, TX, USA*
Catherine A. Stapf • *Immersion Science Program, Fox Chase Cancer Center, Philadelphia, PA, USA*
Jennifer V. Stark • *Immersion Science Program, Fox Chase Cancer Center, Philadelphia, PA, USA*
Lindsay A. Steele • *Immersion Science Program, Fox Chase Cancer Center, Philadelphia, PA, USA*
Todd I. Strochlic • *Immersion Science Program, Fox Chase Cancer Center, Philadelphia, PA, USA; Department of Biochemistry and Molecular Biology, Drexel University, Philadelphia, PA, USA*
Jianjun Sun • *Department of Physiology & Neurobiology, University of Connecticut, Storrs, CT, USA; Institute for Systems Genomics, University of Connecticut, Storrs, CT, USA*
Christina I. Swanson • *Biology Department, Arcadia University, Glenside, PA, USA*
Anikó Szabó • *Eötvös Loránd Research Network (ELKH), Biological Research Centre, Szeged, Hungary*

Szilárd Szikora • *Eötvös Loránd Research Network (ELKH), Biological Research Centre, Szeged, Hungary*
Tarang Tondapu • *Immersion Science Program, Fox Chase Cancer Center, Philadelphia, PA, USA*

TINA L. TOOTLE • *Department of Anatomy and Cell Biology, Carver College of Medicine, University of Iowa, Iowa City, IA, USA*
CHRISTINA TSIOBIKAS • *Immersion Science Program, Fox Chase Cancer Center, Philadelphia, PA, USA*
MARIA K. VARTIAINEN • *Institute of Biotechnology, University of Helsinki, Helsinki, Finland*
ERIN M. VENTRESCA • *Immersion Science Program, Fox Chase Cancer Center, Philadelphia, PA, USA; Albright College, Reading, PA, USA*
PÉTER VILMOS • *Eötvös Loránd Research Network (ELKH), Biological Research Centre, Szeged, Hungary*
LESLEY N. WEAVER • *Department of Biology, Indiana University, Bloomington, IN, USA*
MICHAEL A. WELTE • *Department of Biology, University of Rochester, Rochester, NY, USA*
ROGER P. WHITE • *Department of Biology, University of Rochester, Rochester, NY, USA*
ANNA E. WILLIAMS • *Department of Biology, East Carolina University, Greenville, NC, USA; Biochemistry, Cell & Developmental Biology Graduate Program, Emory University, Atlanta, GA, USA*
AUDREY MILLER WILLIAMS • *Department of Molecular Genetics and Cell Biology, The University of Chicago, Chicago, IL, USA*
ALLISON L. ZAJAC • *Department of Molecular Genetics and Cell Biology, The University of Chicago, Chicago, IL, USA*
BO ZHANG • *School of Life Science and Technology, ShanghaiTech University, Shanghai, China; Institute of Biochemistry and Cell Biology, Shanghai Institutes for Biological Sciences, Chinese Academy of Sciences, Shanghai, China; University of Chinese Academy of Sciences, Beijing, China*
YUANBING ZHANG • *School of Life Science and Technology, ShanghaiTech University, Shanghai, China; Institute of Biochemistry and Cell Biology, Shanghai Institutes for Biological Sciences, Chinese Academy of Sciences, Shanghai, China; University of Chinese Academy of Sciences, Beijing, China*

Chapter 1

The Vast Utility of *Drosophila* Oogenesis

Michelle S. Giedt and Tina L. Tootle

Abstract

In this chapter, we highlight examples of the diverse array of developmental, cellular, and biochemical insights that can be gained by using *Drosophila melanogaster* oogenesis as a model tissue. We begin with an overview of ovary development and adult oogenesis. Then we summarize how the adult *Drosophila* ovary continues to advance our understanding of stem cells, cell cycle, cell migration, cytoplasmic streaming, nurse cell dumping, and cell death. We also review emerging areas of study, including the roles of lipid droplets, ribosomes, and nuclear actin in egg development. Finally, we conclude by discussing the growing conservation of processes and signaling pathways that regulate oogenesis and female reproduction from flies to humans.

Key words *Drosophila*, Oogenesis, Stem cells, Reproduction, Migration, Cell cycle, Cell death, Lipid droplets, Ovulation

1 Overview

Drosophila oogenesis has been, and continues to be, a useful tool to answer questions in cell biology, development, tissue homeostasis, and disease. Further, while morphologically distinct from vertebrate ovaries, many of the cellular and signaling mechanisms driving oogenesis in mammals also take place in the fly, making it an emerging model for studying female reproduction [1].

The utility of the *Drosophila* ovary lies in the presence of multiple cell types, processes, and an organized ovarian structure that permits simultaneous observation of spatiotemporal events across the whole process of oogenesis. Additionally, oogenesis depends upon basic cell biological processes including cell division, signaling, cell migration, and morphogenesis to create a mature egg capable of fertilization. This variety of cell types and biological processes makes the fly ovary a tremendous resource for investigating an array of questions. Further benefits of the system come from its low cost, ease of maintenance, the wide availability of mutants, and robust genetic tools, including genetic mosaic analysis and the

Michelle S. Giedt and Tina L. Tootle (eds.), *Drosophila Oogenesis: Methods and Protocols*,
Methods in Molecular Biology, vol. 2626, https://doi.org/10.1007/978-1-0716-2970-3_1,

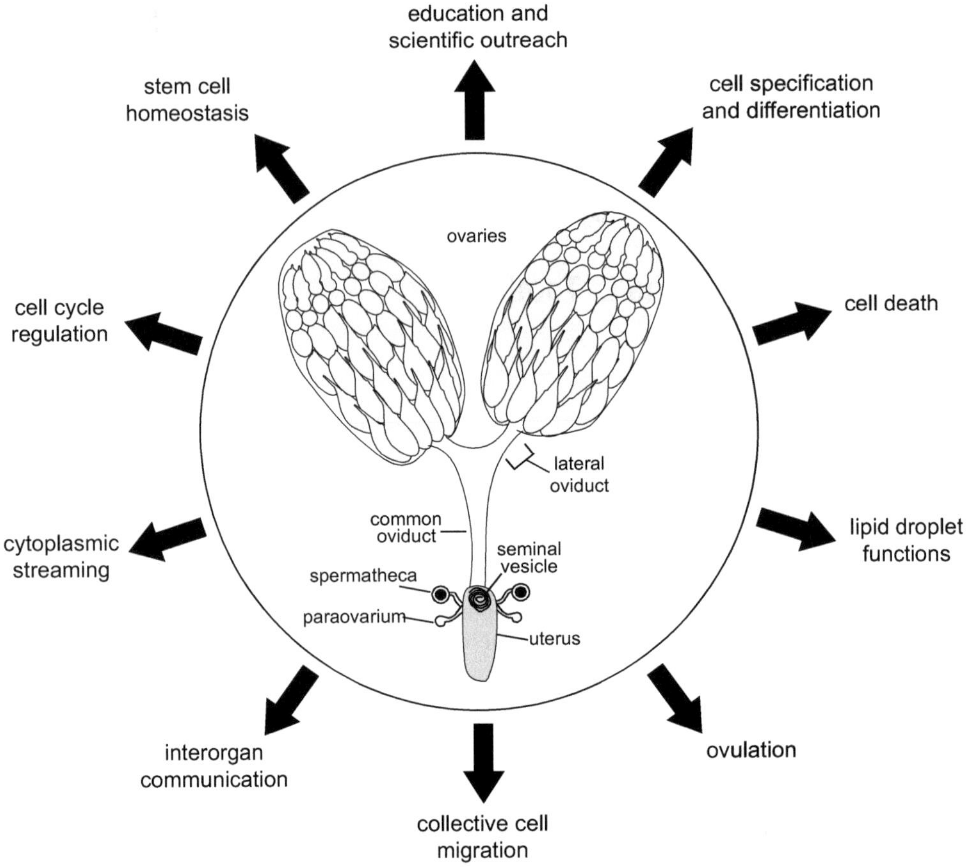

Fig. 1 The utility of the *Drosophila* ovary as a model tissue. Schematic illustrating the *Drosophila* female reproductive tissues and the numerous conserved cellular processes used during *Drosophila* oogenesis. Thus, this tissue is a valuable tool for advancing our understanding of and uncovering the underlying mechanisms driving these processes

ability to overexpress or knockdown genes in specific cells at particular times during development. Additionally, techniques including long-term live imaging, Clustered Regularly Interspaced Short Palindromic Repeats (CRISPR) gene tagging and mutations, single cell RNA sequencing (scRNA-seq), chromatin immunoprecipitation sequencing (ChIP-seq), and proteomic studies have successfully been used to advance understanding of the processes during *Drosophila* oogenesis.

In this chapter, we highlight the utility of *Drosophila* oogenesis as a model (Fig. 1). First, we provide an overview of early ovary development from embryonic gonad coalescence through larval ovary morphogenesis to pupal ovary formation. Then we describe the key events during the 14 stages of adult follicle development. Next, we summarize how *Drosophila* oogenesis has been used to advance the field's understanding of: stem cell maintenance and differentiation using the germline stem cells (GSCs) as an example;

a diverse array of cell cycles from mitosis and meiosis to endocycling and gene amplification; collective cell migrations, including follicle elongation, border cell migration, and dorsal appendage formation; cytoplasmic streaming to establish embryonic polarity; nurse cell dumping; and cell death. Further, we discuss emerging areas of study, including the roles of lipid droplets, ribosome biogenesis and heterogeneity, and nuclear actin in follicle development. We end by highlighting how, in the last decade, *Drosophila* oogenesis has become a recognized model for understanding conserved principles of female reproduction, such as interorgan communication and ovulation.

2 Early Ovary Development

The cell types and events leading to the production of a mature ovary are conserved between flies and mammals [1–3]. Both require the specification and migration of special germline and somatic cell populations. These cells come together and, as development proceeds, undergo rearrangements to form the adult ovary.

The origins of the *Drosophila* ovary are determined during oogenesis in the mother, when a highly specialized, maternally supplied cytoplasm is sequestered at the posterior of the oocyte. This cytoplasm, termed germplasm, is required for germ cell specification and contains transcription factors, proteins, and mRNAs necessary for germline specification such as *oskar*, *vasa*, *tudor*, and *aubergine* [4, 5]. The germplasm is crucial during embryogenesis for specification of the primordial germ cells (PGCs).

Two classes of cells, the PGCs and somatic gonadal precursors (SGPs), are necessary for ovary development and oogenesis [4, 6–8]. These cells arise at different times and locations during embryogenesis. The PGCs appear in the germplasm during stages 4–5 of embryogenesis. The PGCs migrate along and through the posterior midgut until they reach the mesoderm. In contrast, the SGPs develop as clusters of cells in the mesodermal segments during stage 11 of embryogenesis. At stage 12, the PGCs migrate toward the SGPs where they form two nascent gonads on either side of the embryo. Migration and coalescence of primordial germline and somatic cell populations also occur during mammalian development [9, 10].

The *Drosophila* ovary proper begins to form during larval development [7, 11]. Initially, the PGCs and SGPs proliferate, growing the nascent gonads over ~2 days, with some somatic cells taking on the fate of intermingled cells, which begin to wrap around the germ cells [11]. Additional somatic cell types arise throughout larval development. Indeed, scRNA-seq analysis reveals that the third instar larval ovary is comprised of seven types of ovary progenitor cells [12]. In addition to germline cells located at the center of the larval ovary, a group of sheath cells are present at the anterior tip and eventually give rise to the ovarioles in the adult ovary [7]. Two other groups of cells, the terminal filament and cap cells, form the future GSC niche [13]. The fifth cell type is the intermingled cells which wrap the germline cells and regulate larval germline proliferation [14]. The swarm cells (also known as the basal cells) are a group of cells of unknown function that initially form at the anterior of the gonad and then migrate to the posterior of the developing ovary [11, 15]. The seventh group of ovary progenitor cells identified by scRNA-seq is a pool of follicle stem cell (FSC) precursors, which occupy the region between the intermingled cells and the swarm cells [12].

During larval ovary development, the nascent gonad goes from round to oval and ovariole specification begins (Fig. 2a) [7, 11]. The first step in ovariole formation is the differentiation of the terminal filament cells. Stacks of 8–10 of these cells form the terminal filament (TF) at the anterior end of each ovariole, producing 16–20 TFs per ovary. The terminal filament cells induce

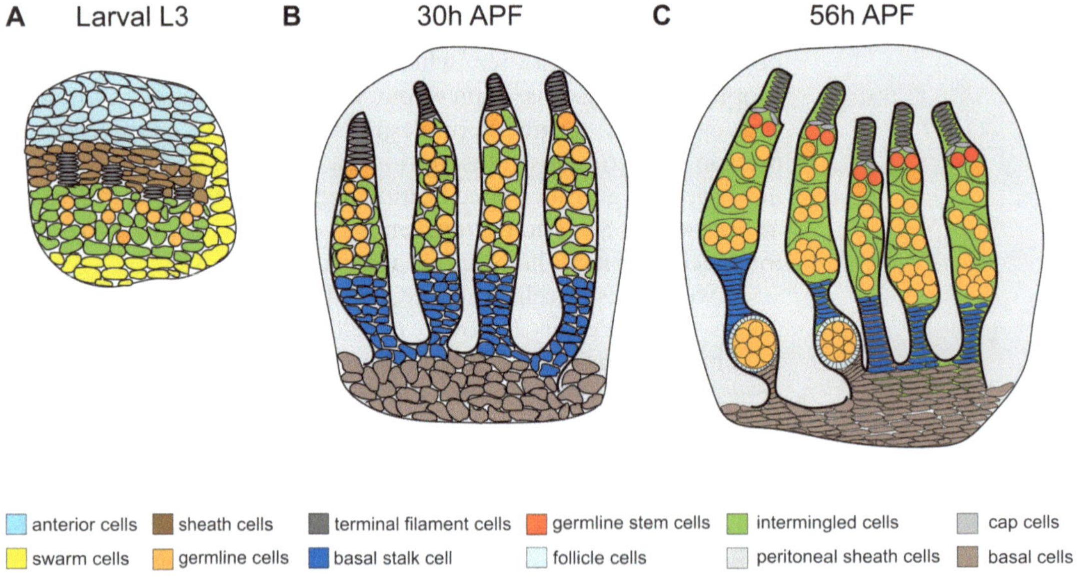

Fig. 2 Larval and pupal ovary development. (**a**) Schematic of a third instar larval ovary. Anterior cells = light blue. Sheath cells = brown. Terminal filament cells = dark gray. Germline cells = orange. Intermingled cells = green. Swarm cells = yellow. (**b**, **c**) Schematics of pupal ovaries at 30 and 56 h (h) after pupal formation (APF). Terminal filament cells = dark gray. Cap cells = light gray. GSCs = dark orange. Germline cells = orange. Intermingled cells = green. Basal stalk cells = dark blue. Basal cells = brown. Follicle cells = light blue. Peritoneal sheath cells are represented by light gray fill. In the third instar larvae, each ovary is a ball of cells that begin to undergo morphogenesis into the adult structure; specifically, the terminal filaments begin to form, delineating the future ovarioles (**a**). These ovarioles continue to form throughout the pupal stage. By 30 h APF, ovarioles are distinct and separated by the tunica propria (**b**). At 56 h APF, the germaria are more defined and the first follicles arise at the posterior of individual germarium, and the space between ovarioles begins to decrease (**c**)

intermingled cells to differentiate into cap cells, The cap cells then recruit the adjacent PGCs to differentiate into GSCs. PGCs distant from the TF and cap cells differentiate to produce the first germline cystoblasts. Apical cells then migrate posteriorly between TFs and their attached GSCs, cystoblasts, and associated somatic cells and lay down a basement membrane called the tunica propria; this separates the gonad into tube structures that will become the adult ovarioles.

The larval ovary undergoes morphogenesis during the pupal stage (Fig. 2b, c). Between 12 and 24 h after pupal formation germaria develop, including the specification of somatic lineages including the escort cells (ECs), FSCs, and differentiating follicle cells [7, 16]. For details on the structure of the germarium, see below and Fig. 3. As pupation progresses, oogenesis proceeds, and by 56 h after the onset of pupation, the first follicle (also referred to as an egg chamber) has budded off the germarium (Fig. 2c). Follicles continue to bud off individual germarium and mature throughout pupal development. For detail on the events occurring during pupal ovariole morphogenesis leading to the formation of the first follicles, please *see* Refs. [7, 16].

3 The Adult Ovary

Drosophila adult females possess a pair of ovaries located in the abdomen. Each ovary is connected to the uterus by oviducts, through which mature eggs pass. Attached to the uterus are sperm storage organs—spermatheca and the seminal vesicle—and secretory organs termed paraovaria (*see* Fig. 1). For more information, please refer to the following paper and reviews [17–19]. Each ovary is composed of 16–20 ovarioles, which are chains of sequentially maturing egg chambers or follicles. In each ovariole, a germarium, which houses the stem cell populations, is at the anterior, and chains of follicles are arranged in order of increasing maturity (Fig. 4) [20]. Oogenesis consists of 14 stages of follicle development. Each follicle is composed of a layer of somatic epithelial cells, termed follicle cells, surrounding sixteen germline-derived cells; fifteen of these are support cells, termed nurse cells, and the other is the oocyte [20].

The germarium, a specialized region at the anterior of each ovariole, contains two stem cell populations, the GSCs and the FSCs that are in separate niches and are required for follicle formation (Fig. 3a). The germarium has four regions—region 1, region 2a, region 2b, and region 3—going from its anterior to its posterior. Region 1 contains the GSC niche which is composed of somatic terminal filament and cap cells. The GSCs contact the cap cells and divide asymmetrically to form a self-renewing GSC and a daughter cell termed a cystoblast [20, 21]. Contributing to this

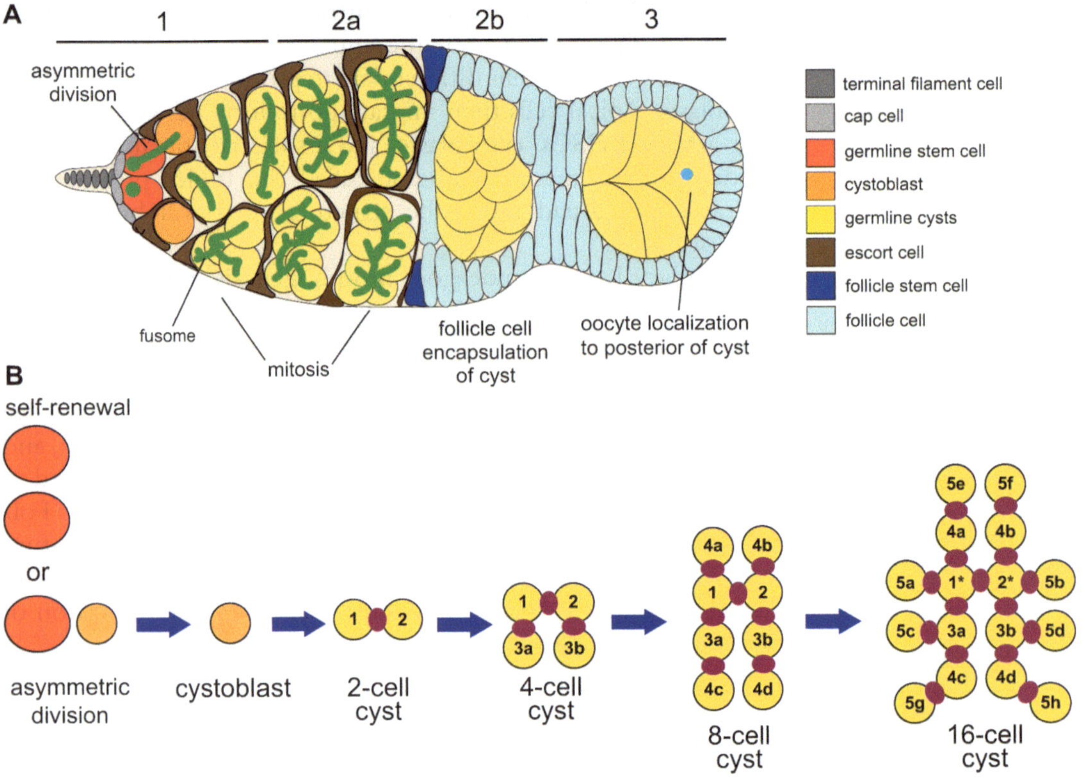

Fig. 3 The germarium and cyst division. (**a**) Diagram of a germarium, with key structures and events indicated. (**b**) Diagram of GSC and cyst division. Terminal filament cells = dark gray. Cap cells = light gray. Escort cells = brown. Germline stem cells (GSCs) = dark orange. Cystoblast = lighter orange. Cysts = yellow. Spectrosome and fusome = green. Cystoblast = lighter orange and cysts = lighter shades of orange. Follicle stem cells (FSCs) = dark blue. Follicle cells = light blue. Ring canals = maroon. The germarium possesses four regions (**a**). Region 1 extends from the GSC niche through the 4-cell cysts. Region 2a includes the 8- and 16-cell cysts. Throughout regions 1 and 2a, each cyst is surrounded by projections from the escort cells. At the 2a/2b boundary, the escort cells handoff the cysts to the follicle cells, which are produced by the two FSCs. In region 2b, the follicle cells encapsulate the 16-cell cyst and the oocyte is specified. In region 3, the oocyte localizes to the posterior of the cyst. Stem cells can either divide symmetrically to maintain the stem cell pool or undergo asymmetric division to produce a daughter cell, termed a cystoblast (**b**). The cystoblast undergoes four rounds of mitosis to produce a 16-cell cyst of interconnected cells. The cyst cells are connected via structures termed ring canals. The number of ring canals per germline cell varies from 1 to 4, with only two cells within the cyst possessing four ring canals (asterisks in 16-cell cyst). Of these two cells, one will be specified as the oocyte

asymmetric division is the GSC spectrosome, a membrane and cytoskeletal-based structure. During mitosis, the spectrosome localizes to the niche side of the mitotic spindle [22], such that after division one cell remains in contact with the cap cells, retains a larger portion of the spectrosome, and maintains stem cell identity, while the other cell orients away from the niche and differentiates (Fig. 3a). The differentiating germ cell then undergoes a series of four synchronous but incomplete mitotic divisions to ultimately

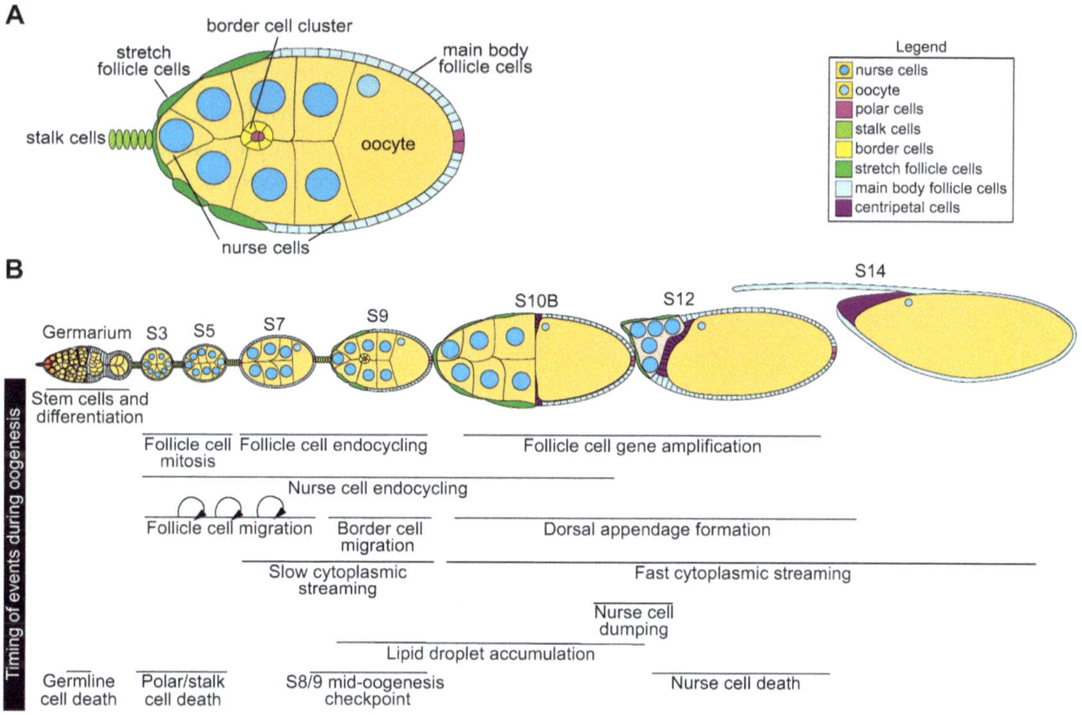

Fig. 4 The *Drosophila* follicle and an ovariole illustrating the developmental timing of events occurring during oogenesis. (**a**) A schematic of a stage 9 follicle, illustrating the various cell types that comprise a *Drosophila* follicle. (**b**) Schematic of a *Drosophila* ovariole and the developmental timing of the cellular events discussed in this chapter. Germline stem cells (GSCs) = dark orange. Nurse cells = yellow with blue nuclei. Oocytes = yellow with cyan nuclei. Escort cells = brown. Follicle stem cells (FSCs) = dark blue. Follicle cells = light blue. Stalk cells = neon green. Polar cells = pink. Border cells = bright yellow. Stretch follicle cells = green. Centripetal cells = purple. For details on events and cell-types within the germarium, please refer to Fig. 3. Each *Drosophila* follicle is comprised of 16 germline cells—15 nurse cells and one oocyte—and a layer of somatic follicle cells (**a**). The follicle cells differentiate into specific subtypes, including stalk cells, polar cells, border cells, squamous stretch follicle cells, and main body follicle cells. *Drosophila* oogenesis requires multiple cell biological processes at specific developmental stages to produce a mature egg (**b**). This temporal precision makes this system a robust model to study both the regulation of and the underlying mechanisms driving these processes

produce a cyst of sixteen interconnected germline cells by region 2a (Fig. 3) [20, 21]. Connecting the cysts cells are ring canals, remnants of the cytokinetic furrows, which permit communication between the cyst cells [23, 24]. Extending through the ring canals is the fusome, the differentiated and branched extension of the spectrosome [25–28]. By region 2b, the 16-cell cyst differentiates into fifteen nurse cells and one oocyte, which is meiotic [20]. This oocyte arises from one of the two cyst cells with four ring canals. The oocyte contains more fusome material than the other germline cells [25, 29]. Associating with both the ring canals and the fusome are proteins and mRNAs that specify the oocyte, including oo18 RNA-binding protein (Orb) and *oskar*. In region 3, the oocyte becomes localized to the posterior end of the cyst [20]. For more information on cyst differentiation and oocyte specification, please refer to [20, 21, 29, 30].

Like the germline cells, there are multiple populations of somatic cells in the germarium that have region-specific functions (Fig. 3a) [21]. In regions 1 and 2a are escort cells (ECs), which encapsulate the developing cysts. At the boundary of region 2a and 2b, the ECs hand-off the cysts to the follicle cells. Follicle cells arise from the two FSCs at the 2a/2b boundary [31, 32]. One FSC resides on each side of the germarium, asymmetrically dividing to maintain one FSC and producing proliferative follicle cells that either migrate across the anterior of the 16-cell cyst where they can either differentiate into follicle cells or replace the other FSC, or the daughter follicle cell can move posteriorly to generate the follicle cells that will surround the cyst [32, 33]. In regions 2b and 3, the follicle cells differentiate into multiple subclasses required for the formation of the follicle [34]. In region 2b, the first group of follicle cells specified are the polar cells; additional polar cells arise during stages 1–2 [31, 33]. The polar cells, as their name implies, signal to establish anterior and posterior identities within the developing follicle and ultimately, the oocyte [35]. In region 3, the anterior polar cell induces a second group of follicle cells to differentiate into the stalk cells which separate each developing follicle from the next [36]. The remaining follicle cell precursors will give rise to the main body follicle cells, which will differentiate into multiple types later in follicle development.

Fully formed follicles bud off the germarium, resulting in a circular follicle composed of a layer of follicle cells surrounding 15 nurse cells and one oocyte (Fig. 4). The follicle then dramatically increases in size by a combination of follicle cell proliferation and increasing nurse cell size by endocyling [20]. At the same time, the follicle goes from round to oval, increasing its length to width ratio; this elongation is driven by collective cell migration and is discussed later in this chapter [37]. Multiple follicle cell-types also arise and contribute to follicle development. For example, at stage 9 the anterior polar cells induce 4–6 surrounding follicle cells to differentiate into border cells (Fig. 4a) [38–40]. The border cells delaminate from the surrounding epithelium and undergo a collective cell migration between the nurse cells to the oocyte, where the cluster then migrates dorsally. Border cell migration is required for the formation of the micropyle, a cone-shaped eggshell structure through which sperm enter to fertilize the egg [41]; border cell migration is discussed in more detail below. Coinciding with border cell migration, the ~50 anterior follicle cells go from cuboidal to squamous and differentiate into stretch follicle cells; by stage 10A, the stretch cells cover the nurse cells, while cuboidal main body follicle cells cover the oocyte [36]. At stage 10B, 30–40 of these follicle cells differentiate into centripetal cells and begin to migrate between the nurse cells and the oocyte to ultimately seal the anterior of the oocyte [36]. As this is occurring (stage 11), the nurse cells transfer their content into the oocyte in a process termed nurse cell dumping [42]. During stage 12, the nurse cells undergo cell death, leaving the follicle to be composed of only the oocyte and the surrounding somatic cells [42, 43]. During late oogenesis (stages 10B–13), the follicle cells secrete the vitelline envelope and eggshell [36], with the centripetal cells producing the operculum and a specialized group of dorsal follicle cells producing the dorsal appendages, paddle-shaped structures that oxygenate the developing embryo [44]. Together, these developmental processes result in the production of a mature, ready to ovulate stage 14 egg.

4 *Drosophila* GSCs: A Model for Stem Cell Maintenance and Differentiation

Stem cells are pluripotent cells that divide asymmetrically to both maintain the stem cell pool and produce cells that will differentiate and maintain tissue homeostasis [45]. *Drosophila* oogenesis has been, and continues to be, a powerful model for uncovering the mechanisms regulating stem cell maintenance and differentiation [46–49]. The power of this system comes from the genetic tools, including generating mosaic clones using the FLP/FRT system and cell-specific gene knockdown and overexpression [50, 51]. Further, the germarium houses two stem cell populations: the GSCs and FSCs. Here we focus on the GSCs, their cellular niche, and the roles of adhesion, nuclear architecture, and signaling in stem cell maintenance and differentiation; for detailed reviews on FSCs, please *see* Refs. [48, 49].

The GSC niche is located at the anterior end of the germarium and is composed of several cell types, the terminal filament cells, cap cells, and ECs (Fig. 3a) [47, 52]. Specifically, at the germarium tip is a stack of 8–10 terminal filament cells, the posterior of which contacts 5–7 cap cells. Attached to the cap cells and lining the outside of the germarium are 4–6 anterior ECs. These ECs extend protrusions which surround the GSCs [13, 53, 54]. Together, these three cell-types comprise the niche and maintain a population of 2–4 GSCs.

Bidirectional communication between the niche and the GSCs is required to both maintain the GSC population and control differentiation. Indeed, direct DE-cadherin cell-cell adhesions attach the GSCs to the cap cells [55, 56]; this keeps the GSCs in contact with the source of the signals required for maintaining stem cell fate. For example, Bone Morphogenetic Protein (BMP) family members Decapentaplegic (Dpp) and Glass bottom boat (Gbb) are secreted from the niche and signal to the GSCs [57–59]. This signal causes the transcriptional repression of the differentiation factor Bag of marbles (Bam) in the GSCs. When the GSC divides, one cell maintains its contact with the niche and its stem-cell promoting signals, and the other cell, the cystoblast, moves away from the niche, which drives the expression of Bam (Fig. 3a) [59–61]. Loss of either Dpp or Gbb leads to depletion of the GSC pool [57–59], whereas loss of Bam leads to germline tumors of undifferentiated cells [62–64].

Multiple mechanisms restrict signaling to the GSCs, including the extracellular matrix (ECM) secreted by the niche cells. This ECM not only provides physical support but restricts the diffusion of signals by sequestering signaling molecules [47, 65, 66]. For example, Division Abnormally Delayed (Dally), a heparin-sulfate proteoglycan (HSPG) within the ECM, restricts the diffusion of Dpp [66, 67]. Intriguingly, the expression of *dally* is repressed in the ECs by Epidermal Growth Factor (EGF) signaling coming from the GSCs. Decreased EGFR expression within the niche expands the expression domain of *dally*, altering the distribution of Dpp and increasing the number of GSCs [68]. The ECM also regulates adhesion of the GSCs to the niche, as the perlecan homolog Terribly reduced optic lobes (Trol) is required for the DE-Cadherin adhesions between the GSCs and the cap cells, in addition to its earlier role in establishing the niche [65]. Another means of restricting the niche signaling to the GSCs is by the Janus kinase (Jak)-signal transducer and activator of transcription (Stat) pathway [69–71]. Indeed, increased Jak/Stat activity in the niche cells expands the number of GSCs [71]. Thus, the niche is a critical regulator of the *Drosophila* GSCs. Similarly, interactions between niche cells, ECM, and stem cells play important roles in stem cell maintenance and differentiation in other organisms and tissues, including hematopoietic stem cells, hair follicle stem cells, intestinal stem cells, and muscle satellite cells [45].

Dynamic crosstalk between the germline and the ECs is also required for GSC maintenance and differentiation. The germline activates Jak/STAT and EGFR signaling in the ECs to drive their protrusions [72, 73]. These protrusions encapsulate the germline cysts, promoting the differentiation of the germline [74, 75]. In particular, the ECs downregulate the stem cell factor Dpp, which drives the pro-differentiation factor Bam [76]. For more in-depth reviews of GSC regulation, we refer readers to [47, 52, 77].

In addition to extrinsic mechanisms regulating stem cell maintenance and differentiation, there are also intrinsic factors. Here we discuss one intrinsic regulator of the *Drosophila* GSCs, the nuclear lamina; for a more in-depth review, please *see* Ref. [78]. The nuclear lamina consists of a protein network of lamins and hundreds of lamin-binding proteins on the inside of the nuclear envelope and regulates both nuclear structure and chromatin organization [79, 80]. In relation to structure, the type of lamin controls nuclear stiffness, with higher levels of Lamin A increasing stiffness [81]. Across organisms and systems, stem cell nuclei are softer and stiffness increases with differentiation. In *Drosophila*, the GSCs express only the B-type Lamin (Lamin B, *Drosophila* Lamin Dm0), and its level decreases as differentiation proceeds; Lamin A (*Drosophila* Lamin C) is present starting from the 16-cell cysts in region 2b [78].

This differential lamin expression also results in a patterned expression of interacting proteins, including the Lap2-emerin-MAN1 domain (LEM-D) family member Emerin (D-emerin/Otefin). D-emerin is highly expressed in the GSCs and decreases with differentiation [78]. D-emerin, like other LEM-D proteins, connects the nuclear lamina to Barrier-to-Autointegration Factor (BAF), which in turn binds chromatin [82, 83]. Loss of either D-emerin or BAF activates kinases of the DNA damage checkpoint, blocking differentiation and causing GSC loss [84–86]. Further, the GSC nuclear lamina plays a critical role in chromatin organization necessary for maintaining stemness, as heterochromatin coalesces in *d-emerin* mutant GSCs [84]. D-emerin and the nuclear lamina also regulate GSC mitosis, see below for more details [87]. Importantly, alterations in the nuclear lamina are linked to patient stem cell dysfunction in the progression of laminopathies [88, 89], making the *Drosophila* GSCs a robust model for uncovering the mechanisms by which specific nuclear lamina components act in stem cell maintenance and tissue homeostasis.

5 Cell Cycle Regulation

Drosophila oogenesis provides a robust system to define the mechanisms controlling cell cycle transitions. In addition to being a model for studying meiosis, during follicle development multiple cell types undergo mitosis, including a noncanonical mitosis in the GSCs. Other mitotic cells transition to endocycling, and follicle cells also undergo gene amplification. This diversity of cell cycle events has made *Drosophila* oogenesis a key system for advancing the field's understanding of cell cycle regulation (Fig. 4b).

Recent work by the Geyer Lab discovered that GSCs undergo a noncanonical mitosis. Indeed, while most eukaryotic cells examined breakdown their nuclear envelope during mitosis, the *Drosophila* ovarian GSCs retain a permeable but intact nuclear envelope [87]. This noncanonical mitosis is highly sensitive to perturbations in the nuclear lamina, with defects resulting in chromosome segregation errors and activation of checkpoints that drive GSC loss. It is tempting to speculate that this noncanonical mitosis is a more broadly used mechanism of assessing stem cell quality across tissues and organisms [87].

The early adult germ cells also provide a system to study how cells transition from mitosis to meiosis. The differentiating germ cells undergo 4 mitotic divisions to produce a 16-cell cyst (Fig. 3b).

In region 2a, the two pro-oocytes initiate prophase I of meiosis, with only one remaining in meiosis by region 3 [90, 91]. At stage 5, the oocyte arrests in diplotene of prophase I and remains arrested until stage 13 when meiosis resumes. In stage 14, metaphase I arrest occurs and is maintained until egg activation during ovulation and transport through the oviduct induces the resumption and completion of meiosis [92, 93].

The germline-derived nurse cells exit mitosis in the germarium and endocycle through stage 10. Endocycling cells go from gap (G) to synthesis (S) phases without intervening mitosis or cytokinesis, generating polyploid cells; this allows for cell and tissue growth without disruptions in tissue integrity due to mitosis. In the nurse cells, oscillating levels of Cyclin E drive 10–12 endocycles over just a few days [94]. This results in massive nurse cell growth, and thereby, contributes to the growth of the follicle. Further, the increased DNA content allows the nurse cells to rapidly synthesize contents for the future egg. Thus, nurse cell endocycling facilitates rapid oogenesis.

The *Drosophila* follicle cells also transition from mitosis to endocycling [95]. From region 3 of the germarium to stage 6 of follicle development, the follicle cells are mitotic, producing ~650 cells. Signaling from the nurse cells then activates Notch within the follicle cells, inducing the switch from mitosis to endocycling [96–101]. The follicle cells undergo 3 rounds of endocycling during stage 7-10A, supporting the rapid growth of the follicle.

Endocycling is found across tissues and organisms. Indeed, many mammalian cells undergo endocycling as part of developmental programs [102, 103]. For example, placental trophoblast giant cells require their polyploid size to form and maintain a barrier between the maternal and embryonic tissues [104–106]. In the liver, hepatocyte polyploidization occurs during both development and in response to stress. This stress response allows hepatocytes to survive genetic instability and to rapidly regenerate the liver in response to injury, aging, or hepatectomy [103, 107, 108]. Thus, the nurse and follicle cells of the *Drosophila* ovary provide a robust system to uncover the underlying mechanisms and functions of endocycling, providing insights that can be used to improve human health.

During stage 10B, the *Drosophila* follicle cells undergo another cell cycle transition, switching from endocycling to gene amplification; this transition is regulated by ecdysone signaling [98]. Gene amplification is when only specific regions of chromosomes are replicated. In the follicle cells, there are 6 sites of gene amplification that exhibit distinct temporal patterns of initiation and elongation of the amplified regions [109–111]. The amplified sites contain genes encoding eggshell components and are required for the temporally controlled, rapid expression, and production of the eggshell at the end of oogenesis. Thus, the multiple cell cycle transitions that follicle cells undertake during oogenesis are required to produce a viable egg and thereby influence fertility. Further, gene amplification in the follicle cells provides a system to study the mechanisms controlling the different steps in DNA replication, from origin recognition, to initiation, to elongation [109, 110, 112]. Additionally, gene amplification also occurs in mammalian cells in response to stress and is routinely observed in cancer [113, 114].

6 Collective Cell Migration

Cell migration is critical for life. In multicellular organisms, cell migration drives not only the morphogenic movements during development, including organ formation, but it is also critical for organ homeostasis and repair, wound healing, and diseases, such as cancer [115–117]. While some cellular migrations occur as single cells moving independently, most in vivo cell migrations occur as groups of cells adhering to each other and migrating as a cohesive cluster, termed collective cell migration [118]. Cell migration is controlled by both signaling molecules and properties of the microenvironment, including stiffness [119–121]. The fly ovary is an excellent model for studying the mechanisms controlling collective cell migration as specific populations of follicle cells undergo developmentally timed collective cell migrations to drive follicle elongation, micropyle formation, and dorsal appendage formation (Fig. 4b). These migratory events, in combination with temporal and cell-specific gene manipulation and advances in live imaging, make *Drosophila* oogenesis a great, in vivo model to understand the mechanisms controlling collective cell migration, and how such migrations mediate morphogenic events across organisms.

From the moment follicles detach from the germarium through stage 8, the outer follicle cells collectively migrate on the surrounding ECM (also referred to as the basement membrane) to lengthen the follicle [37, 122, 123]. All the cells within the follicle cell layer coordinate their behavior and form a polarized actin network. On the basal surface, which faces the outside of the follicle, lamellopodial and filopodial leading edge protrusions and parallel arrays of stress fibers attach to integrin-based adhesions on the ECM to drive the collective migration [122, 124–129]. Genetic perturbations of the actin cytoskeleton, the ECM or its cellular receptors, and the signaling pathways involved result in a round egg phenotype and ultimately sterility. Importantly, similar epithelial collective cell migrations drive tissue remodeling during other morphogenic events, including wound healing and cancer metastasis [115, 116, 118].

During stage 9, another collective cell migration, termed border cell migration, occurs. The border cell cluster ultimately contributes to the formation of the micropyle, the eggshell structure through which sperm enter to fertilize the egg [41], and thus, border cell migration is required for fertility. For detailed reviews on border cell migration, *see* Refs. [38–40]. At the beginning of stage 9, the anterior polar cells specify a group of 4–6 surrounding follicle cells to become border cells. These border cells surround the polar cells and delaminate from the follicular epithelium and migrate as a cohesive cluster between the nurse cells to the oocyte (Fig. 4a). Once at the oocyte (late stage 9, early stage 10A), the border cells migrate dorsally.

Border cell migration is directed by both Platelet-Derived Growth Factor (PDGF) and EGF signaling originating from the oocyte [130, 131] and the physical environment [132]. In particular, the stiffness of the nurse cells, the substrate on which the border cells migrate, is a critical regulator of migration. Increasing the stiffness of the nurse cells impedes migration [133, 134]. Recent work suggests that the border cells modulate the nurse cell stiffness, perhaps creating a durotactic gradient to promote their own migration or altering the nurse cells to mediate later developmental events [135]. Additionally, many signaling pathways control the polarity of the cluster, driving its directed migration by modulating the actin cytoskeleton and its regulators within the border cell cluster [38–40]. For example, prostaglandin signaling regulates both border cell migration and cluster cohesion [136]. In summary, the conserved processes observed in border cell migration—delamination, invasion, and collective migration—along with the ability to visualize migration dynamics live [130, 137, 138] have made border cell migration an ideal model for understanding the conserved signals, microenvironmental properties, and cellular changes driving collective cell migration, including cancer invasion and metastasis.

The formation of the dorsal appendages, respiratory eggshell structures, also depends on collective cell migration (*see* Ref. [44] for a detailed review). During stage 10, the oocyte initiates both EGFR activation and inhibition to specify two separate populations of dorsal follicle cells [139–142]. Further specification of the dorsal-appendage primordia along the anterior-posterior axis comes from Dpp signaling [143–147]. Ultimately, 65–70 cells undergo migration and synthesize the tubular dorsal appendages. Each tube is made from two subpopulations, the cells that produce the roof and sides (termed roof cells) and those that produce the floor [44]. During stage 11, dorsal appendage morphogenesis begins, with the apical constriction of the roof cells, and the floor cells migrating underneath them [148, 149]. In stages 12–13, these cells migrate anteriorly over the stretch follicle cells. Throughout their migration, they secrete the eggshell. By stage 13, the migration stops, and the distal cells shorten along the apical-basal axis to produce the flattened paddle of the appendages [148–150]. Defects in patterning and collective migration result in aberrant dorsal appendages, including a single dorsal appendage, altered separation of the two appendages, and abnormal appendage morphologies, including an "antler" appearance [44]. Thus, dorsal appendage formation is a great model for studying the mechanisms driving tubulogenesis.

7 Cytoplasmic Streaming

A reoccurring tool used to mediate development is fluid flow. For example, fluid flow is required for the development of the vasculature, lungs, and kidneys in vertebrates [151]. Intracellular fluid flow, termed streaming, plays critical roles in oogenesis across organisms [152]. Notably, streaming is generally observed in large cells and is implicated in long-distance transport and compartmentalization. *Drosophila* oogenesis provides a robust system to uncover the mechanisms driving and the functions of cytoplasmic streaming; *see* Ref. [152] for a detailed review.

Two types of streaming occur during *Drosophila* oocyte development (Fig. 4b): slow streaming during mid-oogenesis (stages 7–10A) and fast streaming during late oogenesis (stages 10B–14) [153–155]. Both streaming events depend on microtubules and the plus-end-directed motor protein, Kinesin-1 [155–157]. During slow streaming, the oocyte is filled with an actin mesh, and disassembly of this is required for the transition to fast streaming [158]. Thus, cytoplasmic streaming within the *Drosophila* oocyte is a model for studying the mechanisms regulating both microtubule and actin dynamics. One function of streaming in the *Drosophila* oocyte is to help establish embryonic polarity [152, 159, 160]. The nurse cells transcribe mRNAs that are transported via the ring canals into the oocyte; these mRNAs are delivered to specific locations within the oocyte, and this localization is then maintained. Briefly, head and thorax identity are specified by mRNAs, including *bicoid*, localizing at the anterior end of the oocyte, whereas abdominal and germline identity are controlled by mRNAs such as *nanos* at the posterior end.

8 Nurse Cell Dumping

A conserved cytoplasmic streaming event during oogenesis from flies to mice is when sister germ cells, often called nurse cells, within a germline cyst transfer their cytoplasmic contents into the oocyte, driving its growth as the sister cells shrink and ultimately die [42, 161, 162]. In mice, this process is mediated by both microtubule-dependent transport and actin fibers [161, 162]. Conversely, in late *Drosophila* oogenesis, the actin cytoskeleton is primarily responsible. Specifically, during stage 10B, actin filament bundles initiate at the nurse cell plasma membranes and extend to the nuclei, forming a cage-like structure surrounding each nucleus [163, 164]. The cortical actin also thickens [165]. At stage 11, the nurse cells rapidly transfer mRNAs, proteins, and organelles through the ring canals into the oocyte; this is termed nurse cell dumping.

Recent live-imaging and mathematical modeling work reveals nurse cell dumping occurs in two phases [166]. The first phase is driven by fluid flow from smaller cells into larger cells, with the nurse cells immediately adjacent to the oocyte dumping their contents first, and the more anterior nurse cells transferring their contents through the posterior nurse cells. The second phase of dumping occurs when the nurse cell cytoplasmic volume is comparable to its nuclear volume and is driven by actomyosin contractility of the cortical actin. If nurse cell dumping fails, follicle morphogenesis is disrupted, including failure of the nurse cells to die during stage 12, generation of short eggs, and abnormal dorsal appendage formation. Nurse cell dumping defects can be classified based on their distinct outcomes. In the first case, nurse cell nuclei are observed plugging the ring canals into the oocyte, and thereby, blocking dumping; this phenotype indicates defects in actin bundle formation and/or organization [167–169]. In the second case, dumping fails but no nuclei plug the ring canals; such a phenotype is observed when myosin activity is impaired [165]. Thus, nurse cell dumping during *Drosophila* oogenesis is widely used to study actin cytoskeletal dynamics.

While the functions of numerous actin-binding proteins have been uncovered using nurse cell dumping as a model, the mechanisms initiating this process have largely remained elusive. As discussed above, one means of driving nurse cell dumping is simply cellular size and fluid force [166]. Whether this serves as a mechanical signal to drive actin remodeling events is unknown. Additionally, prostaglandin signaling is required for nurse cell dumping [170]. Loss of prostaglandin signaling results in a severe reduction and often complete loss of actin bundle formation, and cortical actin breakdown [171]. Strikingly, even though the bundles are defective, nurse cell nuclei do not plug the ring canals, suggesting prostaglandins also regulate nurse cell contraction. While the detailed prostaglandin signaling pathway remains to be determined, multiple actin-binding proteins—Fascin (*Drosophila* Singed), Enabled, and Non-muscle myosin II—have been identified as downstream effectors of prostaglandins [172–174]. Whether fluid flow and prostaglandin signaling are conserved mechanisms used to control germline cyst cytoplasmic transfer across organisms is unknown.

The primary purpose of *Drosophila* nurse cell dumping is to provide the oocyte with everything it needs for successfully completing embryogenesis, which occurs outside the mother. During the first ~2 h of embryogenesis, there is no zygotic transcription, and therefore, the embryo is completely dependent on maternally supplied factors [175]. Indeed, roughly 60% of the genome is supplied as maternal RNA, and regulation of its translation, stability, and localization within the embryo are essential for embryogenesis. In the first 2 h, the embryo undergoes 13 rapid nuclear divisions in a shared cytoplasm; this rapid cell cycle depends on maternally supplied proteins. During nuclear cycle 14, the embryo cellularizes to package maternally supplied organelles into individual cells and begins zygotic transcription. One maternally supplied organelle that plays critical roles both early in embryogenesis and after cellularization is the lipid droplet [176]. Lipid droplets consist of a neutral lipid core surrounded by a phospholipid monolayer decorated with an array of proteins. Across organisms, lipid droplet accumulation, composition, and localization are dynamic during oocyte maturation [177–179], and changes in lipids droplets are associated with infertility [180–182]. These lipid droplets are known to support the metabolic needs of the embryo, both when embryonic development occurs outside the mother, such as in *Drosophila*, and prior to implantation in placental mammals [183, 184]. Additionally, in *Drosophila*, maternally supplied lipid droplets sequester and buffer the histone supply prior to cellularization [185–188]; whether this is conserved in higher eukaryotes remains to be determined.

9 Lipid Droplets: Roles in Oogenesis

The conservation of lipid droplet accumulation during oogenesis raises the possibility that in addition to mediating early embryogenesis lipid droplets play roles in follicle development. During *Drosophila* mid-oogenesis, large numbers of lipid droplets are generated in the nurse cells [176]. Lipid droplet accumulation begins at stage 8, such that by stage 10B the nurse cell cytoplasm is packed with lipid droplets (Fig. 4b). This temporal accumulation of lipid droplets is driven by steroid hormone signaling activating the Sterol regulatory-element-binding protein (SREBP) transcription factor, which promotes the expression of lipophorin receptors (LpR1/2) [189]. LpR1/2 take up lipoproteins from the circulating hemolymph, providing the lipids needed for lipid droplet assembly [190, 191]. Diacylglycerol O-acyl transferase 1 (DGAT1, *Drosophila* Midway) incorporates these lipids into triglycerides, driving lipid droplet formation; in its absence, nurse cells and oocytes lack lipid droplets.

Genetic studies on DGAT1 and LpR1/2 suggest lipid droplets play critical roles in oogenesis. Loss of DGAT1 results in a loss of lipid droplet formation and follicles arrest at stage 8, display premature actin bundles, and trigger the mid-oogenesis checkpoint, resulting in follicle death [192]. Thus, DGAT1 is required for follicle development. However, LpR1/2 germline clones also lack lipid droplets, but these have been reported to develop past the mid-oogenesis checkpoint [190]. One difference in these mutants is that *LpR1/2* mutants will fail to take up fatty acids from the hemolymph, while *DGAT1* mutants will have excess free fatty acids in the nurse cells that cannot be sequestered into lipid droplets. A number of free fatty acids are cytotoxic, including arachidonic acid [193, 194]. Indeed, arachidonic acid is found in triglycerides in ovary lysates and loss of the lipid droplet-associated adipose triglyceride lipase (ATGL, also called Brummer) increases such triglycerides. Further, exogenous arachidonic acid inhibits the development of *Drosophila* follicles ex vivo, and reducing ATGL suppresses this toxicity [195]. These data lead to the idea that lipid droplets sequester free fatty acids, including arachidonic acid, to prevent lipotoxicity.

A recent study uncovered another role of lipid droplets in *Drosophila* oogenesis—mediating prostaglandin signaling [195]. Genetic and pharmacological studies support the model that during stage 10B, ATGL releases arachidonic acid from lipid droplet triglycerides. This arachidonic acid is then used as the substrate for prostaglandin production. These prostaglandins activate signaling cascades that drive the actin remodeling events necessary for nurse cell dumping and late-stage follicle morphogenesis. Thus, *Drosophila* oogenesis is emerging as a system to decipher the roles of lipid droplets in oocyte development.

10 Cell Death

Cell death is an essential process for eliminating defective cells and mediating development across organisms. *Drosophila* oogenesis has been widely used to understand the mechanisms and different types of cell death, as cell death is restricted to specific cells and periods of follicle development (Fig. 4b).

In early oogenesis there are three cell death events [43]. Germline cysts can undergo cell death in region 2 of the germarium. Such death, which is a combination of autophagy and apoptosis, is triggered by a checkpoint that can be activated by defects in the cyst or in the environment, such as a shortage of nutrients [196–198]. There are also two types of somatic cell death in early oogenesis. Initially, there are 3–6 polar cells at each end of the follicle, and prior to stage 5, the number is reduced to 2 by apoptosis [199, 200]. Similarly, if there are too many stalk cells, they undergo apoptosis during stages 2–8 [201].

After the follicle has the correct number of polar and stalk cells, it then undergoes a final assessment of both follicle health and the fly's environment prior to the energy-intensive process of vitellogenesis; this is termed the mid-oogenesis checkpoint [43, 196, 202, 203]. Failure to pass this checkpoint results in the elimination of the whole stage 8/9 follicle; this is a mechanism to avoid wasting energy on building an oocyte that is unlikely to produce a viable offspring. This follicle death occurs by all the nurse cell nuclei simultaneously condensing and losing their nuclear integrity. The surrounding follicle cells then act as phagocytes and engulf the dying germline, and subsequently undergo cell death.

In late oogenesis, the 15 nurse cells within the follicle undergo a phagocyte-dependent cell death, leaving only the oocyte and the surrounding follicle cells [42, 43]. After the completion of nurse cell dumping (stage 12), the stretch follicle cells invade, but do not engulf, the nurse cells; this initiates nurse cell death. During stage 13, the nurse cell nuclei become acidified by the stretch follicle cells and then degrade asynchronously, ultimately resulting in the mature stage 14 egg [204–206]. During ovulation, the follicle cell layer is removed, allowing the oocyte, encased by its eggshell, to move through the oviduct. These follicle cells will later die and be phagocytized by the epithelial cells of the lateral oviduct [207, 208].

11 *Drosophila* Oogenesis: A Model for Other Cell Biology Processes

Drosophila oogenesis is a robust model for studying a diverse array of cell biology, from understanding the functions of organelles to uncovering the development roles of understudied factors using both genetics and more recently proteomics. Here we illustrate the utility of *Drosophila* oogenesis to study the roles of ribosomes, nuclear actin, and metabolic filaments.

Control of protein synthesis, at the level of ribosomal activity, is a critical conserved regulator of stem cell function, differentiation, and tissue homeostasis, including during *Drosophila* oogenesis [209]. Indeed, GSCs have higher levels of ribosomal RNA (rRNA) transcription by RNA Polymerase I (RNAPI) than their differentiating daughters but decreased protein synthesis [78, 210, 211]. Inhibiting RNAPI activity results in GSC loss, whereas increasing it drives differentiation [210]. Further, an unbiased RNAi screen discovered that ribosome assembly impacts GSC cytokinesis, and differentiation requires increased ribosome biogenesis and protein synthesis [211]. Recent work indicates three DExD/H-box proteins are required to sense ribosome biogenesis levels to control the GSC cell cycle, and thus, differentiation [212]. Ribosomal activity also plays critical roles in later stages of oogenesis. Two studies analyzing paralogs of ribosome protein S5 (RpS5) found that loss of RpS5b but not RpS5a results in numerous defects in oogenesis, including disruption of vitellogenesis, posterior follicle cell hyperplasia, and mid-oogenesis checkpoint arrest [213, 214]. Contributing to these defects is both increased translation and altered mRNA translation efficiency, resulting in striking transcriptional changes including in metabolic processes, yolk granule proteins, and cytoskeletal regulators. These findings, along with growing evidence of germline-specific differential ribosomal paralog expression [215], suggest that ribosomal specialization is particularly critical for germline function. Together, these studies establish *Drosophila* oogenesis as a model for not only understanding how ribosomal activity controls stemness and differentiation, but how ribosomal protein paralogs drive ribosomal heterogeneity and tissue specialization.

Drosophila oogenesis is also becoming a model to study the functions of actin in the nucleus. While actin has been extensively studied for its cytoskeletal roles, actin localizes to the nucleus in a highly regulated manner where it has numerous functions, including binding to and regulating all three RNA polymerases, functioning in chromatin remodeling complexes, mediating DNA repair, and contributing to nuclear architecture [216–219]. How these different functions of nuclear actin are used to drive development remains poorly understood, and *Drosophila* oogenesis is poised to contribute to this area. The first evidence suggesting nuclear actin has a role in *Drosophila* oogenesis came from the observation that germline expression of tools for imaging actin cytoskeletal dynamics results in a stage-dependent accumulation of thick nuclear actin filaments, often called nuclear actin rods [220]. As these tools stabilize endogenous actin filaments, this finding led to the idea that there was endogenous nuclear actin during *Drosophila* oogenesis.

Subsequent work uncovered three tools that label both overlapping and distinct pools of endogenous nuclear actin [221, 222]. Specifically, monomeric nuclear actin labeled with DNase I is found in the nucleolus of every cell during oogenesis. The C4 actin antibody colocalizes with DNase I in a subset of cells during early oogenesis, but it also labels distinct nucleoplasmic regions in the early germ cells and the oocyte. The AC15 actin antibody labels the chromatin in all cells starting around stage 6 and increases in intensity through stage 10. Additionally, during stages 9–10B, puncta of nuclear actin within the nurse cell nucleoli are labeled with the AC15 actin antibody. Thus, all pools of endogenous nuclear actin exhibit some localization to the nucleolus, suggesting nuclear actin may play a critical role there.

Nuclear actin also regulates transcription during *Drosophila* oogenesis. Using ChIP-seq of whole ovary lysates, the Vartiainen Lab found actin colocalizes with RNAPII on most gene promoters and along the gene body of highly expressed genes, including eggshell genes [223]. Impairing the nuclear import of actin results in decreased eggshell gene expression, supporting that nuclear actin plays a critical role in transcription during oogenesis. Future work using *Drosophila* oogenesis is expected to advance the mechanistic understanding of nuclear actin in development and tissue homeostasis.

The power of using *Drosophila* oogenesis to study cell biology is amplified by the biochemistry capabilities of the ovary. Indeed, due to the ease of obtaining large quantities of either individual follicles or whole ovaries, *Drosophila* oogenesis is becoming more widely used to identify protein-protein interactions. For example, both proximity-dependent biotinylation using the APEX enzyme and tandem affinity purification followed by mass spectrometry were used to uncover the protein interaction partners of the ring canal protein Kelch and its molecular functions [224, 225]. TurboID-mediated biotinylation has also been used to understand the interaction partners of cytidine triphosphate (CTP) synthase when it can or cannot form filamentous cytophidia [226]. Cytophidia are conserved intracellular compartments observed from yeast to humans. Such filamentous structures have been observed with multiple metabolic enzymes and evidence suggests the purpose is to modulate enzymatic activity [227]. As cytophidia are present in both the nurse and follicle cells during *Drosophila* oogenesis, this model has played a critical role in both the discovery and understanding of this cellular structure [227, 228].

12 *Drosophila* as a Female Reproductive Model

While the organization and morphology of the *Drosophila* ovary are distinct from those found in mammals, growing evidence supports that numerous processes during oogenesis and their regulation are conserved. As discussed above, interconnected germline cysts are common, and recent work supports that mammalian germline cells act like *Drosophila* nurse cells, providing contents to the surviving oocyte and then undergoing cell death [42, 161, 162]. Lipid droplet accumulation is also a conserved process during oogenesis [177–179], although it remains unknown whether the roles of lipid droplets in regulating both prostaglandin signaling and lipotoxicity observed in flies extend to higher eukaryotes [195]. Further, across organisms, prostaglandin signaling regulates female reproduction, including follicle maturation [229–232]. However, whether, like in *Drosophila* [136, 171–174], prostaglandins regulate nurse cell dumping, actin cytoskeletal remodeling, and cell migration during mammalian oogenesis remains to be determined. Here we discuss two additional examples—interorgan communication and ovulation—of how *Drosophila* oogenesis serves as a model to uncover conserved mechanisms regulating female reproduction.

Across organisms, female reproduction—oogenesis, ovulation, and fertilization—depends on the status of the whole organism, and therefore, it is critical for the reproductive system to respond to signals from other organs; this is termed interorgan communication. *Drosophila* is a robust model for discovering how interorgan communication controls oogenesis due to the ability to use the UAS/GAL4 system to drive RNA interference (RNAi) knockdown in specific organs and cell-types [233, 234].

One conserved environmental factor driving interorgan control of oogenesis is nutrition. In *Drosophila*, a protein poor diet results in decreased GSC proliferation, GSC loss, increased mid-oogenesis checkpoint death, and a block in ovulation, resulting in increased numbers of stage 14 follicles [43, 196, 197, 235]. The nutritional status of the fly is conveyed to the ovary by a number of organs. For example, upon feeding, the brain secretes insulin-like peptides to signal to the germline, promoting GSC proliferation, follicle growth, and progression through vitellogenesis [236, 237]. Adipose tissue, known as the fat body in *Drosophila*, also plays a critical role in regulating oogenesis in response to nutrition. In adipocytes, insulin signaling promotes GSC maintenance, early germline cyst survival, and vitellogenesis [238]. Adipocytes also use the amino acid response pathway to promote germline cyst survival and progression through vitellogenesis [239]. Further, Target of rapamycin (TOR)-dependent amino acid sensing in adipocytes promotes ovulation [239]. Genetic studies also support that fatty acid oxidation in the fat body is required for GSC maintenance [240].

Another conserved means of regulating oogenesis is through nuclear receptors, ligand-dependent transcription factors. For example, loss of the nuclear receptor *seven up* (*svp*) in adipocytes results in increased GSC loss and early germline cyst death, while loss in oenocytes increases follicle death at the mid-oogenesis checkpoint [241]. The orphan nuclear receptor, Hormone Receptor 4 (Hr4), also acts in multiple tissues to regulate oogenesis [242]. Within the ovary, Hr4 acts in the niche and GSCs, whereas in the muscle Hr4 is required for GSC maintenance and follicle growth. Hr4 acts in other, currently undefined tissues to promote early germline cyst survival and vitellogenesis. Thus, multiple organs use diverse mechanisms to control key aspects of *Drosophila* oogenesis, allowing it to avoid wasting energy producing eggs when the organism is undergoing a stress.

Over the past decade, *Drosophila* has also emerged as a model for studying the conserved mechanisms controlling ovulation. Ovulation in mammals is induced by luteinizing hormone (LH) signaling to the granulosa cells of the preovulatory follicles [243]. This activates multiple pathways, including EGF, progesterone, and prostaglandins, to drive meiotic resumption, somatic cell expansion, and proteolytic breakdown of the follicle wall [244–246]. In *Drosophila*, octopamine signals to the follicle cells on mature stage 14 follicles, activating NADPH oxidase [247] and matrix metalloprotease 2 [248]. The latter, in combination with ecdysone signaling, leads to the breakdown of the follicle wall

[249]. Specifically, the posterior follicle cells are degraded, allowing the mature oocyte to rupture from the follicle cells and be released into the oviduct. The rest of the follicle cells remain at the end of the ovariole to form a corpus luteum-like endocrine structure similar to that in mammals [250]. Further illustrating the utility of *Drosophila* as a model for understanding ovulation, this system has been used to screen for nonsteroidal contraceptive compounds [251]. Thus, *Drosophila* is poised to uncover new and conserved mechanisms regulating the poorly understood process of ovulation.

13 Conclusion

While *Drosophila* oogenesis has already taught us much about fundamental principles of biology and development, with the expanse of technological and genetic tools it remains a model that continues to advance multiple fields, from stem cells and tissue homeostasis to conserved mechanisms controlling female reproduction. Further, the breadth of processes that can be studied, ease of obtaining large amounts of tissue, and simple readouts of defects in follicle development (i.e., fertility and follicle morphology) make *Drosophila* oogenesis poised to transform science education, from outreach activities and classroom exercises for K-12 to undergraduate course-based research experiences [252]. For excellent examples of how-to bring *Drosophila* oogenesis into the classroom, please *see* Refs. [253, 254].

Acknowledgments

We thank the Tootle lab for helpful discussions and careful review of the manuscript. The following sources provided funding for this project: National Institutes of Health R35 GM144057 (T.L.T.) and National Science Foundation MCB 2017797 (T.L.T.). M.S.G. is supported by NIH T32 CA078586 Free Radical and Radiation Biology, University of Iowa.

References

1. Cooley L (1995) Oogenesis: variations on a theme. Dev Genet 16(1):1–5. https://doi.org/10.1002/dvg.1020160103
2. Elis S, Desmarchais A, Cardona E, Fouchecourt S, Dalbies-Tran R, Nguyen T, Thermes V, Maillard V, Papillier P, Uzbekova S, Bobe J, Couderc JL, Monget P (2018) Genes involved in Drosophila melanogaster ovarian function are highly conserved throughout evolution. Genome Biol Evol 10(10):2629–2642. https://doi.org/10.1093/gbe/evy158
3. Doherty CA, Amargant F, Shvartsman SY, Duncan FE, Gavis ER (2022) Bidirectional communication in oogenesis: a dynamic conversation in mice and Drosophila. Trends Cell Biol 32(4):311–323. https://doi.org/10.1016/j.tcb.2021.11.005
4. Santos AC, Lehmann R (2004) Germ cell specification and migration in Drosophila

and beyond. Curr Biol 14(14):R578–R589. https://doi.org/10.1016/j.cub.2004.07.018

5. Ephrussi A, Lehmann R (1992) Induction of germ cell formation by Oskar. Nature 358(6385):387–392
6. Jemc JC (2011) Somatic gonadal cells: the supporting cast for the germline. Genesis 49(10):753–775. https://doi.org/10.1002/dvg.20784
7. King RC, Aggarwal SK, Aggarwal U (1968) The development of the female Drosophila reproductive system. J Morphol 124(2): 143–165. https://doi.org/10.1002/jmor.1051240203
8. Boyle M, DiNardo S (1995) Specification, migration and assembly of the somatic cells of the Drosophila gonad. Development 121(6):1815–1825. https://doi.org/10.1242/dev.121.6.1815
9. Saitou M, Yamaji M (2012) Primordial germ cells in mice. Cold Spring Harb Perspect Biol 4(11). https://doi.org/10.1101/cshperspect.a008375
10. Zhang H, Risal S, Gorre N, Busayavalasa K, Li X, Shen Y, Bosbach B, Brännström M, Liu K (2014) Somatic cells initiate primordial follicle activation and govern the development of dormant oocytes in mice. Curr Biol 24(21): 2501–2508. https://doi.org/10.1016/j.cub.2014.09.023
11. Gilboa L (2015) Organizing stem cell units in the Drosophila ovary. Curr Opin Genet Dev 32:31–36. https://doi.org/10.1016/j.gde.2015.01.005
12. Slaidina M, Banisch TU, Gupta S, Lehmann R (2020) A single-cell atlas of the developing Drosophila ovary identifies follicle stem cell progenitors. Genes Dev 34(3–4):239–249
13. Xie T, Spradling AC (2000) A niche maintaining germ line stem cells in the *Drosophila* ovary. Science 290(5490):328–330. https://doi.org/10.1126/science.290.5490.328
14. Li MA, Alls JD, Avancini RM, Koo K, Godt D (2003) The large Maf Factor Traffic Jam controls gonad morphogenesis in Drosophila. Nat Cell Biol 5(11):994–1000. https://doi.org/10.1038/ncb1058
15. Gilboa L, Forbes A, Tazuke SI, Fuller MT, Lehmann R (2003) Germ line stem cell differentiation in Drosophila requires gap junctions and proceeds via an intermediate state. Development 130(26):6625–6634. https://doi.org/10.1242/dev.00853
16. Reilein A, Kogan HV, Misner R, Park KS, Kalderon D (2021) Adult stem cells and niche cells segregate gradually from common precursors that build the adult Drosophila ovary during pupal development. elife 10: e69749. C69741 – elife 62021;69710: e69749. https://doi.org/10.7554/eLife.69749
17. Allen AK, Spradling AC (2008) The Sf1-related nuclear hormone receptor Hr39 regulates Drosophila female reproductive tract development and function. Development 135(2):311–321. https://doi.org/10.1242/dev.015156
18. Wolfner MF (2011) Precious essences: female secretions promote sperm storage in Drosophila. PLoS Biol 9(11):e1001191. https://doi.org/10.1371/journal.pbio.1001191
19. Schnakenberg SL, Siegal ML, Bloch Qazi MC (2012) Oh, the places they'll go: female sperm storage and sperm precedence in Drosophila melanogaster. Spermatogenesis 2(3): 224–235. https://doi.org/10.4161/spmg.21655
20. Spradling AC (1993) Developmental genetics of oogenesis. In: Martinez-Arias B (ed) The development of Drosophila melanogaster. Cold Spring Harbor Laboratory Press, pp 1–70
21. de Cuevas M (2015) Drosophila oogenesis. In: eLS, pp 1–7. https://doi.org/10.1002/9780470015902.a0001502.pub2
22. Deng W, Lin H (1997) Spectrosomes and fusomes anchor mitotic spindles during asymmetric germ cell divisions and facilitate the formation of a polarized microtubule array for oocyte specification in Drosophila. Dev Biol 189(1):79–94. https://doi.org/10.1006/dbio.1997.8669
23. Yue L, Spradling AC (1992) hu-li tai shao, a gene required for ring canal formation during Drosophila oogenesis, encodes a homolog of adducin. Genes Dev 6(12b):2443–2454. https://doi.org/10.1101/gad.6.12b.2443
24. Robinson DN, Cant K, Cooley L (1994) Morphogenesis of Drosophila ovarian ring canals. Development 120(7):2015–2025. https://doi.org/10.1242/dev.120.7.2015
25. Hinnant TD, Merkle JA, Ables ET (2020) Coordinating proliferation, polarity, and cell fate in the drosophila female germline. Front Cell Dev Biol 8. https://doi.org/10.3389/fcell.2020.00019
26. Lighthouse DV, Buszczak M, Spradling AC (2008) New components of the Drosophila fusome suggest it plays novel roles in signaling and transport. Dev Biol 317(1):59–71. https://doi.org/10.1016/j.ydbio.2008.02.009

27. Röper K, Brown NH (2004) A spectraplakin is enriched on the fusome and organizes microtubules during oocyte specification in Drosophila. Curr Biol 14(2):99–110. https://doi.org/10.1016/j.cub.2003.12.056
28. Snapp EL, Iida T, Frescas D, Lippincott-Schwartz J, Lilly MA (2004) The fusome mediates intercellular endoplasmic reticulum connectivity in Drosophila ovarian cysts. Mol Biol Cell 15(10):4512–4521. https://doi.org/10.1091/mbc.e04-06-0475
29. Huynh J-R, St Johnston D (2004) The origin of asymmetry: early polarisation of the Drosophila germline cyst and oocyte. Curr Biol 14(11):R438–R449. https://doi.org/10.1016/j.cub.2004.05.040
30. Bastock R, St Johnston D (2008) Drosophila oogenesis. Curr Biol 18(23):R1082–R1087. https://doi.org/10.1016/j.cub.2008.09.011
31. Margolis J, Spradling A (1995) Identification and behavior of epithelial stem cells in the Drosophila ovary. Development 121(11):3797–3807. https://doi.org/10.1242/dev.121.11.3797
32. Nystul T, Spradling A (2007) An epithelial niche in the Drosophila ovary undergoes long-range stem cell replacement. Cell Stem Cell 1(3):277–285. https://doi.org/10.1016/j.stem.2007.07.009
33. Nystul T, Spradling A (2010) Regulation of epithelial stem cell replacement and follicle formation in the Drosophila ovary. Genetics 184(2):503–515. https://doi.org/10.1534/genetics.109.109538
34. Rust K, Byrnes LE, Yu KS, Park JS, Sneddon JB, Tward AD, Nystul TG (2020) A single-cell atlas and lineage analysis of the adult Drosophila ovary. Nat Commun 11(1):5628. https://doi.org/10.1038/s41467-020-19361-0
35. Grammont M, Irvine KD (2002) Organizer activity of the polar cells during Drosophilaoogenesis. Development 129(22):5131–5140. https://doi.org/10.1242/dev.129.22.5131
36. Wu X, Tanwar PS, Raftery LA (2008) Drosophila follicle cells: morphogenesis in an eggshell. Semin Cell Dev Biol 19(3):271–282. https://doi.org/10.1016/j.semcdb.2008.01.004
37. Horne-Badovinac S (2014) The Drosophila egg chamber-a new spin on how tissues elongate. Integr Comp Biol 54(4):667–676. https://doi.org/10.1093/icb/icu067
38. Montell DJ (2003) Border-cell migration: the race is on. Nat Rev Mol Cell Biol 4(1):13–24. https://doi.org/10.1038/nrm1006
39. Montell DJ, Yoon WH, Starz-Gaiano M (2012) Group choreography: mechanisms orchestrating the collective movement of border cells. Nat Rev Mol Cell Biol 13(10):631–645. https://doi.org/10.1038/nrm3433
40. Saadin A, Starz-Gaiano M (2016) Circuitous genetic regulation governs a straightforward cell migration. Trends Genet 32(10):660–673. https://doi.org/10.1016/j.tig.2016.08.001
41. Horne-Badovinac S (2020) The Drosophila micropyle as a system to study how epithelia build complex extracellular structures. Philos Trans R Soc Lond Ser B Biol Sci 375(1809):20190561. https://doi.org/10.1098/rstb.2019.0561
42. Buszczak M, Cooley L (2000) Eggs to die for: cell death during drosophila oogenesis. Cell Death Differ 7(11):1071–1074. https://doi.org/10.1038/sj.cdd.4400755
43. Lebo DPV, McCall K (2021) Murder on the ovarian express: a tale of non-autonomous cell death in the Drosophila ovary. Cell 10(6). https://doi.org/10.3390/cells10061454
44. Berg CA (2005) The Drosophila shell game: patterning genes and morphological change. Trends Genet 21(6):346–355. https://doi.org/10.1016/j.tig.2005.04.010
45. Gattazzo F, Urciuolo A, Bonaldo P (2014) Extracellular matrix: a dynamic microenvironment for stem cell niche. Biochim Biophys Acta Gen Subj 1840(8):2506–2519. https://doi.org/10.1016/j.bbagen.2014.01.010
46. Kahney EW, Snedeker JC, Chen X (2019) Regulation of Drosophila germline stem cells. Curr Opin Cell Biol 60:27–35. https://doi.org/10.1016/j.ceb.2019.03.008
47. Losick VP, Morris LX, Fox DT, Spradling A (2011) Drosophila stem cell niches: a decade of discovery suggests a unified view of stem cell regulation. Dev Cell 21(1):159–171. https://doi.org/10.1016/j.devcel.2011.06.018
48. Sahai-Hernandez P, Castanieto A, Nystul TG (2012) Drosophila models of epithelial stem cells and their niches. Wiley Interdiscip Rev Dev Biol 1(3):447–457. https://doi.org/10.1002/wdev.36
49. Rust K, Nystul T (2020) Signal transduction in the early Drosophila follicle stem cell lineage. Curr Opin Insect Sci 37:39–48. https://doi.org/10.1016/j.cois.2019.11.005

50. Laws KM, Drummond-Barbosa D (2015) Genetic mosaic analysis of stem cell lineages in the Drosophila ovary. Methods Mol Biol 1328:57–72. https://doi.org/10.1007/978-1-4939-2851-4_4
51. McLaughlin JM, Bratu DP (2015) Drosophila melanogaster oogenesis: an overview. Methods Mol Biol 1328:1–20. https://doi.org/10.1007/978-1-4939-2851-4_1
52. Eliazer S, Buszczak M (2011) Finding a niche: studies from the Drosophila ovary. Stem Cell Res Ther 2(6):45. https://doi.org/10.1186/scrt86
53. Kirilly D, Xie T (2007) The Drosophila ovary: an active stem cell community. Cell Res 17(1):15–25. https://doi.org/10.1038/sj.cr.7310123
54. Morris LX, Spradling AC (2011) Long-term live imaging provides new insight into stem cell regulation and germline-soma coordination in the Drosophila ovary. Development 138(11):2207–2215. https://doi.org/10.1242/dev.065508
55. Song X, Xie T (2002) DE-cadherin-mediated cell adhesion is essential for maintaining somatic stem cells in the *Drosophila* ovary. Proc Natl Acad Sci 99(23):14813–14818. https://doi.org/10.1073/pnas.232389399
56. Song X, Zhu C-H, Doan C, Xie T (2002) Germline stem cells anchored by adherens junctions in the *Drosophila* ovary niches. Science 296(5574):1855–1857. https://doi.org/10.1126/science.1069871
57. Xie T, Spradling AC (1998) Decapentaplegic is essential for the maintenance and division of germline stem cells in the Drosophila ovary. Cell 94(2):251–260. https://doi.org/10.1016/S0092-8674(00)81424-5
58. Wharton KA, Thomsen GH, Gelbart WM (1991) Drosophila 60A gene, another transforming growth factor beta family member, is closely related to human bone morphogenetic proteins. Proc Natl Acad Sci U S A 88(20):9214–9218. https://doi.org/10.1073/pnas.88.20.9214
59. Chen D, McKearin D (2003) Dpp signaling silences bam transcription directly to establish asymmetric divisions of germline stem cells. Curr Biol 13(20):1786–1791. https://doi.org/10.1016/j.cub.2003.09.033
60. Chen D, McKearin DM (2003) A discrete transcriptional silencer in the bam gene determines asymmetric division of the Drosophila germline stem cell. Development 130(6):1159–1170. https://doi.org/10.1242/dev.00325
61. Song X, Wong MD, Kawase E, Xi R, Ding BC, McCarthy JJ, Xie T (2004) Bmp signals from niche cells directly repress transcription of a differentiation-promoting gene, bag of marbles, in germline stem cells in the Drosophila ovary. Development 131(6):1353–1364. https://doi.org/10.1242/dev.01026
62. McKearin D, Ohlstein B (1995) A role for the Drosophila bag-of-marbles protein in the differentiation of cystoblasts from germline stem cells. Development 121(9):2937–2947. https://doi.org/10.1242/dev.121.9.2937
63. McKearin DM, Spradling AC (1990) bag-of-marbles: a Drosophila gene required to initiate both male and female gametogenesis. Genes Dev 4(12B):2242–2251. https://doi.org/10.1101/gad.4.12b.2242
64. Ohlstein B, Lavoie CA, Vef O, Gateff E, McKearin DM (2000) The Drosophila cystoblast differentiation factor, benign gonial cell neoplasm, is related to DExH-box proteins and interacts genetically with bag-of-marbles. Genetics 155(4):1809–1819. https://doi.org/10.1093/genetics/155.4.1809
65. Díaz-Torres A, Rosales-Nieves AE, Pearson JR, Santa-Cruz Mateos C, Marín-Menguiano M, Marshall OJ, Brand AH, González-Reyes A (2021) Stem cell niche organization in the Drosophila ovary requires the ECM component Perlecan. Curr Biol 31(8):1744–1753.e1745. https://doi.org/10.1016/j.cub.2021.01.071
66. Wilcockson SG, Sutcliffe C, Ashe HL (2017) Control of signaling molecule range during developmental patterning. Cell Mol Life Sci 74(11):1937–1956. https://doi.org/10.1007/s00018-016-2433-5
67. Guo Z, Wang Z (2009) The glypican Dally is required in the niche for the maintenance of germline stem cells and short-range BMP signaling in the Drosophilaovary. Development 136(21):3627–3635. https://doi.org/10.1242/dev.036939
68. Liu M, Lim TM, Cai Y (2010) The Drosophila female germline stem cell lineage acts to spatially restrict DPP function within the niche. Sci Signal 3(132):ra57. https://doi.org/10.1126/scisignal.2000740
69. Wang L, Li Z, Cai Y (2008) The JAK/STAT pathway positively regulates DPP signaling in the Drosophila germline stem cell niche. J Cell Biol 180(4):721–728. https://doi.org/10.1083/jcb.200711022
70. Decotto E, Spradling AC (2005) The Drosophila ovarian and testis stem cell niches:

similar somatic stem cells and signals. Dev Cell 9(4):501–510. https://doi.org/10.1016/j.devcel.2005.08.012
71. Lopez-Onieva L, Fernandez-Minan A, Gonzalez-Reyes A (2008) Jak/Stat signalling in niche support cells regulates dpp transcription to control germline stem cell maintenance in the Drosophila ovary. Development 135(3):533–540. https://doi.org/10.1242/dev.016121
72. Gilboa L, Lehmann R (2006) Soma-germline interactions coordinate homeostasis and growth in the Drosophila gonad. Nature 443(7107):97–100. https://doi.org/10.1038/nature05068
73. Maimon I, Popliker M, Gilboa L (2014) Without children is required for Stat-mediated zfh1 transcription and for germline stem cell differentiation. Development 141(13):2602–2610. https://doi.org/10.1242/dev.109611
74. Kirilly D, Wang S, Xie T (2011) Self-maintained escort cells form a germline stem cell differentiation niche. Development 138(23):5087–5097. https://doi.org/10.1242/dev.067850
75. Antel M, Inaba M (2020) Modulation of cell–cell interactions in Drosophila oocyte development. Cell 9(2):274
76. Liu Z, Zhong G, Chai PC, Luo L, Liu S, Yang Y, Baeg G-H, Cai Y (2015) Coordinated niche-associated signals promote germline homeostasis in the Drosophila ovary. J Cell Biol 211(2):469–484. https://doi.org/10.1083/jcb.201503033
77. Ting X (2013) Control of germline stem cell self-renewal and differentiation in the Drosophila ovary: concerted actions of niche signals and intrinsic factors. WIREs Dev Biol 2(2):261–273. https://doi.org/10.1002/wdev.60
78. Duan T, Green N, Tootle TL, Geyer PK (2020) Nuclear architecture as an intrinsic regulator of Drosophila female germline stem cell maintenance. Curr Opin Insect Sci 37:30–38. https://doi.org/10.1016/j.cois.2019.11.007
79. Burke B, Stewart CL (2013) The nuclear lamins: flexibility in function. Nat Rev Mol Cell Biol 14(1):13–24. https://doi.org/10.1038/nrm3488
80. Vahabikashi A, Adam SA, Medalia O, Goldman RD (2022) Nuclear lamins: structure and function in mechanobiology. APL Bioeng 6(1):011503. https://doi.org/10.1063/5.0082656
81. Swift J, Ivanovska IL, Buxboim A, Harada T, Dingal PC, Pinter J, Pajerowski JD, Spinler KR, Shin JW, Tewari M, Rehfeldt F, Speicher DW, Discher DE (2013) Nuclear lamin-A scales with tissue stiffness and enhances matrix-directed differentiation. Science 341(6149):1240104. https://doi.org/10.1126/science.1240104
82. Segura-Totten M, Wilson KL (2004) BAF: roles in chromatin, nuclear structure and retrovirus integration. Trends Cell Biol 14(5): 261–266. https://doi.org/10.1016/j.tcb.2004.03.004
83. Sears RM, Roux KJ (2020) Diverse cellular functions of barrier-to-autointegration factor and its roles in disease. J Cell Sci 133(16). https://doi.org/10.1242/jcs.246546
84. Barton LJ, Duan T, Ke W, Luttinger A, Lovander KE, Soshnev AA, Geyer PK (2018) Nuclear lamina dysfunction triggers a germline stem cell checkpoint. Nat Commun 9(1): 3960. https://doi.org/10.1038/s41467-018-06277-z
85. Barton Lacy J, Pinto Belinda S, Wallrath Lori L, Geyer Pamela K (2013) The *Drosophila* nuclear lamina protein otefin is required for germline stem cell survival. Dev Cell 25(6):645–654. https://doi.org/10.1016/j.devcel.2013.05.023
86. Duan T, Kitzman SC, Geyer PK (2020) Survival of Drosophila germline stem cells requires the chromatin-binding protein barrier-to-autointegration factor. Development 147(9). https://doi.org/10.1242/dev.186171
87. Duan T, Cupp R, Geyer PK (2021) Drosophila female germline stem cells undergo mitosis without nuclear breakdown. Curr Biol 31(7): 1450–1462. e1453. https://doi.org/10.1016/j.cub.2021.01.033
88. Tatli M, Medalia O (2018) Insight into the functional organization of nuclear lamins in health and disease. Curr Opin Cell Biol 54: 72–79. https://doi.org/10.1016/j.ceb.2018.05.001
89. Shin JY, Worman HJ (2022) Molecular pathology of laminopathies. Annu Rev Pathol 17:159–180. https://doi.org/10.1146/annurev-pathol-042220-034240
90. Bosco G, Orr-Weaver TL (2002) The cell cycle during oogenesis and early embryogenesis in Drosophila. In: Advances in developmental biology and biochemistry, vol 12. Elsevier, pp 107–154. https://doi.org/10.1016/S1569-1799(02)12026-0

91. Smith PA, King RC (1968) Genetic control of synaptonemal complexes in Drosophila melanogaster. Genetics 60(2):335–351. https://doi.org/10.1093/genetics/60.2.335
92. Von Stetina JR, Orr-Weaver TL (2011) Developmental control of oocyte maturation and egg activation in metazoan models. Cold Spring Harb Perspect Biol 3(10):a005553. https://doi.org/10.1101/cshperspect.a005553
93. Mahowald AP, Goralski TJ, Caulton JH (1983) In vitro activation of Drosophila eggs. Dev Biol 98(2):437–445. https://doi.org/10.1016/0012-1606(83)90373-1
94. Lilly MA, Spradling AC (1996) The Drosophila endocycle is controlled by cyclin E and lacks a checkpoint ensuring S-phase completion. Genes Dev 10(19):2514–2526. https://doi.org/10.1101/gad.10.19.2514
95. Klusza S, Deng WM (2011) At the crossroads of differentiation and proliferation: precise control of cell-cycle changes by multiple signaling pathways in Drosophila follicle cells. BioEssays 33(2):124–134. https://doi.org/10.1002/bies.201000089
96. Lopez-Schier H, St Johnston D (2001) Delta signaling from the germ line controls the proliferation and differentiation of the somatic follicle cells during Drosophila oogenesis. Genes Dev 15(11):1393–1405. https://doi.org/10.1101/gad.200901
97. Deng WM, Althauser C, Ruohola-Baker H (2001) Notch-Delta signaling induces a transition from mitotic cell cycle to endocycle in Drosophila follicle cells. Development 128(23):4737–4746. https://doi.org/10.1242/dev.128.23.4737
98. Sun J, Smith L, Armento A, Deng WM (2008) Regulation of the endocycle/gene amplification switch by Notch and ecdysone signaling. J Cell Biol 182(5):885–896. https://doi.org/10.1083/jcb.200802084
99. Calvi BR, Lilly MA, Spradling AC (1998) Cell cycle control of chorion gene amplification. Genes Dev 12(5):734–744. https://doi.org/10.1101/gad.12.5.734
100. Sun J, Deng WM (2007) Hindsight mediates the role of notch in suppressing hedgehog signaling and cell proliferation. Dev Cell 12(3):431–442. https://doi.org/10.1016/j.devcel.2007.02.003
101. Sun J, Deng WM (2005) Notch-dependent downregulation of the homeodomain gene cut is required for the mitotic cycle/endocycle switch and cell differentiation in Drosophila follicle cells. Development 132(19): 4299–4308. https://doi.org/10.1242/dev.02015
102. Edgar BA, Zielke N, Gutierrez C (2014) Endocycles: a recurrent evolutionary innovation for post-mitotic cell growth. Nat Rev Mol Cell Biol 15(3):197–210. https://doi.org/10.1038/nrm3756
103. Lee HO, Davidson JM, Duronio RJ (2009) Endoreplication: polyploidy with purpose. Genes Dev 23(21):2461–2477. https://doi.org/10.1101/gad.1829209
104. Cross JC (2005) How to make a placenta: mechanisms of trophoblast cell differentiation in mice--a review. Placenta 26(Suppl A):S3–S9. https://doi.org/10.1016/j.placenta.2005.01.015
105. Rossant J, Cross JC (2001) Placental development: lessons from mouse mutants. Nat Rev Genet 2(7):538–548. https://doi.org/10.1038/35080570
106. Watson ED, Cross JC (2005) Development of structures and transport functions in the mouse placenta. Physiology (Bethesda) 20: 180–193. https://doi.org/10.1152/physiol.00001.2005
107. Lazzerini Denchi E, Celli G, de Lange T (2006) Hepatocytes with extensive telomere deprotection and fusion remain viable and regenerate liver mass through endoreduplication. Genes Dev 20(19):2648–2653. https://doi.org/10.1101/gad.1453606
108. Fox DT, Duronio RJ (2013) Endoreplication and polyploidy: insights into development and disease. Development 140(1):3–12. https://doi.org/10.1242/dev.080531
109. Claycomb JM, Benasutti M, Bosco G, Fenger DD, Orr-Weaver TL (2004) Gene amplification as a developmental strategy: isolation of two developmental amplicons in Drosophila. Dev Cell 6(1):145–155. https://doi.org/10.1016/s1534-5807(03)00398-8
110. Claycomb JM, Orr-Weaver TL (2005) Developmental gene amplification: insights into DNA replication and gene expression. Trends Genet 21(3):149–162. https://doi.org/10.1016/j.tig.2005.01.009
111. Kim JC, Nordman J, Xie F, Kashevsky H, Eng T, Li S, MacAlpine DM, Orr-Weaver TL (2011) Integrative analysis of gene amplification in Drosophila follicle cells: parameters of origin activation and repression. Genes Dev 25(13):1384–1398. https://doi.org/10.1101/gad.2043111
112. Nordman J, Orr-Weaver TL (2012) Regulation of DNA replication during development. Development 139(3):455–464. https://doi.org/10.1242/dev.061838
113. Shoshani O, Brunner SF, Yaeger R, Ly P, Nechemia-Arbely Y, Kim DH, Fang R, Castillon GA, Yu M, Li JSZ, Sun Y, Ellisman MH,

Ren B, Campbell PJ, Cleveland DW (2021) Chromothripsis drives the evolution of gene amplification in cancer. Nature 591(7848): 137–141. https://doi.org/10.1038/s41586-020-03064-z

114. Matsui A, Ihara T, Suda H, Mikami H, Semba K (2013) Gene amplification: mechanisms and involvement in cancer. Biomol Concepts 4(6):567–582. https://doi.org/10.1515/bmc-2013-0026

115. Friedl P, Gilmour D (2009) Collective cell migration in morphogenesis, regeneration and cancer. Nat Rev Mol Cell Biol 10(7): 445–457. https://doi.org/10.1038/nrm2720

116. Ilina O, Friedl P (2009) Mechanisms of collective cell migration at a glance. J Cell Sci 122(Pt 18):3203–3208. https://doi.org/10.1242/jcs.036525

117. De Pascalis C, Etienne-Manneville S (2017) Single and collective cell migration: the mechanics of adhesions. Mol Biol Cell 28(14):1833–1846. https://doi.org/10.1091/mbc.E17-03-0134

118. Mayor R, Etienne-Manneville S (2016) The front and rear of collective cell migration. Nat Rev Mol Cell Biol 17(2):97–109. https://doi.org/10.1038/nrm.2015.14

119. Aguilar-Cuenca R, Juanes-Garcia A, Vicente-Manzanares M (2014) Myosin II in mechanotransduction: master and commander of cell migration, morphogenesis, and cancer. Cell Mol Life Sci 71(3):479–492. https://doi.org/10.1007/s00018-013-1439-5

120. Gasparski AN, Ozarkar S, Beningo KA (2017) Transient mechanical strain promotes the maturation of invadopodia and enhances cancer cell invasion in vitro. J Cell Sci 130(11): 1965–1978. https://doi.org/10.1242/jcs.199760

121. Eble JA, Niland S (2019) The extracellular matrix in tumor progression and metastasis. Clin Exp Metastasis 36(3):171–198. https://doi.org/10.1007/s10585-019-09966-1

122. Cetera M, Ramirez-San Juan GR, Oakes PW, Lewellyn L, Fairchild MJ, Tanentzapf G, Gardel ML, Horne-Badovinac S (2014) Epithelial rotation promotes the global alignment of contractile actin bundles during Drosophila egg chamber elongation. Nat Commun 5(1):5511. https://doi.org/10.1038/ncomms6511

123. Haigo SL, Bilder D (2011) Global tissue revolutions in a morphogenetic movement controlling elongation. Science 331(6020): 1071–1074. https://doi.org/10.1126/science.1199424

124. Gutzeit HO (1991) Organization and in vitro activity of microfilament bundles associated with the basement membrane of Drosophila follicles. Acta Histochem Suppl 41:201–210

125. Gutzeit HO (1990) The microfilament pattern in the somatic follicle cells of mid-vitellogenic ovarian follicles of Drosophila. Eur J Cell Biol 53(2):349–356

126. Gutzeit HO, Eberhardt W, Gratwohl E (1991) Laminin and basement membrane-associated microfilaments in wild-type and mutant Drosophila ovarian follicles. J Cell Sci 100(Pt 4):781–788. https://doi.org/10.1242/jcs.100.4.781

127. Sherrard KM, Cetera M, Horne-Badovinac S (2021) DAAM mediates the assembly of long-lived, treadmilling stress fibers in collectively migrating epithelial cells in Drosophila. elife 10. https://doi.org/10.7554/eLife.72881

128. Bateman J, Reddy RS, Saito H, Van Vactor D (2001) The receptor tyrosine phosphatase Dlar and integrins organize actin filaments in the Drosophila follicular epithelium. Curr Biol 11(17):1317–1327. https://doi.org/10.1016/s0960-9822(01)00420-1

129. Delon I, Brown NH (2009) The integrin adhesion complex changes its composition and function during morphogenesis of an epithelium. J Cell Sci 122(Pt 23):4363–4374. https://doi.org/10.1242/jcs.055996

130. Bianco A, Poukkula M, Cliffe A, Mathieu J, Luque CM, Fulga TA, Rorth P (2007) Two distinct modes of guidance signalling during collective migration of border cells. Nature 448(7151):362–365. https://doi.org/10.1038/nature05965

131. Poukkula M, Cliffe A, Changede R, Rorth P (2011) Cell behaviors regulated by guidance cues in collective migration of border cells. J Cell Biol 192(3):513–524. https://doi.org/10.1083/jcb.201010003

132. Dai W, Guo X, Cao Y, Mondo JA, Campanale JP, Montell BJ, Burrous H, Streichan S, Gov N, Rappel WJ, Montell DJ (2020) Tissue topography steers migrating Drosophila border cells. Science 370(6519):987–990. https://doi.org/10.1126/science.aaz4741

133. Aranjuez G, Burtscher A, Sawant K, Majumder P, McDonald JA (2016) Dynamic myosin activation promotes collective morphology and migration by locally balancing oppositional forces from surrounding tissue. Mol Biol Cell 27(12):1898–1910. https://doi.org/10.1091/mbc.E15-10-0744

134. Majumder P, Aranjuez G, Amick J, McDonald JA (2012) Par-1 controls myosin-II

activity through myosin phosphatase to regulate border cell migration. Curr Biol 22(5): 363–372. https://doi.org/10.1016/j.cub.2012.01.037

135. Lamb MC, Kaluarachchi CP, Lansakara TI, Mellentine SQ, Lan Y, Tivanski AV, Tootle TL (2021) Fascin limits Myosin activity within Drosophila border cells to control substrate stiffness and promote migration. elife 10:e69836. C69831 – elife 62021;69810: e69836. https://doi.org/10.7554/eLife.69836

136. Fox EF, Lamb MC, Mellentine SQ, Tootle TL (2020) Prostaglandins regulate invasive, collective border cell migration. Mol Biol Cell 31(15):1584–1594. https://doi.org/10.1091/mbc.E19-10-0578

137. Prasad M, Montell DJ (2007) Cellular and molecular mechanisms of border cell migration analyzed using time-lapse live-cell imaging. Dev Cell 12(6):997–1005. https://doi.org/10.1016/j.devcel.2007.03.021

138. Prasad M, Wang X, He L, Montell DJ (2011) Border cell migration: a model system for live imaging and genetic analysis of collective cell movement. Methods Mol Biol 769:277–286. https://doi.org/10.1007/978-1-61779-207-6_19

139. Sapir A, Schweitzer R, Shilo BZ (1998) Sequential activation of the EGF receptor pathway during Drosophila oogenesis establishes the dorsoventral axis. Development 125(2):191–200. https://doi.org/10.1242/dev.125.2.191

140. Peri F, Bokel C, Roth S (1999) Local Gurken signaling and dynamic MAPK activation during Drosophila oogenesis. Mech Dev 81(1–2):75–88. https://doi.org/10.1016/s0925-4773(98)00228-7

141. Wasserman JD, Freeman M (1998) An autoregulatory cascade of EGF receptor signaling patterns the Drosophila egg. Cell 95(3): 355–364. https://doi.org/10.1016/s0092-8674(00)81767-5

142. Neuman-Silberberg FS, Schupbach T (1994) Dorsoventral axis formation in Drosophila depends on the correct dosage of the gene gurken. Development 120(9):2457–2463. https://doi.org/10.1242/dev.120.9.2457

143. Dequier E, Souid S, Pal M, Maroy P, Lepesant JA, Yanicostas C (2001) Top-DER- and Dpp-dependent requirements for the Drosophila fos/kayak gene in follicular epithelium morphogenesis. Mech Dev 106(1–2): 47–60. https://doi.org/10.1016/s0925-4773(01)00418-x

144. Peri F, Roth S (2000) Combined activities of Gurken and decapentaplegic specify dorsal chorion structures of the Drosophila egg. Development 127(4):841–850. https://doi.org/10.1242/dev.127.4.841

145. Queenan AM, Ghabrial A, Schupbach T (1997) Ectopic activation of torpedo/Egfr, a Drosophila receptor tyrosine kinase, dorsalizes both the eggshell and the embryo. Development 124(19):3871–3880. https://doi.org/10.1242/dev.124.19.3871

146. Deng WM, Bownes M (1997) Two signalling pathways specify localised expression of the Broad-Complex in Drosophila eggshell patterning and morphogenesis. Development 124(22):4639–4647. https://doi.org/10.1242/dev.124.22.4639

147. Twombly V, Blackman RK, Jin H, Graff JM, Padgett RW, Gelbart WM (1996) The TGF-beta signaling pathway is essential for Drosophila oogenesis. Development 122(5): 1555–1565. https://doi.org/10.1242/dev.122.5.1555

148. Dorman JB, James KE, Fraser SE, Kiehart DP, Berg CA (2004) bullwinkle is required for epithelial morphogenesis during Drosophila oogenesis. Dev Biol 267(2): 320–341. https://doi.org/10.1016/j.ydbio.2003.10.020

149. Ward EJ, Berg CA (2005) Juxtaposition between two cell types is necessary for dorsal appendage tube formation. Mech Dev 122(2):241–255. https://doi.org/10.1016/j.mod.2004.10.006

150. Tran DH, Berg CA (2003) bullwinkle and shark regulate dorsal-appendage morphogenesis in Drosophila oogenesis. Development 130(25):6273–6282. https://doi.org/10.1242/dev.00854

151. Cartwright JH, Piro O, Tuval I (2009) Fluid dynamics in developmental biology: moving fluids that shape ontogeny. HFSP J 3(2): 77–93. https://doi.org/10.2976/1.3043738

152. Quinlan ME (2016) Cytoplasmic streaming in the Drosophila oocyte. Annu Rev Cell Dev Biol 32:173–195. https://doi.org/10.1146/annurev-cellbio-111315-125416

153. Gutzeit H, Koppa R (1982) Time-lapse film analysis of cytoplasmic streaming during late oogenesis of Drosophila. Development 67(1):101–111. https://doi.org/10.1242/dev.67.1.101

154. Quinlan ME (2013) Direct interaction between two actin nucleators is required in Drosophila oogenesis. Development

140(21):4417–4425. https://doi.org/10.1242/dev.097337

155. Serbus LR, Cha BJ, Theurkauf WE, Saxton WM (2005) Dynein and the actin cytoskeleton control kinesin-driven cytoplasmic streaming in Drosophila oocytes. Development 132(16):3743–3752. https://doi.org/10.1242/dev.01956
156. Gutzeit HO (1986) The role of microfilaments in cytoplasmic streaming in Drosophila follicles. J Cell Sci 80:159–169. https://doi.org/10.1242/jcs.80.1.159
157. Palacios IM, St Johnston D (2002) Kinesin light chain-independent function of the Kinesin heavy chain in cytoplasmic streaming and posterior localisation in the Drosophila oocyte. Development 129(23):5473–5485. https://doi.org/10.1242/dev.00119
158. Dahlgaard K, Raposo AA, Niccoli T, St Johnston D (2007) Capu and Spire assemble a cytoplasmic actin mesh that maintains microtubule organization in the Drosophila oocyte. Dev Cell 13(4):539–553. https://doi.org/10.1016/j.devcel.2007.09.003
159. Becalska AN, Gavis ER (2009) Lighting up mRNA localization in Drosophila oogenesis. Development 136(15):2493–2503. https://doi.org/10.1242/dev.032391
160. Kugler JM, Lasko P (2009) Localization, anchoring and translational control of oskar, gurken, bicoid and nanos mRNA during Drosophila oogenesis. Fly (Austin) 3(1):15–28. https://doi.org/10.4161/fly.3.1.7751
161. Lei L, Spradling AC (2016) Mouse oocytes differentiate through organelle enrichment from sister cyst germ cells. Science 352(6281):95–99. https://doi.org/10.1126/science.aad2156
162. Niu W, Spradling AC (2022) Mouse oocytes develop in cysts with the help of nurse cells. Cell 185:2576. https://doi.org/10.1016/j.cell.2022.05.001
163. Guild GM, Connelly PS, Shaw MK, Tilney LG (1997) Actin filament cables in Drosophila nurse cells are composed of modules that slide passively past one another during dumping. J Cell Biol 138(4):783–797. https://doi.org/10.1083/jcb.138.4.783
164. Huelsmann S, Ylänne J, Brown Nicholas H (2013) Filopodia-like actin cables position nuclei in association with perinuclear actin in Drosophila nurse cells. Dev Cell 26(6): 604–615. https://doi.org/10.1016/j.devcel.2013.08.014
165. Wheatley S, Kulkarni S, Karess R (1995) Drosophila nonmuscle myosin II is required for rapid cytoplasmic transport during oogenesis and for axial nuclear migration in early embryos. Development 121(6):1937–1946
166. Imran Alsous J, Romeo N, Jackson JA, Mason FM, Dunkel J, Martin AC (2021) Dynamics of hydraulic and contractile wave-mediated fluid transport during Drosophila oogenesis. Proc Natl Acad Sci U S A 118(10). https://doi.org/10.1073/pnas.2019749118
167. Cooley L, Verheyen E, Ayers K (1992) chickadee encodes a profilin required for intercellular cytoplasm transport during Drosophila oogenesis. Cell 69(1):173–184. https://doi.org/10.1016/0092-8674(92)90128-y
168. Cant K, Knowles BA, Mooseker MS, Cooley L (1994) Drosophila singed, a fascin homolog, is required for actin bundle formation during oogenesis and bristle extension. J Cell Biol 125(2):369–380. https://doi.org/10.1083/jcb.125.2.369
169. Gates J, Nowotarski SH, Yin H, Mahaffey JP, Bridges T, Herrera C, Homem CC, Janody F, Montell DJ, Peifer M (2009) Enabled and Capping protein play important roles in shaping cell behavior during Drosophila oogenesis. Dev Biol 333(1):90–107. https://doi.org/10.1016/j.ydbio.2009.06.030
170. Tootle TL (2013) Genetic insights into the in vivo functions of prostaglandin signaling. Int J Biochem Cell Biol 45(8):1629–1632. https://doi.org/10.1016/j.biocel.2013.05.008
171. Tootle TL, Spradling AC (2008) Drosophila Pxt: a cyclooxygenase-like facilitator of follicle maturation. Development 135(5):839–847. https://doi.org/10.1242/dev.017590
172. Groen CM, Spracklen AJ, Fagan TN, Tootle TL (2012) Drosophila Fascin is a novel downstream target of prostaglandin signaling during actin remodeling. Mol Biol Cell 23(23): 4567–4578. https://doi.org/10.1091/mbc.E12-05-0417
173. Spracklen AJ, Kelpsch DJ, Chen X, Spracklen CN, Tootle TL (2014) Prostaglandins temporally regulate cytoplasmic actin bundle formation during Drosophila oogenesis. Mol Biol Cell 25(3):397–411. https://doi.org/10.1091/mbc.E13-07-0366
174. Spracklen AJ, Lamb MC, Groen CM, Tootle TL (2019) Pharmaco-genetic screen to uncover actin regulators targeted by prostaglandins during Drosophila oogenesis. G3 (Bethesda) 9(11):3555–3565. https://doi.org/10.1534/g3.119.400704
175. Hamm DC, Harrison MM (2018) Regulatory principles governing the maternal-to-zygotic transition: insights from Drosophila

melanogaster. Open Biol 8(12):180183. https://doi.org/10.1098/rsob.180183
176. Welte MA (2015) As the fat flies: the dynamic lipid droplets of Drosophila embryos. Biochim Biophys Acta 1851(9):1156–1185. https://doi.org/10.1016/j.bbalip.2015.04.002
177. Brusentsev EY, Mokrousova VI, Igonina TN, Rozhkova IN, Amstislavsky SY (2019) Role of lipid droplets in the development of oocytes and preimplantation embryos in mammals. Russ J Dev Biol 50(5):230–237. https://doi.org/10.1134/S1062360419050102
178. Dunning KR, Russell DL, Robker RL (2014) Lipids and oocyte developmental competence: the role of fatty acids and beta-oxidation. Reproduction 148(1):R15–R27. https://doi.org/10.1530/REP-13-0251
179. Ami D, Mereghetti P, Natalello A, Doglia SM, Zanoni M, Redi CA, Monti M (2011) FTIR spectral signatures of mouse antral oocytes: molecular markers of oocyte maturation and developmental competence. Biochim Biophys Acta 1813(6):1220–1229. https://doi.org/10.1016/j.bbamcr.2011.03.009
180. Cardozo E, Pavone ME, Hirshfeld-Cytron JE (2011) Metabolic syndrome and oocyte quality. Trends Endocrinol Metab 22(3):103–109. https://doi.org/10.1016/j.tem.2010.12.002
181. Jungheim ES, Schoeller EL, Marquard KL, Louden ED, Schaffer JE, Moley KH (2010) Diet-induced obesity model: abnormal oocytes and persistent growth abnormalities in the offspring. Endocrinology 151(8):4039–4046. https://doi.org/10.1210/en.2010-0098
182. Marei WFA, Smits A, Mohey-Elsaeed O, Pintelon I, Ginneberge D, Bols PEJ, Moerloose K, Leroy J (2020) Differential effects of high fat diet-induced obesity on oocyte mitochondrial functions in inbred and outbred mice. Sci Rep 10(1):9806. https://doi.org/10.1038/s41598-020-66702-6
183. Prates EG, Nunes JT, Pereira RM (2014) A role of lipid metabolism during cumulus-oocyte complex maturation: impact of lipid modulators to improve embryo production. Mediat Inflamm 2014:692067. https://doi.org/10.1155/2014/692067
184. Khajeh M, Rahbarghazi R, Nouri M, Darabi M (2017) Potential role of polyunsaturated fatty acids, with particular regard to the signaling pathways of arachidonic acid and its derivatives in the process of maturation of the oocytes: contemporary review. Biomed Pharmacother 94:458–467. https://doi.org/10.1016/j.biopha.2017.07.140
185. Cermelli S, Guo Y, Gross SP, Welte MA (2006) The lipid-droplet proteome reveals that droplets are a protein-storage depot. Curr Biol 16(18):1783–1795. https://doi.org/10.1016/j.cub.2006.07.062
186. Li Z, Thiel K, Thul Peter J, Beller M, Kühnlein Ronald P, Welte Michael A (2012) Lipid droplets control the maternal histone supply of Drosophila embryos. Curr Biol 22(22):2104–2113. https://doi.org/10.1016/j.cub.2012.09.018
187. Li Z, Johnson Matthew R, Ke Z, Chen L, Welte Michael A (2014) Drosophila lipid droplets buffer the H2Av supply to protect early embryonic development. Curr Biol 24(13):1485–1491. https://doi.org/10.1016/j.cub.2014.05.022
188. Johnson MR, Stephenson RA, Ghaemmaghami S, Welte MA (2018) Developmentally regulated H2Av buffering via dynamic sequestration to lipid droplets in Drosophila embryos. elife 7:e36021. C36021 – elife 32018;36027:e36021. https://doi.org/10.7554/eLife.36021
189. Sieber MH, Spradling AC (2015) Steroid signaling establishes a female metabolic state and regulates SREBP to control oocyte lipid accumulation. Curr Biol 25(8):993–1004. https://doi.org/10.1016/j.cub.2015.02.019
190. Parra-Peralbo E, Culi J (2011) Drosophila lipophorin receptors mediate the uptake of neutral lipids in oocytes and imaginal disc cells by an endocytosis-independent mechanism. PLoS Genet 7(2):e1001297. https://doi.org/10.1371/journal.pgen.1001297
191. Rodriguez-Vazquez M, Vaquero D, Parra-Peralbo E, Mejia-Morales JE, Culi J (2015) Drosophila lipophorin receptors recruit the lipoprotein LTP to the plasma membrane to mediate lipid uptake. PLoS Genet 11(6):e1005356. https://doi.org/10.1371/journal.pgen.1005356
192. Buszczak M, Lu X, Segraves WA, Chang TY, Cooley L (2002) Mutations in the midway gene disrupt a Drosophila acyl coenzyme a: diacylglycerol acyltransferase. Genetics 160(4):1511–1518. https://doi.org/10.1093/genetics/160.4.1511
193. Pompeia C, Lima T, Curi R (2003) Arachidonic acid cytotoxicity: can arachidonic acid be a physiological mediator of cell death? Cell Biochem Funct 21(2):97–104. https://doi.org/10.1002/cbf.1012

194. Zhang N, Wang L, Luo G, Tang X, Ma L, Zheng Y, Liu S, Price C, Jiang Z (2019) Arachidonic acid regulation of intracellular signaling pathways and target gene expression in bovine ovarian granulosa cells. Animals (Basel) 9(6). https://doi.org/10.3390/ani9060374
195. Giedt MS, Thomalla JM, Johnson MR, Lai ZW, Tootle TL, Welte MA (2022) Adipose triglyceride lipase promotes prostaglandin-dependent actin remodeling by regulating substrate release from lipid droplets. bioRxiv:2021.2008.2002.454724. https://doi.org/10.1101/2021.08.02.454724
196. Drummond-Barbosa D, Spradling AC (2001) Stem cells and their progeny respond to nutritional changes during Drosophila oogenesis. Dev Biol 231(1):265–278. https://doi.org/10.1006/dbio.2000.0135
197. Hou YC, Chittaranjan S, Barbosa SG, McCall K, Gorski SM (2008) Effector caspase Dcp-1 and IAP protein Bruce regulate starvation-induced autophagy during Drosophila melanogaster oogenesis. J Cell Biol 182(6):1127–1139. https://doi.org/10.1083/jcb.200712091
198. Nezis IP, Lamark T, Velentzas AD, Rusten TE, Bjorkoy G, Johansen T, Papassideri IS, Stravopodis DJ, Margaritis LH, Stenmark H, Brech A (2009) Cell death during Drosophila melanogaster early oogenesis is mediated through autophagy. Autophagy 5(3):298–302. https://doi.org/10.4161/auto.5.3.7454
199. Besse F, Pret AM (2003) Apoptosis-mediated cell death within the ovarian polar cell lineage of Drosophila melanogaster. Development 130(5):1017–1027. https://doi.org/10.1242/dev.00313
200. Khammari A, Agnes F, Gandille P, Pret AM (2011) Physiological apoptosis of polar cells during Drosophila oogenesis is mediated by Hid-dependent regulation of Diap1. Cell Death Differ 18(5):793–805. https://doi.org/10.1038/cdd.2010.141
201. Borensztejn A, Mascaro A, Wharton KA (2018) JAK/STAT signaling prevents excessive apoptosis to ensure maintenance of the interfollicular stalk critical for Drosophila oogenesis. Dev Biol 438(1):1–9. https://doi.org/10.1016/j.ydbio.2018.03.018
202. Pritchett TL, McCall K (2012) Role of the insulin/Tor signaling network in starvation-induced programmed cell death in Drosophila oogenesis. Cell Death Differ 19(6):1069–1079. https://doi.org/10.1038/cdd.2011.200
203. Chao S, Nagoshi RN (1999) Induction of apoptosis in the germline and follicle layer of Drosophila egg chambers. Mech Dev 88(2):159–172. https://doi.org/10.1016/s0925-4773(99)00183-5
204. Timmons AK, Mondragon AA, Schenkel CE, Yalonetskaya A, Taylor JD, Moynihan KE, Etchegaray JI, Meehan TL, McCall K (2016) Phagocytosis genes nonautonomously promote developmental cell death in the Drosophila ovary. Proc Natl Acad Sci U S A 113(9):E1246–E1255. https://doi.org/10.1073/pnas.1522830113
205. Yalonetskaya A, Mondragon AA, Hintze ZJ, Holmes S, McCall K (2020) Nuclear degradation dynamics in a nonapoptotic programmed cell death. Cell Death Differ 27(2):711–724. https://doi.org/10.1038/s41418-019-0382-x
206. Mondragon AA, Yalonetskaya A, Ortega AJ, Zhang Y, Naranjo O, Elguero J, Chung WS, McCall K (2019) Lysosomal machinery drives extracellular acidification to direct non-apoptotic cell death. Cell Rep 27(1):11–19 e13. https://doi.org/10.1016/j.celrep.2019.03.034
207. Nezis IP, Stravopodis DJ, Margaritis LH, Papassideri IS (2006) Autophagy is required for the degeneration of the ovarian follicular epithelium in higher Diptera. Autophagy 2(4):297–298. https://doi.org/10.4161/auto.2858
208. Nezis IP, Stravopodis DJ, Margaritis LH, Papassideri IS (2006) Programmed cell death of follicular epithelium during the late developmental stages of oogenesis in the fruit flies Bactrocera oleae and Ceratitis capitata (Diptera, Tephritidae) is mediated by autophagy. Develop Growth Differ 48(3):189–198. https://doi.org/10.1111/j.1440-169X.2006.00856.x
209. Buszczak M, Signer RA, Morrison SJ (2014) Cellular differences in protein synthesis regulate tissue homeostasis. Cell 159(2):242–251. https://doi.org/10.1016/j.cell.2014.09.016
210. Zhang Q, Shalaby NA, Buszczak M (2014) Changes in rRNA transcription influence proliferation and cell fate within a stem cell lineage. Science 343(6168):298–301. https://doi.org/10.1126/science.1246384
211. Sanchez CG, Teixeira FK, Czech B, Preall JB, Zamparini AL, Seifert JR, Malone CD, Hannon GJ, Lehmann R (2016) Regulation of ribosome biogenesis and protein synthesis controls germline stem cell differentiation. Cell Stem Cell 18(2):276–290. https://doi.org/10.1016/j.stem.2015.11.004

212. Martin ET, Blatt P, Nguyen E, Lahr R, Selvam S, Yoon HAM, Pocchiari T, Emtenani S, Siekhaus DE, Berman A, Fuchs G, Rangan P (2022) A translation control module coordinates germline stem cell differentiation with ribosome biogenesis during drosophila oogenesis. Dev Cell 57(7): 883–900 e810. https://doi.org/10.1016/j.devcel.2022.03.005
213. Kong J, Han H, Bergalet J, Bouvrette LPB, Hernandez G, Moon NS, Vali H, Lecuyer E, Lasko P (2019) A ribosomal protein S5 isoform is essential for oogenesis and interacts with distinct RNAs in Drosophila melanogaster. Sci Rep 9(1):13779. https://doi.org/10.1038/s41598-019-50357-z
214. Jang S, Lee J, Mathews J, Ruess H, Williford AO, Rangan P, Betran E, Buszczak M (2021) The Drosophila ribosome protein S5 paralog RpS5b promotes germ cell and follicle cell differentiation during oogenesis. Development 148(19). https://doi.org/10.1242/dev.199511
215. Hopes T, Norris K, Agapiou M, McCarthy CGP, Lewis PA, O'Connell MJ, Fontana J, Aspden JL (2022) Ribosome heterogeneity in Drosophila melanogaster gonads through paralog-switching. Nucleic Acids Res 50(4): 2240–2257. https://doi.org/10.1093/nar/gkab606
216. Green NM, Kimble GC, Talbot DE, Tootle TL (2021) Nuclear actin. In: eLS. Wiley, pp 958–967. https://doi.org/10.1002/9780470015902.a0028471
217. Kelpsch DJ, Tootle TL (2018) Nuclear actin: from discovery to function. Anat Rec (Hoboken) 301(12):1999–2013. https://doi.org/10.1002/ar.23959
218. Misu S, Takebayashi M, Miyamoto K (2017) Nuclear actin in development and transcriptional reprogramming. Front Genet 8:27. https://doi.org/10.3389/fgene.2017.00027
219. Klages-Mundt NL, Kumar A, Zhang Y, Kapoor P, Shen X (2018) The nature of actin-family proteins in chromatin-modifying complexes. Front Genet 9:398. https://doi.org/10.3389/fgene.2018.00398
220. Spracklen AJ, Fagan TN, Lovander KE, Tootle TL (2014) The pros and cons of common actin labeling tools for visualizing actin dynamics during Drosophila oogenesis. Dev Biol 393(2):209–226. https://doi.org/10.1016/j.ydbio.2014.06.022
221. Wineland DM, Kelpsch DJ, Tootle TL (2018) Multiple pools of nuclear actin. Anat Rec (Hoboken) 301(12):2014–2036. https://doi.org/10.1002/ar.23964
222. Kelpsch DJ, Groen CM, Fagan TN, Sudhir S, Tootle TL (2016) Fascin regulates nuclear actin during Drosophila oogenesis. Mol Biol Cell 27(19):2965–2979. https://doi.org/10.1091/mbc.E15-09-0634
223. Sokolova M, Moore HM, Prajapati B, Dopie J, Meriläinen L, Honkanen M, Matos RC, Poukkula M, Hietakangas V, Vartiainen MK (2018) Nuclear actin is required for transcription during Drosophila oogenesis. iScience 9:63–70. https://doi.org/10.1016/j.isci.2018.10.010
224. Hudson AM, Mannix KM, Gerdes JA, Kottemann MC, Cooley L (2019) Targeted substrate degradation by Kelch controls the actin cytoskeleton during ring canal expansion. Development 146(1). https://doi.org/10.1242/dev.169219
225. Mannix KM, Starble RM, Kaufman RS, Cooley L (2019) Proximity labeling reveals novel interactomes in live Drosophila tissue. Development 146(14). https://doi.org/10.1242/dev.176644
226. Zhang B, Zhang Y, Liu J-L (2021) Highly effective proximate labeling in drosophila. G3 Genes|Genomes|Genetics 11(5). https://doi.org/10.1093/g3journal/jkab077
227. Liu J-L (2016) The cytoophidium and its kind: filamentation and compartmentation of metabolic enzymes. Annu Rev Cell Dev Biol 32(1):349–372. https://doi.org/10.1146/annurev-cellbio-111315-124907
228. Liu JL (2010) Intracellular compartmentation of CTP synthase in Drosophila. J Genet Genomics 37(5):281–296. https://doi.org/10.1016/S1673-8527(09)60046-1
229. Lim H, Paria BC, Das SK, Dinchuk JE, Langenbach R, Trzaskos JM, Dey SK (1997) Multiple female reproductive failures in cyclooxygenase 2-deficient mice. Cell 91(2): 197–208. https://doi.org/10.1016/s0092-8674(00)80402-x
230. Langenbach R, Loftin C, Lee C, Tiano H (1999) Cyclooxygenase knockout mice: models for elucidating isoform-specific functions. Biochem Pharmacol 58(8):1237–1246. https://doi.org/10.1016/s0006-2952(99)00158-6
231. Takahashi T, Morrow JD, Wang H, Dey SK (2006) Cyclooxygenase-2-derived prostaglandin E(2) directs oocyte maturation by differentially influencing multiple signaling pathways. J Biol Chem 281(48): 37117–37129. https://doi.org/10.1074/jbc.M608202200
232. Pall M, Friden BE, Brannstrom M (2001) Induction of delayed follicular rupture in the

human by the selective COX-2 inhibitor rofecoxib: a randomized double-blind study. Hum Reprod 16(7):1323–1328. https://doi.org/10.1093/humrep/16.7.1323

233. Droujinine IA, Perrimon N (2016) Interorgan communication pathways in physiology: focus on Drosophila. Annu Rev Genet 50: 539–570. https://doi.org/10.1146/annurev-genet-121415-122024
234. Armstrong AR (2020) Drosophila melanogaster as a model for nutrient regulation of ovarian function. Reproduction 159(2): R69–R82. https://doi.org/10.1530/REP-18-0593
235. Barth JM, Szabad J, Hafen E, Kohler K (2011) Autophagy in Drosophila ovaries is induced by starvation and is required for oogenesis. Cell Death Differ 18(6): 915–924. https://doi.org/10.1038/cdd.2010.157
236. LaFever L, Drummond-Barbosa D (2005) Direct control of germline stem cell division and cyst growth by neural insulin in Drosophila. Science 309(5737):1071–1073. https://doi.org/10.1126/science.1111410
237. Ikeya T, Galic M, Belawat P, Nairz K, Hafen E (2002) Nutrient-dependent expression of insulin-like peptides from neuroendocrine cells in the CNS contributes to growth regulation in Drosophila. Curr Biol 12(15): 1293–1300. https://doi.org/10.1016/s0960-9822(02)01043-6
238. Armstrong AR, Drummond-Barbosa D (2018) Insulin signaling acts in adult adipocytes via GSK-3beta and independently of FOXO to control Drosophila female germline stem cell numbers. Dev Biol 440(1):31–39. https://doi.org/10.1016/j.ydbio.2018.04.028
239. Armstrong AR, Laws KM, Drummond-Barbosa D (2014) Adipocyte amino acid sensing controls adult germline stem cell number via the amino acid response pathway and independently of Target of Rapamycin signaling in Drosophila. Development 141(23): 4479–4488. https://doi.org/10.1242/dev.116467
240. Matsuoka S, Armstrong AR, Sampson LL, Laws KM, Drummond-Barbosa D (2017) Adipocyte metabolic pathways regulated by diet control the female germline stem cell lineage in Drosophila melanogaster. Genetics 206(2):953–971. https://doi.org/10.1534/genetics.117.201921
241. Weaver LN, Drummond-Barbosa D (2019) The nuclear receptor seven up functions in adipocytes and oenocytes to control distinct steps of Drosophila oogenesis. Dev Biol 456(2):179–189. https://doi.org/10.1016/j.ydbio.2019.08.015
242. Weaver LN, Drummond-Barbosa D (2021) Hormone receptor 4 is required in muscles and distinct ovarian cell types to regulate specific steps of Drosophila oogenesis. Development 148(5). https://doi.org/10.1242/dev.198663
243. Espey LL, Richards JS (2006) Chapter 11 – Ovulation. In: Neill JD (ed) Knobil and Neill's physiology of reproduction, 3rd edn. Academic Press, St Louis, pp 425–474. https://doi.org/10.1016/B978-012515400-0/50016-6
244. Duffy DM, Ko C, Jo M, Brannstrom M, Curry TE Jr (2019) Ovulation: parallels with inflammatory processes. Endocr Rev 40(2): 369–416. https://doi.org/10.1210/er.2018-00075
245. Fan H-Y, Liu Z, Mullany LK, Richards JS (2012) Consequences of RAS and MAPK activation in the ovary: the good, the bad and the ugly. Mol Cell Endocrinol 356(1): 74–79. https://doi.org/10.1016/j.mce.2011.12.005
246. Robker RL, Hennebold JD, Russell DL (2018) Coordination of ovulation and oocyte maturation: a good egg at the right time. Endocrinology 159(9):3209–3218. https://doi.org/10.1210/en.2018-00485
247. Li W, Young JF, Sun J (2018) NADPH oxidase-generated reactive oxygen species in mature follicles are essential for *Drosophila* ovulation. Proc Natl Acad Sci 115(30): 7765–7770. https://doi.org/10.1073/pnas.1800115115
248. Deady LD, Shen W, Mosure SA, Spradling AC, Sun J (2015) Matrix metalloproteinase 2 is required for ovulation and corpus luteum formation in Drosophila. PLoS Genet 11(2): e1004989
249. Knapp E, Sun J (2017) Steroid signaling in mature follicles is important for Drosophila ovulation. Proc Natl Acad Sci U S A 114(4): 699–704. https://doi.org/10.1073/pnas.1614383114
250. Deady LD, Sun J (2015) A follicle rupture assay reveals an essential role for follicular adrenergic signaling in Drosophila ovulation. PLoS Genet 11(10):e1005604
251. Jiang K, Zhang J, Huang Y, Wang Y, Xiao S, Hadden MK, Woodruff TK, Sun J (2021) A platform utilizing Drosophila ovulation for nonhormonal contraceptive screening. Proc Natl Acad Sci U S A 118(28). https://doi.org/10.1073/pnas.2026403118

252. Tootle TL, Hoffmann DS, Allen AK, Spracklen AJ, Groen CM, Kelpsch DJ (2019) Research and teaching: mini-course-based undergraduate research experience: impact on student understanding of STEM research and interest in STEM programs. J Coll Sci Teach. 48(6)
253. Cortes JA, Swanson CI (2023) Using Drosophila oogenesis in the classroom to increase student participation in biomedical research. In: Tootle TL, Giedt MS (eds) Drosophila oogenesis. Methods in molecular biology. Springer Nature, New York
254. Ruiz-Whalen DM, P. Aichele CP, Dywon ER et al (2023) Drosophila oogenesis. Methods in Molecular Biology. Springer Nature, New York

Chapter 2

Visualization and Quantification of *Drosophila* Larval Ovaries

Alicia E. Rosales-Nieves, Miriam Marín-Menguiano, Alejandro Campoy-Lopez, and Acaimo González-Reyes

Abstract

The morphogenesis of the ovarian germline stem cell (GSC) niche during larval stages in *Drosophila* provides the initial cellular and molecular basis for female gamete production in the adult. During larval instars, the *Drosophila* female gonad matures gradually from a round structure enclosing primordial germ cells (PGCs) and somatic cells into a functional organ containing GSC populations in their niches that later in adult stages support oogenesis. In this chapter, we describe a technique for dissecting, staining, and analyzing gonads from female *Drosophila* larvae and early pupae, offering the possibility of a direct view of the morphogenesis of an ovarian niche.

Key words *Drosophila*, Larvae, Gonad, Ovary, Dissection, Staining, Image analysis

1 Introduction

Gamete production in the *Drosophila* ovary relies on populations of somatic and germline stem cells (GSCs) housed in the germarium, a well-studied structure present at the tip of adult ovaries. Somatic support cells constitute the cellular component of the GSC niche and are essential for ovary homeostasis and gamete production, emphasizing their importance for proper female reproduction.

This well-characterized ovarian niche develops from somatic and germline precursors that proliferate during larval instars and populate the larval gonad. The future GSC niche forms during the third larval instar (L3), starting at ~4 days after the egg was laid. In addition to the primordial germ cells (PGCs) or germline

Alicia E. Rosales-Nieves is a lead contact to this chapter.

Supplementary Information The online version contains supplementary material available at https://doi.org/10.1007/978-1-0716-2970-3_2.

Michelle S. Giedt and Tina L. Tootle (eds.), *Drosophila Oogenesis: Methods and Protocols*, Methods in Molecular Biology, vol. 2626, https://doi.org/10.1007/978-1-0716-2970-3_2,

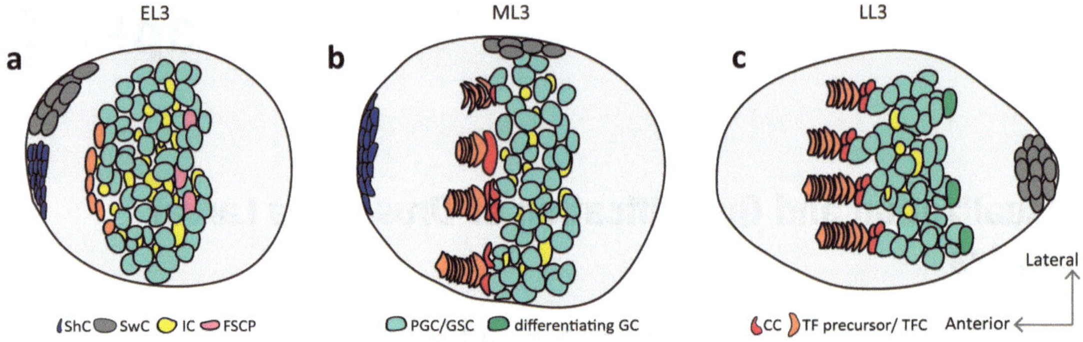

Fig. 1 *Drosophila* larval ovary development. The future germline stem cell niche forms during third larval instar (L3) beginning at ~4 days after egg laying. (**a**–**c**) Drawings of female gonads from (**a**) an early third instar larva (EL3), (**b**) a middle third instar larva (ML3) and, (**c**) a late third instar larva (LL3). Some of the cell types found in the developing gonads are labeled by different colors. These include the somatic sheath cells (ShC) in dark blue, swarm cells (SwC) in gray, intermingled cells (IC) in yellow, follicle stem cell precursors (FSCP) in pink, cap cells (CpC) and terminal filament precursors (TF) and terminal filament cells (TFC) in orange; and the germline-derived primordial germ cells (PGC), germline stem cells (GSC), both in light blue, and differentiating germline cells in green

precursors, the main somatic cell types present in early L3 (EL3) gonads are the anterior sheath cells (ShCs), follicle stem cell precursors (FSCPs), intermingled cells (ICs), and swarm cells (SwCs) (Fig. 1a). As gonads develop into the middle (ML3) and late L3 (LL3) stages, six somatic cell types are found, which include terminal filament cells (TFCs), cap cells (CpCs), ShCs, SwCs, FSCPs, and ICs (Fig. 1b–c) [1–7].

These cell types are organized into structured units that will function as GSC niches by a number of signaling and cellular events. First, a wave of morphogenetic cell rearrangements and cell recruitment is directed by an early ecdysone peak starting between EL3 and ML3. As a consequence, individual terminal filaments are gradually formed along the medial-lateral axis. In addition, 2–3 ICs located juxtaposed to the base of the terminal filaments are incorporated into each of the units and adopt a CpC fate. Next, the PGCs that directly associate with the developing niches become GSCs and begin to self-renew. However, PGCs not in direct contact with CpCs differentiate by a process controlled by hormone pulses and cell signaling [2, 8]. Importantly, for adult gametogenesis, a number of the cell types specified in larval stages are maintained into adulthood, including GSCs, TFCs, CpCs, FSCPs, and ICs, which will give rise to escort cells.

In this protocol, we describe a method for dissecting and staining *Drosophila* third instar larval and pre-pupal gonads. In addition, we provide some morphological cues that can assist in the classification of the developmental stage of the gonads.

Specifically, third instar gonads are grouped into EL3, ML3, or LL3 based on the presence and numbers of mature terminal filaments and associated CpC clusters and GSCs (Fig. 1).

Finally, because the larval gonad is a disk-shaped structure with hundreds of cells arranged in precise patterns that change with time, an accurate description of the different cell types and their organization could be cumbersome. To aid in this, we have developed an ImageJ-based method to quantify cell numbers in fixed gonads. This method, which requires immunofluorescence staining and gives consistent results, can be adapted to different fluorescent signals and cellular localizations of antigens, thus providing a versatile and efficient tool to analyze larval gonads.

2 Materials

Diligently follow all rules and regulations for waste disposal.

2.1 Gonad Dissection

1. Ringer's solution: 128 mM NaCl, 2 mM KCl, 1.8 mM $CaCl_2$, 4 mM $MgCl_2$, 35.5 mM Sucrose, 5 mM Hepes pH 6.9.
2. # 5 forceps.
3. Dissecting microscope (*see* **Note 1**).
4. Fly culture with larvae of the desired age. Avoid crowded cultures to improve the yield of well-fed larvae; (Fig. 2a; *see* **Note 2**).
5. Dissecting dish: 9-well glass plates (85 mm × 100 mm, 22 mm O.D. × 7 mm deep; Fig. 2b).

2.2 Gonad Staining

1. 6-well plastic plates (127.76 mm × 85.46 mm; Fig. 2c).
2. 40 μm cell strainer (25 mm diameter; Fig. 2c).
3. 5% formaldehyde in Ringer's solution, diluted from 10% formaldehyde, methanol-free, electron microscopy grade.
4. Orbital rotator.
5. 1% PBT solution: 1% Triton X-100 in phosphate buffered saline (PBS).
6. 0.3% PBTB (blocking solution): 0.3% Triton X-100 and 1% bovine serum albumin (BSA) in PBS.
7. Fetal bovine serum (FBS).
8. 0.3% PBT: 0.3% Triton X-100 in PBS.
9. Primary and secondary antibodies.
10. Nutator mixer or mini-shaker.
11. DNA stain (*see* **Note 3**).
12. Pipettes and pipette tips.

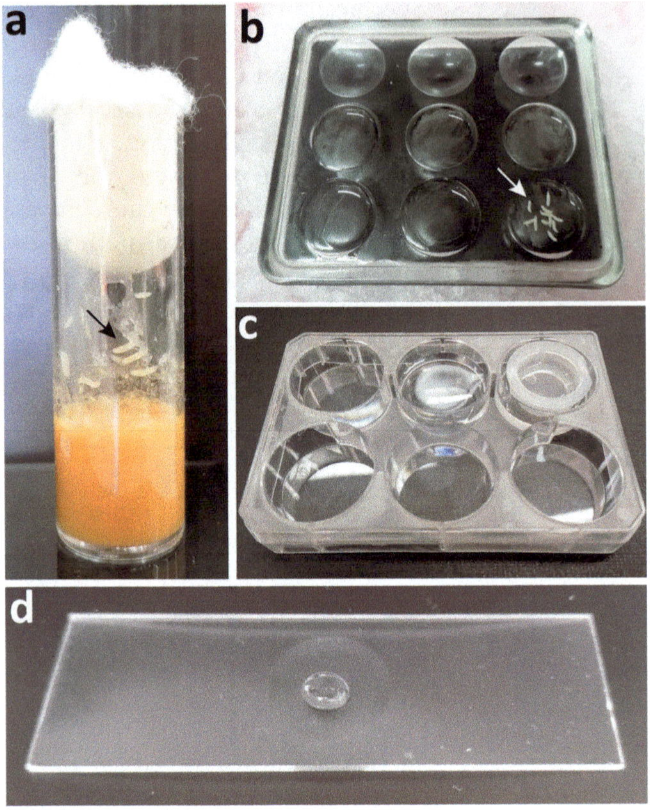

Fig. 2 Specialized material. (**a**) Fly culture vial containing wandering larvae and pre-pupa larvae. (**b**) Dissecting dish: 9-well glass plate. (**c**) 6-well plate with 40 μm cell strainer. (**d**) Shallow dissecting dish with a drop of liquid

2.3 Gonad Mounting

1. Shallow dissecting dish: Microscope slides with cavities 76 mm × 26 mm, 15 mm O. D. × 0.8 mm deep (Fig. 2d).
2. Nickel plated pin holder and stainless steel minutien pins 0.15 mm diameter, 12 mm long.
3. Microscope slides (25 mm × 75 mm non-treated surface).
4. Coverslips (22 mm × 22 mm, 0.13–0.16 mm thick).
5. Mounting medium (*see* **Note 4**).
6. Nail polish.

2.4 Gonad Counting

1. ImageJ.
2. Workstation.

3 Methods

3.1 Dissect Larval Fat Body

1. Collect larvae of desired stage (*see* **Note 2**) by either flooding a food vial with tap water or by directly picking the larvae with forceps from the walls of the vial (Fig. 2a).
2. Place the selected larvae in a dissecting dish containing cold Ringer's solution (Fig. 2b).
3. Under a dissecting microscope, separate female from male larvae. Larval gonads are clear spheres embedded in fat body and located about one-third from the posterior end of the larvae (Fig. 3a, a'). Female gonads can be distinguished because they are significantly smaller than male gonads (Fig. 3a').
4. Transfer the selected larvae into a shallow dissecting well (Fig. 2d) containing Ringer's solution and remove the head by pulling with one pair forceps while holding with another (Fig. 4a). Grasp the remaining of the larva with one pair of forceps and slowly turn the cuticle inside out from the posterior

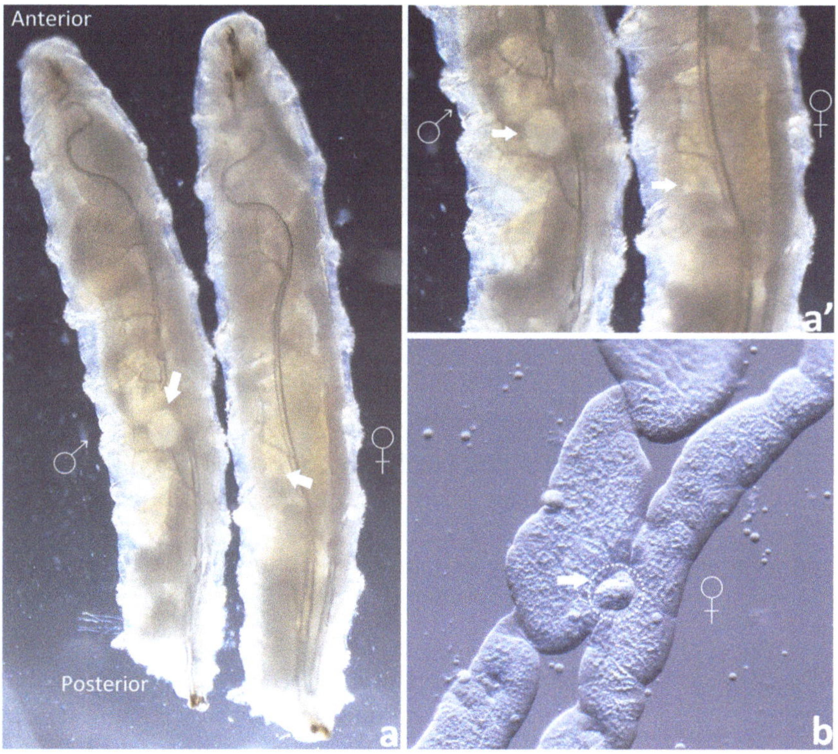

Fig. 3 Details of wandering third instar larvae. (**a**) male and female larvae. White arrows point at the gonads (magnified in **a'**). (**b**) Fat body from a female larva. The white arrow points at the female gonad imbedded in the fat body (dashed line)

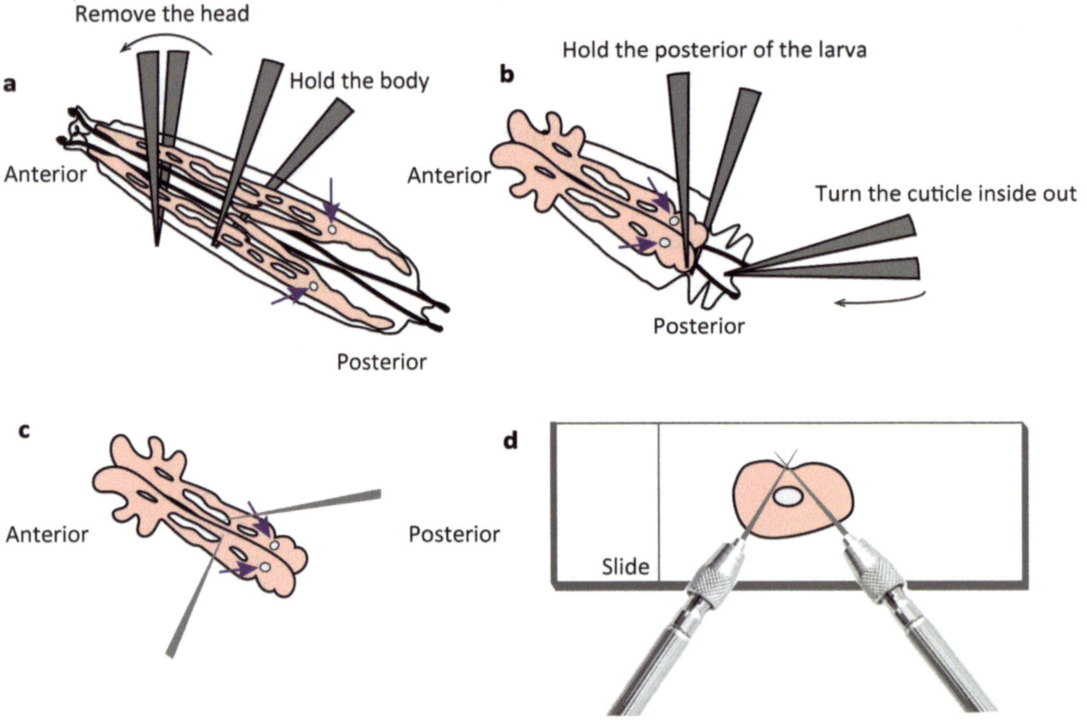

Fig. 4 Schematic illustration of the larval fat body/gonad dissection procedure. To dissect the fat body, (**a**) first remove the larval head with a pair of forceps. (**b**) Next, hold the remaining of the larva with one pair of forceps and slowly turn, from the posterior spiracles, the cuticle inside out with the second pair. The black arrows indicate the movement of the forceps. (**c**) This should expose the fat body ready for isolation, use two minutien pins to separate the fat body surrounding the gonads from the rest. (**d**) When mounting the gonads for microscopic observation, remove the excess of fat body surrounding the gonad with the two pins. The gonads are translucent spheres surrounded by a "flower-shaped" fat body (purple arrows). The anterior and posterior ends of the larva are indicated

spiracles toward the anterior end with the other pair of forceps (Fig. 4b). Ideally, the now exposed fat body should be recovered intact (*see* **Note 5**).

5. Transfer all the collected fat bodies to a cell strainer placed in a 6-well plate filled with Ringer's solution (Fig. 2c).

3.2 Stain Larval Fat Body

Transfer the cell strainer containing the fat bodies into new wells with the appropriate solutions for fixation and washing steps (*see* **Note 6**).

1. Fix the dissected larval fat bodies in 5% formaldehyde in Ringer's solution for 20 min on an orbital rotator at 35 revolutions per minute (rpm) at room temperature (RT).
2. Wash for 5, 10, and 45 min with 1% PBT (35 rpm, RT).
3. Block with 0.3% PBTB for 1 h (35 rpm, RT).

4. After blocking, with a cropped pipette tip, transfer the fat bodies to 0.2 mL tubes and incubate with the desired primary antibody diluted in 0.3% PBTB overnight at 4 °C with gentle agitation on a nutator or mini-shaker (*see* **Note 7**).
5. The following day, transfer with a cropped pipette tip the fat bodies back to the cell strainer and wash three times, 30 min each, with 0.3% PBTB (35 rpm, RT).
6. Block with 0.3% PBTB supplemented with 5% fetal bovine serum for 1 h (35 rpm, RT).
7. After blocking, transfer with a cropped pipette tip fat bodies to 0.2 mL tubes and incubate with the secondary antibodies for at least 2 h in blocking solution with gentle agitation at RT (*see* **Note 8**) on a nutator or mini-shaker.
8. Using a cropped pipette tip, return fat bodies to the cell strainer. Wash three times, 30 min each, with 0.3% PBT (*see* **Note 9**).
9. In case of staining the gonads with the DNA dye Hoechst, add Hoechst (1/1000 dilution of a 5 mg/mL stock) after 20 min in the third wash in 0.3% PBT, and incubate for 10 min.
10. Wash three times, 5 min each, in 0.3% PBT (*see* **Note 9**).

3.3 Mount Gonads

1. Transfer 2–3 fat bodies to a shallow dissecting dish with a cropped 0.2 mL pipette and examine thoroughly each of the fat bodies looking for the gonads. They are translucent spheres surrounded by a "flower-shaped" fat body (Fig. 3b). Exert extreme caution with the tweezers to avoid damaging the gonads during their manipulation.
2. Separate the fat body area containing the gonad from the main fat body using the minutien pins (Fig. 4c). Be careful to leave enough fat body still attached to the gonad to facilitate transfer to a mounting slide.
3. Add a drop of mounting medium to a slide and transfer the gonads. With the help of two pin holders, remove excess fat body surrounding the gonad (Fig. 4d). This can be done by shearing the fat body with the two minutien pins (Fig. 4d).
4. Repeat **steps 1–3** until all the processed gonads are in the mounting medium with the fat bodies removed.
5. Position the cleaned gonads at the center of the drop of mounting medium and remove the fat body debris using the pins (we use the same slide from the previous step). Once all the gonads are clustered together, carefully place the coverslip on top of the drop. Avoid squishing the gonads with the coverslip.
6. Seal the preparation with nail polish and store at 4 °C (or at −20 °C for long-term storage).

3.4 Method to Count Germline Cells in Larval Gonads

We developed a reliable and easy-to-use method to automatically quantify the number of germline cells present in larval gonads. Because the most common marker for germline cells is the cytoplasmic protein Vasa, it cannot be used as an approach for scoring positively labeled nuclei. Thus, we developed a method that renders the non-stained nuclei of Vasa-positive germline cells as labeled spheres. In the example below, we quantify the number of germline cells present in fixed gonads stained with a rabbit anti-Vasa primary antibody visualized with an Alexa Fluor 488-conjugated secondary antibody (*see* **Note 10**).

3.4.1 Acquire Images

1. Obtain z-stacks of the gonads using a 40× oil immersion objective (Numerical Aperture 1.15) on a confocal microscope at 1.7 zoom and 1024 × 1024 resolution to achieve a final pixel size of 158 × 158 nm (*see* **Note 11**). A scanning frequency of 600 Hz results in a pixel dwell of 480 ns for the final 1024 × 1024 pixels image size.
2. Set the pinhole aperture to <1 Arbitrary Units (AU) to increase the resolution along the z-axis and adjust the number of slices to capture the complete gonad (approximately 60 μm thick) using 0.6 μm z-steps. This thickness ensures that all germline cells are imaged.
3. Use excitation and emission settings specific to the fluorophore utilized. In the case of the AF488 fluorophore, we used a 488 nm laser and a photomultiplier detector filtering the emission between 505 and 536 nm. Adjust the laser power and sensor gain to maximize the dynamic range of the images without reaching pixel saturation.

3.4.2 Analyze Images

Since the Vasa protein is confined to the cytoplasm and its localization signal does not allow single cell resolution, we quantify the Vasa-negative, “black” nuclei. All steps are performed on each image independently using FIJI [9]. Complete z-stacks encompassing entire gonads can be analyzed using a macro script (*see* Appendix I and Electronic Supplementary Material).

1. Apply a *3D Gaussian Blur* filter to smooth the signal and remove noise (Fig. 5a, b).
2. Increase and equalize the contrast across the image by applying *Contrast Limited Adaptive Histogram Equalization* [10] using the *Enhance Local Contrast (CLAHE)* plugin (Fig. 5c). This enhances the contrast of the weak signal in middle sections due to limited laser penetration in thick samples. In addition, acting locally on the contrast results in enhanced contrast differences between the Vasa signal and the negative nuclei. For our images, we normally utilize the following parameters: Block Size 50 pixels (7.9 μm, larger than our object of interest), Histogram 200, Slope 4.

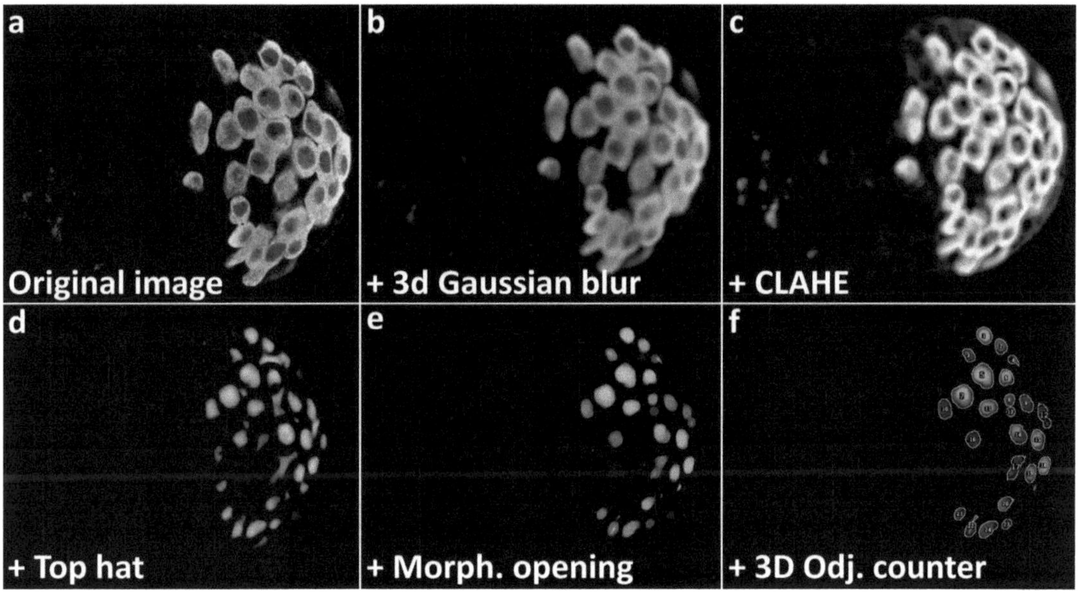

Fig. 5 A method to count germline cells in larval gonads using FIJI. (**a**) Original confocal image. (**b**) Same image after applying a *3D Gaussian Blur* filter. (**c**) Image after applying *Contrast Limited Adaptive Histogram Equalization* using the *Enhance Local Contrast (CLAHE)* plugin. (**d**–**f**) The use of two *Morphological Filtering* techniques, the *Top Hat* filter (**d**) and the *Morphological Opening* method (**e**), rendered image that is then segmented using the native FIJI 3D Object Counter plugin (**f**)

3. Reveal the black nuclei by using two *Morphological Filtering* techniques [11] (https://imagej.net/plugins/morpholibj). First, run the *Top Hat filter* in the *Gray Scale Attribute Filtering* plugin with a size parameter of 1800 pixels (45 μm^2) (Fig. 5d). This ought to identify small dark structures and to turn them bright. Thus, the output is an image where the black nuclei now appear as bright spheres. Second, apply the *Morphological Opening* method using a sphere radius of 12 pixels (1.8 μm) to get rid of nonspherical structures (Fig. 5e).
4. Segment the resulting images using the native FIJI *3D Object Counter* plugin to count the elements (bright spheres). The plugin thus gives back the total number of nuclei in the 3D gonad (Fig. 5f).

4 Notes

1. We use a stereomicroscope with a 1× plan-apochromatic objective (working distance: 60 mm) and a maximum magnification of 125×.
2. To allow for a precise timing of larval gonad development, adults should lay eggs for 2–4 h and the (ideally uncongested) fly cultures kept at 25 °C. In 4–5 days, the larvae should reach the non-wandering third instar stage (ML3).

3. We use Hoechst. However, any fluorescent DNA dye can be used.
4. We use VECTASHIELD, but any mounting medium for preserving fluorescence can be used.
5. To prevent fat body damage, discard the cuticle, trachea, and gut.
6. Avoid leaving the dissected, unfixed larvae in the Ringer's solution for more than 30 min, as this may affect tissue integrity.
7. To reduce the incubation volume, to facilitate antibody penetration, and to avoid fat body damage, transfer the fat bodies from the cell strainer to individual 0.2 mL PCR tubes. We recommend using a cropped 0.2 mL pipette tip to move the tissue.
8. Incubate the samples with the secondary antibody for at least 2 h at RT. However, in our experience, overnight incubation at 4 °C gives better results.
9. The fat bodies in 0.3% PBT can be stored at 4 °C up to a week.
10. We use anti-Vasa [8] and Alexa-fluor 488-conjugated secondary antibody, but any other suitable combination that renders strong signal might be used as well.
11. 1024 × 1024 image size, although not critical for PGC quantification, will result in a bigger, higher-quality final image.

Appendix I: Macro Script to Process Entire z-stacks (*See* Electronic Supplementary Material)

- Author: Alejandro Campoy-López
- Microscopy Unit (CABD)
- Date: 10.11.2021
- Description: Counting negative nuclei
- Input: Folder containing z-stacks to be processed
- Output: Processed images and csv file with counted elements

References

1. Slaidina M, Banisch TU, Gupta S, Lehmann R (2020) A single-cell atlas of the developing Drosophila ovary identifies follicle stem cell progenitors. Genes Dev 34:239–249
2. Banisch TU, Slaidina M, Gupta S, Ho M, Gilboa L, Lehmann R (2021) A transitory signaling center controls timing of primordial germ cell differentiation. Dev Cell 56:1742–1755 e1744
3. Gancz D, Gilboa L (2013) Insulin and target of rapamycin signaling orchestrate the development of ovarian niche-stem cell units in Drosophila. Development 140:4145–4154

4. Gancz D, Lengil T, Gilboa L (2011) Coordinated regulation of niche and stem cell precursors by hormonal signaling. PLoS Biol 9: e1001202
5. Sarikaya DP, Extavour CG (2015) The hippo pathway regulates homeostatic growth of stem cell niche precursors in the Drosophila ovary. PLoS Genet 11:e1004962
6. Yatsenko AS, Shcherbata HR (2018) Stereotypical architecture of the stem cell niche is spatiotemporally established by miR-125-dependent coordination of Notch and steroid signaling. Development 145:dev159178
7. Zhu C, Xie T (2003) Clonal expansion of ovarian germline stem cells during niche formation in Drosophila. Development 130:2579–2588
8. Gilboa L, Lehmann R (2004) Repression of primordial germ cell differentiation parallels germ line stem cell maintenance. Curr Biol 14:981–986
9. Schindelin J, Arganda-Carreras I, Frise E, Kaynig V, Longair M, Pietzsch T, Preibisch S, Rueden C, Saalfeld S, Schmid B et al (2012) Fiji: an open-source platform for biological-image analysis. Nat Methods 9: 676–682
10. Zuiderveld K (1994) Contrast limited adaptive histogram equalization. In: Graphics gems IV. Academic Press Professional, Inc., pp 474–485
11. Legland D, Arganda-Carreras I, Andrey P (2016) MorphoLibJ: integrated library and plugins for mathematical morphology with ImageJ. Bioinformatics 32:3532–3534

Chapter 3

Dissection, Fixation, and Standard Staining of Adult *Drosophila* Ovaries

Julie A. Merkle

Abstract

Studies of the *Drosophila* ovary have provided significant insight into the molecular and cellular processes that control cell function, tissue organization, and animal development. To characterize mutants with defects in oogenesis or to observe the distribution of gene products involved in egg production, the ovaries must be carefully extracted and prepared for analysis. This chapter describes the manual dissection of ovaries from adult *Drosophila* females, followed by standard fixation and staining of the isolated tissue. Specifically, this chapter provides procedures for simple DNA and F-actin staining to assess cell and tissue morphology, as well as immunostaining to localize proteins of interest.

Key words *Drosophila*, Oogenesis, Gametogenesis, Ovary, Dissection, Staining, Immunostaining, Immunohistochemistry

1 Introduction

The *Drosophila* ovary is an excellent model system for studying fundamental mechanisms that govern cell structure and function and organ development. In particular, the conserved processes that control cell division, communication, and patterning have been extensively studied in the *Drosophila* ovary and have contributed to a deeper understanding of these molecular events across metazoans (reviewed in [1–5]). The ovary (refer to Fig. 1e) uniquely contains both germline and somatic cell lineages, which signal to one another to generate haploid oocytes, each surrounded by a protective eggshell. While the oocyte is a product of the germline lineage, the follicle cells that produce the layers of the eggshell are somatically derived (refer to Fig. 2a). Much insight has been gained by studying these lineages in a developing organ, from stem cell populations to the differentiated cell types they produce [6–12].

Genetic and biochemical screens have uncovered many gene products critical for oogenesis [13–25]. Such screens typically

Michelle S. Giedt and Tina L. Tootle (eds.), *Drosophila Oogenesis: Methods and Protocols*,
Methods in Molecular Biology, vol. 2626, https://doi.org/10.1007/978-1-0716-2970-3_3,

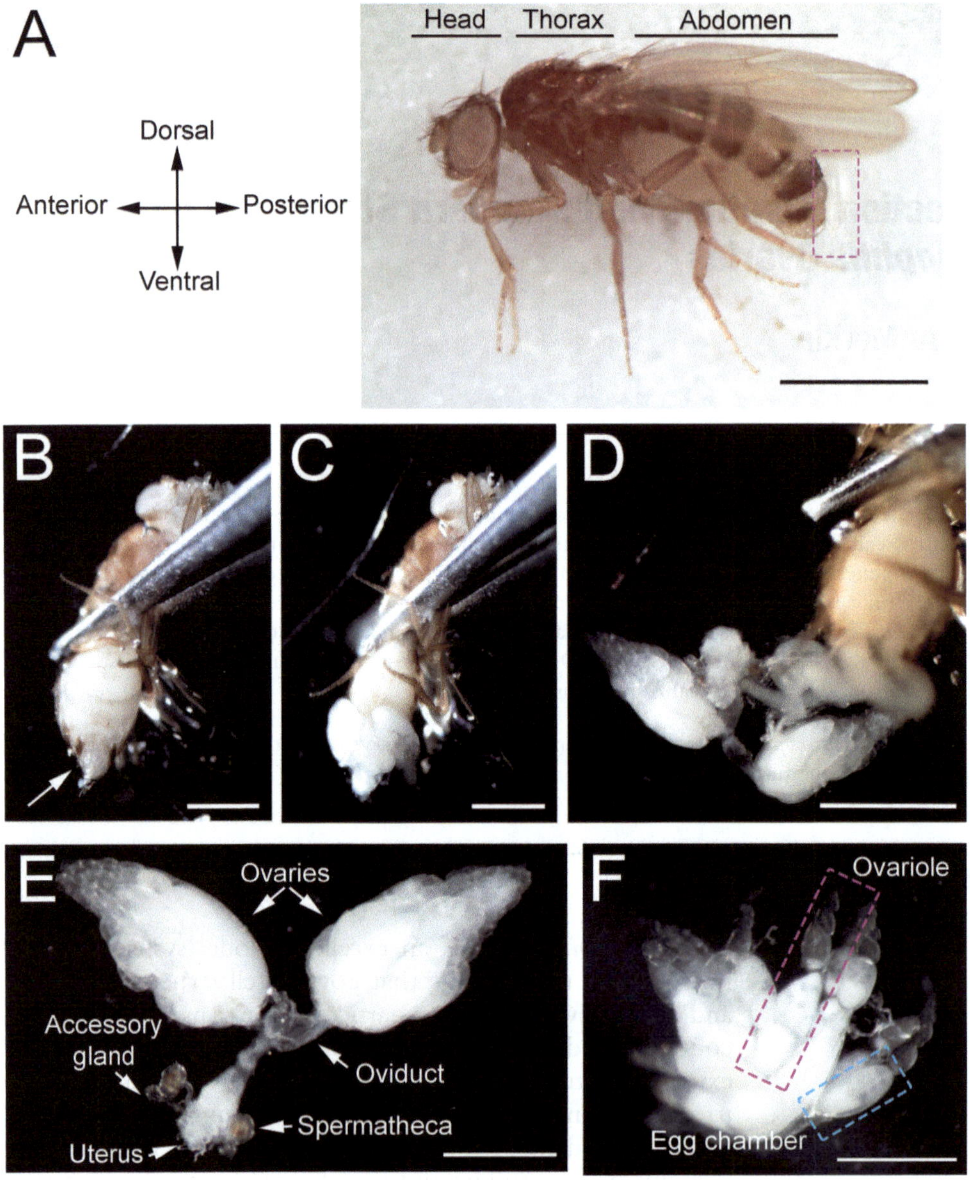

Fig. 1 Ovarian dissection of an adult *Drosophila* female. (**a**) Adult *Drosophila melanogaster* female (genotype *y w*) after being fed yeast for 2 days. The body segments (head, thorax, and abdomen) are labeled. Body axes (anterior, posterior, dorsal, and ventral) are indicated with an orientation reference. Females often lay a mature egg (magenta box) when under anesthesia. (**b–f**) The anterior of the fly or ovary is oriented at the top and the posterior is at the bottom. (**b**, **c**) The ventral (belly) of the fly is facing up. (**b**) A fly is held with forceps at the anterior region of the abdomen, just anterior to the ovaries. Sharper forceps should be used to tug at the most posterior end, just anterior to the ovipositor (arrow). (**c**) Exposed ovaries after initial removal of the posterior end of the fly. Once the ovaries are visible, they can be extracted from the abdominal cavity either by carefully pulling them from the abdomen or by pressing gently on the abdomen, thereby pushing the ovaries from the abdominal cavity. (**d**) Ovaries extracted from the abdominal cavity of a fly, but still attached to other abdominal organs. The female carcass can now be discarded. (**e**) Ovary pair after dissection from an adult *Drosophila* female. Many of the associated structures needed for oogenesis are indicated (spermatheca,

involve the dissection of ovaries from females of isolated genotypes, followed by cytological or molecular analyses, including tissue fixation and immunohistochemistry, live imaging, or RNA or protein isolation. To identify and characterize genes involved in *Drosophila* oogenesis while avoiding artifactual results due to improper techniques, it is critical to follow well-established protocols [26–35]. One main goal of this chapter is to provide detailed guidance on ovary dissection for such experiments. For helpful videos showing manual isolation of ovaries from adult *Drosophila*, refer to Wong and Schedl [34] and Zimmerman et al. [35].

Many genes known to be involved in oogenesis were first identified by either observing mutant phenotypes or visualizing RNA or protein expression in the ovary [17, 18, 22, 23, 36]. To initially characterize a mutant phenotype in oogenesis, it is often informative to use fluorescently labeled probes or dyes for common cellular markers, including DAPI (4′, 6-diamidino-2-phenylindole) or Hoechst to observe nuclei and phallotoxins (i.e., phalloidin) to observe filamentous actin (F-actin) of the cytoskeleton. These markers allow for relatively quick assessment of ovarian morphology. In many instances, nuclear and F-actin staining are sufficient to identify primary architectural defects to inform future analyses. Therefore, this chapter provides procedures for efficient initial phenotypic characterization of oogenesis in wild-type and mutant females using Hoechst and phalloidin stains.

Another focus of this chapter is to provide standard methods to visualize the subcellular localization of proteins in the *Drosophila* ovary. Visualizing the spatial and temporal distribution of a protein of interest allows for the identification of specific cell types and structures in which the protein may be expressed and/or localized. This may be achieved by using antibodies generated against a protein, fusing the protein to a fluorescent protein (e.g., GFP, mCherry) or epitope tag (e.g., HA, Myc), or using tagged transposons serving as in vivo "traps." Transposon insertions are useful to serve as enhancer traps, gene traps, and protein traps to visualize expression patterns of mRNAs or proteins, or to control the expression of the gene with an inducible promoter (e.g., using the *Gal4-UAS* system) [37–40]. Protein traps are especially useful for

Fig. 1 (continued) accessory gland, oviduct, and uterus). It is recommended to keep the ovaries attached at the oviduct, if possible. If the extracted ovaries are smaller and few vitellogenic stages (stages after yolk update that appear white and opaque) are present, it is recommended to keep the ovaries attached to the uterus and other associated structures. (**f**) Ovary after careful separation of the ovarioles (magenta box) at the anterior end of the ovary. This removes the muscle sheath covering each ovariole, allowing for better fixation and staining. Egg chambers (cyan box) within each ovariole are visible after the ovaries are properly teased apart. Images were acquired using a Nikon SMZ1000 stereomicroscope at 20× (**a**–**d**) or 40× (**e**, **f**) magnification. Scale bars, 1 mm (**a**–**d**) or 0.5 mm (**e**, **f**)

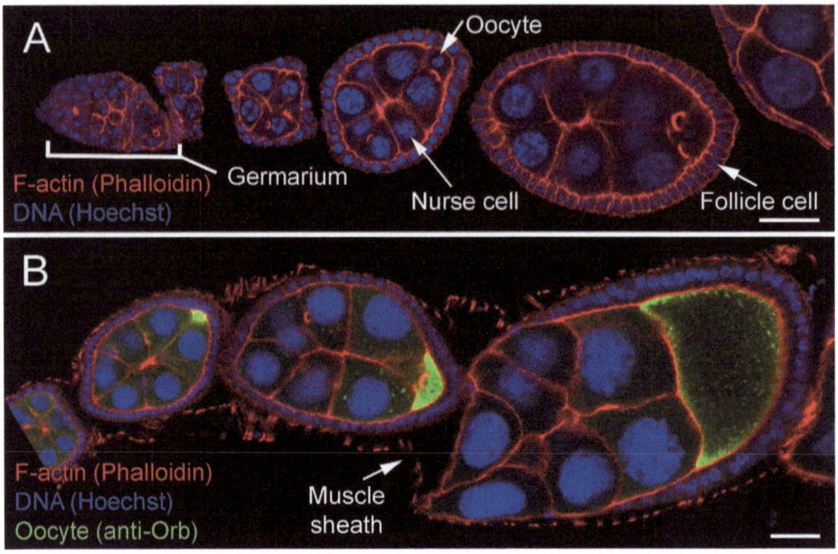

Fig. 2 Visualization of oogenesis after fixation and staining. (**a**, **b**) Nuclei are stained with the dye Hoechst (blue) and filamentous actin (F-actin) of the cytoskeleton is stained with Alexa Fluor 568-conjugated phalloidin (Life Technologies) (red). (**a**) Longitudinal section through an ovariole after fixation and simple DNA and F-actin staining. Egg chamber and ovariole morphology are easily visualized with DNA and F-actin dyes and key structures are indicated. The germarium houses the germline and follicle cell stem cells. Egg chambers are produced from the germarium and consist of 16 germline cells (1 oocyte and 15 nurse cells) surrounded by a monolayer of somatic follicle cells. (**b**) Longitudinal section through an ovariole after fixation, blocking, and immunostaining for the oocyte-specific protein Orb (Oo18 RNA-binding protein) (green). DNA and F-actin staining was performed in combination with secondary antibody incubation. The muscle sheath that surrounds the ovariole was not fully removed and is indicated. Mouse anti-Orb monoclonal antibodies (clones 4H8 and 6H4; Developmental Studies Hybridoma Bank) were each used at a 1:60 dilution. Anti-mouse Alexa Fluor 488-conjugated secondary antibodies (Life Technologies) were used at a dilution of 1:500. Images were acquired using a 40×, 1.30 NA Nikon Apo Plan Fluor objective on a Nikon A1 laser scanning confocal microscope. Scale bars, 20 μm

imaging endogenous protein distribution in living tissue [41, 42]. However, since fluorescent proteins may produce a weak signal due to photo-instability, antibodies (e.g., anti-GFP) can be used to amplify weak fluorescence in fixed tissue. In addition to single protein observation, co-localization analyses may be used to study molecular interactions in vivo, thereby providing insight into cellular pathways important for tissue development. Some of these approaches may require genetic crosses to first be performed, followed by an analysis of the progeny.

This chapter provides standard procedures for ovary dissection, tissue fixation, staining for nuclear and cellular architecture, and immunostaining for localization studies. The methods described herein can be used to examine most stages of egg production. These procedures may be adapted to address specific experimental questions as recommended by published studies. For example, the suggested dissection media, reagent concentrations, fixative, antibody dilutions, and incubation times may vary for tissue fixation, permeabilization, blocking, or antibody incubation. Such alterations usually correspond to optimizations according to the stage of oogenesis to be studied and/or the antibodies or probes to be used. It is, therefore, always advisable to first optimize the experimental conditions before piloting a screen or characterizing phenotypes (*see* **Note 1**).

2 Materials

2.1 Fly Husbandry

1. Flies of appropriate genotypes in vials with standard *Drosophila* medium (*see* **Note 2**).
2. Active or inactive dry baker's yeast or prepared wet yeast paste (*see* **Note 3**).
3. 25 °C incubator (recommended for most conditions; *see* **Note 4**).

2.2 Ovary Dissection

1. Dissecting stereomicroscope.
2. Carbon dioxide (CO_2) and CO_2-supplied fly pad (*see* **Note 5**).
3. Two dissection forceps (*see* **Note 6**).
4. Dissection needles in holders (*see* **Note 7**).
5. Glass concavity slide or depression wells for dissection.
6. 1.5 mL microcentrifuge tubes.
7. Glass Pasteur pipette with a rubber bulb.
8. Kimwipes.
9. 1× Phosphate buffered saline (PBS), pH 7.4: 137 mM NaCl, 2.7 mM KCl, 10 mM Na_2HPO_4, 1.8 mM KH_2PO_4 in dH_2O (*see* **Notes 8** and **9**).

2.3 Tissue Fixation and Washing

1. Fixation solution: 4% Paraformaldehyde (PFA) diluted in 1× PBS (*see* **Notes 8**, **10**, and **11**).
2. 1× PBST: 0.1% Triton-X100 in 1× PBS (*see* **Notes 1** and **9**).
3. Nutating platform rocker.
4. Tube rack.
5. Glass Pasteur pipette with a rubber bulb.
6. p1000 micropipettor.
7. p1000 micropipette tips.
8. Capped waste container for paraformaldehyde waste.

2.4 DNA and F-Actin Staining

1. Hoechst and phalloidin staining solution: Hoechst (1:1000 dilution of 10 mg/mL solution) and phalloidin probe conjugated with a fluorophore (phalloidin solutions should be made according to manufacturer's instructions and then diluted to working solutions) (*see* **Note 12**).
2. 1× PBST: 0.1% Triton-X100 in 1× PBS (*see* **Notes 1** and **9**).
3. p1000 micropipettor.
4. p1000 micropipette tips.
5. Aluminum foil.
6. Nutating platform rocker.
7. Glass Pasteur pipette with a rubber bulb.

2.5 Immunostaining

1. Blocking solution: 5% BSA in 1× PBST (*see* **Note 13**).
2. Primary antibody solution: 1% BSA (optional) and Primary antibody (at recommended concentration) in 1× PBST (*see* **Notes 13** and **14**).
3. Secondary antibody solution: 1% BSA (optional) and Secondary antibody (at recommended concentration) in 1× PBST (*see* **Note 15**).
4. 1× PBST: 0.1% Triton-X100 in 1× PBS (*see* **Notes 1** and **9**).
5. p1000 micropipettor.
6. p1000 micropipette tips.
7. Glass Pasteur pipette with a rubber bulb.
8. Aluminum foil.
9. Nutating platform rocker at room temperature and in a 4 °C refrigerator or walk-in cold room.

2.6 Mounting

1. Mounting medium, such as the self-hardening mountant Aqua-Poly/Mount (Polysciences, Inc.) (*see* **Notes 16** and **17**).
2. Dissection needles in holders (*see* **Note 7**).
3. Glass Pasteur pipette with a rubber bulb.
4. p200 and p1000 micropipettors.
5. p200 and p1000 micropipette tips.
6. Glass slides.
7. Scissors.
8. Kimwipes.
9. Glass coverslips (*see* **Note 18**).
10. Confocal or epifluorescent microscope.
11. Nail polish or plastic sealant (*see* **Notes 16** and **17**).

3 Methods

For standard ovary dissection, females should be selected 1–2 days post-eclosion and then fed yeast for another 1–2 days at the appropriate temperature (*see* **Note 4**). Dissection may take 20–30 min per genotype, depending on experience, the number of females to be dissected, and the size of the ovaries. Fixation and washing take approximately 1 h. Ovaries may then be processed immediately or stored up to a week at 4 °C. If simple probes are used without immunostaining, such as Hoechst and phalloidin stains, then the processes of staining and mounting can be completed in about 1 h. If protein detection relies on the use of antibodies, blocking, immunostaining, and mounting can be completed in as little as 2 days or can be extended to 3–4 days as necessary.

3.1 Fly Husbandry

1. Collect 10–15 females per genotype to be analyzed. If more than 20 females are to be dissected, they should be separated into several vials to avoid overcrowding, which may affect ovary quality.
2. To induce egg-laying and ensure continued egg production, add a few wild-type or fertile sibling males to each vial of females.
3. Feed flies in vials with standard *Drosophila* medium supplemented with dry yeast 24–48 h at 25 °C prior to dissection to increase ovary size (*see* **Notes 4** and **19**).

3.2 Ovary Dissection

1. Anesthetize flies using a fly pad connected to CO_2 (*see* **Note 5**).
2. Add 1× PBS (or other dissection medium) to the depression wells (*see* **Notes 8** and **9**).
3. Pick up a female fly, ventral side up and with the anterior (head) away from the end of the forceps (Fig. 1a). Use the nondominant hand to place the blunter forceps around the anterior end of the abdomen. It is advisable to place the forceps about two-thirds up the abdomen toward the thorax (Fig. 1b). If the forceps are placed higher, the entire abdomen will separate from the thorax. Use enough pressure to securely pick up and hold the fly, being careful not to crush the abdomen (*see* **Note 6**) [34].
4. Submerge the female fly into 1× PBS/media in a depression well (*see* **Note 20**). 20× magnification is typically ideal for dissection.
5. Using sharper forceps, grasp the most posterior end of the abdomen (Fig. 1b, arrow) and tug gently, pulling the posterior end away from the fly to expose the abdominal organs. The ovaries should be attached to the tissue pulled away from the

posterior end. If they are, proceed to **step 7**. If the ovaries are not attached to the tissue separated from the posterior end, proceed to **step 6**.

6. If the ovaries are not attached to the tissue pulled from the anterior end, then they must be extracted from the abdominal cavity. Until the ovaries are successfully extracted from the female, it is important to maintain a firm hold on the female using the blunter forceps in the nondominant hand. It also is critical that the fly and extracted ovaries remain submerged below the surface of PBS/medium.
   (a) *Method 1*: For females with large, opaque abdomens (suggesting that the ovaries are also large), lightly clamp the forceps around the anterior end of the abdomen and then gently slide the forceps along the abdomen posteriorly. This should squeeze the ovaries from the abdominal cavity (like toothpaste squeezed from a tube) (Fig. 1c, d).
   (b) *Method 2*: A similar approach is to lightly press on the ventral side of the abdomen with forceps, starting at the anterior end of the abdomen, and then moving posteriorly using the same repetitive motion. This should slowly push the ovaries from the abdominal cavity.
   (c) *Method 3*: For females with smaller abdomens and ovaries, it is best to use forceps to gently remove tissue from the abdominal cavity until the ovaries are successfully extracted. For the recovery of smaller ovaries, *see* **Notes 21** and **22**.
7. Before discarding the female carcass, locate both ovaries and detach them from other abdominal organs (e.g., intestines, uterus, seminal receptacle, and spermathecae) (Fig. 1d, e). It may help to position two Kimwipes on the table on either side of the microscope. These can be used for swift cleaning of the forceps and/or removal of carcasses during dissection.
8. Using dissecting forceps or needles, carefully tease apart the ovarioles at the anterior end of the ovary (Fig. 1f). Stabilize the ovaries with one pair forceps in the nondominant hand, and then use a sharp dissecting needle in the dominant hand to separate the ovarioles from one another. It is effective to employ a slicing motion between ovarioles, from near the posterior end of the ovary where vitellogenic stages are found toward the anterior end where the germaria are located. This removes the muscle sheath surrounding each ovariole and allows for better penetration of the fixation solution and antibodies. It is helpful to keep the ovarioles attached at the most posterior end so that tissue loss is reduced during future fixation, staining, and washing steps.

9. After dissection, use forceps to transfer the ovaries to a microcentrifuge tube containing 1× PBS/medium (*see* **Notes 8**, **20**, **23**, and **24**). Alternatively, dissected ovaries of the same genotype may be stored in a depression well with fresh PBS/media until all females have been dissected. The ovaries should be transferred to a microcentrifuge tube using a glass Pasteur pipette, being mindful to keep the ovaries within 2–3 cm of the pipette opening (*see* **Note 25**).
10. Once all ovary pairs are dissected and transferred to a microcentrifuge tube, remove the liquid media from the tube with a glass Pasteur pipette attached to a rubber bulb (*see* **Note 26**).
11. Proceed to Subheading 3.3. For more information on working with smaller ovaries and previtellogenic egg chambers, *see* **Notes 21** and **22**.

3.3 Tissue Fixation and Washing

1. Add ~700 μL fixation solution and incubate on a nutating rocker at room temperature for 15 min (*see* **Note 27**). If the solution and ovaries do not move back and forth when rocking, raise the end of the tube at an angle on the platform nutating rocker.
2. Place the tube in a tube rack and allow the ovaries to sink to the bottom of the tube. Flick the tube intermittently and leave the tube undisturbed in a tube rack so that the tissue sinks to the bottom of the tube. To avoid overfixation, it is important that the tissue should not remain in the fixation solution for longer than 20 min.
3. Remove the fixation solution with a glass Pasteur pipette and dispose it of in a capped and labeled waste container, which should remain closed between uses (*see* **Note 26**).
4. Quick wash ovaries 2× with ~1 mL 1× PBST using a wash bottle or micropipettor (*see* **Note 9**). For all quick washes in this protocol, invert the tube several times, wait for the tissue to settle to the bottom, and then carefully remove and discard the PBST. Sufficient washing is necessary after fixation and staining to remove reagents and reduce background fluorescence.
5. Wash ovaries 3× for 10 min in ~1 mL 1× PBST, placing the tubes on a nutating rocker between washes.
6. Fixed ovaries may be processed immediately or stored at 4 °C for up to a week before staining. Proceed to Subheading 3.4 or Subheading 3.5.

3.4 DNA and F-Actin Staining

If ovaries will not be immunostained, this process is to be completed next. If immunostaining is to be performed, proceed to Subheading 3.5. A representative ovariole stained for DNA and F-actin using the described methods is shown in Fig. 2a.

1. Add 500 μL Hoechst and phalloidin staining solution, cover tubes with aluminum foil, and incubate on a nutating rocker for 15 min (*see* **Note 28**).
2. Quick wash ovaries 2× with 1× PBST. Carefully remove and discard the PBST after each wash.
3. Wash ovaries 3× for 10 min in 1× PBST on a nutating rocker.
4. Samples may be stored at 4 °C in the dark for up to 1 week before mounting. Proceed to Subheading 3.6.

3.5 Immunostaining

To study the spatial and temporal distribution of proteins during oogenesis, it is often useful to use primary antibodies raised against a protein of interest or a tagged fusion protein. A representative ovariole immunostained with antibodies that recognize the oocyte-specific protein Oo18 RNA-binding protein (Orb) and then subsequently stained for DNA and F-actin using the described methods is shown in Fig. 2b.

3.5.1 Day 1: Blocking and Primary Antibody Incubation

1. Add 1 mL blocking solution. Rock on a nutating rocker at room temperature for 1–3 h or overnight at 4 °C, as recommended by published methods for the antibodies used and as time permits.
2. Remove the blocking solution with a glass Pasteur pipette and discard.
3. Quick wash ovaries with 1× PBST. Carefully remove and discard the wash solution.
4. Add 300 μL primary antibody solution (*see* **Note 29**). Rock on a nutating rocker at 4 °C overnight.

3.5.2 Day 2: Washing and Secondary Antibody Incubation

1. Carefully remove primary antibody solution using a clean pipette tip and store in a labeled microcentrifuge tube at 4 °C for reuse (*see* **Note 30**).
2. Quick wash 2× with 1× PBST. Carefully remove and discard the PBST after each wash.
3. Wash 3× for 10 min in 1× PBST on a nutating rocker.
4. Add 300 μL of secondary antibody solution (*see* **Notes 15** and **29**). Rock aluminum foil-covered tubes on a platform nutating rocker for 2 h at room temperature or 4 °C overnight. From this point on, ovaries should be kept out of light and covered with aluminum foil as much as possible (*see* **Note 28**).
5. If fluorescently tagged dyes (e.g., Hoechst and phalloidin) are also being used, they can be added to the secondary antibody solution at the beginning of secondary antibody incubation, in the last 15 min, or sequentially if the secondary antibody solution is to be reused.
6. Carefully remove the secondary antibody solution.

7. Quick wash ovaries 2× with 1× PBST. Carefully remove and discard the PBST after each wash.
8. Wash ovaries 3× for 10 min in 1× PBST on a nutating rocker.
9. Samples may be stored at 4 °C in the dark for up to 1 week before mounting. Proceed to Subheading 3.6.

3.6 Mounting

3.6.1 Manual Separation of Ovarioles

This approach is recommended in most cases. While it takes more time and patience than Subheading 3.6.2 Mechanical disruption, it ensures less tissue loss and results in fewer phenotypic artifacts.

1. Label a glass slide.
2. Remove all but ~200 μL of liquid from the sample.
3. Using a glass Pasteur pipette, transfer the ovaries and 1× PBST to the slide (Fig. 3a).
4. Use dissecting needles to collect the ovaries to the center of the liquid mound. 20× or 40× magnification is ideal for mounting.
5. Carefully wick away the remaining PBST with a Kimwipe by placing it to the edge of the liquid. Do not place the Kimwipe too close to the tissue or allow it to touch the egg chambers.

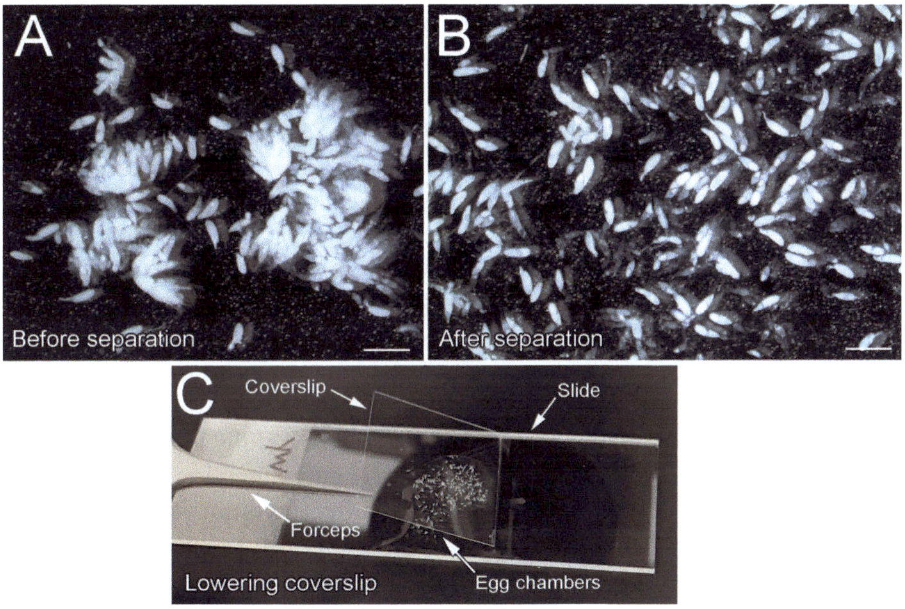

Fig. 3 Whole mounting of fixed egg chambers. (**a**–**c**) Ovarian tissue in mounting media on a glass slide after fixation and staining during whole mounting. (**a**) Before separation, the ovarioles remain connected at the posterior end of the ovary. (**b**) During manual separation, ovarioles and egg chambers are carefully separated from each other and other tissues (e.g., muscle sheath, oviducts) using a slicing motion with sharp dissecting needles. (**c**) After the egg chambers are sufficiently separated, they are spread out in mounting media on a glass slide and a glass coverslip is lowered at an angle onto the tissue. Forceps may aid in preventing air bubbles in the mounting media. (**a**, **b**) Images were acquired using a Nikon SMZ1000 stereomicroscope at 20× magnification. Scale bars, 1 mm

Previtellogenic egg chambers can easily be wicked away or become desiccated. On the other hand, it is important to remove as much of the PBST as possible, as remaining buffer will interfere with the polymerization of self-hardening mounting media, if used (*see* **Notes 16** and **17**).

6. Add 2 drops of mounting medium to the slide (*see* **Note 17**). It is ideal to minimize the amount of medium added as too much medium will seep out from the coverslip and may damage objectives, especially if the medium has not completely polymerized or the coverslip is not completely sealed.
7. Carefully separate ovarioles with dissecting needles, and carefully remove any attached muscle sheath (refer to Fig. 2b) and extraneous tissue. It may be easiest to stabilize the ovaries with forceps or a dissecting needle with the nondominant hand, and then to use a sharp dissecting needle in the dominant hand to separate the ovarioles from one another (*see* **Note** 7). It is effective to employ a slicing motion between ovarioles from the posterior end of the ovary (vitellogenic stages) toward the anterior end. The goal is to separate the egg chambers, remove remaining muscle sheath, and organize the egg chambers in a single layer on the slide (Fig. 3b). If egg chambers are stacked on top of each other, they will be difficult to properly image.
8. Using dissecting forceps or needles, gently spread out the ovarioles and mounting medium to an area just smaller than the dimensions of the coverslip. It is also recommended to eliminate bubbles at this time. Bubbles can be popped by poking the center of each bubble with forceps or dissecting needles. Another effective approach is to drag the bubbles to the edge of the mounting medium, away from the tissue, where they can be segregated from the area to be covered with a coverslip.
9. Carefully place a coverslip on the tissue and mounting medium. Lower the coverslip at an angle to avoid bubbles. Some find it helpful to use forceps or to carefully lean the coverslip on the end of a paintbrush as the coverslip is lowered (refer to Fig. 3c). If self-hardening mounting medium is not used, coverslips should be sealed with nail polish or plastic sealant (*see* **Notes 16** and **17**).
10. Dry slides overnight at room temperature on a flat surface (such as in a drawer) in the dark. Once the self-hardening mounting medium or sealant has set, slides should be stored at 4 °C to preserve fluorescence. Most fluorophores will be reliably stable in the dark at 4 °C for up to 1 month.
11. Visualize using a confocal or epifluorescent microscope (Fig. 2).

3.6.2 Mechanical Disruption of Ovarioles

This method is intended for use when ovary tissue is abundant, as it results in more tissue loss. It is also advisable to use this method with discretion when analyzing follicle cell morphology and Flp/FRT-induced clones, as mechanical disruption can result in tissue damage and false clones [29].

1. After the last wash, remove ~500 μL of 1× PBST and use a p1000 micropipettor to slowly pipet up and down the remaining buffer and tissue, separating ovarioles and individual egg chambers from each other and the overlying muscle sheath. Bubbles are inevitable with this method, but should be avoided as much as possible.
2. Allow ovaries to sit undisturbed for 10–15 min. Egg chambers should sink to the bottom of the tube. Previtellogenic egg chambers may not easily sink and are easily caught in bubbles. Flick the tube several times to dislodge egg chambers from the bubbles to recover as many stages as possible.
3. Remove as much liquid as possible, first with a glass Pasteur pipette and then with a p200 pipette.
4. Carefully add two drops of mounting medium to the microcentrifuge tube (*see* **Notes 16** and **17**).
5. Label a glass slide.
6. Cut off the end of a p200 pipette tip using scissors or a razor blade to produce a wider bore.
7. Slowly and carefully transfer ovaries in mounting medium to the glass slide, being mindful not to transfer bubbles. It is advisable to only transfer a small volume (~20 μL) at a time to avoid tissue loss, repeatedly adding tissue and medium to the slide until most of the medium and tissue has been transferred. Additionally, it helps to first transfer as much of the tissue as possible (it often clumps together in the medium), and then to transfer mounting medium in subsequent attempts. This will help to dislodge tissue that gets stuck on the inner walls of the pipette tip.
8. Using dissecting forceps or needles, gently spread out the egg chambers and mounting medium to an area just smaller than the dimensions of the coverslip. Remove any remaining accessory tissue, including muscle sheath. It is also recommended to eliminate bubbles at this time. Bubbles can be popped by poking the center of each bubble with forceps or dissecting needles. Another effective approach is to drag the bubbles to the edge of the mounting medium, away from the tissue, where they can be segregated from the area to be covered with a coverslip.
9. Carefully place a coverslip on the tissue and mounting medium. Lower the coverslip at an angle to avoid bubbles. Some find it

helpful to use forceps or to carefully lean the coverslip on the end of a paintbrush as the coverslip is lowered (refer to Fig. 3c). If self-hardening mounting medium is not used, coverslips should be sealed with nail polish or plastic sealant (*see* **Notes 16** and **17**).

10. Dry slides overnight at room temperature on a flat surface (such as in a drawer) in the dark. Once the self-hardening mounting medium or sealant has set, slides should be stored at 4 °C to preserve fluorescence. Most fluorophores will be reliably stable in the dark at 4 °C for up to 1 month.
11. Visualize using a confocal or epifluorescent microscope (Fig. 2).

4 Notes

1. Researchers are encouraged to refer to published methods for isolation, fixation, blocking, and staining when using previously reported reagents. For example, this protocol uses paraformaldehyde, which is recommended for most applications, however, is not ideal when visualizing microtubules and may not be compatible with all antibodies. In such cases, fixatives such as methanol or glutaraldehyde may be recommended. Moreover, immunostaining of stage 14 oocytes requires the removal of the vitelline membrane using heptane during fixation. This protocol also uses 0.1% Triton-X100 in 1× PBS for permeabilization, while other methods use higher concentrations of Triton-X100 or other detergents such as NP-40 to allow for better permeabilization of early-stage egg chambers. However, higher concentrations of Triton-X100 are not recommended if using GFP-tagged proteins, as this may result in GFP denaturation and reduced fluorescence.
2. Nutri-Fly Premixed Fly Food can be purchased from Genesee Scientific and prepared according to the manufacturer's instructions. Prepared food should be supplemented with preservatives (0.15% w/v p-Hydroxy-benzoic acid methyl ester (Tegosept) in 95% ethanol or 0.059 M propionic acid) before pouring into bottles or vials.
3. Wet yeast paste at a consistency of peanut butter (approximately 1:1 mixture of dry baker's yeast dissolved in H_2O) may also be used to feed flies by placing it on the side of the vial. Avoid placing the paste directly on the food as the flies can easily get stuck in it and die. Store dry yeast powder and prepared yeast paste at 4 °C.
4. If other temperatures are recommended for phenotypic analysis, as with temperature-sensitive *Gal80* (*tsGal80*) expression,

Gal4 expression, or other genetic manipulations, the time prior to dissection may be adjusted. For example, females may be maintained for 8 days at 18 °C or 4 or more days at 29 °C to allow for egg maturation as necessary [2, 5].

5. Alternatively, flies may be anesthetized using ice or chemicals (e.g., FlyNap). Ice packs layered with a Kimwipe and a Petri dish are an effective way to immobilize flies if CO_2 is not feasible.
6. Dumont Inox #5 forceps are recommended (available from Fine Science Tools, Inc. or other vendors). It is advisable to use two slightly different forceps. One pair should be heavier and slightly less sharp. These forceps are mainly used to maintain a grip on the fly and are held in the nondominant hand. The other pair, which should be finely sharpened, are held in the dominant hand and used to manipulate the tissue during dissection.
7. Tungsten carbide needles and nickel-plated pin holders are available from Fine Science Tools, Inc. or other vendors. Needles can be sharpened chemically in sodium nitrite or electrolytically in potassium hydroxide.
8. Ovaries can be dissected into 1× PBS, or more specialized dissection medium for ex vivo culture (e.g., Grace's Insect media, Schneider's *Drosophila* media, or *Drosophila* Ringer's solution). When a medium other than PBS is used, this should also be used to make the fixation solution. Subsequent steps should still use 1× PBS or 1× PBST, as indicated.
9. 1× PBS and 1× PBST can be kept in plastic wash bottles for quick and efficient dispensing. This can be used to fill depression wells or conduct washes.
10. 16%, 20%, or 32% paraformaldehyde solutions (electron microscopy-grade and methanol free) are available in 10 mL ampules from vendors such as Electron Microscopy Sciences.
11. PFA can be stored in the dark at 4 °C for 1 week. Alternatively, aliquots of diluted PFA can be stored at −20 °C for several months. Do not subject PFA to multiple freeze-thaw cycles. If frozen and thawed PFA precipitates, warm the solution in a 37 °C water bath, inverting intermittently, to dissolve.
12. Alexa Fluor phalloidin probes from Life Technologies are recommended due to their photostability. We prepare the stock solution according to the manufacturer's instructions by resuspending the probes in 1.5 mL of 100% methanol. Stock solutions should be stored in the dark at −20 °C. For staining, the stock solution is further diluted in 1× PBST to the following dilutions: 1:500 for Alexa Fluor 546 or Alexa Fluor 488, or 1:200 for Alexa Fluor 647. Diluted solutions should be stored in the dark at 4 °C.

13. Some methods involve the use of additional serum, such as 1% fetal bovine serum or normal goat serum, in blocking and/or antibody solutions. Such additions depend on the specific primary antibodies used and should be used as recommended by published methods.
14. To prevent microbial growth, sodium azide can be added to primary and secondary antibody solutions to a final concentration of 0.02% (w/v). Sodium azide is toxic, should be treated with caution, and should not be used for live imaging.
15. Alexa Fluor-conjugated secondary antibodies are recommended due to their photostability and reduced photobleaching rates. In most instances, Alexa Fluor fluorescent secondary antibodies should be resuspended following the manufacturer's instructions and diluted to a working stock of 1:500–1:1000. Diluted solutions should be stored in the dark at 4 °C.
16. ProLong Glass Antifade Mountant (Invitrogen) and Vectashield Antifade Mounting Medium (Vector Laboratories) are other popular choices. Vectashield and ProLong are available in formulas with and without DAPI (4′, 6-diamidino-2-phenylindole). If using these DAPI-containing formulas, Hoechst staining is not necessary. Some mounting media, such as Vectashield, require the coverslip to be sealed using nail polish or a plastic sealant.
17. Due to the viscosity of Aqua-Poly/Mount and other mounting media, it is advisable to store the bottle at 4 °C, capped with the dropper end down. To avoid the nozzle getting clogged, wipe the dropper with a Kimwipe before replacing the lid after each use. Note that after mounting, Aqua-Poly/Mount and ProLong Glass Antifade Mountant do not require the slide to be sealed with nail polish due to their polymerization property.
18. For most microscope objectives, it is recommended to use #1.5 thickness (0.17 mm or 0.0067 in) coverslips. It is advisable to consult the manufacturer's specifications for the objectives used.
19. If flies are fed yeast for longer than 3 days, late-stage egg chambers will accumulate and few previtellogenic stages (i.e., stages 1–8) may be recovered.
20. It is critical to keep dissected tissue covered with liquid solution at all times from this point on (even after fixation). Just a few seconds are enough to desiccate and destroy the tissue.
21. When dissecting smaller ovaries, it is often difficult to distinguish the ovaries from other structures. It is particularly helpful to use the oviducts, to which the ovaries are attached, as a guide when locating smaller ovaries. It may be helpful to keep associated organs (i.e., oviducts and uterus) attached to the ovaries

during dissection and staining to better see the tissue in solution so that ovaries are not discarded. If the abdomen is filled with an abundance of fat body or if the intestine is punctured, this will spill out of the abdominal cavity and make the dissection media very cloudy. It is advisable to carefully transfer the tissue to a well with fresh medium to separate the ovaries from connecting structures.

22. Ovaries dissected from females mutant for genes required for oogenesis may exhibit significantly smaller ovaries. The following tips are intended to reduce loss of smaller ovaries and previtellogenic stage egg chambers: (1) it is recommended to leave ~50 μL of liquid at the bottom of the tube during fixation and staining; (2) smaller ovaries and previtellogenic egg chambers do not sink easily or quickly and require extra patience and attention when removing solutions during fixation and staining.
23. Some protocols involve an additional step to first coat microcentrifuge tubes, and subsequently glass Pasteur pipettes upon removal of the solution, with 3% BSA in H_2O to prevent ovaries from sticking to the sides of the plastic or glass.
24. To preserve molecular interaction and cellular structures, ovaries may be stored in tubes on ice until fixation or one is ready for follow-up procedures (i.e., homogenization for protein or RNA isolation). For visualization or recovery of microtubules or microtubule-associated RNAs and proteins, however, the tissue should be kept at room temperature until fixation. Microtubules and actin filaments are susceptible to cold-induced depolymerization.
25. Ovaries easily stick to glass or plastic tips, pipettes, and tubes. Therefore, it is imperative to be careful when transferring ovaries. For example, if ovaries are pipetted above the neck of a glass Pasteur pipette, they may not be recovered. It is advisable to minimize the distance at which ovaries travel up the shaft of the pipette to minimize tissue loss. If ovaries stick to a tube or pipette, it may be possible to carefully pipette up and down with 1× PBST to dislodge the ovaries.
26. When removing solutions from the ovaries in a tube, carefully inspect the solution as it is being pipetted. It may be helpful to briefly hold the tube and Pasteur pipette up to the light as solutions are removed from the tube and before discarding them. Previtellogenic egg chambers and smaller ovaries are rather transparent and can easily be unseen and discarded without careful observation.
27. Some protocols use a higher concentration of PFA (e.g., 5% or 5.3%) and fix for a shorter amount of time (e.g., 13 min). To avoid over-fixation, it is important to not allow the tissue to stay in fixative for longer than 20 min before washing.

28. Tubes should be protected from light after the addition of fluorescent dyes or secondary antibodies to avoid photobleaching. Instead of covering individual tubes, it is advisable to cover the entire platform rocker with a layer of aluminum foil. Alternatively, a nontransparent freezer box lid or a transparent pipette tip box lid covered in aluminum foil can be placed over the tubes.
29. If co-localization studies are being performed, it is important to be mindful of the animals in which the primary antibodies were raised. For example, if one primary antibody was raised in mice and the other is raised in rabbits, then these primary antibodies will be distinguishable upon the use of secondary antibodies raised against those animals (i.e., anti-mouse and anti-rabbit). However, if two primary antibodies were raised in rabbits, then secondary antibodies cannot distinguish the two primary antibodies as separate antigens. This should be avoided except in specific circumstances where this is the desired outcome.
30. Many antibodies can be reused several times. In fact, some antibodies work better after pre-absorption in which nonspecific antibodies in the serum or supernatant are reduced, resulting in lower background staining. However, as antibodies are depleted from the solution, detection may diminish and become undetectable. It is advisable to optimize antibody use, including the number of times a specific antibody lot can be reused at the recommended dilution.

Acknowledgments

The author thanks Olivier Devergne, Makayla Gomperts, Julia Wittes, and McKenzie Young for their contributions to this protocol.

References

1. Hinnant TD, Merkle JA, Ables ET (2020) Coordinating proliferation, polarity, and cell fate in the Drosophila female germline. Front Cell Dev Biol 8:19. https://doi.org/10.3389/fcell.2020.00019
2. Mahowald A, Kambysellis M (1980) Oogenesis. In: Ashburner M, Wright T (eds) The genetics and biology of Drosophila, 2nd edn. Academic Press, pp 141–225
3. McLaughlin JM, Bratu DP (2015) Drosophila melanogaster oogenesis: an overview. In: Bratu DP, McNeil GP (eds) Drosophila oogenesis. Springer New York, New York, pp 1–20
4. Merkle JA, Wittes J, Schüpbach T (2020) Signaling between somatic follicle cells and the germline patterns the egg and embryo of Drosophila. In: Current topics in developmental biology. Elsevier, pp 55–86
5. Spradling A (1993) Developmental genetics of oogenesis. In: Bate M, Martinez-Arias A (eds) The development of Drosophila melanogaster. Cold Spring Harbor Press, Cold Spring Harbor, pp 1–70
6. Blatt P, Martin ET, Breznak SM, Rangan P (2020) Post-transcriptional gene regulation regulates germline stem cell to oocyte

transition during Drosophila oogenesis. In: Current topics in developmental biology. Elsevier, pp 3–34. https://doi.org/10.1016/bs.ctdb.2019.10.003
7. Drummond-Barbosa D (2019) Local and physiological control of germline stem cell lineages in *Drosophila melanogaster*. Genetics 213:9–26. https://doi.org/10.1534/genetics.119.300234
8. Eliazer S, Buszczak M (2011) Finding a niche: studies from the Drosophila ovary. Stem Cell Res Ther 2:45. https://doi.org/10.1186/scrt86
9. Flora P, McCarthy A, Upadhyay M, Rangan P (2017) Role of chromatin modifications in Drosophila germline stem cell differentiation. In: Arur S (ed) Signaling-mediated control of cell division. Springer International Publishing, pp 1–30
10. Riechmann V (2017) In vivo RNAi in the Drosophila follicular epithelium: analysis of stem cell maintenance, proliferation, and differentiation. In: Zhang B (ed) RNAi and small regulatory RNAs in stem cells. Springer New York, New York, pp 185–206
11. Slaidina M, Gupta S, Banisch TU, Lehmann R (2021) A single-cell atlas reveals unanticipated cell type complexity in *Drosophila* ovaries. Genome Res 31:1938–1951. https://doi.org/10.1101/gr.274340.120
12. Wu X, Tanwar PS, Raftery LA (2008) Drosophila follicle cells: morphogenesis in an eggshell. Semin Cell Dev Biol 19:271–282. https://doi.org/10.1016/j.semcdb.2008.01.004
13. Ables ET, Hwang GH, Finger DS et al (2016) A genetic mosaic screen reveals ecdysone-responsive genes regulating *Drosophila* oogenesis. G3 Bethesda 6:2629–2642. https://doi.org/10.1534/g3.116.028951
14. Czech B, Preall JB, McGinn J, Hannon GJ (2013) A transcriptome-wide RNAi screen in the Drosophila ovary reveals factors of the germline piRNA pathway. Mol Cell 50:749–761. https://doi.org/10.1016/j.molcel.2013.04.007
15. Gans M, Audit C, Masson M (1975) Isolation and characterization of sex-linked female-sterile mutants in *Drosophila melanogaster*. Genetics 81:683–704. https://doi.org/10.1093/genetics/81.4.683
16. Hayashi R, Wainwright SM, Liddell SJ et al (2014) A genetic screen based on *in vivo* RNA imaging reveals centrosome-independent mechanisms for localizing *gurken* transcripts in *Drosophila*. G3 Bethesda 4:749–760. https://doi.org/10.1534/g3.114.010462
17. Jagut M, Mihaila-Bodart L, Molla-Herman A et al (2013) A mosaic genetic screen for genes involved in the early steps of Drosophila oogenesis. G3 Bethesda. https://doi.org/10.1534/g3.112.004747
18. Jambor H, Surendranath V, Kalinka AT et al (2015) Systematic imaging reveals features and changing localization of mRNAs in Drosophila development. eLife 4:e05003. https://doi.org/10.7554/eLife.05003
19. Jia D, Soylemez M, Calvin G et al (2015) A large-scale in vivo RNAi screen to identify genes involved in Notch-mediated follicle cell differentiation and cell cycle switches. Sci Rep 5:12328. https://doi.org/10.1038/srep12328
20. Lee M-C, Skora AD, Spradling AC (2017) Identification of genes mediating *Drosophila* follicle cell progenitor differentiation by screening for modifiers of GAL4::UAS variegation. G3 Bethesda 7:309–318. https://doi.org/10.1534/g3.116.036038
21. Sanchez CG, Teixeira FK, Czech B et al (2016) Regulation of ribosome biogenesis and protein synthesis controls germline stem cell differentiation. Cell Stem Cell 18:276–290. https://doi.org/10.1016/j.stem.2015.11.004
22. Schüpbach T, Wieschaus E (1989) Female sterile mutations on the second chromosome of Drosophila melanogaster. I. Maternal effect mutations. Genetics 121:101–117. https://doi.org/10.1093/genetics/121.1.101
23. Schüpbach T, Wieschaus E (1991) Female sterile mutations on the second chromosome of Drosophila melanogaster. II. Mutations blocking oogenesis or altering egg morphology. Genetics 129:1119–1136
24. Tootle TL, Williams D, Hubb A et al (2011) Drosophila eggshell production: identification of new genes and coordination by Pxt. PLoS One 6:e19943. https://doi.org/10.1371/journal.pone.0019943
25. Wittes J, Schüpbach T (2019) A gene expression screen in *Drosophila melanogaster* identifies novel JAK/STAT and EGFR targets during oogenesis. G3 Bethesda 9:47–60. https://doi.org/10.1534/g3.118.200786
26. Abbaszadeh EK, Gavis ER (2016) Fixed and live visualization of RNAs in Drosophila oocytes and embryos. Methods 98:34–41. https://doi.org/10.1016/j.ymeth.2016.01.018
27. Cetera M, Lewellyn L, Horne-Badovinac S (2016) Cultivation and live imaging of Drosophila ovaries. In: Dahmann C (ed) Drosophila. Springer New York, New York, pp 215–226

28. Dai W, Montell DJ (2016) Live imaging of border cell migration in Drosophila. In: Jin T, Hereld D (eds) Chemotaxis. Springer New York, New York, pp 153–168
29. Haack T, Bergstralh DT, St Johnston D (2013) Damage to the *Drosophila* follicle cell epithelium produces "false clones" with apparent polarity phenotypes. Biol Open 2:1313–1320. https://doi.org/10.1242/bio.20134671
30. Hudson AM, Cooley L (2014) Methods for studying oogenesis. Methods 68:207–217. https://doi.org/10.1016/j.ymeth.2014.01.005
31. Peters NC, Berg CA (2016) In vitro culturing and live imaging of Drosophila egg chambers: a history and adaptable method. In: Nezis IP (ed) Oogenesis. Springer New York, New York, pp 35–68
32. Shalaby NA, Buszczak M (2017) Live-cell imaging of the adult Drosophila ovary using confocal microscopy. In: Buszczak M (ed) Germline stem cells. Springer New York, New York, pp 85–91
33. Thompson L, Randolph K, Norvell A (2015) Basic techniques in Drosophila ovary preparation. In: Bratu DP, McNeil GP (eds) Drosophila oogenesis. Springer New York, New York, pp 21–28
34. Wong LC, Schedl P (2006) Dissection of Drosophila ovaries. J Vis Exp 52. https://doi.org/10.3791/52-v
35. Zimmerman SG, Peters NC, Altaras AE, Berg CA (2013) Optimized RNA ISH, RNA FISH and protein-RNA double labeling (IF/FISH) in Drosophila ovaries. Nat Protoc 8:2158–2179. https://doi.org/10.1038/nprot.2013.136
36. Hsu H-J, Drummond-Barbosa D (2017) A visual screen for diet-regulated proteins in the Drosophila ovary using GFP protein trap lines. Gene Expr Patterns 23–24:13–21. https://doi.org/10.1016/j.gep.2017.01.001
37. Buszczak M, Paterno S, Lighthouse D et al (2007) The carnegie protein trap library: a versatile tool for Drosophila developmental studies. Genetics 175:1505–1531. https://doi.org/10.1534/genetics.106.065961
38. Morin X, Daneman R, Zavortink M, Chia W (2001) A protein trap strategy to detect GFP-tagged proteins expressed from their endogenous loci in Drosophila. Proc Natl Acad Sci 98:15050–15055. https://doi.org/10.1073/pnas.261408198
39. Quiñones-Coello AT, Petrella LN, Ayers K et al (2007) Exploring strategies for protein trapping in Drosophila. Genetics 175:1089–1104. https://doi.org/10.1534/genetics.106.065995
40. Rørth P (1998) Gal4 in the Drosophila female germline. Mech Dev 78:113–118. https://doi.org/10.1016/S0925-4773(98)00157-9
41. He L, Wang X, Montell DJ (2011) Shining light on Drosophila oogenesis: live imaging of egg development. Curr Opin Genet Dev 21:612–619. https://doi.org/10.1016/j.gde.2011.08.011
42. Pokrywka NJ (2013) Live imaging of GFP-labeled proteins in Drosophila oocytes. J Vis Exp:50044. https://doi.org/10.3791/50044

Chapter 4

Utilizing the FLP-Out System for Clonal RNAi Analysis in the Adult *Drosophila* Ovary

Daniel N. Phipps, Amanda M. Powell, and Elizabeth T. Ables

Abstract

The ability to conduct spatially controlled RNA interference (RNAi) for gene knockdown using the UAS/Gal4 system is among the most appealing techniques available for analysis of gene function in the *Drosophila* ovary. While gene knockdown experiments in somatic cells in the developing organism (i.e., embryos and larvae) are effectively and commonly performed, the use of RNAi in adult ovarian cells can be hampered by the unintended deleterious effects of Gal4 expression in "off-target" developing tissues. Mosaic analysis overcomes these problems by imparting temporal and spatial control over gene manipulation, providing a useful tool to compare manipulated cells with wild-type cells in the same tissue. Here, we provide a method to utilize the UAS/Gal4 system in combination with the Flippase (FLP)-Flippase Recognition Target (FRT) system to generate positively labeled "FLP-Out" clones expressing a chosen RNAi in both the germline and the soma in the *Drosophila* ovary. This protocol outlines each step of the generation of clones and the selection of appropriate fly stocks and reagents, providing a guide to this powerful tool in the *Drosophila* genetic toolbox. These techniques allow for RNAi analysis within a specific cell type, providing an opportunity to study a variety of unique aspects of cell function that would not be possible in more traditional RNAi-based experiments.

Key words Germline stem cell, Cyst, Oocyte, Germarium, Germline, Follicle cell

1 Introduction

1.1 Harnessing the Power of Genetic Mosaics in Ovarian Cell Analyses

The *Drosophila melanogaster* ovary provides an elegant model for the study of many key aspects of cell biology, including stem cell self-renewal, cell cycle regulation, cell-cell signaling, and tissue morphogenesis. Among its many positive attributes as an experimental model, one of the best reasons for using the *Drosophila* ovary to address cell biological questions is the wealth of cell type-specific tools available for manipulating gene expression in space and time. One particularly elegant example is the ability to create genetic mosaics: clonal populations of genetically distinct

Daniel N. Phipps and Amanda M. Powell contributed equally.

Michelle S. Giedt and Tina L. Tootle (eds.), *Drosophila Oogenesis: Methods and Protocols*,
Methods in Molecular Biology, vol. 2626, https://doi.org/10.1007/978-1-0716-2970-3_4,

cells that can be compared within a shared microenvironment [1, 2]. Mosaic and chimeric experiments have been central to the study of cell and developmental biology, allowing powerful methods to examine lineage relationships, test fate specification and induction, and compare cellular morphology and connectivity. Genetic mosaics allow rigorous analyses comparing protein or reporter transgene levels, as cells with different genetic backgrounds can be compared in the same tissue, having undergone the same histological preparation. Genetic mosaics also offer the ability to analyze cells carrying homozygous mutations or insertions that might otherwise be deleterious to organismal development. The utility and technical advantages of genetic mosaics have made them essential for the analysis of ovarian cells.

The structural arrangement and distinct lineages of the ovary make mosaic analyses particularly amenable (Fig. 1a). First, the ovary is composed of both germ cells and somatic cells, each with its own unique gene expression and thus, unique genetic toolkit. Second, adult germ cells (nurse cells and oocytes) and somatic cells (follicle cells) arise from mitotically active stem cells, making it possible to utilize genetic mosaic systems that require a mitotic event to be activated. Ovarian germline stem cells (GSCs) arise from primordial germ cells during development and reside in the germarium at the anterior of each ovariole, in a somatic niche that promotes their self-renewal (Fig. 1b, c) [3, 4]. Asymmetric division of a GSC results in the production of another GSC (self-renewal) and a daughter cell (cystoblast) committed to differentiation. The cystoblast divides mitotically, ultimately forming a 16-cell cyst that becomes encapsulated (forming an egg chamber; Fig. 1d, e) by an epithelial layer of somatic follicle cells, the descendants of follicle stem cells (FSCs).

1.2 Selecting a Genetic Mosaic Generation System

At the center of genetic mosaic generation techniques in *Drosophila* lie two critical exogenous gene expression systems: the Upstream Activating Sequence (UAS)/GAL4 system and the Flippase (FLP)/Flippase Recognition Target (FRT) system. These have been reviewed recently in great depth, and we encourage the reader to use those resources to learn more about each system [1, 2, 5–7]. We focus here instead on how these systems have been adapted for ovarian cells. Indeed, while both systems work well for experiments in somatic cells, there are caveats to their use in germ cells, particularly for the UAS/Gal4 system [8].

The UAS/Gal4 system utilizes the yeast transcription factor Gal4, which binds to a specific target DNA sequence (the UAS) [9, 10]. Transgenic constructs can be engineered with UAS sequences upstream of a variety of tools, including fluorescent reporters such as green fluorescent protein (GFP), gene coding regions for overexpression or misexpression analyses, or hairpin sequences for the induction of short interfering RNAs (RNAi; *see*

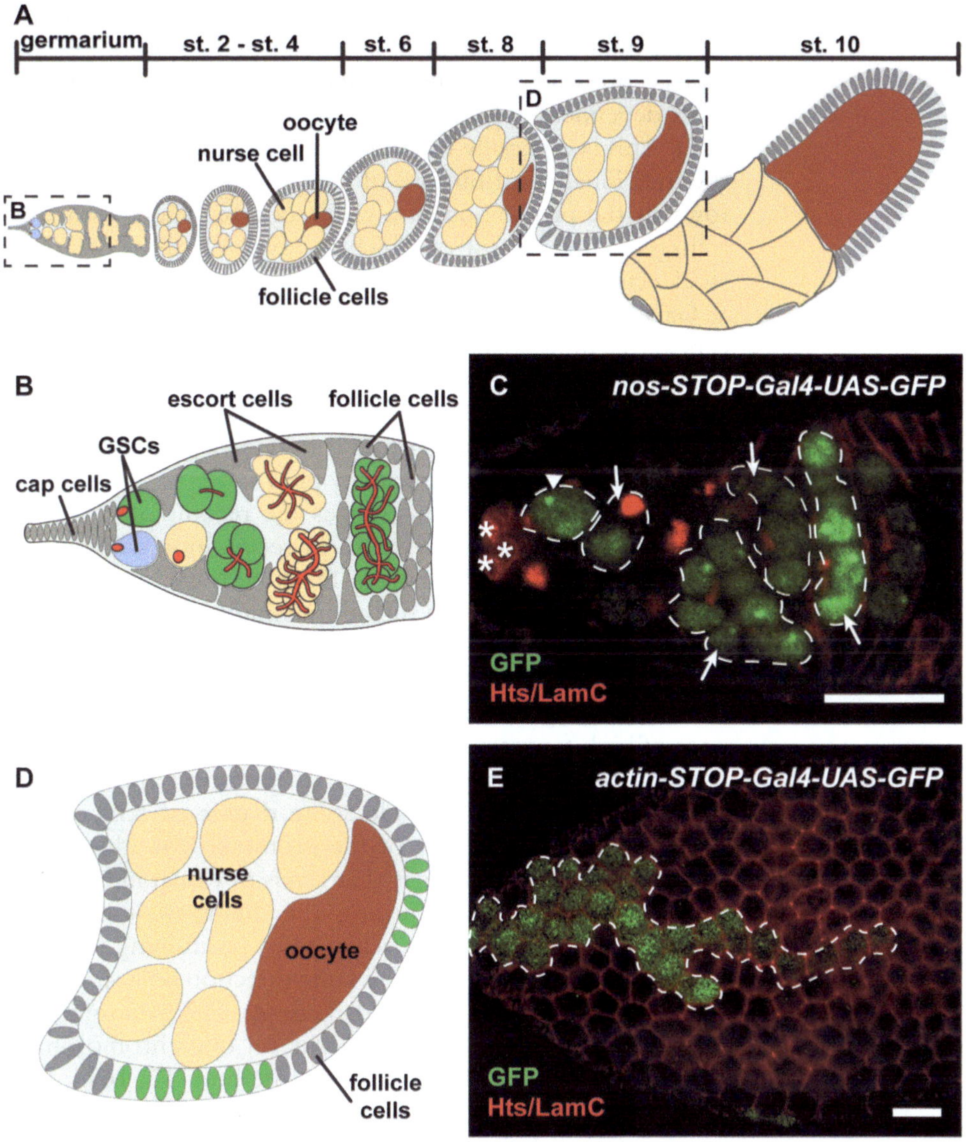

Fig. 1 Utilizing FLP-Out RNAi in the *Drosophila* ovary for genetic analysis of somatic and germ cells. *Drosophila* ovaries offer a linear array of developing oocytes. (**a**) A single ovariole is composed of egg chambers, each harboring a developing oocyte (brown) in progressively older stages, denoted 1–14. Somatic cells (gray) surround the germ cells (beige). Oogenesis begins in the germarium, where germ cells divide and egg chambers are created. Germline stem cells (GSCs; light blue) divide and form a cystoblast (beige) which will then mitotically divide four times to create a 16-cell cyst. The boxed areas are magnified in B and D (below). (**b**) In the germarium, germ cells are interconnected by ring canals and a fusome (red). As they divide, cysts move posteriorly and become encapsulated by follicle cells. Here, a positively labeled GSC (green) has divided to make three cysts (green). (**c**) Germarium from a *hsFLP; nos* > *STOP* > *Gal4, UASp-H4-EGFP* ovary following clone induction, labeled with GFP (green) to mark clones and Hts and LamC (red) to mark fusomes, cap cell nuclear membranes, and follicle cell plasma membranes. Dashed lines indicate GFP+ GSC (arrowhead) and cystoblast/cysts (arrows). Asterisks mark cap cells. (**d**–**e**) Schematic (**d**) and representative image (**e**) of a stage 8 egg chamber from a *hsFLP;actin5c* > *CD2* > *Gal4, UAS-GFP* ovary showing positively labeled follicle cell clones (green), co-labeled with Hts to mark follicle cell plasma membranes (red). Dashed lines indicate GFP+ follicle cells. Scale bars represent 10 μm

below). The power of this system lies in the ability of Gal4 to be placed under the control of a desired tissue-specific promoter or native promoter/enhancer elements, allowing for targeted UAS-dependent expression in a specific cell type of interest [11]. This is especially useful in the study of oogenesis to separate the individual contributions of somatic and germ cell populations towards the production of a viable oocyte. There are many Gal4 transgenic lines (most of which are available from the *Drosophila* stock centers) that have been established over the last two decades, as Gal4 was randomly inserted downstream of endogenous enhancers [12]. An important caveat, however, is that the full expression pattern of most Gal4-containing transgenes has not been described in all tissues across all stages of development. Thus, careful examination of the expression pattern is necessary prior to drawing conclusions about cell type specificity [13].

The FLP/FRT system utilizes the recombinase protein FLP with a series of two conserved short recognition sequences known as FRTs [2, 14]. FLP catalyzes DNA cleavage and chromosomal exchange, but it has been further manipulated to precisely control clonal induction both temporally and spatially [2, 12]. Temporal control can be achieved by use of the heat-inducible *hsp70* promoter, allowing for FLP protein to be produced only in the presence of high temperature [14]. Spatial control can be imparted by use of a tissue-specific promoter (or enhancer trap) upstream of *FLP*, which may also provide some degree of temporal control depending on the activity of the promoter. An added benefit of the *hsp70*-based FLP lines is that levels of clonal induction can be easily manipulated by adjusting the duration of heat exposure [15]. Spatial control can be achieved by positioning two FRT sites flanking a stop sequence in the same direction on a single chromosome downstream of a desired tissue-specific promoter, allowing for a recombination event to occur on one chromosome, commonly referred to as the "FLP-Out" method [12, 16].

A wide variety of techniques have been developed that use combinations of the UAS/Gal4 and/or FLP/FRT strategies to induce genetic mosaics. These techniques have been broadly reviewed elsewhere, and we point the reader to those resources for in-depth discussion [2, 12]. Broadly speaking, genetic mosaic generation strategies fall into two categories: negative labeling strategies (FLP/FRT) and positive labeling strategies (MARCM, FLP-Out, FINGR) [2, 12, 17]. Both rely on the ability to mark clones with an easily detectable, nonlethal, cell autonomous indelible marker that can be inherited by daughter cells after mitotic division. The bacterial *lacZ* gene and GFP or other fluorescent proteins fit all these requirements and are commonly used in ovarian genetic mosaics. While both negative and positive labeling strategies work well in the ovary (*see* [18, 19]), the negative labeling strategy has predominated in germ cell analyses. This is largely because ovarian germ cells block expression of the *hsp70* promoter incorporated in UASt-based transgenes, which are the most commonly available in the stock centers [6, 8]. To circumvent this problem, other transgenes using the germline-specific UASp and VALIUM22 or the ubiquitous and highly effective UASz promoters were developed and are much more suitable for germ cell analyses [8, 20, 21]. As these are not widely available in the stock centers, these transgenes must be custom developed using molecular biology techniques and available vectors (*see* https://dgrc.bio.indiana.edu/vectors/Catalog) for your specific gene of interest. Understanding the properties of the vector used to build a UAS-containing transgene is essential to determine the compatibility of the transgene with the targeted cell type in ovarian experiments.

1.3 The FLP-Out Method: Combining UAS/Gal4 and FLP/FRT with RNAi to Knockdown Gene Function in Germ Cells and Somatic Cells in the Adult Ovary

One of the most widely used tools in *Drosophila* to experimentally block gene expression is RNAi. This naturally occurring system, first observed in *C. elegans,* relies on the generation of a double-stranded RNA molecule which is targeted for degradation by an endonuclease [22, 23]. The cleaved dsRNA is then complexed with an Argonaut protein to form the RNA Induced Silencing Complex (RISC), which will seek out and degrade a complementary mRNA sequence, resulting in a reduction of the expression of a given gene [24]. RNAi has been highly effective in the *Drosophila* model due to its compatibility with the UAS/Gal4 system, which allows for spatial specificity in RNAi knockdown experiments. The ability to spatially restrict gene knockdown to a single tissue or cell type helps to reduce the risk of embryonically lethal phenotypes, while elucidating cell autonomous gene functions in adults. The UAS-RNAi system, however, carries the same limitations of the UAS-Gal4 system itself. For the study of adult ovarian cells, the most notable is the lack of temporal control over RNAi induction. Although a temperature-sensitive version of the Gal80 repressor (Gal80^{ts}) can be used to suppress Gal4 expression [25, 26], incubation of flies at the permissive temperature (29 °C) for a week or more can result in nonspecific germline cell death independent of the RNAi, especially if the temperature is not closely regulated.

The FLP-Out technique combines the UAS/Gal4, FLP/FRT, and RNAi methods to provide temporal and spatial control over gene knockdown, making it an exceptional tool for the study of cells in the adult ovary (Fig. 1b–d). In the FLP-Out system, Gal4 is preceded by a stop cassette flanked by two FRT sites (Fig. 2). Heat shock-inducible FLP recognizes the FRT sites, catalyzing removal of the stop cassette and activation of Gal4 transcription. With the Gal4 protein active, any UAS-controlled transgene can then be expressed. Like other mosaic techniques, the FLP-Out system includes an indelible fluorescent reporter controlled by a UAS promoter, thus positively labeling any cells in which a

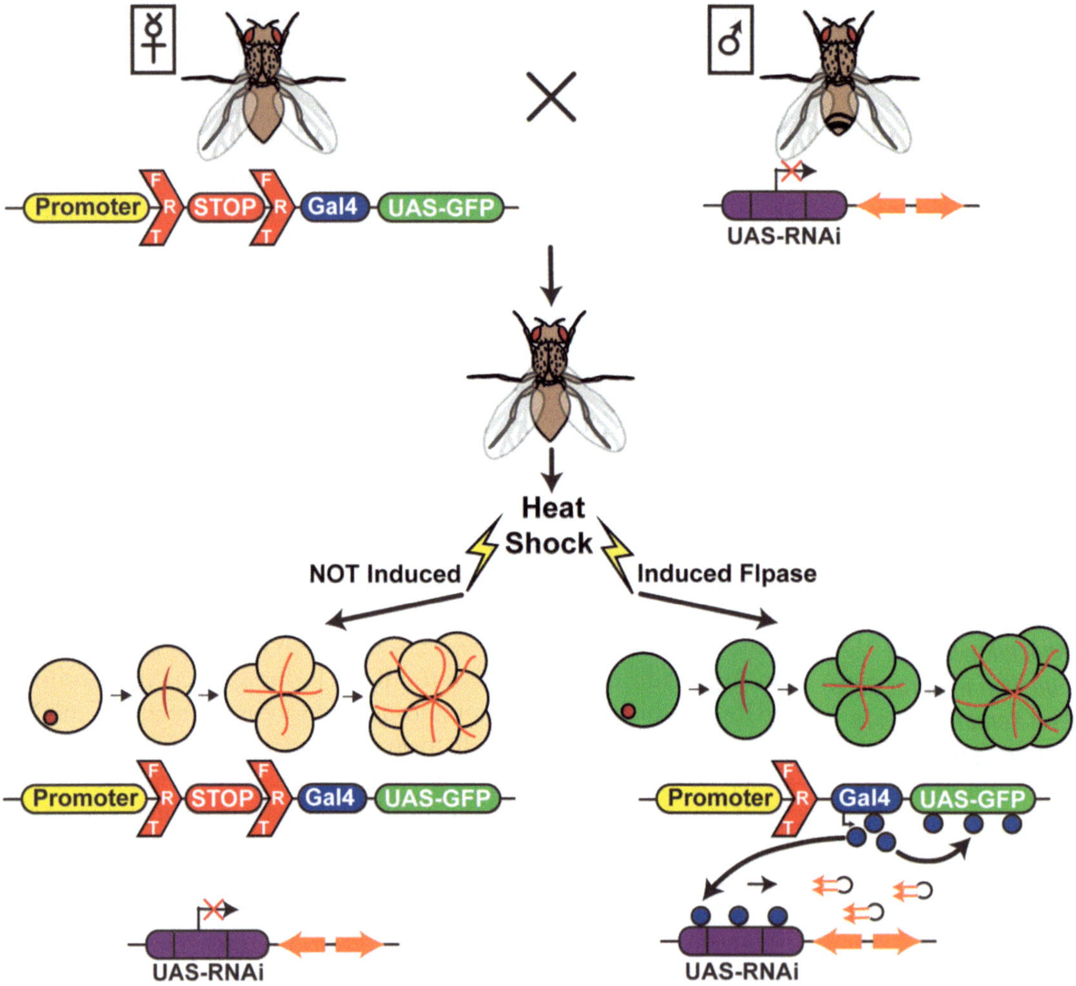

Fig. 2 Overview of FLP-Out RNAi. To create mosaic clones, FLP-Out Gal4 virgin females (containing a tissue-specific promoter driving the expression of Gal4) are crossed to males carrying compatible UAS-controlled RNAi. Heat shock-inducible FLP (hs-FLP) catalyzes excision of a stop sequence flanked by two FRT sites. This allows for Gal4 protein to be produced, which will then bind to UAS-GFP and UAS-RNAi, positively labeling any cell in which RNAi is expressed. This positive label will be maintained in any progeny produced by a dividing cell, allowing for RNAi knockdown to be observed in both stem cells and differentiated progeny. If FLP is not induced, cells and their progeny remain wild type

recombination event occurs, as well as any progeny produced from that cell. Use of an *actin5C* promoter and UASt-based fluorescent reporters provides reliable clonal induction and labeling in somatic cells (Fig. 1d, e) [17] and is a useful tool for a variety of assays [27–29]; however, it does not provide reliable clone generation in ovarian germ cells. To circumvent this problem, the Xie, Lin, and Yamashita labs modified the FLP-Out system for use in germ cells by substituting the germline-specific *nanos* promoter upstream of Gal4 and including a UASp-driven GFP [30]. This allows clones to be generated specifically in the germline, with no activation of the system in somatic populations (Fig. 1b, c). The germline FLP-Out is active in GSCs and dividing cysts, and in combination with germline-compatible UASp, UASz, or VALIUM22 transgenes, is a versatile and effective means for positively labeled mosaic studies in the germline that have not previously been possible.

In this protocol, we describe our method for using FLP-Out systems to knockdown gene expression in adult ovarian cells. We describe the husbandry necessary to use the *nanos* FLP-Out or *actin5C* FLP-Out transgenes to generate positively labeled genetic mosaics in germ cells or somatic cells, respectively (Fig. 2). We then provide step-by-step methods for clone induction, ovary dissection, immunostaining, and whole-mount preparation of ovaries for confocal microscopy. This versatile protocol can be adapted for a wide variety of cell biology analyses (i.e., cell proliferation, cell death, gene expression) to suit the needs of the researcher with the rigor of side-by-side cellular analysis.

2 Materials

2.1 Drosophila *Strains and Culture*

1. *Drosophila* strains (*see* **Note 1**).
   (a) FLP-Out Gal4:
      (i) For mosaic RNAi analysis in the germline: *hsFLP; nos > STOP > Gal4, UASp-H4-EGFP/CyO* [30] (*see* **Note 2**).
      (ii) For mosaic RNAi analysis in somatic cells: *hsFLP; actin5c > CD2 > Gal4, UAS-GFP.*
   (b) UAS-RNAi: When selecting an RNAi line for a target gene to use, be careful to select lines with the appropriate UAS construct. While most UAS-containing RNAi lines are compatible with somatic expression, you must select UASp, VALIUM20, VALIUM22, or UASz transgenes for germ cell expression.
2. Standard fly culture medium (cornmeal/molasses/yeast/agar) in bottles and vials (*see* **Note 3**).
3. Dry Active Yeast (*see* **Note 3**).
4. Yeast paste: 1–1 mixture of dry active yeast (*see* **Note 3**) and distilled water, mixed to the consistency of creamy peanut butter. Store covered with Parafilm at 4 °C to prevent drying (*see* **Note 4**).
5. Dissecting stereomicroscope with dual goose-neck illumination, 0.6–4× magnification range, and a flat black dissection base, equipped with a fly pad and air needle to deliver CO_2 (*see* **Note 5**).

2.2 Materials for Heat Shock to Induce Clones

1. Large water bath set to exactly 37 °C (confirmed with a reliable glass thermometer).
2. Heat shock vials: use standard *Drosophila* vials (without food), placing tightly folded Kimwipe in the bottom to create a reasonably flat surface, topped with a light layer of dry yeast (Fig. 3a, b).
3. Weighted rings (e.g., round covered lead flask rings) or blocks.
4. Flat surface to hold weights on top of vials (lid of a pipette tip box works well).
5. Waterproof vial rack with enough space to distribute vials evenly.

2.3 Ovary Dissection and Immunostaining

1. 1.5 mL microfuge tubes, pre-coated with 3% Bovine Serum Albumin (BSA) prepared in water (*see* **Note 6**).
2. 15 mL polypropylene conicals.
3. 50 mL polypropylene conicals.
4. Glass or plexiglass dissection dish.
5. Kimwipes.
6. Glass Pasteur pipettes and bulbs.
7. Two pairs of #5 dissection forceps (INOX, Dumont #5, Biologie point).
8. Two 27 × 1¼ gauge needles with 1 mL syringes.
9. Nutating mixer.
10. Grace's Insect Medium (discard if cloudy or if liquid has a strange smell) (*see* **Note 7**).
11. 0.1% Triton-X-100 in Phosphate Buffered Saline (PBS).
12. 5.3% formaldehyde fixative solution diluted in Grace's medium (*see* **Note 8**).
    - Prepared fresh before each dissection using EM Grade 16% formaldehyde opened for no more than 4 days.
    - For each sample, add 300 μL 16% formaldehyde and 600 μL Grace's medium. Keep on ice (*see* **Note 9**).
13. Blocking Solution: 5% BSA, 5% normal goat serum, 0.1% Triton-X-100 in PBS. Prepare using sterile technique, and store at 4 °C. Discard if cloudy (*see* **Note 10**).
14. Primary Antibodies: Both the described germline and somatic clonal systems utilize a positive labeling system by way of a UAS-GFP element. This will identify all cells in which the Gal4 protein has been activated, indicating single cells in which the RNAi is being expressed. For a standard immunostaining experiment in the ovary, labeling cell membranes and fusome structure is extremely helpful for identification of different cell

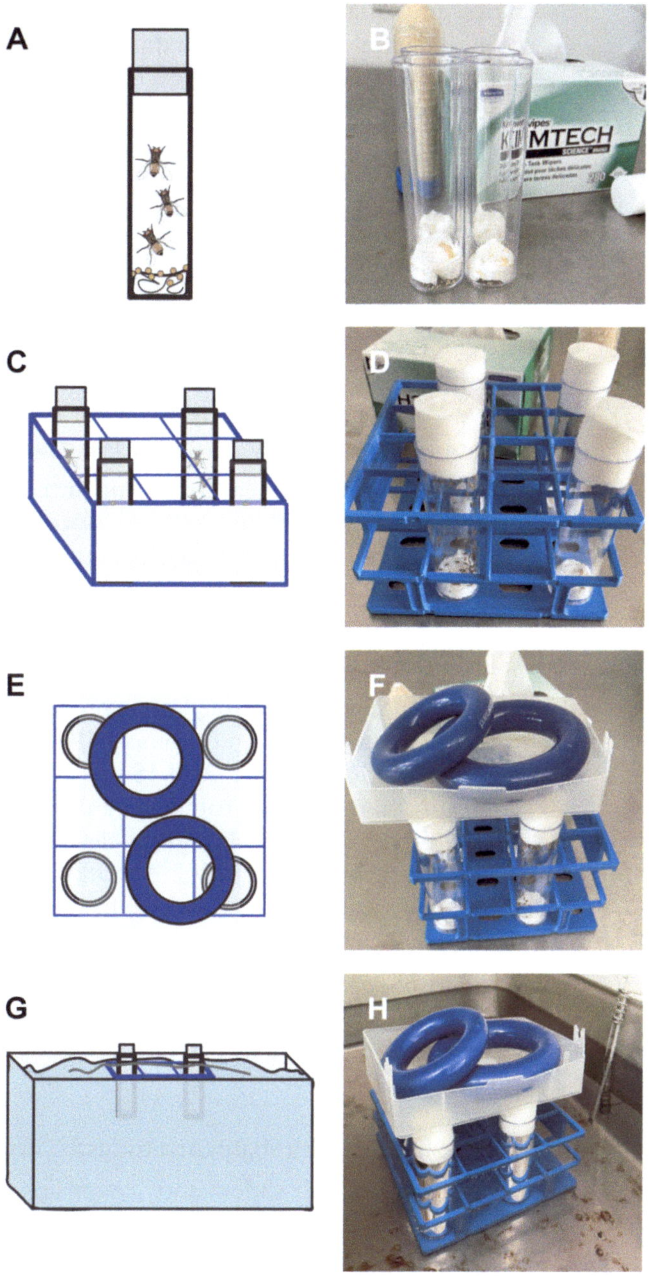

Fig. 3 Heat shock setup. (**a**, **b**) When preparing heat shock vials, ensure that a tightly folded Kimwipe is pressed firmly to the bottom, creating a reasonably flat surface. This prevents flies from becoming entangled in the wipe and dying during heat shock. Add dry yeast for optimal fly nutrition during the heat shock. (**c**, **d**) Ensure that vials are spaced out evenly in the rack, avoid placing vials directly beside each other. Especially in the absence of a circulating water bath, this keeps heat distribution even among vials, allowing for optimal clonal induction within each sample. (**e**, **f**) Weights and flat surface should be carefully placed to ensure that each vial is equally pressed down into the water bath. If a vial is not adequately exposed to the weight and the vial floats up, this will result in little to no clonal induction in that vial. (**g**) It is crucial to ensure that the waterline is above the bottom line of the plug, without being high enough to touch the exposed plug. If the waterline is too low, flies will escape to the cooler region at the top of the vial. If the water touches the exposed plug, this could result in flies becoming wet and drowning

types and quantifications, including locating germline stem cells. The following combination of antibodies is extremely versatile and makes for an excellent preliminary experiment when testing this system (*see* **Notes 11** and **12**):

- Labeling FLP-Out positive cells: chicken anti-GFP (ab13970; Abcam), 1:2000 dilution.
- Cell membrane and fusome marker: mouse anti-Hts (1B1-s; Developmental Studies Hybridoma Bank), 1:10 dilution.
- Nuclear envelope marker: mouse anti-LaminC (LC28.26-s; Developmental Studies Hybridoma Bank), 1:100 dilution.

15. Secondary Antibodies: AlexaFluor-conjugated secondary antibodies matching the respective host species of the corresponding primary antibody. For example:
    - For the detection of anti-Hts and anti-LaminC, we use goat anti-mouse-AlexaFluor568 (*see* **Note 13**), diluted to a final concentration of 1:200 in blocking solution.
16. 4′,6-Diamidino-2-phenylindole (DAPI): For working concentration, prepare a 1:500 dilution of a 5.0 mg/mL DAPI stock solution in 0.1% Triton-X-100 in PBS (*see* **Note 14**).
17. Mounting Medium: 20 mg/mL n-propyl gallate in glycerol. In a 50 mL conical tube, combine 1.0 g n-propyl gallate with 5 mL PBS; vortex to mix. Add 45 mL 100% glycerol. Cover the conical tube in foil and rotate on a nutator at room temperature overnight. Media should be clear but very viscous. Store protected from light at 4 °C. A working stock can be poured into an opaque dropper bottle for ease of application (*see* **Note 15**).
18. Dissecting stereomicroscope with dual goose-neck illumination, 0.6–4× magnification range, and a flat black dissection base, equipped with a fly pad and air needle to deliver CO_2 (*see* **Note 5**).

2.4 Sample Preparation and Confocal Microscopy

1. Needle holders with sharpened tungsten needles (*see* **Note 16**).
2. Glass microscope slides and coverslips (1 μm thickness, 22 × 22 mm).
3. Steel weight, measuring approximately 250 g.
4. Fingernail polish (ideally of a relatively thick consistency to prevent running).
5. Laser scanning confocal microscope equipped with 63× oil immersion lens (n.a. = 1.4) and 1.5–3.0× optical zoom (*see* **Note 17**).
6. Confocal image acquisition and analysis software (e.g., Zeiss Zen) (*see* **Note 18**).

3 Methods

3.1 Drosophila Husbandry (See Note 19)

1. Since low-level *hsFLP* expression can induce premature recombination in the *actin-Gal4* FLP-out parent line (*w⁻; +; actin5c > CD2 > Gal4, UAS-GFP*) (Bloomington #4780, #30558, or similar), we do not maintain the *actin-Gal4* FLP-Out parent line with *hsFLP* in our laboratory stocks. Prior to starting your somatic FLP-Out experiment, use standard husbandry to create a temporary *hsFLP; actin5c > CD2 > Gal4, UAS-GFP* recombinant stock. For germ cell FLP-Out mosaics, the *hsFLP; nos > STOP > Gal4, UASp-H4-EGFP/CyO* [30] already contains a hsFLP element; however, test the line frequently to ensure that the STOP has not been prematurely deleted.
2. Expand at least two bottles of the appropriate FLP-Out Gal4 line to ensure an adequate supply of virgin females for crosses. Maintain expanded bottles at room temperature to prevent flies from being exposed to elevated temperature before heat shock.
3. Expand at least one bottle each of the somatic RNAi line and a control line (e.g., y^1w^1, *UAS-lacZ*, or Canton-S) (*see* **Note 20**).
4. Collect FLP-Out Gal4 virgin females, maintaining them for no longer than 4 days at room temperature before setting crosses (*see* **Note 21**).
5. Set Crosses in two to four replicates with the addition of dry yeast to promote optimal egg laying conditions (Fig. 2). Maintain vials at room temperature transferring adults to new vials every 3 days. New flies should emerge 10–14 days after the setup date (*see* **Note 22**).
   - Set 5 virgin females of the *hsFLP* driver with 4–5 males from a control line (e.g., *UAS-lacZ*).
   - Set 5 virgin females of the *hsFLP* driver with 4–5 males from your RNAi line.
6. Collect progeny upon eclosion and prepare for heat shock.

3.2 Heat Shocking Flies to Induce FLP-Out

1. Place 15 pairs of progeny into prepared heat shock vials with Kimwipes and dry yeast (Fig. 3a, b) (*see* **Note 23**).
2. Place vials into a waterproof vial rack (Fig. 3c, d) (*see* **Note 24**).
3. Place flat surface with weights on top of vials (Fig. 3e, f) (*see* **Note 25**).
4. Submerge vials in 37 °C water bath with weights for 1 hour (Fig. 3g, h) (*see* **Note 26**).
5. At the end of the hour, remove vials from water bath and place flies in vials with standard fly food and fresh wet yeast paste at room temperature.

6. Perform **steps 1–5**, 2 times per day for 3 days, for a total 6 heat shock events (*see* **Note 27**). Flies should be well-fed with fresh wet yeast paste each day prior to dissection.

3.3 Ovary Dissection and Immunostaining

1. Use CO_2, ice, or FlyNap to anesthetize flies. Remove males and discard. Dissect ovaries from female flies in ice cold Grace's medium (to preserve tissue integrity) in a dissecting dish. To dissect ovaries (*see* **Note 28**), grasp the anterior end of the abdomen with forceps in the nondominant hand. Using forceps in the dominant hand, pinch the end of the abdomen, and pull away from the body. The forceps will grab the cuticle, urogenital system, and frequently the ovaries. Repeat the dissection for all the females in that vial and continuing in the cold Grace's medium, break the outer muscle sheath surrounding the ovarioles to tease apart individual ovarioles using your forceps and a needle. Using a glass Pasteur pipet, remove BSA from labeled, pre-coated microfuge tube and suction up and down to coat the pipet and discard BSA. Using the same pipet, move the dissected ovaries in Grace's medium to the microfuge tube. Place labeled tube back on ice.
2. Repeat dissection with additional vials of flies, making sure to keep genotypes separated and labeled. Fix samples (*see* **step 3**) within an hour of dissection of the first female (*see* **Note 29**).
3. Allow ovaries to settle to the bottom of the microfuge tube by gravity. Remove as much Grace's medium as possible (*see* **Note 30**) and add 900 μL of fixative. Invert the microfuge tube and rotate on a nutator for 13 minutes at room temperature.
4. Remove fixative to an appropriate waste container and quickly add 1.0 mL of 0.1% PBS-Triton (wash solution). To rinse: invert the tube several times and remove the wash to a formaldehyde waste container. Add 1.0 mL of fresh wash solution and repeat rinse two more times. After final removal, add fresh wash solution, suspend ovaries, and rotate on a nutator for 10 minutes.
5. Discard solution and wash ovaries two more times in 1.0 mL wash solution for greater than 10 minutes each.
6. Discard solution and add 1.0 mL of blocking solution, place ovaries on the nutator for 3 hours at room temperature.
7. Discard solution and add 500 μL of first primary antibody mixture (*see* **Notes 31** and **32**). We achieve the highest signal with lowest background fluorescence by incubating ovaries for one or two nights at 4 °C rotating on a nutator.
8. Discard antibody (*see* **Note 33**) and wash ovaries in 1.0 mL wash solution four times for at least 30 minutes each on a rotating nutator.

9. Discard solution and add 500 μL of second primary GFP antibody mixture. Incubate ovaries for one night at 4 °C rotating on a nutator.
10. Discard antibody and wash ovaries in 1.0 mL wash solution four times for at least 30 minutes each on a rotating nutator.
11. Discard the final wash solution and add 500 μL of secondary antibody mixture. Incubate ovaries for 1–2 hours at room temperature on a rotating nutator. Incubate ovaries from this step forward in the dark to minimize the loss of fluorescence.
12. Discard antibody and wash ovaries in 1.0 mL wash solution four times for at least 30 minutes each on a rotating nutator.
13. Discard the final wash and add 500 μL of DAPI. Incubate ovaries for 15 minutes on a nutator at room temperature.
14. Discard DAPI and wash ovaries two times in 1.0 mL wash solution for 10 minutes each on nutator.
15. Discard the final wash, removing as much liquid as possible and add at least four drops of mounting medium. Samples can be stored upright at 4 °C for up to a month before mounting.

3.4 Sample Preparation and Confocal Imaging

1. Prepare samples on glass slides for imaging no more than 24 hours before imaging (*see* **Note 34**).
2. Using a glass Pasteur pipet, move ovaries and mounting media onto a slide. Remove excess mounting media using your pipet. Separate individual ovarioles using forceps and needles. For optimal imaging, remove all large egg chambers (*see* **Note 35**). Spread ovarioles out to avoid clumping.
3. Add 1–2 drops of mounting media back to the slide and slowly drop a glass coverslip to the center of the slide, allow capillary action to spread the mounting media across the coverslip. If any air remains, use a Pasteur pipet to slowly add more mounting media to the edges of the coverslip.
4. Place Kimwipes and a steel weight on top of the coverslip/slide to flatten the ovarioles. After 5–10 minutes, remove the weight and seal the coverslip using fingernail polish. Allow slides to dry before imaging. Store in a slide box protected from light at 4 °C.
5. Use a confocal microscope with a 63× oil immersion lens and optical zoom (1.5–3×) to image samples. Images are collected as confocal z-stacks with 1 μm optical sections (or the optimal step size for the objective being used). Each germarium on the slide should be imaged for analysis, usually totaling 50–200 germaria per slide (depending on analysis).

3.5 Data Analysis, Quantification, and Next Steps

The first major step of data analysis using this system will be determining the appropriate number of days after heat shock to dissect. This will depend on the type of analysis that is being done. If observing GSC loss over time using either the somatic or germline drivers, this will require dissections to be done at multiple time points over days to track loss. Our lab usually uses 1, 3, 6, and 10 days after heat shock. If observing early cyst divisions or cyst formation, a time point of approximately 3 days after heat shock would likely be sufficient. To observe the effectiveness of a chosen RNAi line using immunofluorescence, an image processing software can be used to determine fluorescence intensity of a GFP-positive cell directly adjacent to a wild-type GFP-negative cell, providing an excellent internal control to determine that mRNA and thus protein levels are, in fact, reduced in cells actively expressing RNAi. This method is effective in both the germline and somatic cell types. This system also provides opportunities for additional staining sets such as cell cycle markers, mitotic markers, and other indicators of cellular processes within a single RNAi-positive cell in a background of wild-type cells.

4 Notes

1. Many RNAi lines can be obtained from the Bloomington Drosophila Stock Center (http://flystocks.bio.indiana.edu), the Vienna Drosophila Resource Center (http://stockcenter.vdrc.at), or the Kyoto Drosophila Stock Center (http://www.dgrc.kit.ac.jp).
2. The *hsFLP; nos > STOP > Gal4, UASp-H4-EGFP/CyO* strain [30] is available by request from the Xie Lab.
3. We recommend Nutri-Fly MF mixed using the manufacturer's instructions and supplemented with Tegosept and propionic acid to prevent mold growth. Nutri-Fly MF is a molasses-based media formulation available from Genesee Scientific (https://geneseesci.com) that is easily mixed using an immersion blender and convection hot plate. Because it can be made in relatively small batches, it is a good option for small *Drosophila* labs. Fly culture medium is best when used fresh (withing 3–4 weeks). We use 6 oz. square bottom polypropylene bottles and narrow polystyrene vials both sealed with cotton flugs.
4. We use Fleischmann's instant dry yeast, available in one-pound packages from commercial grocery stores, mixed at a 1:1 ratio with water until a peanut-butter consistency is achieved.
5. *Drosophila* husbandry and anesthetic supplies can be purchased from numerous suppliers, including Genesee Scientific.

6. Tubes are pre-coated in a 3% BSA solution [diluted from a 30% solution in distilled water] to prevent the ovaries from sticking to the side of the tubes. Add 250 μL of 3% BSA in microfuge tubes. Place tubes on a nutator at room temperature for 1 hour to ensure complete coverage of inside of tube. Store tubes at 4 ° C with BSA still in the tube. Coated tubes are good unless BSA becomes cloudy or discolored.
7. We make 50 mL sterile aliquots of Grace's medium and store at 4 °C until ready for dissection. A fresh batch should be used if the aliquot appears cloudy, as this indicates it is contaminated.
8. Optimal fixative concentration and fixation time may vary between primary antibodies and should be experimentally determined. We have found that the concentration and incubation time given here are optimal for the primary and secondary antibodies listed in this protocol.
9. Formaldehyde rapidly degrades when in contact with air. For best results, we order 10 mL ampules of ultrapure 16% formaldehyde, transfer the liquid to a 15 mL conical tube, and store for use no longer than 1 week at 4 °C.
10. Ideal blocking solutions help to reduce background staining and may vary with primary and secondary antibodies. It is also important to choose a serum for blocking solution that it matches the host species of the secondary antibodies used in the immunostaining reaction.
11. We found that some concentrated aliquots of LC28.26 will also label the nuclear envelopes of GSCs, germ cells, and, occasionally, follicle cells.
12. If the protein that is being targeted for RNAi knockdown has a reliable antibody readily available, it is important to perform a separate experiment to test how effective the RNAi is at depleting the protein of interest. Protein levels should be compared in adjacent clones to ensure equivalent histological preparation. In the absence of an antibody, RNA levels should be examined by reverse transcriptase polymerase chain reaction.
13. Upon arrival, we add an equal volume of 100% glycerol to each commercial secondary antibody, mix well, and store at −20 °C in 100 μL aliquots.
14. Prepare a 5 mg/mL DAPI stock solution according to a manufacturer's instructions in deionized water. Store at 4 °C in the dark.
15. While a variety of commercially available glycerol-based antifade mounting media are available, we prefer to make our own solution. We have found that it is less expensive and preserves fluorescence intensity as well as commercially available brands. We have previously used mounting media directly

supplemented with DAPI, but have found that these formulations do not penetrate well into the ovary, resulting in uneven nuclear staining. To stabilize fluorescence on slides over multiple days of imaging, we have also found some commercially available hard-set media to work well.

16. Members of our lab separate ovarioles ("tease") using different methods. Some decide to use two forceps, while others use one pair of forceps and a tungsten needle. However, we find it easiest to have one needle with an "L" or "shepherd's crook" shape that helps pull larger follicles off the ovarioles in a left-right motion (think of a cane pulling someone off the stage in old movies).
17. We recommend the Zeiss LSM800 system, particularly for small labs with routine use.
18. We use Zeiss image acquisition and analysis software. Many common analyses can also be conducted in ImageJ, which is freely available (http://imagej.nih.gov/ij/).
19. Methods of *Drosophila* culture maintenance and husbandry have been described elsewhere.
20. y^1w^1 is frequently used as an inbred isotype control because of its prevalent use in generating transgenic flies. These, as well as *UASp-LacZ* and Canton-S, are available from Bloomington Stock Center (#1495, #1776, #64349).
21. Our lab uses female virgins aged no longer than 4 days in order to promote optimal egg laying output.
22. Experimental crosses and their controls should always have the same diet, ambient temperature, and be age-matched.
23. When preparing heat shock vials (*see* Fig. 3), make sure that the Kimwipe is as flat as you can get it so that flies do not get stuck between the folds. Adding dry yeast provides a small food source that will not be affected by the heat.
24. When placing heat shock vials in vial rack, make sure that heat will be distributed evenly. Placing vials too close together makes "cool spots" between vials.
25. We use an old pipet tip box top (*see* Fig. 3) with two ring weights on top. It is important to make sure that the weight is evenly distributed to avoid any mishaps when placing in the water bath.
26. Ensure that the water line is above the bottom of the flug line to confirm that flies are constantly exposed to the heat source for the duration of the hour. Flies will seek out space away from the heat if vials are not submerged correctly, which will reduce clone induction. Make sure not to submerge the exposed part of the flugs. If the plug gets wet, water will seep into the vials and can kill the flies.

27. Ideally, spread out heat shocks by several hours (one in the morning and one in the afternoon) to ensure that flies are not exposed to heat in quick succession and thereby, reduce unwanted stress. The density of clones is determined by the number and duration of heat shocks; therefore, this step can be titrated to determine the optimal density of clones needed for each specific experiment.
28. A variety of reviews have been recently published with excellent descriptions of ovary dissection techniques; we refer readers there who want additional photographs/videos of ovary dissection [31, 32] and *see* "*Dissection, Fixation, and Standard Staining of Adult Drosophila Ovaries*" by Dr. Julie Merkle (also in this edition). We have also found that the YouTube video from Dr. Scott Ferguson (https://www.youtube.com/watch?v=oYzjE_zqc1w) is helpful for first-time ovary dissection.
29. Fixing ovaries within 1 hour prevents tissue degradation.
30. At each liquid change throughout the rest of the protocol, allow a few seconds for ovaries to fall to the bottom of the microfuge tube by gravity before attempting to remove the solution. With exception of the fixative and primary antibody solution, we use a vacuum trap flask quipped with a thin pipet tip to remove liquid from the ovaries. We typically leave about 50–100 μL of solution on dissected ovaries at any given time in this protocol. This helps to avoid accidental suction of the ovaries into waste and prevent the samples from drying.
31. We achieve vastly decreased background staining by separately incubating primary antibodies with different host species overnight, separated by extensive wash steps (4 times 30 minutes each). In effect, primary antibodies are layered on top of the sample. In contrast, secondary antibodies against different species are typically grouped together into a single solution.
32. If two antibodies are raised in the same species, but it is necessary to image them separately, apply one antibody followed by washes and then the appropriate secondary. Then wash ovaries five times (30–45 minutes each) in a wash solution. Re-block for 1 hour and apply the second primary antibody, following by incubation, proceed with wash, and then the secondary (with a different fluorophore).
33. The mixture of anti-Hts and anti-LaminC at the concentration given can typically be collected, stored at 4 °C, and reused for one additional staining experiment.
34. Samples on slides will dry out and lose their fluorescence intensity over time. Since image acquisition is frequently the bottleneck to these experiments, we store all samples in mounting media until no more than 24 hours become imaging.

35. Stage 10 follicles are easily identified under a dissection stereomicroscope, because the size of the oocyte approximates one-half the size of the whole follicle. We remove all follicles larger than a Stage 10 if we are imaging GSCs because larger egg chambers can tilt the coverslip, allowing germaria to move and create blurry images.

Acknowledgments

We are grateful to Dr. Ting Xie and the *Drosophila* community for sharing reagents and protocols and to members of the Ables lab past and present for helpful comments on this manuscript. This work was supported by National Institutes of Health (NIH) grant R15-GM117502 (E.T.A.) and an East Carolina University Undergraduate Research and Creative Activity Award (A.M.P.).

References

1. Griffin R, Binari R, Perrimon N (2014) Genetic odyssey to generate marked clones in Drosophila mosaics. Proc Natl Acad Sci U S A 111(13):4756–4763. https://doi.org/10.1073/pnas.1403218111
2. Germani F, Bergantinos C, Johnston LA (2018) Mosaic analysis in Drosophila. Genetics 208(2):473–490. https://doi.org/10.1534/genetics.117.300256
3. Hinnant TD, Merkle JA, Ables ET (2020) Coordinating proliferation, polarity, and cell fate in the Drosophila female germline. Front Cell Dev Biol 8:19. https://doi.org/10.3389/fcell.2020.00019
4. McLaughlin JM, Bratu DP (2015) Drosophila melanogaster Oogenesis: An overview. Methods Mol Biol 1328:1–20. https://doi.org/10.1007/978-1-4939-2851-4_1
5. Caygill EE, Brand AH (2016) The GAL4 system: a versatile system for the manipulation and analysis of gene expression. Methods Mol Biol 1478:33–52. https://doi.org/10.1007/978-1-4939-6371-3_2
6. Duffy JB (2002) GAL4 system in Drosophila: a fly geneticist's Swiss army knife. Genesis (New York, NY: 2000) 34(1–2):1–15. https://doi.org/10.1002/gene.10150
7. Hales KG, Korey CA, Larracuente AM, Roberts DM (2015) Genetics on the fly: a primer on the Drosophila model system. Genetics 201(3):815–842. https://doi.org/10.1534/genetics.115.183392
8. DeLuca SZ, Spradling AC (2018) Efficient expression of genes in the drosophila germline using a UAS promoter free of interference by Hsp70 piRNAs. Genetics 209(2):381–387. https://doi.org/10.1534/genetics.118.300874
9. Laughon A, Gesteland RF (1984) Primary structure of the Saccharomyces cerevisiae GAL4 gene. Mol Cell Biol 4(2):260–267. https://doi.org/10.1128/mcb.4.2.260-267.1984
10. Fischer JA, Giniger E, Maniatis T, Ptashne M (1988) GAL4 activates transcription in Drosophila. Nature 332(6167):853–856. https://doi.org/10.1038/332853a0
11. Brand AH, Perrimon N (1993) Targeted gene expression as a means of altering cell fates and generating dominant phenotypes. Development 118(2):401–415
12. Hafezi YaN TG (2012) Advanced techniqus for cell lineage labeling in drosophila in eLS (Ed). https://doi.org/10.1002/9780470015902.a0022539
13. Weaver LN, Ma T, Drummond-Barbosa D (2020) Analysis of Gal4 expression patterns in adult Drosophila females. G3 (Bethesda) 10(11):4147–4158. https://doi.org/10.1534/g3.120.401676
14. Golic KG, Lindquist S (1989) The FLP recombinase of yeast catalyzes site-specific recombination in the Drosophila genome. Cell 59(3):499–509. https://doi.org/10.1016/0092-8674(89)90033-0
15. Solomon JM, Rossi JM, Golic K, McGarry T, Lindquist S (1991) Changes in hsp70 alter

thermotolerance and heat-shock regulation in Drosophila. New Biol 3(11):1106–1120
16. Basler K, Struhl G (1994) Compartment boundaries and the control of Drosophila limb pattern by hedgehog protein. Nature 368(6468):208–214. https://doi.org/10.1038/368208a0
17. Pignoni F, Zipursky SL (1997) Induction of Drosophila eye development by decapentaplegic. Development 124(2):271–278
18. Laws KM, Drummond-Barbosa D (2015) Genetic mosaic analysis of stem cell lineages in the Drosophila ovary. Methods Mol Biol 1328:57–72. https://doi.org/10.1007/978-1-4939-2851-4_4
19. Rubin T, Huynh JR (2015) Mosaic analysis in the Drosophila melanogaster ovary. Methods Mol Biol 1328:29–55. https://doi.org/10.1007/978-1-4939-2851-4_3
20. Ni JQ, Zhou R, Czech B, Liu LP, Holderbaum L, Yang-Zhou D, Shim HS, Tao R, Handler D, Karpowicz P, Binari R, Booker M, Brennecke J, Perkins LA, Hannon GJ, Perrimon N (2011) A genome-scale shRNA resource for transgenic RNAi in Drosophila. Nat Methods 8(5):405–407. https://doi.org/10.1038/nmeth.1592
21. Rørth P (1998) Gal4 in the Drosophila female germline. Mech Dev 78(1–2):113–118. https://doi.org/10.1016/s0925-4773(98)00157-9
22. Fire A, Xu S, Montgomery MK, Kostas SA, Driver SE, Mello CC (1998) Potent and specific genetic interference by double-stranded RNA in Caenorhabditis elegans. Nature 391(6669):806–811. https://doi.org/10.1038/35888
23. Kennerdell JR, Carthew RW (2000) Heritable gene silencing in Drosophila using double-stranded RNA. Nat Biotechnol 18(8):896–898. https://doi.org/10.1038/78531
24. Ameres SL, Zamore PD (2013) Diversifying microRNA sequence and function. Nat Rev Mol Cell Biol 14(8):475–488. https://doi.org/10.1038/nrm3611
25. McGuire SE, Mao Z, Davis RL (2004) Spatiotemporal gene expression targeting with the TARGET and gene-switch systems in Drosophila. Sci STKE 2004 (220):pl6. https://doi.org/10.1126/stke.2202004pl6
26. Blake AJ, Finger DS, Hardy VL, Ables ET (2017) RNAi-based techniques for the analysis of gene function in Drosophila germline stem cells. Methods Mol Biol 1622:161–184. https://doi.org/10.1007/978-1-4939-7108-4_13
27. Takemura M, Bowden N, Lu YS, Nakato E, O'Connor MB, Nakato H (2021) Drosophila MOV10 regulates the termination of midgut regeneration. Genetics 218(1). https://doi.org/10.1093/genetics/iyab031
28. Mika K, Cruchet S, Chai PC, Prieto-Godino LL, Auer TO, Pradervand S, Benton R (2021) Olfactory receptor-dependent receptor repression in Drosophila. Sci Adv 7(32). https://doi.org/10.1126/sciadv.abe3745
29. Gunawan F, Arandjelovic M, Godt D (2013) The Maf factor Traffic jam both enables and inhibits collective cell migration in Drosophila oogenesis. Development 140(13):2808–2817. https://doi.org/10.1242/dev.089896
30. Ma X, Wang S, Do T, Song X, Inaba M, Nishimoto Y, Liu LP, Gao Y, Mao Y, Li H, McDowell W, Park J, Malanowski K, Peak A, Perera A, Li H, Gaudenz K, Haug J, Yamashita Y, Lin H, Ni JQ, Xie T (2014) Piwi is required in multiple cell types to control germline stem cell lineage development in the Drosophila ovary. PLoS One 9(3):e90267. https://doi.org/10.1371/journal.pone.0090267
31. Prasad M, Wang X, He L, Cai D, Montell DJ (2015) Border cell migration: a model system for live imaging and genetic analysis of collective cell movement. Methods Mol Biol 1328:89–97. https://doi.org/10.1007/978-1-4939-2851-4_6
32. Thompson L, Randolph K, Norvell A (2015) Basic techniques in Drosophila ovary preparation. Methods Mol Biol 1328:21–28. https://doi.org/10.1007/978-1-4939-2851-4_2

Chapter 5

Analysis of Physiological Control of Adult *Drosophila* Oogenesis by Interorgan Communication

Lesley N. Weaver

Abstract

Tissue homeostasis is dependent on the interaction between various organs within an organism in response to physiological inputs. The adult *Drosophila melanogaster* ovary is sensitive to environmental challenges and has recently been shown to be regulated by signaling from peripheral organs. To dissect the intricate coordination between overall organism health and reproduction, it is necessary to meticulously characterize both experimental tools and oogenesis processes. This chapter provides a guide for the careful analysis of interorgan communication in regulating oogenesis in adult *Drosophila melanogaster*.

Key words Oogenesis, *Drosophila*, *Gal4/UAS*, Germline stem cell, Follicle stem cell, Follicle cell, Germline cyst, Physiology

1 Introduction

Organs relay their status to peripheral tissues to maintain organismal tissue homeostasis in response to different physiological cues. The *Drosophila* ovary responds to systemic changes in physiology or in the environment (e.g., changes in diet [1] and aging [2]) to maintain proper egg production. The energy-intensive process of oogenesis requires a concerted effort between organs to coordinate progeny production with the physiological status of adult *Drosophila* females [3]. Recent studies have used the ovary as a model to understand how tissues such as adipocytes [4–8], the central nervous system [9], and muscle [10] communicate their physiological status to the ovary to influence distinct steps of oogenesis.

The *Drosophila* ovary is a powerful model for understanding how different organs communicate their physiological status due to well-established methods for studying oogenesis and the wide variety of transgenic tools available to dissect tissue-specific requirements regulating overall physiology [3, 11–14]. For example, the *Gal4/UAS/Gal80* [15] and *LexA/LexOp* [16] systems are

Michelle S. Giedt and Tina L. Tootle (eds.), *Drosophila Oogenesis: Methods and Protocols*,
Methods in Molecular Biology, vol. 2626, https://doi.org/10.1007/978-1-0716-2970-3_5,

commonly used to perform temporally restricted and tissue-specific manipulations in *Drosophila*. The *Gal4/UAS/Gal80* system uses the yeast Gal4 transcription factor that is under control of a specific promoter sequence. Gal4 binds to an *Upstream Activating Sequence* (*UAS*) that resides upstream of a target gene or sequence of interest, in turn inducing expression of the transgene [15]. Gal80 is an inhibitor of Gal4 [17] that can be used in combinations to provide temporal control. For example, a temperature-sensitive mutation of *Gal80* (*$Gal80^{ts}$*) can be used in combination with a tissue-specific *Gal4-driver* to allow for temporal transgene expression by rearing animals at the permissive *$Gal80^{ts}$* temperature (18 °C) and then shifting the animals to 29 °C (the $Gal80^{ts}$ restrictive temperature) at the desired developmental time point [18]. Together, the *$Gal4/UAS/Gal80^{ts}$* system is beneficial for both temporal and spatial transgene expression. Use of this system in *Drosophila* has revolutionized understanding of how proteins regulate tissue homeostasis in a cell-specific manner.

However, to properly understand how a gene functions within a specific cell type or tissue to influence the activity of a peripheral tissue requires extreme care in the selection of tools used for the experiment. A recent analysis of commonly used *Gal4* drivers in adult *Drosophila* females revealed that many "tissue-specific" drivers are also expressed in additional tissues beyond their commonly reported tissue of interest [19]. The use of tools that have expression in additional tissues or that change their expression pattern in response to a physiological manipulation can greatly confound interpretation of how interorgan communication influences distinct steps of oogenesis and other physiological processes. Therefore, when attempting to infer how one tissue influences the behavior of another, it is imperative that *Gal4* (or *LexA*) drivers used for transgene expression are clean (i.e., only expressed in the tissue of interest to be manipulated) and proper controls are performed.

The focus of this chapter is to provide guidance on how to study the role of interorgan communication in regulating adult *Drosophila* oogenesis. My work [7, 8, 10, 20] and the work of others [4–6] using these techniques have shown how signaling in peripheral tissues regulates distinct and overlapping processes that occur during oogenesis. The described protocol represents a detailed reference to check *Gal4* strains of interest, perform experiments in a temporal and adult-specific manner, dissect and immunostain ovaries, and analyze data. It should be noted that initial selection of *Gal4* driver and controls described in this protocol can be used to perform tissue-specific analysis to understand additional processes including interorgan communication regulating spermatogenesis, gonad formation, and somatic tissue homeostasis.

2 Materials

2.1 Drosophila Strains and Culture

1. Combination of *Gal4* drivers (*see* [19] for expression patterns of commonly used drivers in adult *Drosophila* females) to be used in the study and *UAS*-marker lines (*see* **Note 1**) for crosses. *Gal4* drivers can be used in combination with a temperature-sensitive *Gal80* construct ($tubPGal80^{ts}$ [18]) or an auxin-inducible degradable *Gal80* ($tubGal80^{AID}$ [21]) to add temporal control of *Gal4* expression.
2. Standard fly culture medium (*see* **Note 2**) in plugged bottles and vials.
3. Incubators set at 18 °C and 29 °C (*see* **Note 3**).
4. Active or inactive dry yeast.
5. Wet yeast paste without Auxin: 3.2 g of active or inactive, dry yeast mixed with 4.5 mL H_2O (*see* **Note 4**).
6. Wet yeast paste with 10 mM Auxin (optional): 3.2 g of inactive, dry yeast mixed with 10 mM Auxin in 0.01% NaOH (*see* **Note 4**).
7. Molasses/agar plates: 10% molasses, 22 g/L agar, 0.05% tegosept. Pour approximately 2 mL of molasses/agar solution into the lid of 60 mm petri dishes (*see* **Note 5**).
8. Perforated plastic fly bottles for egg count and hatching percentage analyses (*see* **Note 5**).

2.2 Dissection, Immunostaining, and Slide Preparation

1. Stereomicroscope.
2. 1.5 mL microcentrifuge tubes.
3. Multi-well dissection dish.
4. Kimwipes.
5. Dumont #5 forceps.
6. 0.5 mm tungsten needles.
7. Glass Pasteur pipettes and bulbs.
8. Nutator.
9. Grace's Insect medium.
10. Phosphate buffered saline (PBS).
11. 3% Bovine Serum Albumin (BSA) in water.
12. Fixation solution: 5.3% formaldehyde in Grace's Insect Medium (*see* **Note 6**).
13. Washing solution: 0.1% Triton-X100 in 1× PBS.
14. Blocking solution: 0.1% Triton-X100, 5% BSA, 5% normal goat serum in 1× PBS.
15. Primary antibodies (*see* **Note 7**):

- For checking Gal4 expression patterns: chicken anti-GFP or chicken anti-β-gal.
- For assaying the effect of interorgan cross-talk on oogenesis: mouse anti-alpha spectrin (3A9, Developmental Studies Hybridoma Bank), mouse anti-Lamin C (LC28.26, Developmental Studies Hybridoma Bank), rat anti-vasa (anti-vasa, Developmental Studies Hybridoma Bank, *see* **Note 8**), mouse anti-Fasciclin III (7G10, Developmental Studies Hybridoma Bank), mouse anti-hts RC (Developmental Studies Hybridoma Bank), rabbit anti-pHH3 ser10 (Millipore Sigma, [10]), rabbit anti-Dcp1(Cell Signaling, [4]).

16. Secondary antibodies (*see* **Note 9**):
    - For checking Gal4 expression patterns: anti-chicken Alexa Fluor 488.
    - For assaying the effect of interorgan cross-talk on oogenesis: anti-mouse IgG Alexa Fluor 568, anti-rat IgM Alexa Fluor 488, anti-rabbit Alexa Fluor 488.
17. Click-iT EdU Imaging Kit Alexa Fluor 568 (Fisher Scientific, *see* **Note 10**).
18. ApopTag Fluorescein Direct In Situ Apoptosis Detection Kit (Fisher Scientific, *see* **Note 10**).
19. Mounting, anti-fade medium (*see* **Note 11**).
20. Slides (1 mm thickness).
21. Coverslips (18 × 18 mm for checking Gal4 expression and 18 × 40 mm for processes of oogenesis analyses).
22. 200 g weight.
23. Clear nail polish.

2.3 Image Acquisition and Analysis

1. Fluorescent light microscope (confocal or equivalent).
2. Image analysis software (such as ImageJ).
3. Data analysis software (such as Microsoft Excel or GraphPad Prism).

3 Methods

3.1 Checking Gal4 Expression Patterns

This step is important for determining the expression pattern of the *Gal4-driver* to be used for analysis of interorgan communication. The user should set up crosses so that the number of progeny generated will be a sufficient sample size to dissect all organs of the adult female (brain, thoracic muscle, gut and Malpighian tubules, ovary, and carcass) at two time points. Typically, 10 well-dissected samples per time point provide enough tissue for analysis. To ensure this number is obtained, collect 20 females per time point. The time points used should include 0 days of transgene expression (flies grown at 18 °C for *Gal80*^{ts} or those grown at 25 °C without Auxin exposure for *Gal80*^{AID}) and 5–7 days of transgene expression (flies that have been shifted from 18 °C to 29 °C or fed 10 mM Auxin).

1. Determine driver specificity by generating flies expressing a *UAS* driven reporter [e.g., *UAS-GFP, UAS-lacZ, UAS-m-Cherry*] under control of desired *Gal4* driver (*see* **Notes 12** and **13**).
2. Collect at least 40 0-to-2-day old adult females of progeny from **step 1** (enough for two dissection time points) plus an equivalent number of control (*y w* or *Oregon-R*) males. Feed couples wet yeast paste daily for 2 days at 18 °C (for Gal80^{ts}) or 25 °C without Auxin (for Gal80^{AID}). This step will allow for remodeling of larval organs [22].
3. Dissect 15–20 females collected from Subheading 3.1 of **step 2** for t = 0 days of transgene induction controls.
4. For the t = 5–7 days of transgene expression time point, shift remaining 20 flies from **step 2** to 29 °C (for Gal80^{ts}) or feed wet yeast plus 10 mM Auxin [21] at 25 °C (for Gal80^{AID}) (*see* **Note 14**).
5. Transfer remaining flies to fresh vial with wet yeast paste with or without 10 mM Auxin accordingly each day.

3.2 Dissection and Immunostaining

For analysis of interorgan communication in adult *Drosophila* females, it is important that *UAS-transgenes* are not expressed during development. Therefore, the expression pattern of a control sample in which transgene expression is prevented (i.e., reared at 18 °C or without Auxin) should be checked and compared to a time period in which transgene expression has occurred (29 °C or 10 mM Auxin).

1. Prepare 1.5 mL microcentrifuge tubes for each dissection sample by filling with 3% BSA/H_2O solution. Nutate overnight or for at least for 1 hour at room temperature to coat tubes to prevent tissues from sticking to the tubes.
2. Transfer Grace's Insect Medium to dissecting dish wells using a sterile pipet tip or Pasteur pipette.
3. Anesthetize flies and select females for dissection.
4. Dissect organs (brain, thoracic muscle, abdominal carcass/fat body, gut + Malpighian tubules, and ovary) from at least 10 females. Place each organ type in an individual well of multi-well dissection dish (Fig. 1).
   - For the brain, carefully pick females up by the thorax with nondominant hand. Separate the head from the rest of the

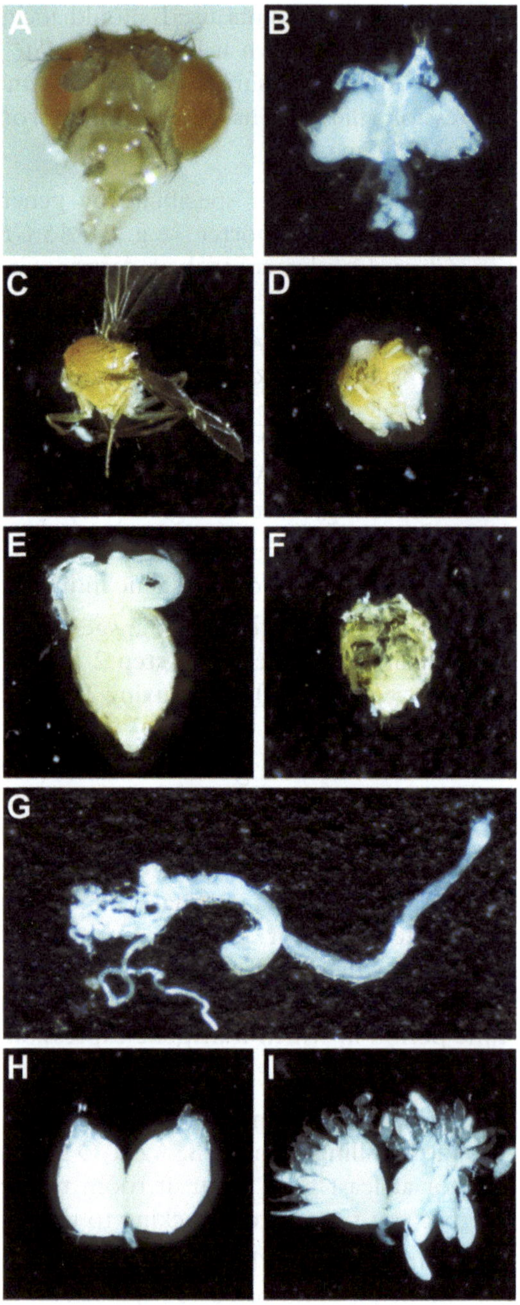

Fig. 1 Dissection of adult female tissues for immunofluorescence. The adult female head is separated from the thorax (**a**) and the brain is removed from the cuticle with the ventral side up (**b**). The thorax is further separated from the carcass (**c**), which contains the ovary and the gut. The wings and legs are removed from the thorax (**d**), and the tissue can be ripped open to fix and stain for immunofluorescence. The abdominal carcass ventral side (**e**) is filleted, and the gut (**g**) and ovaries (**h**) are removed from the carcass (**f**). The carcass (**f**) can be fixed and processed for immunofluorescence with the fat tissue attached. The ovarioles of the ovary are teased apart (**i**) to disrupt the muscle sheath prior to fixation and processing for image analysis

body by severing at the base of the head and thorax with forceps and place the head in a separate well (Fig. 1a). Isolate the brain from the head cuticle using forceps (Fig. 1b, [23]).

- For muscle samples, separate the thorax from the head and abdominal cuticle using forceps. Place the thorax in a separate well of the dissection dish (Fig. 1c). Remove the wings and legs from the thorax (Fig. 1d). Separate the thorax into at least four pieces using forceps to allow for antibody absorption [24].
- For the carcass, gut, and ovary samples, fillet open the abdomen with forceps and carefully remove the gut and ovaries (Fig. 1e–h). Place each organ in a separate well of the dissection dish.
- Tease ovarian ovarioles apart at the anterior end (pre-vitellogenic stages) using tungsten needles to remove muscle sheath (compare Fig. 1h–i).

5. Remove BSA solution from the microcentrifuge tubes (**step 1**) using a Pasteur pipette. Use a coated pipette to transfer organs to the appropriate tubes to prevent tissues from sticking to the glass.
6. Fix ovaries (13 minutes), brain (20 minutes), thoracic muscle (20 minutes), abdominal carcass (20 minutes), and gut (1 hour) in 5.3% fixation solution at room temperature with rotation on a nutator.
7. Remove fixation solution to organic waste and quickly rinse samples 3× in washing solution by repeatedly changing the buffer after allowing samples to settle to the bottom of the tube.
8. Wash samples with washing solution for an additional 3 × 15 minutes each wash on a nutator at room temperature (*see* **Note 15**).
9. Block samples in Blocking Solution for at least 3 hours at room temperature or overnight at 4 °C on a nutator (*see* **Note 16**).
10. Stain organs with primary antibodies diluted in blocking solution overnight at 4 °C on a nutator: anti-GFP or anti-β-gal (*see* **Note 17**).
11. Wash samples 3 × 15 minutes each wash in washing solution on a nutator.
12. Stain organs with secondary antibody (anti-chicken Alexa Fluor 488) diluted 1:200 in blocking solution. Protect sample from light by covering with foil and incubate on the nutator at room temperature for 2 hours (*see* **Note 18**).
13. Wash samples 3 × 15 minutes each wash in washing solution on a nutator covered with foil.

14. Remove washing solution and add two drops of mounting medium of choice (e.g., Vectashield with 10 μg/mL 4′, 6-diamidino-2-phenylindole [DAPI]) overnight at 4 °C to allow tissues to clear (*see* **Note 19**).
15. To mount the samples, transfer the samples mixed with mounting medium onto a glass slide under a stereomicroscope.
    - Gut samples: Ensure that all sections of the gut are observable and are not overlapping.
    - Brain samples: Mount the brain with either the ventral or dorsal side up [25] and overlay coverslip.
    - Ovary samples: Carefully separate the large late-stage egg chambers (stages 11–14) from the ovarioles using a tungsten needle and remove them from the slide. This will ensure that the sample is not too thick for imaging due to mature egg chambers. Make sure ovarioles are clearly separated from each other. To mount late-stage egg chambers, carefully place egg chambers on glass slide and ensure that there is no overlap.
16. Slowly place a glass coverslip on top of sample, avoiding bubbles. Cover with a Kimwipe and gently apply pressure (excluding the brain, gut, and late egg chamber samples) by overlaying a 200 g weight on the coverslip for 1–5 minutes to flatten tissue for image analysis (*see* **Note 20**). Seal coverslips with nail polish and store at 4 °C in the dark.
17. Using a fluorescent microscope, analyze each organ for expression of *UAS-transgene* marker. If there is expression in additional tissues than the tissue of interest, the *Gal4* driver can be paired with a tissue-specific *Gal80* (for the nonintended tissue of interest).

3.3 Assaying Interorgan Cross-Talk on Oogenesis

To understand how manipulation of pathways in one tissue may influence oogenesis, it is important to have a *UAS-transgene* control (e.g., RNAi or overexpression construct for *Luciferase*, *GFP*, or *mCherry*) to compare to experimental samples. Addition of a control transgene will determine if observed phenotypes are due to activation of RNAi machinery. In addition, it is recommended that at least two *UAS-RNAi* lines be used for each targeted mRNA to ensure that phenotypes observed are not due to off-target effects.

1. Generate flies of control and experimental genotypes through standard crosses. Collect 0–2-day old females of appropriate genotypes and mate with equivalent number of control (*y w* or *Oregon-R*) males. Collect at least 10 females per time point ($t = 0, 5, 10, 15$ days of transgene induction) in the same biological replicate (*see* **Note 21**).

2. Incubate couples of all genotypes at 18 °C for 3 days (to allow remodeling of fat body and other tissues [22]) and feed wet yeast paste each day to promote oogenesis.
3. Separate out $t = 0$ days of transgene expression (3–5 day old) females for dissection. Transfer remaining vials of control and experimental couples to 29 °C for transgene induction (or add 10 mM Auxin to experimental samples incubated at 25 °C using $Gal80^{AID}$). Dissect females on $t = 5$, 10, and 15 days after transgene induction (*see* **Note 22**).
4. Dissect ovaries according to **steps 1–9** in Subheading 3.2 (For proliferation assays including EdU incorporation, *see* **Note 23**).
5. Stain ovaries with primary antibodies diluted in blocking solution overnight at 4 °C:
   - For germline stem cell (GSC) and cap cell maintenance: anti-α-spectrin (1:20, fusome), anti-Lamin C (1:50, nuclear lamina), and anti-Vasa (1:10, germ cells).
   - For GSC, follicle stem cell (FSC), and follicle cell proliferation: anti-α-spectrin (1:10, fusome), anti-Lamin C (1:50, nuclear lamina), and anti-pHH3 (1:200, mitotic cells) prior to performing Click-iT reaction to detect EdU (S-phase cells, *see* **Note 24**).
   - For early germline cyst survival: anti-α-spectrin (1:10, fusome), anti-Lamin C (1:50, nuclear lamina) prior to performing incorporation with ApopTag. Alternatively, stain with anti-alpha-spectrin (1:10, fusome), anti-Lamin C (1:50, nuclear lamina), and anti-Dcp1 (1:100, dying cells).
   - For early germline cyst distribution: anti-α-spectrin (1:20, fusome) and anti-Lamin C (1:50, nuclear lamina) (*see* **Note 25**).
   - For survival of vitellogenic egg chambers: anti-α-spectrin (1: 20, fusome) and anti-Lamin C (1:50, nuclear lamina) (*see* **Note 26**).
6. Wash samples 3 × 15 minutes each wash in washing solution on a nutator.
7. Stain ovaries with secondary antibodies diluted 1:200 in blocking solution protected from light:
   - For GSC and cap cell maintenance: anti-mouse Alexa Fluor 568 and anti-rat IgM Alexa Fluor 488.
   - For GSC, FSC, and follicle cell proliferation: anti-mouse Alexa Fluor 568 and anti-rabbit Alexa Fluor 488.

- For early germline cyst survival: anti-mouse Alexa Fluor 568 and anti-rabbit Alexa Fluor 488 (if using anti-Dcp1 primary antibody).
- For early germline cyst distribution: anti-mouse Alexa Fluor 568 (*see* **Note 25**).
- For survival of vitellogenic egg chambers: anti-mouse Alexa Fluor 568.

8. Proceed with **steps 13–16** from Subheading 3.2 above.

3.4 Data Analysis and Image Acquisition

Observing differences in the number of eggs laid per female can indicate defects in oogenesis (*see* **Note 27**).

3.4.1 Number of Laid Eggs

1. To determine the number of eggs laid per female per day, place 5 couples per bottle (experimental females with control males) in a perforated bottle topped with a molasses plate (or a similar cage used for embryo collections) supplemented with wet yeast paste (*see* **Note 28**). Set up 5 bottles for each genotype (25 couples total for the experiment).
2. Bottles should be kept in a humid chamber or incubator with at least 80% humidity to ensure maximum egg production.
3. Change plates at least twice per day over the course of 2–3 weeks and count the number of eggs laid over a 24 hour period using a stereomicroscope. At least 5 bottles (25 couples total) per genotype are sufficient for reliable measurements. Data should be reported as the average of the total egg number divided by the number of females per day from the five bottles.

3.4.2 Hatching Percentage Analysis

Analysis of the percentage of viable embryos can indicate defects in the maternal load requirement for proper embryogenesis and development.

1. Measure the percentage of hatched embryos by mating 5 females per genotype with 5 control (*y w* or *Oregon R*) males in a perforated bottle topped with a molasses plate.
2. Keep plates in a humid chamber or incubator with at least 80% humidity to ensure the best hatching conditions.
3. Collect the embryos (~100–150 for control genotypes; however, experimental genotypes may have fewer) and place in groups of 10 around a molasses agar plate with a dollop of wet yeast in the center using a paintbrush dipped in water to prevent damage of the embryo.
4. Count the number of hatched embryos at 24 and 48 hours using a stereomicroscope to determine the percentage of hatched embryos (total number of hatched eggs divided by the number of eggs placed on the plate).

3.4.3 Block in Ovulation

A block in ovulation results in increased numbers of stage 14 egg chambers within an ovariole.

1. Whole ovaries dissected in Grace's media should be thoroughly fixed by incubating in 5.3% fixation solution for 1 hour at room temperature in a 1.5 mL Eppendorf tube.
2. Using a dissection scope, carefully separate each ovariole and count the number of mature oocytes (stage 14 egg chambers with dorsal appendages) as well as the number of ovarioles per ovary.
3. Graph the average number of ovarioles per ovary as well as the average number of mature oocytes/ovariole in each condition (*see* **Note 29**).

3.4.4 GSC and Cap Cell Maintenance

Reduction of cap cell (the niche) or stem cell number over time can impair the number of eggs laid by a female.

1. For each time point desired for analysis, dissect 10 females to generate sufficient sample sizes (~100 germaria analyzed per time point per genotype) and stain according to **steps 5–8** in Subheading 3.3. Ensure that all females are collected at the same time for each biological replicate (~50 females per genotype).
2. To quantify the number of GSCs and cap cells over time (acquired from ovaries dissected in Subheading 3.3), count the number of GSCs and cap cells from at least 100 germaria per time point per biological replicate for each genotype. GSCs can be distinguished by fusome morphology [26] and cap cells can be identified by bright LamC staining and the oval structure of their nuclei (Fig. 2a).
3. To assess the starting point for GSC and cap cell numbers in each genotype, it is imperative that GSCs and cap cells be counted at day zero of transgene induction. Subsequent time points for analysis of the change in GSC and cap cell number can occur at 5 day intervals for up to 3 weeks.

3.4.5 GSC and FSC Proliferation

Defects in cell cycle progression of GSCs or FSCs can influence oogenesis. At least 700–1000 GSCs or FSCs should be counted from at least three biological replications to ensure a sufficient sample size for statistics.

1. Collect 20–30 females for zero days and 7 days of transgene induction time points (10–15 females per time point), which will generate sufficient sample sizes (~100 germaria analyzed per time point per genotype) per biological replicate. Stain samples according to **steps 5–8** in Subheading 3.3.

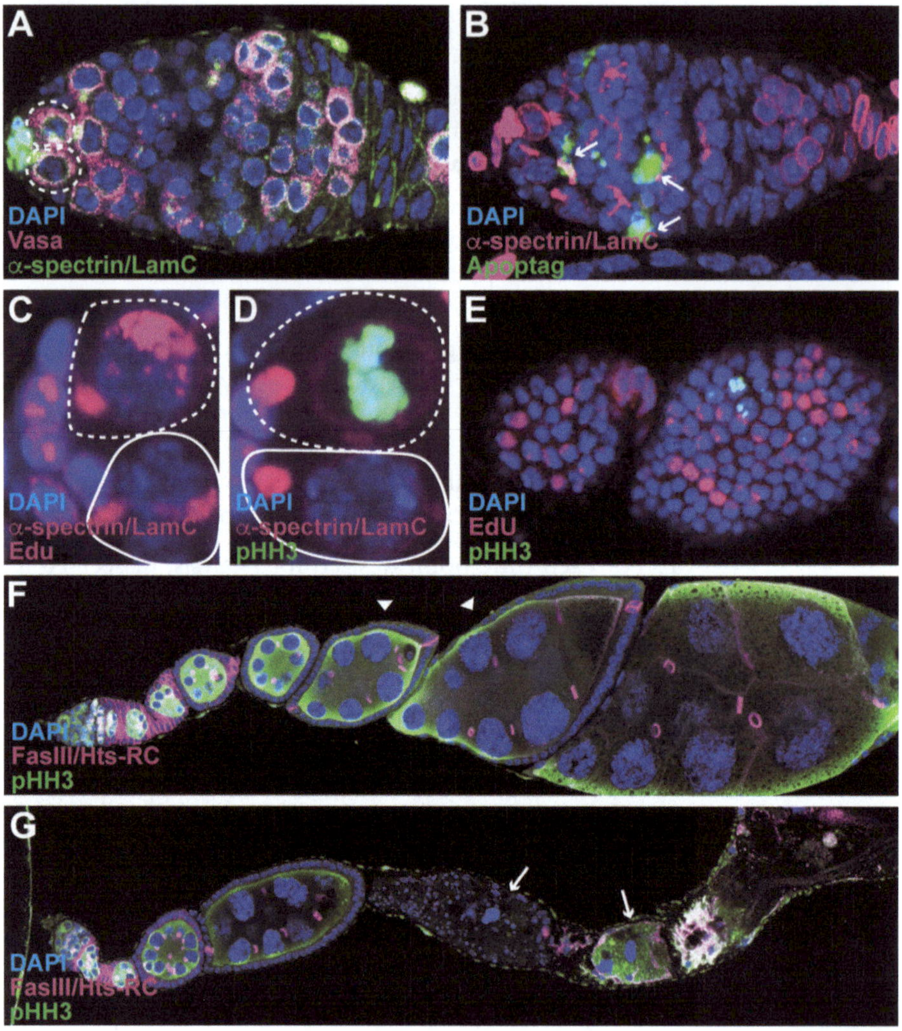

Fig. 2 Processes of oogenesis influenced by changes in physiology. (**a**) Germarium stained with Vasa (magenta, germ cells), α-spectrin (green, fusome), LamC (green, cap cell nuclear lamina), and DAPI (blue, DNA). GSCs are outlined by white dashed line. (**b**) Germarium labeled with ApopTag (green, dying cells), α-spectrin (magenta, fusome), LamC (magenta, cap cell nuclear lamina), and DAPI (blue, DNA). Arrows point to dying germline cysts. (**c**, **d**) Anterior region of a germarium labeled with EdU (magenta, C, GSCs in S-phase) or phospho-histone H3 (pHH3, green, D, GSCs in M-phase), and α-spectrin (magenta, fusome), LamC (magenta, cap cell nuclear lamina), and DAPI (blue, nuclei). EdU (**c**) and pHH3 (**d**) GSCs have dashed outlines, unlabeled GSCs have solid outlines. (**e**) Stage 4–6 follicles labeled with EdU (magenta, follicle cells in S-phase), phospho-histone H3 (pHH3, green, follicle cells in M-phase), and DAPI (blue, DNA). (**f**, **g**) Ovarioles containing healthy (**f**) and dying (**g**) vitellogenic egg chambers stained for Fasciclin III (magenta, follicle cells), Hts-RC (magenta, ring canals), Vasa (green, germ cells), and DAPI (blue, DNA). Arrowheads point to healthy vitellogenic egg chambers (**f**), and arrows point to dying vitellogenic egg chambers (**g**)

2. To measure the proportion of GSCs or FSCs in S or M phase, quantify the total number of GSCs/FSCs that have incorporated the thymidine analog EdU or are positively stained with pHH3 (*see* **Note 30**) (Fig. 2c, d).
3. Calculate the percentage of S-phase or M-phase GSCs/FSCs by dividing the total number of EdU incorporated GSCs/FSCs or pHH3+ GSCs/FSCs per genotype by the total number of GSCs/FSCs within the sample set. Using two markers for proliferation within the same sample can help determine whether observed defects are not due to arrest at a specific cell cycle stage.

3.4.6 Follicle Cell Proliferation

Follicle cell proliferation of mitotic egg chambers (stages 2–6) serves as a proxy for egg chamber growth. Perturbations in the cell cycle of follicle cells can alter the number of eggs laid by females.

1. Collect 20 females for zero days and 7 days of transgene induction time points (10 females per time point) to generate 20–50 ovarioles for analysis per biological replicate and stain samples according to **steps 5–8** of Subheading 3.3.
2. Calculate the percentage of proliferating follicle cells by counting the number of EdU+ or pHH3+ follicle cells and dividing by the total number of cells present in mitotic egg chambers (stages 2–6) (Fig. 1e).

3.4.7 Early Germline Cyst Survival

Death of developing germline cysts in the germarium is a readout of early oogenesis defects and can severely impact egg production.

1. Collect 20–30 females for zero days and 10 days of transgene induction time points (10–15 females per time point) to generate at least 100 germaria for analysis per biological replicate.
2. Label developing cysts undergoing apoptosis using the ApopTag In Situ Apoptosis Detection Kit [1] (*see* **Note 10**, Fig. 2b) or by immunostaining using an apoptosis-related antibody [e.g., anti-Dcp1 [4]] according to **steps 5–8** of Subheading 3.3.
3. Calculate the percentage of germaria containing dying germline cysts by determining the fraction of germaria containing at least one ApopTag+ germline cyst to the total number of analyzed germaria (*see* **Note 31**).

3.4.8 Early Germline Cyst Distribution

If defects in early germline cyst survival (calculated in Subheading 3.4.7) are observed, it is possible to determine at which stage(s) early cyst development is affected.

1. Collect 10–20 females for zero days and 10 days of transgene induction time points (5–10 females per time point) to generate at least 25 germaria for analysis per biological replicate.
2. Stain samples according to **steps 5–8** in Subheading 3.3. Count the number of 2-, 4-, 8-, and 16-cell cysts present in each germarium using fusome morphology or by counting the number of ring canals in each germline cyst [26].
3. Calculate the average number of cystoblasts and cysts per GSC per germarium.

3.4.9 Survival of Vitellogenic Egg Chambers

Vitellogenesis is one of the most regulated steps during oogenesis and death of vitellogenic egg chambers indicates defects in oogenesis.

1. Collect 20–30 females for zero days and 10-days of transgene induction time points (10–15 females per time point) to generate at least 100 ovarioles for analysis per biological replicate.
2. Determine the percentage of vitellogenic egg chambers (stages 8-10B) undergoing apoptosis by visualization of pyknotic nuclei by DAPI staining (Fig. 2f, g).
3. Determine the percentage of ovarioles with dying egg chambers by calculating the fraction of ovarioles containing at least one dying/apoptotic vitellogenic egg chamber to the total number of ovarioles with vitellogenic egg chambers.

4 Notes

1. Common *UAS-marker* lines available in the Bloomington *Drosophila* Stock Center include *UAS-mCherry, UAS-GFP.nls, UAS-LacZ* (https://bdsc.indiana.edu/stocks/gfp/gfp_uas_bycolor.html). Gal4 expression should be monitored at multiple time points indicative of the intended experimental framework.
2. My lab currently uses the Bloomington *Drosophila* Stock Center Cornmeal Food, which consists of 15.9 g/L inactive yeast, 9.2 g/L soy flour, 67 g/L yellow cornmeal, 42.4 g/L light malt extract, 5.3 g/L agar, 7% light corn syrup, 0.06 M propionic acid. Alternatively, Nutri-Fly Bloomington Formulation pre-mixed food can be purchased from Genessee (https://geneseesci.com/) and prepared according to manufacturer's instructions.
3. My lab currently uses the temperature-sensitive mutation of *Gal80* (*tubPGal80*^{ts}) for adult-specific manipulation. Alternatively, the auxin-inducible degradable *Gal80* construct (*tubGa-l80*^{AID}) allows experiments to be performed at 25 °C or room

temperature. Using *tubGal80*^{AID} is desirable as it accelerates the growth time from 18 days (with *tubPGal80*^{ts}) to the normal life cycle from egg to adult (~10 days). In addition, use of *tubGal80*^{AID} prevents toxic stress due to heat, which can impair oogenesis [27].

4. My lab typically makes our wet yeast paste fresh every week with a consistency of smooth peanut butter. Wet yeast paste should be stored at 4 °C and covered with parafilm. If using *tubPGal80*^{AID}, it is important to use inactive yeast and check that Auxin activity does not diminish throughout the course of the week. Yeast paste containing Auxin should be stored in a dark container to protect from light. Auxin activity can be further monitored using a *UAS-GFP.nls* transgene to monitor *Gal4* activity.
5. Alternative bottle and cap combinations (e.g., common embryo chambers available through commercial vendors) can also be used for counting the number of eggs laid per female. If using plastic *Drosophila* bottles as chambers, make sure to poke small holes (perforation) so that air can circulate.
6. I dilute methanol-free 16% formaldehyde to 5.3% in Grace's Insect Medium. Opened bottles of 16% formaldehyde can be stored at 4 °C for 1 week, or alternatively, aliquoted and frozen at −20 °C for extended periods of time.
7. Antibodies from different species can be used that suit the experimenter's need. The antibody combinations described in this protocol are based on those used in my lab.
8. The rat anti-vasa antibody from the Developmental Studies Hybridoma Bank is an IgM antibody and will only be recognized by IgM secondary antibodies.
9. Other secondary fluorescent combinations can be used that suit the experimenter's need for their studies.
10. The Click-iT EdU Imaging and ApopTag In Situ Apoptosis Detection kits are routinely used and optimized in my lab. However, additional kits from other companies can be used.
11. My lab uses Vectashield with and without 4′, 6-diamidino-2-phenylindole (DAPI). However, anti-fade made in house or from different commercial vendors is also applicable.
12. If possible, it is best to use a *UAS* reporter transgene that is at the same genetic location as the transgenes used for the intended experiment.
13. If using *tubPGal80*^{ts} for an experiment, crosses should be set at 18 °C and shifted to 29 °C with adult females. If using *tubP-Gal80*^{AID}, crosses can be set at 25 °C in food lacking auxin during development. Auxin can be supplemented in the food or wet yeast paste once experimental adult flies have eclosed.

14. For initial analysis of Gal4 expression patterns, it is sufficient to perform an early timepoint analysis. However, the Gal4 expression pattern should be monitored throughout the course of the experiment (e.g., combining the Gal4 driver with a *UAS*-driven marker) as changes in age, genetic manipulation, temperature, as well as many additional physiological factors can influence gene expression.
15. After fixation and first quick washing step, samples can be washed for longer periods of time or stored at 4 °C in washing solution for up to 1 month.
16. Samples can be stored in blocking solution for an extended period of time (no longer than 1 year) at 4 °C.
17. The concentration used for each primary antibody will depend on the specific antibody that is used and should be optimized. My lab typically uses a 1:1000 dilution of chicken anti-GFP and chicken anti-β-gal.
18. For the remainder of this protocol, the researcher should ensure that samples are covered with foil and protected from light to prevent fluorescence decay.
19. I find that DAPI incorporates into the sample better if incubated at least overnight at 4 °C. Alternatively, one could incubate samples with 10 μg/mL DAPI during the secondary incubation step and add mounting media without DAPI for mounting and imaging.
20. Longer flattening times for ovary samples are recommended for imaging of epithelial follicle cell layers in egg chambers.
21. For temporal expression control, make sure to use either $tubPGal80^{ts}$ or $tubPGal80^{AID}$.
22. Females should be fed fresh, wet yeast every day throughout the course of the experiment or at least two consecutive days prior to dissection. For experiments using Auxin, females should be fed every day to ensure proper transgene expression throughout the course of the experiment.
23. For GSC, FSC, and follicle cell proliferation, samples should be dissected and incubated with EdU in room temperature Grace's Insect Medium to prevent defects in cell cycle progression due to a decrease in temperature. Ovaries are incubated in EdU solution for 1 hour at room temperature prior to teasing ovarioles apart. Ovarioles should be teased apart after fixation.
24. For EdU analysis, the Click-It reaction should be performed according to manufacturer's instructions immediately after washing in washing solution and prior to incubation in the secondary antibodies. Samples should be covered with foil

during this process. There is an intermediate 3 × 15 minute washing step after the Click-It reaction has been performed preceding secondary antibody incubation.

25. For early germline cyst analysis, an additional stain such as anti-Hts-RC (Developmental Studies Hybridoma Bank) can be used to count the number of ring canals associated with each germline cyst. However, this antibody does not stain well in region 1 of the germarium. Alternatively, an anillin antibody can be used [26] or lines can be generated with y^1w^*; $vsg^{CA07004}$ (Bloomington *Drosophila* Stock Center #50812, GFP protein trap that recognizes Vsg and stains ring canals in all regions of the ovary). Finally, it should be noted that cyst distribution can be analyzed with staining for only the fusome if needed [26].
26. It is possible to analyze defects in the survival of vitellogenic egg chambers using DAPI alone without any additional stain.
27. It should be noted that reliance on the number of eggs laid per female should not be the sole assay to detect defects in oogenesis. Physiological perturbations can have subtle, yet significant, effects on early stages of oogenesis that may not be observed using an egg count assay, as previously reported [8, 28].
28. For experiments using $tubPGal80^{AID}$, ensure that females are eating wet yeast from the supplemented molasses plate to induce transgene expression throughout the course of the experiment. Researchers can add colored food dye to the wet yeast paste to visualize eating by the presence of dye in intestines. In addition, transgene expression can be monitored as described in Subheading 3.1.
29. For severe ovulation defects, it may be beneficial to dissect one female at a time to account for any eggs that "spill" out of the abdomen during dissection. In addition, for more detailed analysis of defects in oogenesis, females can be dissected to observe the presence of eggs in the reproductive tract [29].
30. Samples can be co-labeled for EdU and pHH3 in the same set of samples. To do so, stain cap cells, fusomes, and EdU in the same fluorophore (e.g., Alexa Fluor 594) and pHH3 in a different fluorophore (e.g., Alexa Fluor 488).
31. Samples labeled with ApopTag should be stained for immunofluorescence according to **steps 5–8** in Subheading 3.3 within 30 days after fixation to prevent nonspecific labeling of escort cells.

Acknowledgments

I am grateful to Kaitlin Laws for critical comments on drafts of this manuscript. This work was supported by National Institutes of Health (NIH) grant R00 GM127605.

References

1. Drummond-Barbosa D, Spradling AC (2001) Stem cells and their progeny respond to nutritional changes during Drosophila oogenesis. Dev Biol 231(1):265–278. https://doi.org/10.1006/dbio.2000.0135
2. Ishibashi JR, Taslim TH, Ruohola-Baker H (2020) Germline stem cell aging in the Drosophila ovary. Curr Opin Insect Sci 37:57–62. https://doi.org/10.1016/j.cois.2019.11.003
3. Drummond-Barbosa D (2019) Local and physiological control of germline stem cell lineages in Drosophila melanogaster. Genetics 213(1):9–26. https://doi.org/10.1534/genetics.119.300234
4. Armstrong AR, Drummond-Barbosa D (2018) Insulin signaling acts in adult adipocytes via GSK-3beta and independently of FOXO to control Drosophila female germline stem cell numbers. Dev Biol 440(1):31–39. https://doi.org/10.1016/j.ydbio.2018.04.028
5. Armstrong AR, Laws KM, Drummond-Barbosa D (2014) Adipocyte amino acid sensing controls adult germline stem cell number via the amino acid response pathway and independently of Target of Rapamycin signaling in Drosophila. Development 141(23):4479–4488. https://doi.org/10.1242/dev.116467
6. Matsuoka S, Armstrong AR, Sampson LL, Laws KM, Drummond-Barbosa D (2017) Adipocyte metabolic pathways regulated by diet control the female germline stem cell lineage in Drosophila melanogaster. Genetics 206(2):953–971. https://doi.org/10.1534/genetics.117.201921
7. Weaver LN, Drummond-Barbosa D (2018) Maintenance of proper germline stem cell number requires adipocyte collagen in adult Drosophila females. Genetics 209(4):1155–1166. https://doi.org/10.1534/genetics.118.301137
8. Weaver LN, Drummond-Barbosa D (2019) The nuclear receptor seven up functions in adipocytes and oenocytes to control distinct steps of Drosophila oogenesis. Dev Biol 456(2):179–189. https://doi.org/10.1016/j.ydbio.2019.08.015
9. Sieber MH, Spradling AC (2015) Steroid signaling establishes a female metabolic state and regulates SREBP to control oocyte lipid accumulation. Curr Biol 25(8):993–1004. https://doi.org/10.1016/j.cub.2015.02.019
10. Weaver LN, Drummond-Barbosa D (2021) Hormone receptor 4 is required in muscles and distinct ovarian cell types to regulate specific steps of Drosophila oogenesis. Development 148(5). https://doi.org/10.1242/dev.198663
11. Ables ET, Drummond-Barbosa D (2017) Steroid hormones and the physiological regulation of tissue-resident stem cells: lessons from the Drosophila ovary. Curr Stem Cell Rep 3(1):9–18. https://doi.org/10.1007/s40778-017-0070-z
12. Greenspan LJ, de Cuevas M, Matunis E (2015) Genetics of gonadal stem cell renewal. Annu Rev Cell Dev Biol 31:291–315. https://doi.org/10.1146/annurev-cellbio-100913-013344
13. Hales KG, Korey CA, Larracuente AM, Roberts DM (2015) Genetics on the fly: a primer on the Drosophila model system. Genetics 201(3):815–842. https://doi.org/10.1534/genetics.115.183392
14. Laws KM, Drummond-Barbosa D (2017) Control of germline stem cell lineages by diet and physiology. Results Probl Cell Differ 59:67–99. https://doi.org/10.1007/978-3-319-44820-6_3
15. Brand AH, Perrimon N (1993) Targeted gene expression as a means of altering cell fates and generating dominant phenotypes. Development 118(2):401–415
16. Pfeiffer BD, Ngo TT, Hibbard KL, Murphy C, Jenett A, Truman JW, Rubin GM (2010) Refinement of tools for targeted gene expression in Drosophila. Genetics 186(2):735–755. https://doi.org/10.1534/genetics.110.119917
17. Douglas HC, Hawthorne DC (1966) Regulation of genes controlling synthesis of the galactose pathway enzymes in yeast. Genetics 54(3):911–916

18. McGuire SE, Le PT, Osborn AJ, Matsumoto K, Davis RL (2003) Spatiotemporal rescue of memory dysfunction in Drosophila. Science 302(5651):1765–1768. https://doi.org/10.1126/science.1089035
19. Weaver LN, Ma T, Drummond-Barbosa D (2020) Analysis of Gal4 expression patterns in adult Drosophila females. G3 (Bethesda) 10(11):4147–4158. https://doi.org/10.1534/g3.120.401676
20. Weaver LN, Drummond-Barbosa D (2020) The nuclear receptor seven up regulates genes involved in immunity and xenobiotic response in the adult Drosophila female fat body. G3 (Bethesda) 10(12):4625–4635. https://doi.org/10.1534/g3.120.401745
21. McClure CD, Hassan A, Aughey GN, Butt K, Estacio-Gomez A, Duggal A, Ying Sia C, Barber AF, Southall TD (2022) An auxin-inducible, GAL4-compatible, gene expression system for Drosophila. eLife 11. https://doi.org/10.7554/eLife.67598
22. Nelliot A, Bond N, Hoshizaki DK (2006) Fat-body remodeling in Drosophila melanogaster. Genesis 44(8):396–400. https://doi.org/10.1002/dvg.20229
23. Wu JS, Luo L (2006) A protocol for dissecting Drosophila melanogaster brains for live imaging or immunostaining. Nat Protoc 1(4): 2110–2115. https://doi.org/10.1038/nprot.2006.336
24. Xiao YS, Schock F, Gonzalez-Morales N (2017) Rapid IFM dissection for visualizing fluorescently tagged sarcomeric proteins. Bio Protoc 7(22). https://doi.org/10.21769/BioProtoc.2606
25. Arain U, Valentino P, Islam IM, Erclik T (2021) Dissection, immunohistochemistry and mounting of larval and adult drosophila brains for optic lobe visualization. J Vis Exp 170. https://doi.org/10.3791/61273
26. de Cuevas MSA (1998) Morphogenesis of the Drosophila fusome and its implications for oocyte specification. Development 125(15): 2781–2789
27. Gandara ACP, Drummond-Barbosa D (2022) Warm and cold temperatures have distinct germline stem cell lineage effects during Drosophila oogenesis. Development 149(5). https://doi.org/10.1242/dev.200149
28. Ma T, Matsuoka S, Drummond-Barbosa D (2020) RNAi-based screens uncover a potential new role for the orphan neuropeptide receptor Moody in Drosophila female germline stem cell maintenance. PLoS One 15(12): e0243756. https://doi.org/10.1371/journal.pone.0243756
29. Sun J, Spradling AC (2013) Ovulation in Drosophila is controlled by secretory cells of the female reproductive tract. eLife 2:e00415. https://doi.org/10.7554/eLife.00415

Chapter 6

Immunohistochemical Analysis of Nuclear Lamina Structures in the *Drosophila* Ovary Using CRISPR-Tagged Genes

Tingting Duan, Felipe Rodriguez-Tirado, and Pamela K. Geyer

Abstract

The *Drosophila* ovary represents an outstanding model for investigating tissue homeostasis. Females continuously produce oocytes throughout their lifetime. However, as females age, fecundity declines, in part, due to changes in ovarian niche function and germline stem cell (GSC) homeostasis. Understanding the dynamics of GSC maintenance will provide needed insights into how coordinated tissue homeostasis is lost during aging. Critical regulators of GSC maintenance are proteins that reside in the nuclear lamina (NL), including the NL proteins emerin and Barrier-to-Autointegration Factor (BAF). Continued investigation of how emerin, BAF, and other NL proteins contribute to GSC function depends upon the availability of antibodies for NL proteins, a limiting resource. In this chapter, we discuss strategies for using clustered regularly interspaced short palindromic repeats (CRISPR) genomic editing to produce endogenously tagged NL genes to circumvent this obstacle, using the generation of the *gfp-baf* allele as an example. We describe strategies for validation of tagged alleles. Finally, we outline methods for immunohistochemical analysis of resulting tagged-NL proteins.

Key words CRISPR, Cas9, Gene tagging, GFP, Drosophila, Nuclear lamina, LEM-domain proteins, Drosophila oogenesis

1 Introduction

Tissue homeostasis depends on the replacement of injured and defective cells by proliferation of resident adult stem cells. With age, tissue function declines, largely due to decreases in stem cell number and/or functions that impact tissue renewal. To intervene in the aging process, an improved understanding of mechanisms involved in stem cell maintenance is needed. The *Drosophila* ovary represents an outstanding model for investigation of these processes. Ovaries are divided into 16 to 20 ovarioles, with each carrying an assembly line of maturing oocytes (Fig. 1a). Oogenesis begins in the germarium, a structure located at the tip of each

Michelle S. Giedt and Tina L. Tootle (eds.), *Drosophila Oogenesis: Methods and Protocols*,
Methods in Molecular Biology, vol. 2626, https://doi.org/10.1007/978-1-0716-2970-3_6,

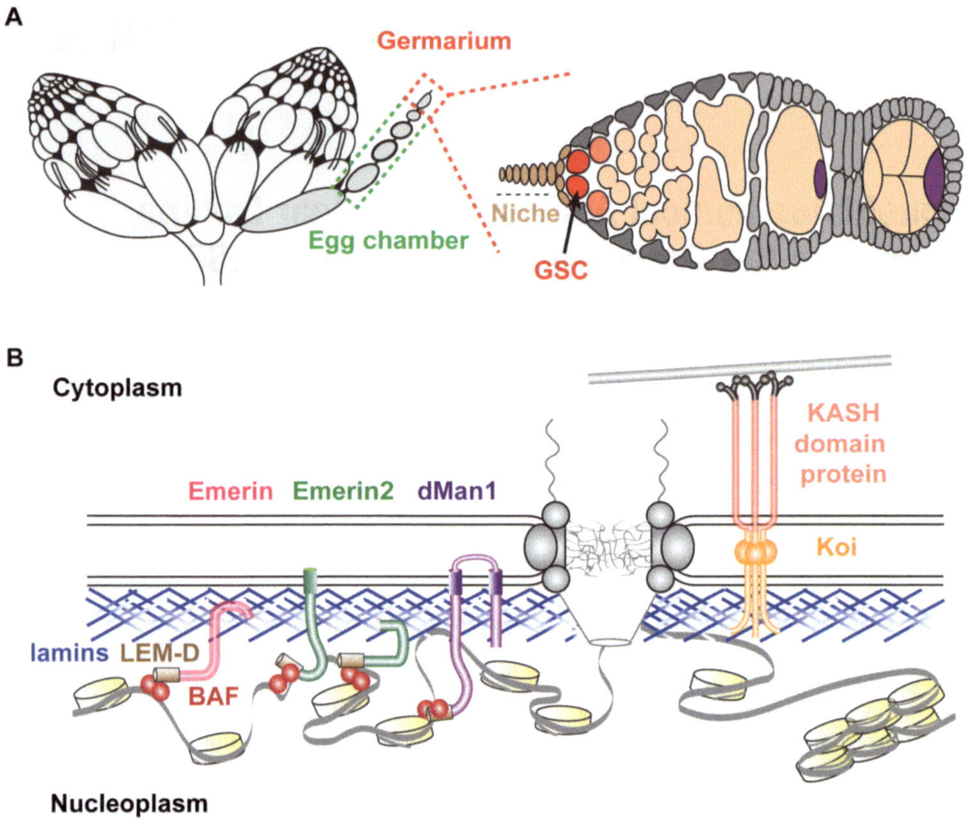

Fig. 1 *Drosophila* oogenesis depends upon proteins in the nuclear lamina (NL). (**a**) Left: Shown is a diagram of a pair of *Drosophila* ovaries that are organized into ovarioles that carry oocytes of advancing stages of differentiation. The germarium (red box) is located at the anterior end of each ovariole. Posterior to the germarium are individual egg chambers (green box) that carry maturing oocytes. Right: Shown is a diagram of the germarium comprised of somatic niche cells (brown), germline stem cells (GSCs; red), differentiating germ cells (peach), somatic support cells (grey), and oocytes with the egg chamber (purple). (**b**) Shown is a schematic presentation of the Drosophila NL. The NL lies underneath the inner nuclear membrane and is composed of lamins (dark blue lines) and multiple lamin-associated proteins. Within this network are three LEM domain (LEM-D, brown) proteins, including emerin (also known as Otefin, pink), emerin2 (also known as Bocksbeutel, green), and dMan1 (magenta). LEM-D proteins share interactions with BAF (red dimer). Other NL proteins include components of the LINC complex, including Klaroid (Koi, orange) and KASH domain proteins (Msp300, peach). LEM-D protein associations with BAF tether the genome (shown as strings of nucleosomes) to the nuclear periphery

ovariole. Each germarium carries two to three germline stem cells (GSCs) that are anchored to somatic niche cells. Asymmetric GSC divisions produce one daughter cell that self-renews and a second daughter that commits to differentiation. Ultimately, an interconnected 16-cell cyst is formed that contains one oocyte and fifteen supporting nurse cells (Fig. 1a). Females continuously produce oocytes throughout their lifetime. However, over time, fecundity drops, due in part to changes of the function of the ovarian stem cell niche. These changes include decreases in GSC numbers, declines in stem cell proliferation, and increases in death of differentiating germ cells [1]. Both extrinsic and intrinsic signals regulate GSC homeostasis [2]. Recent evidence suggests that one intrinsic regulatory factor is nuclear structure [3], contributed by proteins in the nuclear lamina (NL). Comprised of lamins and hundreds of lamin interacting proteins, the NL network lies beneath the nuclear envelope [4] and builds nuclear architecture critical for balancing stem cell self-renewal and differentiation [3]. Notable among NL proteins are the Lap2-emerin-MAN1 domain (LEM-D) proteins (Fig.1b) [5], as these proteins connect the genome to the nuclear periphery through shared interactions with Barrier-to-Autointegration Factor (BAF) [6, 7]. Both emerin and BAF are required for maintenance of *Drosophila* GSCs [8, 9], wherein loss of either NL protein activates kinases of the DNA damage checkpoint that block germ cell differentiation and cause GSC loss [8, 10]. Strikingly, NL dysfunction is linked to failures in stem cell maintenance in progressive human diseases known as laminopathies [5, 11–17]. As such, studies of the NL in stem cell maintenance in the ovary will contribute knowledge relevant to tissue homeostasis in age-enhanced NL-associated human diseases.

1.1 Resources for Immunohistochemical NL Analysis in the Ovary

Antibodies for NL proteins have been central for investigation of the role of NL function in GSC maintenance [18]. Whereas commercially available antibodies against the *Drosophila* A- and B-type lamins are available (Developmental Studies Hybridoma Bank, DSHB), antibodies for LEM-D proteins, BAF, and other NL proteins are either limiting or not available. Although new antibodies for NL proteins could be produced, the generation of such reagents is uncertain, time-consuming, and subject to variable success. An alternative approach is the production of NL alleles that encode proteins tagged with protein segments for which antibodies are already available, such as the epitope tags of Flag, Hemagglutinin (HA), V5, and fluorescent proteins such as Green Fluorescent Protein (GFP). Epitope-tagged proteins provide opportunities for biochemical purification [19]. Additionally, fluorescent tags are advantageous, as they allow application of additional experimental methods to explore gene function, including increased flexibility of imaging that includes fixed and live tissues and opportunities for conditional removal of the gene products using strategies of RNA interference and deGradFP [20–22]. These features emphasize advantages of generating NL knock-in alleles, wherein sequences encoding epitope tags are inserted into the endogenous gene.

Large-scale transposon screens have generated many genes encoding epitope-tagged proteins, including the protein-trap and MiMIC screens [23–25]. Both strategies employ transposons to mediate cassette insertion into introns that are nested within coding exons of genes. As such, genes with multiple, large introns have a greater prospect of being transposon-tagged. Notably, such

Table 1
Availability of endogenously tagged NL proteins

Gene	Function	Tag	Method of tagging	Stock #	Reference
lam DmO	B-type lamin	N/A	N/A	N/A	N/A
lam C	A-type lamin	GFP	Protein trap	Bloomington #6837	PMID:11742088
emerin/ ote	LEM-D protein	N/A	N/A	N/A	N/A
emerin2/ bocks	LEM-D protein	EGFP	CRISPR (Fungene)	N/A	This study
dMan1	LEM-D protein	EGFP	CRISPR (Fungene)	N/A	This study
Baf	Chromatin protein	sfGFP	CRISPR	N/A	PMID:32345742
LBR	Integral inner nuclear membrane protein	N/A	N/A	N/A	N/A
Klar	LINC complex	EGFP. Flag	MiMIC	Bloomington #61652	PMID:25824290
Koi	LINC complex	GFP	Protein Trap	Bloomington #51525	PMID:17194782
msp300	LINC complex	VEnus	Protein Trap	DGRC #115409	PMID:34831284
msp300	LINC complex	EGFP. Flag	MiMIC	Bloomington #59757	PMID:25824290
CG11138	Transcriptional repressor	GFP	Protein Trap	N/A	PMID:23341925

N/A Not applicable

features are uncommon in some *Drosophila NL* genes. For example, the *baf* gene is 1.1 kb and carries a single small 83 bp intron. Even so, transposon-tagged alleles of some *NL* genes have been produced (Table 1) [24, 26].

1.2 Targeted Gene Editing Using CRISPR to Expand NL Tools

Intentional methods can be used to tag genes of interest. Among these, clustered regularly interspaced short palindromic repeats (CRISPR) genome editing stands out due to its versatility and flexibility. This two-component system exploits an RNA-guided Cas9 endonuclease for genome editing [27, 28]. The specificity of Cas9 targeting is provided by a 20-nucleotide spacer sequence within a guide RNA (gRNA) that directs the endonuclease to complementary sites in the genome. CRISPR editing only requires that the spacer sequence lies immediately upstream of a three nucleotide protospacer adjacent motif (PAM) in the genome.

Indeed, even though *baf* is a small gene, CRISPR editing successfully generated an endogenously tagged *gfp-baf* allele (Figs. 2 and 3) [8]. Using this allele, contributions of BAF to mitosis in oogenesis and neurogenesis have been investigated, in fixed and live tissues [8, 29, 30].

Application of CRISPR gene editing requires consideration of many design features. First, decisions are needed about where to place the tag in the protein. This decision reflects observations that tags can affect protein folding, stability, localization, and activity. Additionally, the structure of the edited gene needs to be considered, as alternative transcription start sites and splicing have the potential to produce mRNAs that do not contain coding sequences for the tag. Commonly, GFP tags are integrated into the 5′ or 3′ end of a coding region, generating N- or C-terminal fused proteins. Second, decisions are needed whether to include linker sequences that separate the tag from the protein of interest, as direct tagging might produce nonfunctional fusion proteins. Two types of linkers are common [31–34]. A flexible linker is composed of small, nonpolar (e.g., Gly) or polar (e.g., Ser or Thr) amino acids to enable mobility and flexibility of sequences between joined domains. A rigid linker is composed of alpha helices or Pro-rich sequences that adopt a stiff structure, chosen to efficiently separate joined domains and prevent interactions [34]. Third, sequences of the gRNA need to be chosen. Critically, a unique 18- to 20-nucleotide gRNA spacer sequence needs to be selected that lies adjacent to a PAM (*see* **Note 1**). Spacer sequences should be positioned close to the site of insertion of the tag, as the highest efficiency of incorporation of exogenous DNA occurs near the double-stranded break (DSB) [35]. Efficiency of editing is also affected by complementarity of the spacer sequence to the genome, being highest with complete complementary (*see* **Note 2**). Mismatches of three or more nucleotides greatly reduce cutting efficiency [36]. Polymorphisms between fly strains and the reference genome sequence are common, requiring sequencing of the target gene within the genome of the fly strain to be engineered (Subheading 3.1). Once gRNA sequences are defined, these sequences are used to generate a gRNA plasmid (Subheading 3.2). Fourth, a decision is needed concerning the type of donor sequences to be used. Both oligonucleotide and plasmid donors can be used, depending on size of DNA to be integrated. For small (<200 bp) insertions that might encode HA or Myc tags, single-stranded oligonucleotide donors can be used, with tag sequences surrounded by ~50 bp of homology on either side of the gRNA cut site [37]. When short oligonucleotides are used for editing, positive integration events are screened by PCR [38]. For large (>200 bp) insertions that might encode tags such as a GFP, double-stranded plasmid donors are used, wherein tag sequences are surrounded by ~1-kb of homology on either side of the gRNA cut site to facilitate genome integration by homologous

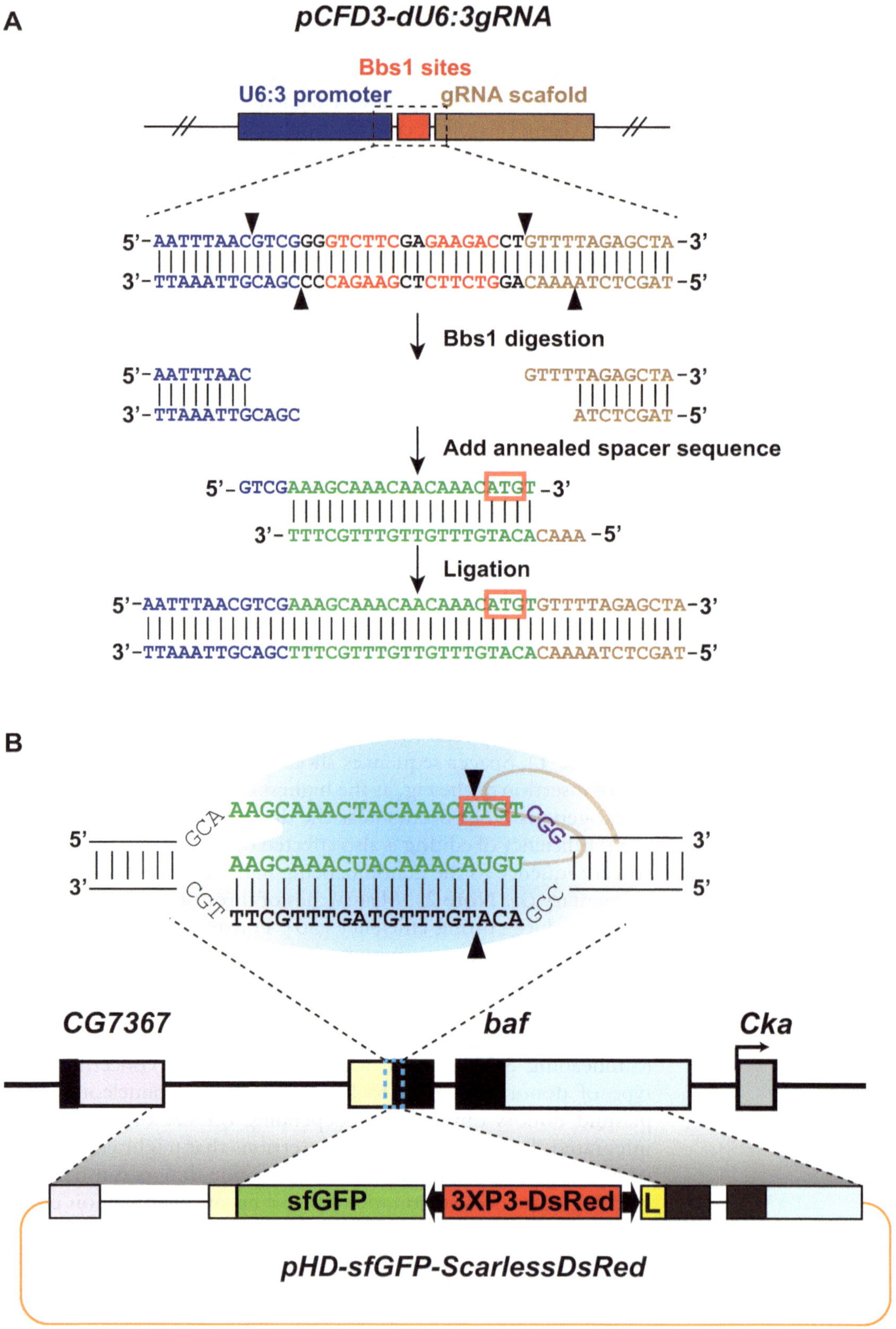

Fig. 2 Generation of the *gfp-baf* allele with the scarless tagging method. (**a**) Structure of the pCFD3-dU6:3gRNA plasmid used for expression of the gRNA. This plasmid contains a U6:3 promoter (blue) for gRNA expression and two recognition sites for the restriction endonuclease BbsI (red). BbsI is a type IIS restriction

recombination. A set of plasmids have been developed to facilitate CRISPR editing involving insertion of tags, including the pHD-ScarlessDsRed plasmids (Subheading 3.3; FlyCRISPR). These donor plasmids carry coding sequences of sfGFP, 3XFLAG, or 2XHA flanked on either side by coding sequences for a flexible linker. Using these donor plasmids, N-terminal, C-terminal, or internally tagged proteins can be made. Additionally, pHD-ScarlessDsRed plasmids carry the dominant selection marker DsRed that allows fast screening of engineered lines and efficient removal by piggyBac transposase.

In this chapter, we describe methods used in CRISPR/Cas9-mediated editing to introduce coding sequences of protein tags into endogenous genes, using the generation of *gfp-baf* allele as an example. We include strategies to validate gene function after tagging. Finally, we describe the generation of two additional CRISPR-tagged NL genes (Table 1). Throughout the protocol, we highlight key considerations for the design and execution of the CRISPR tagging method.

2 Materials

2.1 Fly Media and Stocks

1. Fly media: Every liter of food contains 8.9 g of Agar, 50 mL of yeast, 82 mL of cornmeal, 18 mL of sugar, and 31.5 mL of a 10% Tegosept (methyl-p-hydroxy benzoate) solution in 95% ethanol, used as a mold inhibitor.
2. Fly lines:

Fig. 2 (continued) endonuclease that recognizes asymmetric DNA sequences (5'-GAAGAC-3') and cuts outside of these sequences to generate two unique 4-nucleotide overhangs. Black arrowheads denote the positions of BbsI cleavage. BbsI sites are positioned upstream of the gRNA scaffold sequences (brown), thereby allowing easy cloning of spacer sequences (green) into the expression vector. Following BbsI digestion, the unique 4-nt overhangs can be used to insert oligonucleotides that carry specific spacer sequences. To generate the *gfp-baf* allele, the spacer sequence included the *baf* translation start site (red box). (**b**) Diagram of the *baf* locus, including *baf* and the 5' *CG7367* and 3' *Cka* genes. Coding regions are shown in black, while UTRs are shown in purple, yellow, blue, and grey, respectively. The CRISPR-targeting region is positioned at the *baf* translation start site (red box) and is shown boxed in blue. Above the *baf* locus is shown a detailed view of the association of the gRNA with the *baf* gene, where the targeting spacer sequence that carries the translation start site is shown in green (red box), the PAM sequence (CGG) is shown in purple, and the Cas9 cutting sites are indicated by black arrowheads. Repair of Cas9 cleavage uses the donor plasmid shown at the bottom. This donor plasmid is built from the pHD-sfGFP-ScarlessDsRed vector and contains 1-kb of genomic sequences (grey shades) that immediately flank the Cas9 cleavage site. Other features of the pHD-sfGFP-ScarlessDsRed targeting plasmid include coding sequences for *sfGFP* (green) and *DsRed* (red), that is flanked by two *piggyBac* recognition sites (black arrows). The positions of coding sequences for the flexible linker are shown (L, yellow)

(a) Germline Cas9 injection line: *nos-Cas9* (e.g., Bloomington stock #54591) or *vas-Cas9* (e.g., Bloomington stock #51323).

(b) Ubiquitous Cas9 injection line: *Act5C-Cas9* (e.g., Bloomington stock #54590).

(c) *piggyBac* transposase expression line. For example, *w [1118]; CyO, P[Tub-PBac\T] 2/wg[Sp-1]; l(3)* [*]/TM6B, Tb* [1] (Bloomington stock #8285).

2.2 Molecular Biology Reagents and Equipment

1. Cloning vectors: pCFD3-dU6:3gRNA (Addgene #49410), pHD-sfGFP-ScarlessDsRed (DGRC #1365).
2. *E. coli* DH5α.
3. Luria Broth (LB) agar plates with 100 μg/mL ampicillin.
4. BbsI endonuclease and buffer.
5. Nanodrop spectrophotometer.
6. T4 DNA Ligase.
7. T4 Polynucleotide Kinase.
8. Plasmid Mini and Maxi kit, such as QIAGEN Plasmid Mini and Maxi kit.
9. 1% TAE (Tris base, glacial acetic acid, and EDTA) agarose gels.
10. 4–20% precast protein gels.
11. 1 M Dithiothreitol (DTT).
12. Beta-mercaptoethanol.
13. 0.45 μm Nitrocellulose membrane.
14. 4× Laemmli Sample Buffer: 0.25 M Tris base, pH 6.8, 8% SDS, 40% glycerol, 20% β-Mercapto-ethanol, 4 mg/ml Bromophenol blue
15. Standard PCR reagents.
16. Thermocycler.
17. Standard agarose gel electrophoresis equipment.
18. Scalpel or razor blade.
19. Primers for sequencing region of gene being targeted.
20. Access to DNA sequencing facility.
21. Gibson assembly kit, such as NEB Gibson Assembly Cloning Kit.
22. DpnI endonuclease.
23. PCR gel purification kit, such as QIAquick Gel Extraction Kit.
24. Pellet pestles.
25. Heat block.
26. −20 °C freezer.

27. Phosphate-buffered saline (PBS): 137 mM NaCl, 2.7 mM KCl, 10 mM Na_2HPO_4, 1.8 mM KH_2PO_4, pH = 7.4.
28. PBS-Tween: 0.3% Tween 20 in PBS.
29. 5% milk: powdered skim milk diluted in PBS-Tween.
30. Standard Western Blotting equipment.
31. GFP Antibodies, such as goat anti-GFP (Abcam 5450), goat anti-GFP (Abcam 6673), or rabbit anti-GFP [Life Technologies (A11122)].
32. IRDye secondary antibodies.
33. Odyssey imager.
34. Centrifuge.
35. Stereo microscope fluorescence adapter with dsRed filter.
36. Squish buffer (10 mL): 9.8 mL H_2O, 100 μL 1 M Tris pH 8.0, 20 μL 0.5 M EDTA, 50 μL 5 M NaCl.
37. Proteinase K, 10 mg/mL.
38. Primers used for tagged line validation:
    (a) Primer A (*sfGFP+1F*): 5'- ATGGTGTCCAAGGGC GAGGA-3'
    (b) Primer B (*sfGFP+647R*): 5'-CGCTTCTCGTTGGGGT CCTT-3'
    (c) Primer C (*baf-1310F*): 5'-GTAATCCTCGAAACCGGT TCCAG-3'
    (d) Primer D (*baf+1493R*): 5'- CACTGGTGTTCGCA TATGCCAA-3'

2.3 Immunohistochemical Analyses Reagents and Equipment

1. Dissecting microscope.
2. CO_2 anesthesia equipment.
3. Pyrex 9 depression spot plates.
4. Electron microscopy graded formaldehyde.
5. Extra fine forceps (Tip width 0.5 mm, Tip dimensions: 0.5 × 0.5 mm).
6. Micro scissors (Tip diameter: 0.05 mm, cutting edge: 3 mm).
7. Tungsten needle (tip diameter: 0.001 mm).
8. P200 pipette and tips.
9. PBS.
10. PBST: 0.3% Triton X-100 in PBS.
11. Fixative: 4% paraformaldehyde in PBST.
12. Blocking solution: 5% Bovine serum albumin in PBST.
13. 24-well cell culture plate.
14. Kimwipes.

15. Horizontal rotator.
16. PET Membrane Inserts 8 μm.
17. GFP Antibodies, such as goat anti-GFP (Abcam 5450), goat anti-GFP (Abcam 6673), or rabbit anti-GFP [Life Technologies (A11122)].
18. Lamin-B antibody: DSHB# ADL84.12.
19. Cold room or refrigerator with an electrical outlet.
20. 1 mg/mL Diamidino-2-phenylindole (DAPI).
21. Microscope slides.
22. Cover glass #1.5.
23. Mounting medium (e.g., nonhardening: SlowFade Diamond, VECTASHIELD; hardening: Prolong diamond).
24. Clear nail polish.
25. Confocal microscope.

2.4 Web Resources and Tools

1. Guide RNA (gRNA) and donor plasmids: Addgene (addgene.org/crispr/drosophila), DGRC (dgrc.bio.indiana.edu).
2. CRISPR design protocols: crisprflydesign.org/, flycrispr.org/scarless-gene-editing/
3. Spacer sequence identification and off-target evaluation: flycrispr.org, crispr.mit.edu/, chopchop.rc.fas.harvard.edu/, www.deskgen.com/.
4. CRISPR fly lines design and generation: rainbowgene.com/, thebestgene.com, http://www.fungene.tech/.

3 Methods

3.1 Sequence of the Gene of the Fly Strain Used in Editing

Cas9 binding is best when the gRNA is entirely complementary to the genomic DNA. As such, an essential step is to sequence the target region in the genome of the fly strain being engineered, as polymorphisms between fly strains and the reference genome are common.

1. Place a single fly into a 200 μL PCR tube, freeze at −20 °C until ready to process.
2. Add 1 μL of 10 mg/mL proteinase K per 50 μL of squish buffer.
3. Pipette 50 μL of squish buffer with proteinase K into a P200 tip. Take the buffer-loaded tip and grind the fly against the wall of the PCR tube without releasing the buffer. Once the fly is fully ground, eject the buffer into the tube.

4. Place the PCR tube in the thermocycler, incubate at 37 °C for 30 min, 95 °C for 2 min. The resulting DNA can be stored at – 20 °C for several months in the original PCR tube.
5. PCR with appropriate primers to amplify regions of interest.
6. Run PCR products on a 1% agarose gel.
7. Excise the DNA fragment from the gel with a clean, sharp scalpel.
8. Obtain DNA using a gel extraction kit according to manufacturer's instructions.
9. Determine DNA concentration on a Nanodrop spectrophotometer and send appropriate amount of DNA (usually 10 μL of purified fragment with 1 μL of gene specific primer) for sequencing.

3.2 Generate the gRNA Expression Plasmid

gRNAs can be supplied as RNAs or expressed from plasmids. Here, we describe the generation of a gRNA expression vector based on the vector pCFD3-dU6:3gRNA, which directs gRNA expression under the control of RNA polymerase (Pol) III promoter U6 (Fig. 2a).

1. Order sense and antisense oligonucleotides of the spacer sequence from a company. The sense strand oligonucleotide for a spacer should include the 5'-overhang-spacer sequence-3' (5'-GTCG-spacer squence-3') and the antisense oligonucleotide should include the 5'-overhang-reverse complement of spacer sequence-3' (5'-AAAC- reverse complement-3'). Spacer sequences for the *baf* gRNA include the translation start site (Fig. 2a).
2. Combine 1 μL of the 100 μM sense oligonucleotide and 1 μL of 100 μM antisense oligonucleotide in a buffer containing 1 μL of T4 DNA Ligase buffer (10×), 6 μL of H_2O, and 1 μL of T4 Polynucleotide Kinase (10 U/μL). Incubate the mixture in a thermocycler at 37 °C for 30 min and 95 °C for 5 min, then gradually cool down to 25 °C at a rate of 5 °C/min. Store the double-stranded spacer DNA at −20 °C.
3. Digest 1 μg of the gRNA expression plasmid pCFD3-dU6: 3gRNA (*see* **Note 3**) by mixing 1 unit of BbsI endonuclease, 5 μL of Buffer (10×), and H_2O to a final volume of 50 μL. Incubate the mixture at 37 °C for 1 hr and then heat inactivate BbsI at 65 °C for 20 min (Fig. 2a).
4. To ligate, mix 1 μL of annealed oligos from **step 2**, 50 μg of digested plasmid from **step 3**, 1.5 μL of T4 DNA Ligase buffer (10×), and H_2O to a final volume of 15 μL. Incubate the mixture at room temperature for 30 min.

5. Transform 1 μL of ligation product into 50 μL of DH5α cells according to manufacturer's instructions and plate on LB ampicillin agar plates for selection.
6. Miniprep and send the plasmids to sequence with the following primer: 5'-CAGGAAACAGCTATGAC-3'.
7. After verifying the sequence is correct, maxi prep and store the plasmid at −20 °C before injection into *Drosophila* embryos.

3.3 Preparation of a Donor Plasmid

1. Amplify 1-kb homology arms flanking the site of gRNA cleavage from genomic DNA of the fly strain used for injection (see example for *gfp-baf*, Fig. 2b, *see* **Notes 4–5**).
2. PCR amplify backbones from the pHD-sfGFP-ScarlessDsRed plasmid (*see* **Note 6**).
3. Treat PCR products with the DpnI restriction endonuclease for 30 min at 37 °C to eliminate plasmid template.
4. Run PCR products on a 1% agarose gel and gel purify PCR products according to manufacturer's instructions.
5. Determine the DNA concentration with Nanodrop spectrophotometer and run 2 μL of each product on a gel to verify the concentration (*see* **Note** 7).
6. Define the molar concentration of the PCR-amplified DNA using the following equation: pmols = (molecular weight in ng × 1000)/(base pairs × 650 daltons). Calculate how much of each fragment needs to be added to provide an equal molar mixture.
7. Set up the Gibson assembly reaction on ice by mixing PCR fragments, H_2O, and 2× Gibson master mix.
8. Incubate the reaction in a PCR machine at 50 °C for 1 h.
9. Transform 2 μL of assembled plasmids into 50 μL DH5α cells according to manufacturer's instructions and plate on LB ampicillin agar plates for selection.
10. Mini prep 5 colonies and sequence with the following primer: 5'-CAGGAAACAGCTATGAC-3' found upstream of the piggyBac 5' inverted repeat.
11. Maxi prep one correct clone. The resulting DNA can be stored at −20 °C before injection into *Drosophila* embryos.

3.4 Injection and Isolation of CRISPR-Engineered Lines

Multiple methods can be used for introduction of the Cas9 endonuclease for editing. These include injection of a Cas9 expression plasmid, mRNA or protein, or the use of transgenic flies that express Cas9. Here, we outline steps of generating CRISPR lines using transgenic flies carrying a vector that directs germline expression of Cas9, which yields a high efficiency of engineering.

1. Inject gRNA and donor plasmid into a transgenic line expressing Cas9 or send to a company for injection into *Drosophila* embryos (*see* **Notes 8–10**). If performing the injections in house, move hatched, injected animals to a food vial and incubate at 25 °C for development.
2. Mate emerged injected F0 adults to *w⁻; Tub-piggyBac* flies that carry a balancer chromosome relevant to the location of the engineered gene (e.g., CyO for *gfp-baf*, Bloomington stock #8285; Fig. 3a).
3. Identify edited DsRed+ F1 flies by their mottled, red fluorescent eye phenotype (*see* **Note 11**).
4. Mate single DsRed+ F1 flies to flies carrying a balancer of the same chromosome as the edited allele (2nd chromosome in the example shown) to ensure recovery of any edited, DsRed alleles.
5. Establish stocks.

3.5 Sequence Validation of Tagged Lines

1. Isolate genomic DNA from five edited individuals using the procedure outlined in Subheading 3.1.
2. Use PCR to amplify the edited gene. For amplification, DNA primers should be located both within the tag (A, B primers) and outside of the homology regions (C, D primers) in the donor plasmid (see example for *gfp-baf*, Fig. 3b, c; *see* **Note 12**). Primer sequences will depend on gene of interest. Using this design, only correctly edited events at the endogenous location will produce PCR products when AD or BC primer pairs are used.
3. Sequence resulting DNA to ensure that integration of the tag occurred correctly (*see* **Note 13**). Primer selection will depend on gene of interest.

3.6 Confirm Expression of Tagged Fusion Protein by Western Blot Analysis

1. Dissect 15 pairs of ovaries from newly eclosed flies or 7.5 pairs of 3-day old ovaries using strategies outlined in the Subheading 3.8, **step 1**. Transfer ovaries to a 1.5 mL microcentrifuge tube containing cold PBS on ice to reduce protein degradation.
2. Spin down ovaries at 16,000 g at 4 °C for 1 min, remove the PBS, and freeze ovaries at −80 °C until needed.
3. Prepare sample buffer. Add 30 μL of 1 M Dithiothreitol (DTT) and 15 μL of Beta-Mercaptoethanol into 1 mL of 4× Laemmli sample buffer.
4. Add 150 μL of prepared sample buffer to the frozen ovaries. Grind the tissue with a pellet pestle.
5. Boil protein extracts at 100 °C for 10 min in heat block.
6. Prepare 4–20% gels and electrophoresis equipment.

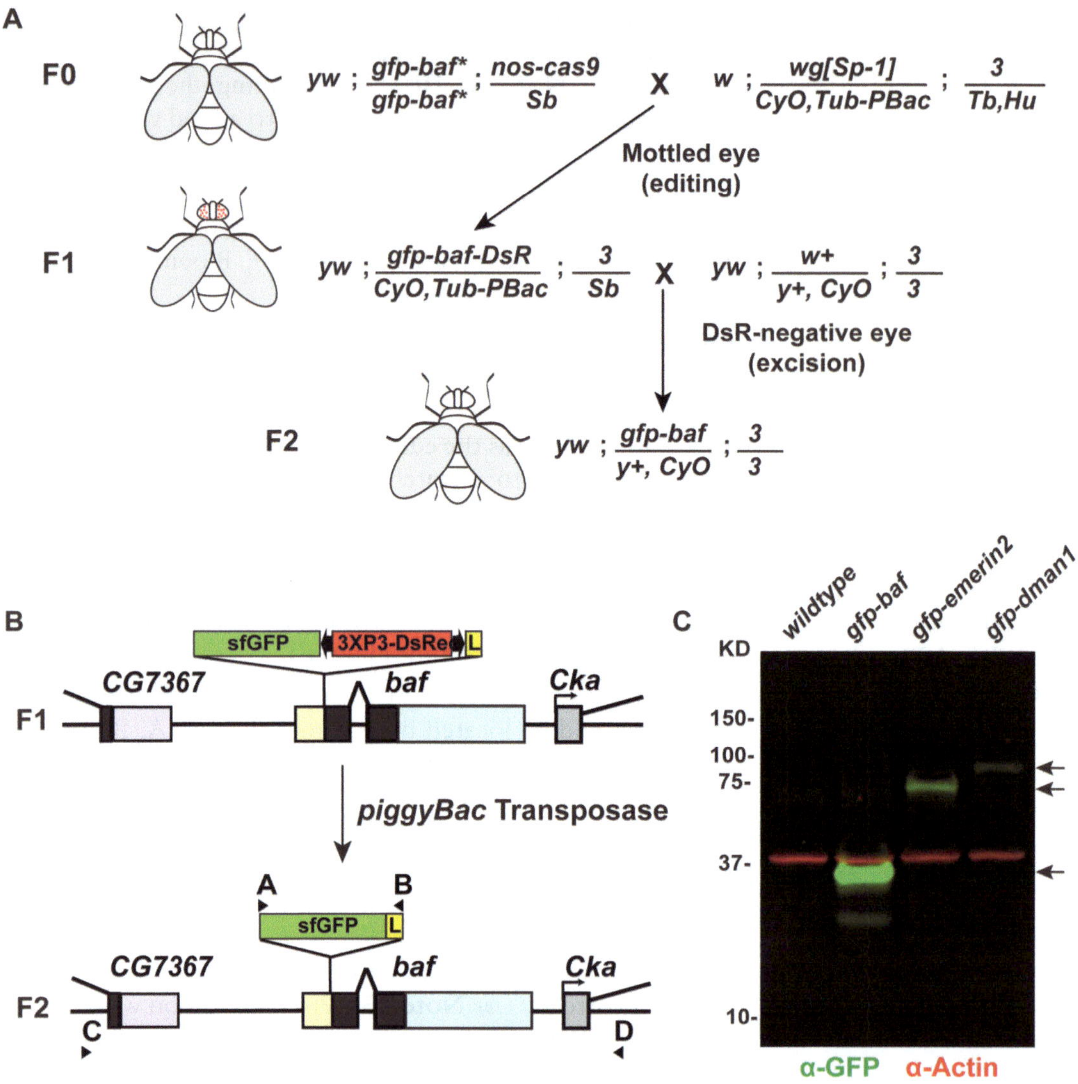

Fig. 3 Strategy for isolating flies carrying the *gfp-baf* allele. (**a**) Injected flies (F0) that have potential editing of the *baf* genes (asterisks) were crossed to flies carrying a balancer chromosome with insertion of a transgene that constitutively expresses the *piggyBac* transposase. Edited F1 progeny were identified by screening for the presence of mottled fluorescent eyes, indicating insertion of DsRed into the genome. In the case of *baf* editing, F1 flies were recovered with mottled eyes. Subsequently, F2 flies carrying edited and DsRed jumped alleles were isolated by screening for DsR-negative eye phenotype. (**b**) Diagrams of edited *gfp-baf* locus in the F1 and F2 generations. Following Cas9 cutting, homology-dependent repair incorporates *sfGFP* and *DsRed* into the *baf* gene at the translation start site (F1). *DsRed* sequences are subsequently removed by the *piggyBac* transposase, resulting in the N-terminal tagging of *baf* with *sfgfp* (F2), which we refer to as *gfp-baf*. Primer positions used in the PCR verification are shown by black arrowheads. (**c**) Western blot of proteins extracted from mature ovaries of the indicated genotypes. Blots were probed with antibodies against GFP (green) to monitor levels of GFP-tagged NL protein production and Actin (red), as a loading control. Arrows point to the bands of expected GFP-NL protein sizes

7. Load 10 μL extracts into gels (*see* **Note 14**).
8. Run gel at 150 Volts for 1 h.
9. Transfer proteins to 0.45 μm nitrocellulose membrane using 70 Volts for 1.5 h.
10. Block the membrane in 5% milk at room temperature for 1 h.
11. Prepare the primary antibody solution by diluting an antibody against GFP and an antibody against a loading control (*see* **Notes 15–16**) into 5% milk [8].
12. Incubate the membrane with the primary antibody solution at 4 °C on a shaker overnight.
13. Wash membrane 5 times with PBS-Tween for 5 min each.
14. Dilute the IRDye fluorescent secondary antibodies into 5% milk.
15. Incubate the membrane with the fluorescent secondary solutions in the dark at room temperature in a shaker for 2 h.
16. Scan the membrane on an Odyssey Imager.

3.7 Assess the Functionality of the Edited Gene

1. Cross flies with tagged alleles to flies that carry a mutation in the tagged gene to obtain F1 progeny corresponding to tagged allele/mutant allele.
2. Assess whether mutant phenotypes remain in F1 heterozygous progeny (*see* **Note 17**).

3.8 Immunohistochemical Analysis of the Tgged Protein

1. Immobilize newly emerged females with anesthesia and remove ovaries under a dissecting scope (*see* **Note 18**). To remove ovaries, grasp females by the thorax with one pair of extra fine forceps and submerge females wings down into a glass well filled with cold PBS to protect against tissue degradation. Next, place a second pair of forceps at the base of the abdomen corresponding to the second to last abdominal stripe. Gently pull the cuticle down and pinch it off, releasing the ovary pair.
2. Remove any remaining cuticle, gut, or surrounding fat tissue from around the ovary, as these affect the quality of antibody staining. Transfer up to 10 ovary pairs to a 24 well plate on ice that has a PET membrane cell culture insert sitting in 1 mL cold PBS (Fig. 4). To prevent degradation of tissue, we recommend that dissections take no longer than 30 min before starting fixation.
3. Fix the tissue by transferring the insert into a second well of the 24-well plate that contains 1 mL of freshly made fixative (Fig. 4, *see* **Note 19**). Before placing the insert into the fixative, blot the bottom with a Kimwipe to ensure that the liquid inside the insert is completely drained. Fix at room temperature (RT) on a horizontal rotator at low speed for 20 min.

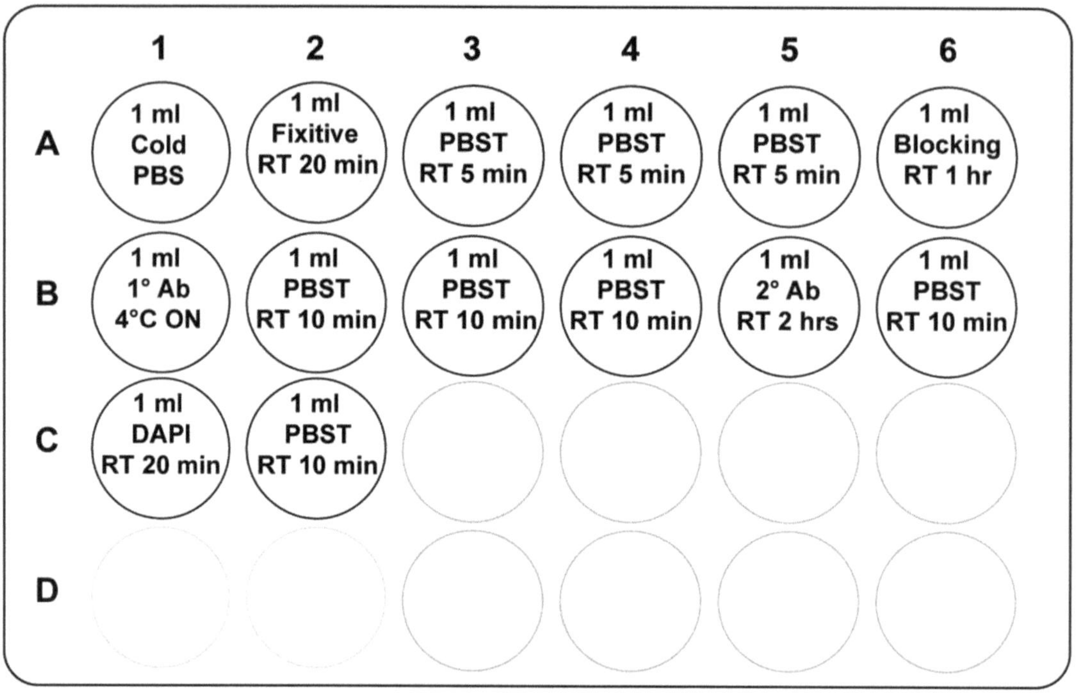

Fig. 4 Schematic of protocol used for antibody staining of ovaries. Example of the layout of a 24 well plate used for immunohistochemical analysis. Solutions and incubation time are labelled inside the well. Empty wells are shown as open circles. In general, one plate is used for staining one sample with one set of antibodies

4. Wash ovaries three times by successive transfer of the insert into wells containing 1 mL of PBST (Fig. 4) for 5 min each at RT on the rotator.
5. Tease apart ovaries to separate ovarioles to improve the penetration of primary antibodies. To do this, transfer ovaries from the insert to a glass well with a P200 pipette with 1 cm cut off the tip. Avoid the introduction of bubbles during the transfer, especially as PBST contains detergent. Under a dissecting microscope, gently use a tungsten needle to scrape the ovary in a posterior to anterior direction until ovarioles are separated.
6. Transfer teased ovaries back into an insert placed in a well containing 1 mL of freshly made blocking solution (Fig. 4). Block ovaries at RT for 1 h or at 4 °C for >6 h, while rotating at low speed.
7. Make antibody mixtures by diluting primary antibody stocks in 1 mL of blocking solution. Commonly, ovaries are incubated with multiple primary antibodies simultaneously (*see* **Note 20**).
8. Transfer ovaries into a well containing 1 mL of primary antibody mixture (Fig. 4). Incubate ovaries with primary antibodies on a rotator at low speed at 4 °C for 16 h (*see* **Note 21**).

9. Remove the primary antibody mixture and save for use in future experiments (*see* **Note 22**).
10. Wash stained ovaries three times by transferring the insert into successive wells of 1 mL of PBST (Fig. 4) for 10 min each at RT on a rotator.
11. During the last wash step, prepare secondary antibodies by diluting in blocking solution.
12. Transfer stained and washed ovaries into 1 mL secondary antibody mixture. Incubate at RT for 2–4 h on the rotator in the dark (Fig. 4, *see* **Note 23**).
13. Wash ovaries by transferring inserts to three successive wells of 1 mL of PBST at RT on a rotator for 10 min each (Fig. 4).
14. DAPI stain ovaries by transferring washed ovaries into 1 mL of PBST to a final concentration of 1 to 5 μg/ml (*see* **Note 24**), and incubate for 20 min at RT on a rotator.
15. Transfer ovaries and inserts to 1 mL of PBST at RT on a rotator for 10 min.
16. Transfer all ovaries from the inserts into a well in a glass plate with a P200 pipette with 1 cm cut off the tip. Under the dissecting microscope, remove unwanted tissue debris and remaining late-stage egg chambers (*see* **Note 18**).
17. Transfer ovaries with forceps onto a clean microscope slide that has a drop (~20–50 μL) of mounting medium. Try to transfer multiple ovaries at once and keep transfers to fewer than three to avoid diluting the mounting media with excess PBST. Mount five to ten ovaries per slide. Check the slide under the scope to ensure ovaries are well separated.
18. Slowly, lower a 0.17 mm cover slip over the mounted ovaries (*see* **Note 25**). Avoid trapping air bubbles. Wipe the edge of coverslip with a kimwipe to remove excess mounting medium.
19. Seal edges of the coverslip with clear nail polish and let it dry for 5 min. Ensure that the coverslip is completely sealed, as exposure of the mounting media to air shortens the shelf life of a slide.
20. Slides can be imaged immediately or stored at 4 °C in the dark (*see* **Note 26**).

3.9 Confocal Imaging of Antibody-Stained Ovaries

Confocal analysis of slides depends upon the microscope used. Shown in Fig. 5 are examples of immunohistochemical staining of ovaries dissected from females carrying three different CRISPR generated *gfp-NL*-tagged genes. These proteins include the GFP fusion proteins GFP-BAF, GFP-emerin2, and GFP-dMAN1 (Fig. 5). To assess GFP-NL protein localization, ovaries were stained with GFP and Lamin-B antibodies. The specificity of the GFP antibody is demonstrated by the absence of staining of the

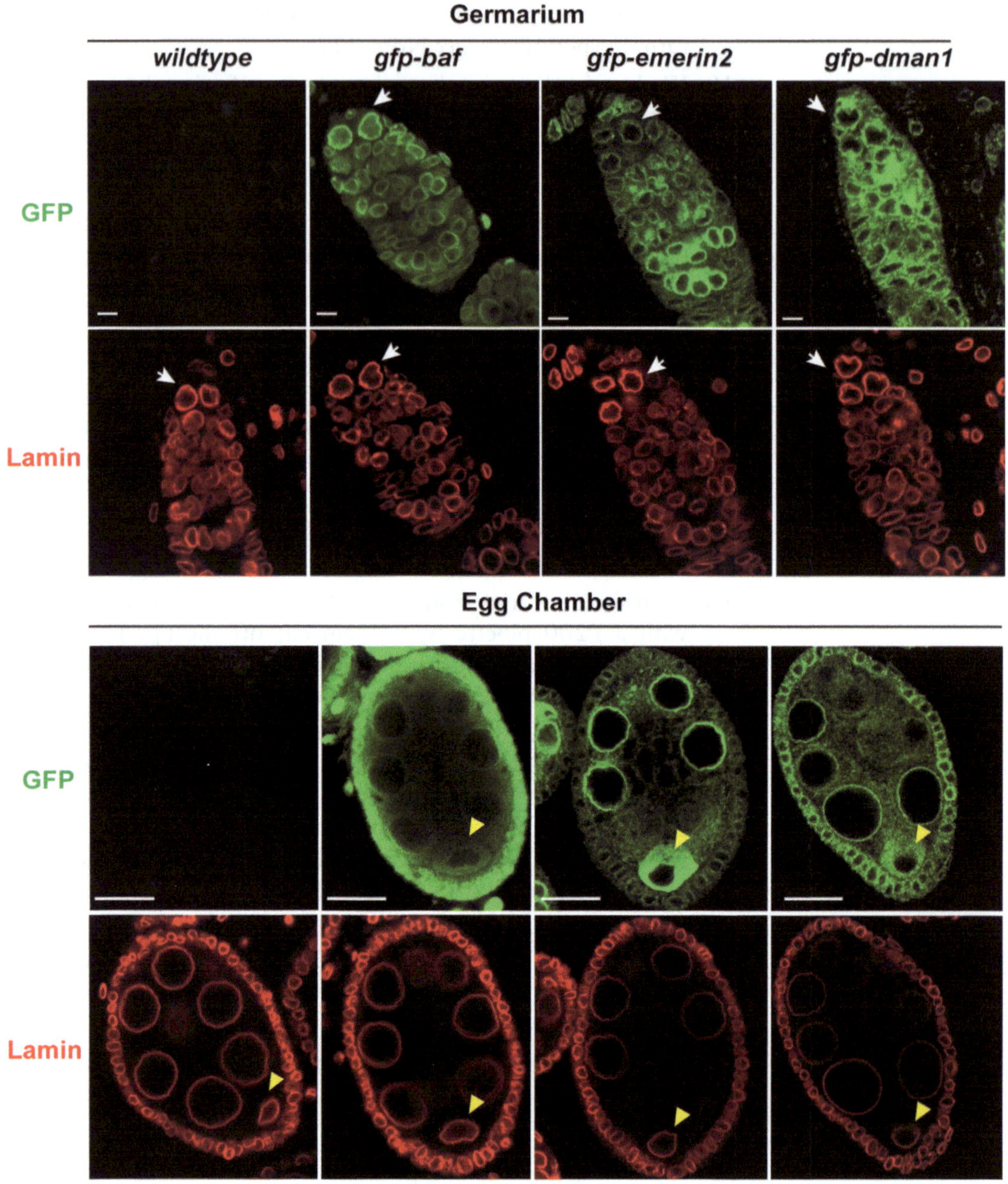

Fig. 5 Immunohistochemical analysis of GFP-tagged NL proteins. Confocal images of representative germaria and egg chambers isolated from females carrying GFP-tagged NL endogenous genes. Ovaries were stained using antibodies against GFP (Abcam 6673, green) and Lamin-B (red, DHSB ADL84.12). Genotypes are noted on top of the images. White arrows point to GSCs. Yellow arrow heads point to oocytes. Patterns of expression GFP-NL proteins differ, revealing high levels of BAF and dMAN1 in the NL of GSCs, whereas levels of emerin2 are lower. With the exception of BAF, all other NL proteins increase during oogenesis and are deposited into the developing oocyte. Scale bar in germaria images: 5 μm; scale bar in egg chamber images: 20 μm

wildtype ovaries and the cell type-specific localization patterns of the different GFP-NL fusion proteins. The generation of these tagged alleles provides opportunities to expand our understanding of the NL network in the ovary and other tissues. As CRISPR technology is versatile and includes readily available resources, endogenously tagged NL alleles can be readily generated by individual investigators.

4 Notes

1. The PAM sequence is 5'-NGG-3' for the commonly used *Streptococcus pyogenes* Cas9.
2. The efficiency of Cas9 cleavage is affected by several parameters. First, complementarity between the spacer sequence and genome is essential, as imperfect matches with gRNAs are cleaved with low efficiency. Notably, low frequency off-target cleavage events can produce mutations. However, such mutations are likely lost during outcrossing, unless the lesions are located close to the edited gene. Possible off-targets events can be predicted using online CRISPR design tools [38]. Second, the GC content and secondary structure of gRNAs affect stability of the DNA-RNA (R-loop) complex and are critical for efficient cleavage [36, 39–41]. Several scoring algorithms are available to evaluate and rank efficiencies of potential spacer sequences [42, 43] (e.g., http://crispr.mit.edu/, https://chopchop.rc.fas.harvard.edu/, https://www.deskgen.com/). Third, Cas9 cleavage is diminished by the presence of multiple PAM sequences nearby a target site [44]. Although target site cleavage is essential for gene editing by CRISPR, high cleavage efficiency might not always be preferred. For example, low efficiency cutting might be advantageous to avoid biallelic editing of genes that are needed for viability or fertility.
3. Three types of gRNA plasmids are available that carry different U6 promoters, corresponding to U6.1, U6.2, and U6.3. U6.3 was found to be the most efficient in driving gRNA expression [45]. As such, pCFD3-dU6:3gRNA was chosen for constructing the *baf* gRNA.
4. If Gibson assembly is used in the following step, then do not use Taq polymerase because this polymerase adds an extra A to the 3' end. Polymerases such as Primestar or Phusion generate blunt ends and can be used.
5. For Gibson assembly, PCR primers need to be designed with overlapping sequences between the adjacent DNA fragments. These primers can be designed with NEBuilder, available at NEBGibson.com.

6. A set of donor plasmids have been developed to facilitate CRISPR editing [46]. These pHD-Scarless DsRed plasmids carry coding sequences for multiple protein tags flanked by sequences encoding a flexible Ser/Gly linker (L), corresponding to IKAGGSGGSGGSGGS. In addition, the pHD-Scarless DsRed plasmids carry the gene encoding the DsRed flanked by piggyBac transposon ends. PiggyBac transposons integrate into genomic sites of TTAA, duplicating these sites upon insertion. PiggyBac excision is accompanied by restoration of the single TTAA site, thereby leaving the genome in its original state. As such, removal of the DsRed selectable marker occurs without scars.
7. Adjust the volume of DNA that is loaded into the gel with water to make sure the amount of DNA is within the range of the ladder.
8. Embryo injections can be done by companies such as BestGene Inc.
9. Optimal concentrations for injection are between 50 and 250 ng/μL of gRNA plasmid and between 250 and 500 ng/μL of donor plasmid [46].
10. As *baf* is an essential gene [8, 47], the fly line chosen for editing provides germline restricted Cas9 expression and easy removal of *Cas9* transgene in subsequent genetic crosses. Two common germline expression lines are available, *nos-Cas9* and *vas-Cas9*. However, *vas-Cas9* directs some somatic expression, as Vasa expression is not restricted to the germline early during *Drosophila* embryogenesis [48]. For this reason, the *nos-Cas9* expression line was chosen. Importantly, germline expressed Cas9 can be maternally deposited into the embryo, which has the potential to modify the embryonic genome in the presence of gRNA [49].
11. Injected F0 flies are crossed to *w*$^-$; *CyO, Tub-PBac* flies (Bloomington stock #8285; Fig. 3a). The resulting F1 flies are mutant for *white* gene (*w*$^-$) to facilitate identification of fluorescent eyes resulting from integration of the donor vector. These flies also carry a second chromosome balancer, *CyO*, with a piggyBac transgene that constitutively produces transposase to generate excision events of the DsRed module in somatic and germ cells. As such, edited F1 flies will show a mottled eyes phenotype due to the removal of DsRed in a subset of photoreceptors. DsRed can be visualized under a fluorescent scope or by adding a fluorescent adaptor to the lab stereo microscope.
12. GFP primer sequences depend on the type of *gfp* construct used. We recommend the following primers: sfGFP fw: ATGGTGTCCAAGGGCGAGGA, sfGFP rev: CGCTTCTC

GTTGGGGTCCTT; and for eGFP fw: ATGGTGAGC AAGGGCGAGG, eGFP rev: GTACAGCTCGTCCATGCCG.

13. For the generation of the *gfp-baf* allele, only two DsRed positive lines were obtained from 47 F0 crosses. This low level of recovery might reflect efficient editing at the translation start site of the *baf* gene without integration of donor sequences, producing mutations in this essential gene that resulted in lethality. For the two DsRed positive lines, DsRed were successfully removed through PiggyBac transposon-mediated excision, but final sequencing found that only one gene was correctly tagged. The other had a deletion of *baf* sequences.
14. The amount of material analyzed depends on the expression level of the NL-tagged genes and might need to be adjusted after the initial western blot analysis to achieve optimal results. For Fig. 3c, three pairs of ovaries were used per lane.
15. A wide variety of anti-GFP antibodies are available that differ in specificity. As a rule of thumb, a negative control for GFP needs to be included in all analyses. Anti-Actin and Anti-Tubulin are commonly used as loading controls. However, Tubulin shows different bands depending upon tissue type due to tissue-specific expression of its isoforms.
16. If an antibody against the endogenous protein is available, the stability of tagged proteins can be assessed in heterozygous backgrounds by comparing intensity of tagged with untagged proteins [50]. This approach addresses whether the protein tag alters the stability of the fusion protein.
17. In the case of *gfp-baf*, flies carrying this allele were crossed to flies carrying a deletion mutation in the *baf* gene. F1 progeny was screened, wherein GFP-BAF was found to rescue viability of the null allele [8].
18. The depth of focus of a confocal oil lens is about 300 nm or less [51]. For optimal imaging of germaria, it is best to use ovaries dissected from newly born females that only contain pre-vitellogenic stages. These ovaries are thinner, allowing closer placement of the germaria to the coverslip for optimal imaging. If both early and late stages of oogenesis are to be imaged, then older females can be used. Then, it is recommended that late-stage egg chambers be removed because the large size of older egg chambers keeps germaria out of focus. To this end, a pair of micro scissors are used to carefully cut off the opaque egg chambers at the posterior end of the ovary, under the dissecting scope. The remaining transparent ovary tip includes the germaria and is about half the size of a newly eclosed ovary. The ovary tip and late-stage egg chambers can be stained together, but must be mounted separately.

19. Formaldehyde is commonly used, as formaldehyde cross-links proteins and nucleic acids and preserves cellular structures [52]. However, formaldehyde crosslinking can sometimes mask antibody recognition sites. An alternative method for fixing tissues is the use of −20 °C methanol (100%) for 5 min followed by permeabilization with −20 °C acetone (100%) for 1 min.
20. For the imaging of NL proteins, antibodies against Lamin-B are helpful as they provide strong, specific marking of the NL. In the case of primary antibodies for GFP, there are many choices. Importantly, the source depends upon whether the fusion protein contains enhanced (eGFP) or superfolder (sfGFP), as antibodies differ in their ability to recognize the different GFP variants. Both Abcam 5450 and Abcam 6673 anti-GFP recognize sfGFP. However, Abcam 6673 recognizes eGFP better than Abcam 5450. In addition, different tissues show distinct background when stained with different GFP antibodies. Therefore, a negative control for GFP needs to be included in all analyses.
21. For weak antibodies the incubation time can be increased. For example, many Actin antibodies require a minimum of 20 h to achieve good staining [53].
22. Reuse of antibodies can be particularly important when antibodies are limiting. If an antibody mixture is to be reused, then, it is important to keep track of the number of uses. Staining intensity typically decreases upon reuse. In our experience, we have found that antibody mixtures can be stored at 4 °C for no longer than six months and reused no more than three times. However, this needs to be empirically determined, as antibody depletion might occur quickly for highly expressed proteins. For quantification of staining intensity, a fresh antibody mixture should be used.
23. Generally, secondary antibodies chosen for GFP are conjugated with Alexa Fluor 488 (green channel), as GFP retains some fluorescence after fixation. We used Alexa Fluor 488 secondary antibodies at 1:500 dilution.
24. A 2 mg/mL DAPI stock solution is used for dilution into PBS-T, wherein a 1:1000 dilution yields a 2 μg/mL solution. DAPI staining can be omitted if a DAPI-containing mounting medium is used. However, we recommend direct DAPI staining if slides are to be used immediately, as the DAPI-containing mounting solution requires ~24 h to fully penetrate tissues.
25. The optimal coverslip is lens specific. However, the #1.5 coverslip is optimal for imaging with all major microscope brands, including Zeiss, Leica, Nikon, and Olympus.

26. If tissues are mounted in nonhardening mounting medium, they can be imaged right away. However, using a hardening mounting medium, such as Prolong diamond, requires a cure step of a minimum of 24 h in the dark. Typically, slides last for 3 months. After that time, DAPI staining starts to fade, but many fluorescence signals are still visible.

Acknowledgments

We thank Lacy Barton and William Zaremba for critically reading this manuscript. Imaging was supported by the University of Iowa Central Microscopy Research Facility and use of the Zeiss LSM710 confocal microscope acquired via National Institutes of Health (NIH) funding (S10 RR025439-01). This work was supported by the NIH R01 (GM087341) and R21 (AG072397) to PKG.

References

1. Miller PB, Obrik-Uloho OT, Phan MH, Medrano CL, Renier JS, Thayer JL, Wiessner G, Bloch Qazi MC (2014) The song of the old mother: reproductive senescence in female drosophila. Fly (Austin) 8(3):127–139. https://doi.org/10.4161/19336934.2014.969144
2. Gleason RJ, Anand A, Kai T, Chen X (2018) Protecting and diversifying the germline. Genetics 208(2):435–471. https://doi.org/10.1534/genetics.117.300208
3. Duan T, Green N, Tootle TL, Geyer PK (2020) Nuclear architecture as an intrinsic regulator of Drosophila female germline stem cell maintenance. Curr Opin Insect Sci 37:30–38. https://doi.org/10.1016/j.cois.2019.11.007
4. Smith ER, Meng Y, Moore R, Tse JD, Xu AG, Xu XX (2017) Nuclear envelope structural proteins facilitate nuclear shape changes accompanying embryonic differentiation and fidelity of gene expression. BMC Cell Biol 18(1):8. https://doi.org/10.1186/s12860-017-0125-0
5. Barton LJ, Soshnev AA, Geyer PK (2015) Networking in the nucleus: a spotlight on LEM-domain proteins. Curr Opin Cell Biol 34:1–8. https://doi.org/10.1016/j.ceb.2015.03.005
6. Segura-Totten M, Wilson KL (2004) BAF: roles in chromatin, nuclear structure and retrovirus integration. Trends Cell Biol 14(5): 261–266
7. Sears RM, Roux KJ (2020) Diverse cellular functions of barrier-to-autointegration factor and its roles in disease. J Cell Sci 133(16). https://doi.org/10.1242/jcs.246546
8. Duan T, Kitzman SC, Geyer PK (2020) Survival of Drosophila germline stem cells requires the chromatin-binding protein barrier-to-autointegration factor. Development 147(9). https://doi.org/10.1242/dev.186171
9. Barton LJ, Pinto BS, Wallrath LL, Geyer PK (2013) The Drosophila nuclear lamina protein otefin is required for germline stem cell survival. Dev Cell 25(6):645–654. S1534-5807(13)00317-1 [pii]. https://doi.org/10.1016/j.devcel.2013.05.023
10. Barton LJ, Duan T, Ke W, Luttinger A, Lovander KE, Soshnev AA, Geyer PK (2018) Nuclear lamina dysfunction triggers a germline stem cell checkpoint. Nat Commun 9(1):3960. https://doi.org/10.1038/s41467-018-06277-z
11. Vlcek S, Foisner R (2007) A-type lamin networks in light of laminopathic diseases. Biochim Biophys Acta 1773(5):661–674
12. Gotzmann J, Foisner R (2006) A-type lamin complexes and regenerative potential: a step towards understanding laminopathic diseases? Histochem Cell Biol 125(1–2):33–41
13. van Steensel B, Belmont AS (2017) Lamina-associated domains: links with chromosome architecture, heterochromatin, and gene repression. Cell 169(5):780–791. https://doi.org/10.1016/j.cell.2017.04.022
14. Tatli M, Medalia O (2018) Insight into the functional organization of nuclear lamins in health and disease. Curr Opin Cell Biol 54:

72–79. https://doi.org/10.1016/j.ceb.2018.05.001
15. Cenni V, Capanni C, Mattioli E, Schena E, Squarzoni S, Bacalini MG, Garagnani P, Salvioli S, Franceschi C, Lattanzi G (2020) Lamin A involvement in ageing processes. Ageing Res Rev 62:101073. https://doi.org/10.1016/j.arr.2020.101073
16. Bitman-Lotan E, Orian A (2021) Nuclear organization and regulation of the differentiated state. Cell Mol Life Sci 78(7):3141–3158. https://doi.org/10.1007/s00018-020-03731-4
17. Shin JY, Worman HJ (2021) Molecular pathology of Laminopathies. Annu Rev Pathol. https://doi.org/10.1146/annurev-pathol-042220-034240
18. Barton LJ, Wilmington SR, Martin MJ, Skopec HM, Lovander KE, Pinto BS, Geyer PK (2014) Unique and shared functions of nuclear lamina LEM domain proteins in Drosophila. Genetics 197(2):653–665. https://doi.org/10.1534/genetics.114.162941
19. Guruharsha KG, Rual JF, Zhai B, Mintseris J, Vaidya P, Vaidya N, Beekman C, Wong C, Rhee DY, Cenaj O, McKillip E, Shah S, Stapleton M, Wan KH, Yu C, Parsa B, Carlson JW, Chen X, Kapadia B, VijayRaghavan K, Gygi SP, Celniker SE, Obar RA, Artavanis-Tsakonas S (2011) A protein complex network of Drosophila melanogaster. Cell 147(3):690–703. https://doi.org/10.1016/j.cell.2011.08.047
20. Caussinus E, Kanca O, Affolter M (2011) Fluorescent fusion protein knockout mediated by anti-GFP nanobody. Nat Struct Mol Biol 19(1):117–121. https://doi.org/10.1038/nsmb.2180
21. Caussinus E, Affolter M (2016) deGradFP: a system to knockdown GFP-tagged proteins. Methods Mol Biol 1478:177–187. https://doi.org/10.1007/978-1-4939-6371-3_9
22. Prole DL, Taylor CW (2019) A genetically encoded toolkit of functionalized nanobodies against fluorescent proteins for visualizing and manipulating intracellular signalling. BMC Biol 17(1):41. https://doi.org/10.1186/s12915-019-0662-4
23. Morin X, Daneman R, Zavortink M, Chia W (2001) A protein trap strategy to detect GFP-tagged proteins expressed from their endogenous loci in Drosophila. Proc Natl Acad Sci U S A 98(26):15050–15055. https://doi.org/10.1073/pnas.261408198
24. Buszczak M, Paterno S, Lighthouse D, Bachman J, Planck J, Owen S, Skora AD, Nystul TG, Ohlstein B, Allen A, Wilhelm JE, Murphy TD, Levis RW, Matunis E, Srivali N, Hoskins RA, Spradling AC (2007) The carnegie protein trap library: a versatile tool for Drosophila developmental studies. Genetics 175(3):1505–1531
25. Venken KJ, Schulze KL, Haelterman NA, Pan H, He Y, Evans-Holm M, Carlson JW, Levis RW, Spradling AC, Hoskins RA, Bellen HJ (2011) MiMIC: a highly versatile transposon insertion resource for engineering Drosophila melanogaster genes. Nat Methods 8(9):737–743. https://doi.org/10.1038/nmeth.1662
26. Rohrbaugh M, Clore A, Davis J, Johnson S, Jones B, Jones K, Kim J, Kithuka B, Lunsford K, Mitchell J, Mott B, Ramos E, Tchedou MR, Acosta G, Araujo M, Cushing S, Duffy G, Graves F, Griffin K, Gurudatta BV, Jackson D, Jaimes D, Jamison K, Jones K, Kelley D, Kilgore M, Laramore D, Le T, Mazhar B, Mazhar MM, McCrary B, Miller T, Moreland C, Mullins A, Munye E, Okoorie S, Pittman E, Roberts N, Rose D, Rowland A, Shagarabi A, Smith J, Stallworth T, Stroud N, Sung E, Sung K, Takenaka N, Torre E, Veira J, Vu K, Wagstaff W, Wood AM, Wu K, Yang J, Corces VG (2013) Identification and characterization of proteins involved in nuclear organization using Drosophila GFP protein trap lines. PLoS One 8(1):e53091. https://doi.org/10.1371/journal.pone.0053091
27. Doudna JA, Charpentier E (2014) Genome editing. The new frontier of genome engineering with CRISPR-Cas9. Science 346(6213):1258096. https://doi.org/10.1126/science.1258096
28. Wang H, La Russa M, Qi LS (2016) CRISPR/Cas9 in genome editing and beyond. Annu Rev Biochem 85:227–264. https://doi.org/10.1146/annurev-biochem-060815-014607
29. Duan T, Cupp R, Geyer PK (2021) Drosophila female germline stem cells undergo mitosis without nuclear breakdown. Curr Biol 31(7):1450–1462. e1453. https://doi.org/10.1016/j.cub.2021.01.033
30. Warecki B, Ling X, Bast I, Sullivan W (2020) ESCRT-III-mediated membrane fusion drives chromosome fragments through nuclear envelope channels. J Cell Biol 219(3). https://doi.org/10.1083/jcb.201905091
31. Amet N, Lee HF, Shen WC (2009) Insertion of the designed helical linker led to increased expression of tf-based fusion proteins. Pharm Res 26(3):523–528. https://doi.org/10.1007/s11095-008-9767-0
32. Bai Y, Shen WC (2006) Improving the oral efficacy of recombinant granulocyte colony-

stimulating factor and transferrin fusion protein by spacer optimization. Pharm Res 23(9): 2116–2121. https://doi.org/10.1007/s11095-006-9059-5

33. Bai Y, Ann DK, Shen WC (2005) Recombinant granulocyte colony-stimulating factor-transferrin fusion protein as an oral myelopoietic agent. Proc Natl Acad Sci U S A 102(20):7292–7296. https://doi.org/10.1073/pnas.0500062102

34. Chen X, Zaro JL, Shen WC (2013) Fusion protein linkers: property, design and functionality. Adv Drug Deliv Rev 65(10):1357–1369. https://doi.org/10.1016/j.addr.2012.09.039

35. O'Brien AR, Wilson LOW, Burgio G, Bauer DC (2019) Unlocking HDR-mediated nucleotide editing by identifying high-efficiency target sites using machine learning. Sci Rep 9(1): 2788. https://doi.org/10.1038/s41598-019-39142-0

36. Ren X, Yang Z, Xu J, Sun J, Mao D, Hu Y, Yang SJ, Qiao HH, Wang X, Hu Q, Deng P, Liu LP, Ji JY, Li JB, Ni JQ (2014) Enhanced specificity and efficiency of the CRISPR/Cas9 system with optimized sgRNA parameters in Drosophila. Cell Rep 9(3):1151–1162. https://doi.org/10.1016/j.celrep.2014.09.044

37. Korona D, Koestler SA, Russell S (2017) Engineering the Drosophila genome for developmental biology. J Dev Biol 5(4). https://doi.org/10.3390/jdb5040016

38. Gratz SJ, Cummings AM, Nguyen JN, Hamm DC, Donohue LK, Harrison MM, Wildonger J, O'Connor-Giles KM (2013) Genome engineering of Drosophila with the CRISPR RNA-guided Cas9 nuclease. Genetics 194(4):1029–1035. https://doi.org/10.1534/genetics.113.152710

39. Xu X, Duan D, Chen SJ (2017) CRISPR-Cas9 cleavage efficiency correlates strongly with target-sgRNA folding stability: from physical mechanism to off-target assessment. Sci Rep 7(1):143. https://doi.org/10.1038/s41598-017-00180-1

40. Liu X, Homma A, Sayadi J, Yang S, Ohashi J, Takumi T (2016) Sequence features associated with the cleavage efficiency of CRISPR/Cas9 system. Sci Rep 6:19675. https://doi.org/10.1038/srep19675

41. Alekseyenko AA, Gorchakov AA, Zee BM, Fuchs SM, Kharchenko PV, Kuroda MI (2014) Heterochromatin-associated interactions of Drosophila HP1a with dADD1, HIPP1, and repetitive RNAs. Genes Dev 28(13):1445–1460. https://doi.org/10.1101/gad.241950.114

42. Moreno-Mateos MA, Vejnar CE, Beaudoin JD, Fernandez JP, Mis EK, Khokha MK, Giraldez AJ (2015) CRISPRscan: designing highly efficient sgRNAs for CRISPR-Cas9 targeting in vivo. Nat Methods 12(10):982–988. https://doi.org/10.1038/nmeth.3543

43. Doench JG, Fusi N, Sullender M, Hegde M, Vaimberg EW, Donovan KF, Smith I, Tothova Z, Wilen C, Orchard R, Virgin HW, Listgarten J, Root DE (2016) Optimized sgRNA design to maximize activity and minimize off-target effects of CRISPR-Cas9. Nat Biotechnol 34(2):184–191. https://doi.org/10.1038/nbt.3437

44. Malina A, Cameron CJF, Robert F, Blanchette M, Dostie J, Pelletier J (2015) PAM multiplicity marks genomic target sites as inhibitory to CRISPR-Cas9 editing. Nat Commun 6:10124. https://doi.org/10.1038/ncomms10124

45. Port F, Chen HM, Lee T, Bullock SL (2014) Optimized CRISPR/Cas tools for efficient germline and somatic genome engineering in Drosophila. Proc Natl Acad Sci U S A 111(29): E2967–E2976. https://doi.org/10.1073/pnas.1405500111

46. Gratz SJ, Harrison MM, Wildonger J, O'Connor-Giles KM (2015) Precise genome editing of Drosophila with CRISPR RNA-guided Cas9. Methods Mol Biol 1311: 335–348. https://doi.org/10.1007/978-1-4939-2687-9_22

47. Furukawa K, Sugiyama S, Osouda S, Goto H, Inagaki M, Horigome T, Omata S, McConnell M, Fisher PA, Nishida Y (2003) Barrier-to-autointegration factor plays crucial roles in cell cycle progression and nuclear organization in Drosophila. J Cell Sci 116(Pt 18): 3811–3823

48. Renault AD (2012) vasa is expressed in somatic cells of the embryonic gonad in a sex-specific manner in Drosophila melanogaster. Biol Open 1(10):1043–1048. https://doi.org/10.1242/bio.20121909

49. Bier E, Harrison MM, O'Connor-Giles KM, Wildonger J (2018) Advances in engineering the fly genome with the CRISPR-Cas system. Genetics 208(1):1–18. https://doi.org/10.1534/genetics.117.1113

50. Glenn SE, Geyer PK (2019) Investigation of the developmental requirements of Drosophila HP1 and insulator protein partner, HIPP1. G3 (Bethesda) 9(2):345–357. https://doi.org/10.1534/g3.118.200705

51. Conchello JA, Lichtman JW (2005) Optical sectioning microscopy. Nat Methods 2(12):

920–931. https://doi.org/10.1038/nmeth815

52. Thavarajah R, Mudimbaimannar VK, Elizabeth J, Rao UK, Ranganathan K (2012) Chemical and physical basics of routine formaldehyde fixation. J Oral Maxillofac Pathol 16(3):400–405. https://doi.org/10.4103/0973-029X.102496

53. Wineland DM, Kelpsch DJ, Tootle TL (2018) Multiple pools of nuclear actin. Anat Rec (Hoboken) 301(12):2014–2036. https://doi.org/10.1002/ar.23964

Chapter 7

Visualizing Fusome Morphology via Tubulin Immunofluorescence in *Drosophila* Ovarian Germ Cells

Anna E. Williams and Elizabeth T. Ables

Abstract

In many species, oocytes are initially formed by the mitotic divisions of germline stem cells and their differentiating daughters. These progenitor cells are frequently interconnected in structures called cysts, which may function to safeguard oocyte quality. In *Drosophila*, an essential germline-specific organelle called the fusome helps maintain and coordinate the mitotic divisions of both germline stem cells and cyst cells. The fusome also serves as a useful experimental marker to identify germ cells during their mitotic divisions. Fusomes are cytoplasmic organelles composed of microtubules, endoplasmic reticulum-derived vesicles, and a meshwork of membrane skeleton proteins. The fusome branches as mitotic divisions progress, traversing the intercellular bridges of germline stem cell/cystoblast pairs and cysts. Here, we provide a protocol to visualize fusome morphology in fixed tissue by stabilizing microtubules and immunostaining for α-Tubulin and other protein constituents of the fusome. We identify a variety of fluorophore-tagged proteins that are useful for visualizing the fusome and describe how these might be combined experimentally. Taken together, these tools provide a valuable resource to interrogate the genetic control of germline stem cell function, oocyte selection, and asymmetric division.

Key words Germline stem cell, Cyst, Oocyte, Germarium, Germline, Microtubules

1 Introduction

To build oocytes, *Drosophila melanogaster* germ cells develop as cysts of undifferentiated cells that remain interconnected during mitotic division [1–3] (Fig. 1a, b). Within each cyst, one cell is ultimately selected as the oocyte and the other 15 become nurse cells. The cells of the cyst remain interconnected from their initial division through the late stages of oogenesis, where nurse cells shift ("dump") their cytoplasmic contents to the oocyte. The "nursing" mechanism of oocyte development requires that cyst cells maintain conduits between cells, called ring canals, that are initially built during cyst cell mitotic divisions. Ring canals arise due to incomplete cytokinesis at each mitotic division, manifested by cleavage furrow arrest and the subsequent formation and growth of stable

Michelle S. Giedt and Tina L. Tootle (eds.), *Drosophila Oogenesis: Methods and Protocols*,
Methods in Molecular Biology, vol. 2626, https://doi.org/10.1007/978-1-0716-2970-3_7,

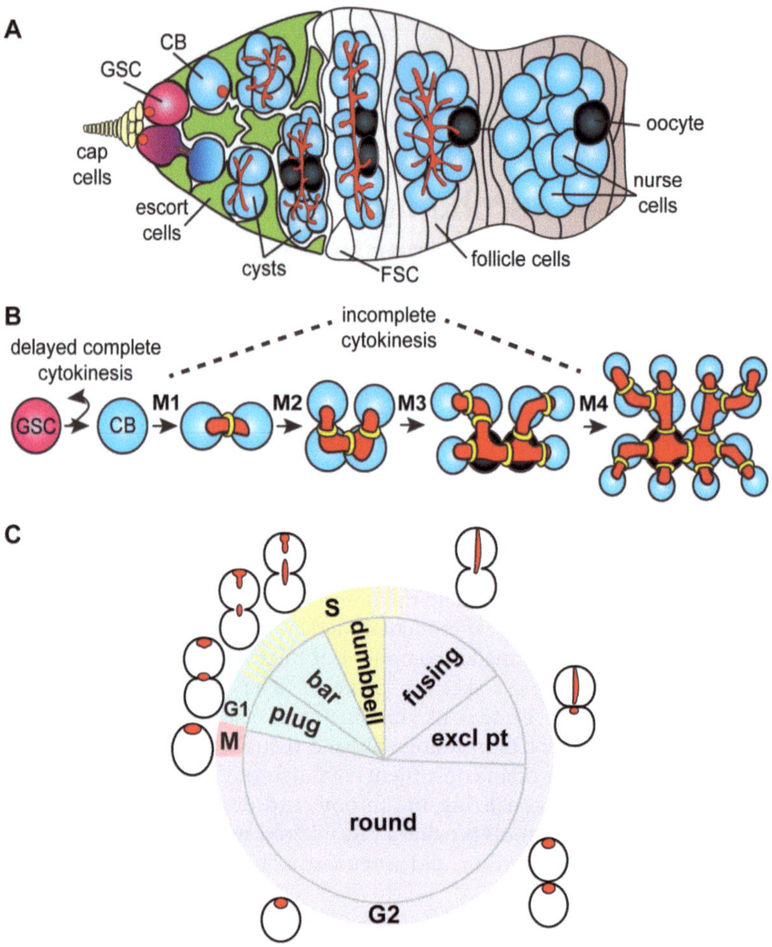

Fig. 1 Germ cell mitotic division in adult *Drosophila melanogaster*. *Drosophila* ovaries are composed of 14 to 16 ovarioles, or strings of progressively older egg chambers. (**a**) Schematic of a germarium. Germline stem cells (GSCs, pink) are located at the anterior tip of the germarium, which resides at the anterior-most tip of each ovariole. GSCs lie adjacent to somatic cap cells (yellow) and escort cells (green) which support GSC self-renewal. GSC division gives rise to another GSC and a differentiating daughter cell, called a cystoblast (CB), which forms posterior to the GSC and continues to divide into 2-cell, 4-cell, 8-cell, and 16-cell cysts (blue). Within each cyst, one germ cell is specified as the oocyte (dark blue), while others become nurse cells. At the posterior of the germarium, cysts are surrounded by somatic follicle cells (tan), which descend from follicle stem cells (FSC). (**b**) Germ cells divide in a stereotypical fashion. GSCs undergo asymmetric mitotic divisions with complete cytokinesis, while cystoblasts divide four times (M1-M4) with incomplete cytokinesis, giving rise to the interconnected cells of the cyst. Individual cells remain connected by small ring canals (yellow), through which the fusome (red) branches. (**c**) Schematic diagram of the GSC cell cycle. GSCs divide approximately every 15 h; however, abscission (the final stage of cytokinesis) is delayed well into the G2 phase of the next cell cycle [9, 14, 23, 24, 26]. Fusome morphology (red) can be used for identifying GSCs generally, and more specifically, as an indirect indicator of the cell cycle stage of the GSC [9, 14, 22]. G1 gap phase 1; S synthesis phase; G2 gap phase 2; M mitosis

structures post-mitosis that permit transport of cytoplasm and organelles [3–5]. Apart from their role in nurse cell to oocyte transport, ring canals also support cell-cell communication and oocyte specification during germ cell mitotic divisions [3–5]. The evolutionary conservation of cyst formation suggests that germ cell interconnectivity is biologically relevant, and in support of this, stable intercellular bridges are required for optimal fertility in multiple phyla [1, 5, 6]. Cyst formation has been postulated to increase cytoplasmic volume, enhance sensitivity to DNA damage, and ensure robust oocyte development [3–7].

In *Drosophila*, cysts arise via four mitotic divisions of the cystoblast, an undifferentiated progenitor cell that is produced from germline stem cells (GSCs) (Fig. 1a, b). In contrast to the incomplete cytokinesis of the dividing cyst cells, GSCs produce cystoblasts via an asymmetric division with complete cytokinesis. GSCs are located at the anterior-most tip of each ovariole (the major subunit of the ovary) adjacent to somatic cap cells (Fig. 1a). GSCs divide asymmetrically every 12–15 h to maintain the GSC population while also forming the cystoblast, which is committed to differentiation [8, 9]. The switch from complete cytokinesis of the GSC/cystoblast pair to incomplete cytokinesis of the dividing cystoblast happens on a rapid timescale, suggesting tight molecular regulation.

Mitotically dividing cysts can be easily distinguished based on the prominent appearance of a germline-specific organelle called the fusome (Fig. 1a, b) [8]. A cytoplasmic vesicular organelle that most closely resembles the endoplasmic reticulum, the fusome core is composed of microtubules, the adducin-like protein Hu-li tai shao (Hts), and the membrane skeletal proteins α- and β-spectrin [10–13]. Other proteins, including cell cycle regulatory proteins and polarity factors, are transiently associated with the fusome. In dividing cyst cells, the fusome extends through ring canals, connecting the cells within a cyst [14–17]. When cells divide, the fusome also grows in a stereotypical pattern. New fusome material accumulates as mitotic spindles are disassembled, resulting in a continuously branched structure [14–17]. Like cysts, GSCs and cystoblasts also contain fusomes, sometimes referred to as spectrosomes due to their prominent dot-like morphology. Although the GSC and cystoblast fusomes are thought to be the precursors of the cyst fusomes, they differ in protein composition, morphology, and temporal dynamics [8, 11, 18]. Despite its variable composition, the function of the fusome is the same: to establish cell polarity, coordinate cyst cell communication, and promote oocyte differentiation [10, 12, 19–21].

Fixed and live-cell imaging analyses demonstrate that morphology of the GSC fusome changes coordinately with the phases of the cell cycle, as new fusome material is distributed to the daughter cystoblast [9, 14, 15, 22–24]. Importantly, delayed cytokinesis between the GSC and the nascent cystoblast keeps the pair connected via a temporary ring canal through S phase of the subsequent cell cycle (Fig. 1c) [25–27]. Through most of G1 and S phase, the GSC fusome appears as a round structure located at the posterior side of the GSC, juxtaposed to the adherens junctions which anchor the GSC to the cap cells (Fig. 1c). As the pair of cells exit S phase and proceed through G2, three events occur: the distribution of new fusome material to the cystoblast, the completion of abscission, and the assembly of the mitotic spindle in the GSC. The cystoblast fusome forms in the nascent ring canal, opposite to the original GSC fusome, as a plug of new fusome material merges with microtubule remnants from the previous cycle's mitotic spindle. The fusome continues to grow during G2, connecting the GSC and cystoblast and forming a thick elongated structure. Abscission closes the ring canal between the two cells, separating the fusome asymmetrically (appears as an "exclamation mark" morphology). As the GSC progresses through G2, the fusome retracts into the round shape, remaining at the posterior of the cell. The round fusome then becomes the anchor point for the mitotic spindle, whose growth is facilitated by the centrosome. This allows the division plane of the GSC to divide largely perpendicular to the adjacent somatic cap cells, facilitating cystoblast formation posteriorly. Despite the well-characterized morphogenesis of the fusome, the molecular control of this process remains largely unknown.

The stereotypical shape and predictable morphogenesis of the fusome, as well as the availability of excellent antisera against fusome core proteins (Hts and α-spectrin) [10, 12, 28, 29], make the fusome a highly useful experimental tool with which to monitor mitotic expansion of the germ cell pool. Here, we provide a protocol to use α-Tubulin immunofluorescence in combination with other fusome core proteins and/or GFP-tagged germ cell resident proteins to visualize GSCs and their daughters during mitotic division. Immunofluorescence for α-Tubulin works particularly well to visualize mitotic spindles and the cytoskeleton in germ cells (Fig. 2) but must be performed carefully to stabilize the microtubules through fixation. Therefore, our protocol is based on the methodology of Grieder and Spradling, who described the fixation conditions and temperature requirements necessary for optimal microtubule stabilization and immunostaining [19]. Our protocol uses Hts localization to both identify germ cells and discern fusome structure, as it works well in co-immunofluorescence with other antibodies, including α-Tubulin (Fig. 2) [12]. We also identify a variety of transgenic fly lines which are useful tools for visualizing different characteristics of GSCs or germ cells during division (Table 1 and Fig. 2). These include GFP-tagged forms of the plasma membrane-associated scaffolding protein Scribble (Fig. 2c, d") and the microtubule-associated protein Jupiter (Fig. 2a), and a UAS-based α-Tubulin that can be driven specifically in germ cells using the *nanos-Gal4::VP16* driver (Fig. 2b).

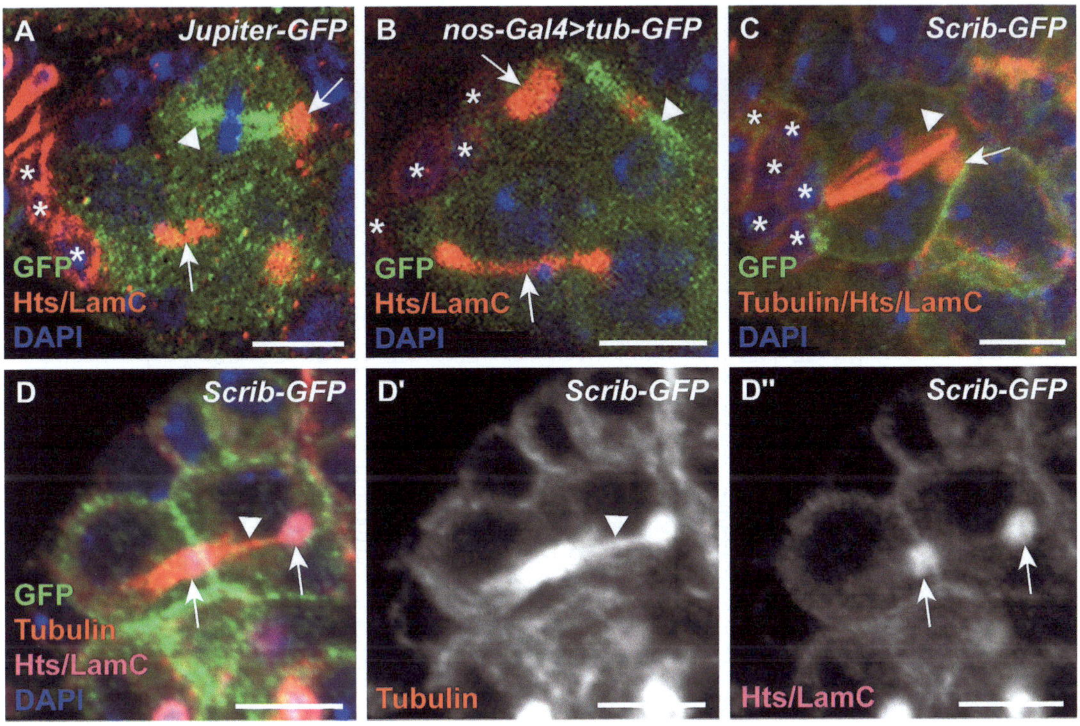

Fig. 2 Visualizing fusome and spindle morphology. (**a**, **b**) GSCs and cystoblasts from *Jupiter-GFP* (**a**) and *nos-Gal4::VP16 > tub-GFP* (**b**) germaria immunostained for GFP (green; mitotic spindle), Hts and LaminC (LamC; red; fusomes and nuclear membrane of cap cells), and DAPI (blue; DNA). (**c**) GSC in metaphase of mitosis from *Scrib-GFP* germarium immunostained for GFP (green; cell membrane), α-Tubulin (red; mitotic spindle), Hts and LamC (red), and DAPI (blue; DNA). (**d-d**") Dividing cystoblast from *Scrib-GFP* germarium co-immunostained for GFP (green; cell membrane), α-Tubulin (magenta; mitotic spindle remnants), Hts and LamC (red; fusome), and DAPI (blue; DNA). Images in **d**'-**d**" display Tubulin (**d**') or Hts/LamC (**d**") channels only. Asterisks indicate cap cells; dashed white lines indicate GSCs; arrows indicate fusomes; arrowheads indicate mitotic spindles or spindle remnants. (Images **a–c** were acquired with a Zeiss LSM 700 laser scanning confocal; images **d-d**" acquired with a Zeiss LSM 800 with Airyscan. Scale bars = 5 μm)

2 Materials

2.1 Fly Strains and Culture

1. Any *Drosophila* strains are conducive for visualizing fusome morphology via tubulin antibody staining, and many lines utilizing endogenous protein tags are available (some useful examples for visualizing germ cells are listed in Table 1). Standard *Drosophila* husbandry can be used to generate progeny with genotypes of interest. Flies can be maintained in bottles or vials on standard cornmeal/molasses/yeast media.

Table 1
Fly stocks useful for visualizing fusome dynamics and/or microtubules in dividing ovarian germ cells

Fluorescent marker	Fly Stock	Description	Localization	Reference
UASp-tubulin-GFP	*P{UASp-GFPS65C-α Tub84B}3* (BDSC #7373 or #7253)	Expresses N-terminal region of αTub84B with a N-terminal GFP tag under UASp control	Fusome, mitotic spindle	[19]
UASp-Hts-mCherry	*P{UASp-hts.mCherry}attP2* (BDSC #66171)	Expresses Hts protein isoform I with a C-terminal mCherry tag under UASp control	Fusome, ring canals, plasma membrane	Tony Harris, personal communication to FlyBase
UASp-KDEL-RFP	*P{UASp-RFP.KDEL}* (BDSC #30909 & #30910)	Expresses DsRed with a N-terminal bovine preprolactin signal sequence and a endoplasmic reticulum retention signal under UASp control	Fusome, endoplasmic reticulum	Jennifer Lippincott-Schwartz, personal communication to FlyBase
Scrib-GFP	*P{PTT-GA}scrib*^{CA07683}	Protein trap	Plasma membrane, fusome	[11, 30]
Jupiter-GFP	*Jupiter*^{CPTI003917} (Kyoto #115457)	Protein trap	Mitotic spindle, cytoplasm	[31]
Rtnl1-GFP	*P{PTT-GA}Rtnl1*^{CA06523}	Protein trap	Fusome	[11, 13, 30, 32]
Par-1-GFP	*P{PTT-GC}par-1*^{CC01981}	Protein trap	Fusome	[11]
Fkbp14-GFP	*Mi{PT-GFSTF.2}Fkbp14*$^{MI04530\text{-}GFSTF.2}$ (BDSC #66358)	Protein trap	Fusome, endoplasmic reticulum	[33]

2. Wet yeast paste: used to supplement flies' diet prior to dissection to increase ovary size aiding in the dissection process. Make a 1:1 mixture of dry active yeast and distilled water; mix until the consistency of smooth peanut butter is reached and store at 4 °C, tightly covered to prevent drying.
3. Dissecting microscope equipped with CO_2 fly pad and tube with air needle to deliver CO_2 to flies in bottles or vials.

2.2 Ovary Dissection

1. 1.5 mL microfuge tubes pre-coated with 3% Bovine Serum Albumin (BSA). Add 500 μL BSA to clean microfuge tubes and incubate at room temperature on a nutating mixer for about 1 h. Pre-coating the tubes helps prevent ovaries from sticking to the sides of the tube. Pre-coated tubes can be stored at 4 °C.
2. Glass or plexiglass dissection dish.
3. Kimwipes.
4. Glass pasteur pipettes and bulbs.
5. Two pairs of #5 dissection forceps (INOX, Dumont #5, Biologie point).
6. Two 27 × 1¼ gauge needles with 1 mL syringes.
7. Nutating mixer.
8. Grace's Insect Medium.

2.3 Immunostaining

1. 15 mL and 50 mL conical tubes.
2. 1× Phosphate Buffered Saline (PBS).
3. PBS-T: 1× PBS with 0.1% Triton-X-100, store at room temperature.
4. Grace's Insect Medium.
5. 16% electron microscopy grade formaldehyde.
6. Pipettors and pipette tips.
7. Blocking reagent: 5% Normal Goat Serum (NGS) in PBS. For 50 mL, add 2.5 mL NGS to 47.5 mL PBS. Store at 4 °C; do not use if solution becomes cloudy or develops an unusual smell.
8. Primary antibodies, diluted in blocking reagent (Table 2):
   - Labeling endogenously tagged proteins: chicken anti-GFP (1:2000 dilution).
   - Cell membrane and fusome marker: mouse anti-Hts (1:10 dilution).
   - Nuclear envelope marker: mouse anti-LaminC (1:100 dilution).
   - Microtubules and fusomes: anti-α-Tubulin-AlexaFluor555 (1:200 dilution).

Table 2
Antibodies useful for visualizing the fusome and/or microtubules in dividing ovarian germ cells

Antigen (Host Species)	Vendor (Catalog)	Description	Localization
Hts (mouse)	DSHB (1B1)	Monoclonal antisera	Fusome
LaminC (mouse)	DSHB (LC28.26)	Monoclonal antisera, useful for identifying cap cells	Nuclear envelope
α-Spectrin (mouse)	DSHB (3A9)	Monoclonal antisera	Fusome
α-Tubulin-AF555 (mouse)	Millipore (#05-829-AF555)	Directly conjugated monoclonal antisera (clone DM1A), can be purchased with AlexaFluor-555 or AlexaFluor-488	Fusome, mitotic spindle
GFP (chicken)	Abcam (ab13970)	Polyclonal antisera, useful for marking GFP-tagged transgenes (see Table 1)	Variable
dsRed (rabbit)	Clontech / Takara (632496)	Polyclonal antisera, useful for marking mCherry- and RFP-tagged transgenes (see Table 1)	Variable

DSHB Developmental Studies Hybridoma Bank

6. Secondary Antibodies, diluted 1:200 in blocking reagent from a stock solution in which the product obtained from the manufacturer was diluted 1:1 with 100% glycerol, conjugated with an AlexaFluor of interest and matching the host species of the corresponding primary antibody. For example:
   - GFP: goat anti-chicken-AlexaFluor488
   - Hts + LaminC: goat anti-mouse-AlexaFluor633
7. 4',6-Diamidino-2-phenylindole (DAPI): dilute 5 mg/mL DAPI stock 1:500 in 0.1% PBS-T; store at 4 °C in dark.
8. Grace's Medium.
9. Mounting medium (*see* **Note 1**).

2.4 Sample Mounting and Microscopy

1. Kimwipes.
2. Glass Pasteur pipettes and bulbs.
3. Fine dissection tools for ovary mounting (forceps and needles).
4. Glass microscope slides and coverslips.
5. Mounting media (*see* **Note 1**).
6. Steel weight, measuring approximately 250 g.
7. Fingernail polish (clear or colors) to seal slide (if necessary).
8. Laser scanning confocal microscope with a 63× oil immersion lens.
9. Image acquisition and analysis software (ex. Zeiss Zen Blue, Zeiss Zen Black, Imaris).

3 Methods

3.1 Fly Preparation

1. Use standard *Drosophila* husbandry to rear flies and/or cross fly lines under UAS control with a Gal4 driver. For analyses of GSCs and their dividing daughters, *nos-Gal4::VP16* is a convenient driver [34]. After 2–3 days of egg laying, remove parents to a new vial, such that only eclosing progeny will be collected for dissection. Progeny takes approximately 10 days to eclose when reared at 25 °C.
2. Age-match flies and feed wet yeast paste (changed daily) prior to dissection. Age, diet, temperature, and genetic background all impact cell cycle dynamics in GSCs; therefore, it is imperative to control these variables.

3.2 Ovary Dissection

1. Remove Grace's Medium, formaldehyde, and 1.5 mL microfuge tubes precoated with BSA from refrigerator. Allow these to come completely to room temperature (RT).
2. Mix fixative. For each batch of flies to be dissected, add 700 μL Grace's Medium plus 400 μL 16% formaldehyde. Keep fixative at RT (do NOT put in ice bucket) to preserve microtubules during fixation.
3. Dissect 10–15 pairs of ovaries in Grace's medium in glass or plastic dissecting dishes, using needles to break open the outer layer of muscle around each ovariole. Teasing apart ovarioles helps fixative and antibodies to reach all germaria and egg chambers more equivalently (*see* **Note 2**).
4. Remove BSA from pre-coated microfuge tube using a glass Pasteur pipet.
5. Add your dissected ovaries in Grace's medium to the pre-coated microfuge tube using the same glass pipet from **step 4** above. Do NOT put tubes on ice.

3.3 Immunostaining

Although there are a variety of ways to utilize immunofluorescence to visualize mitotically active GSCs and germ cells in the germarium, the method below specifically uses four-color immunofluorescence to simultaneously visualize microtubules, a GFP-tagged protein of interest, the fusome, and the DNA-binding fluorescent dye DAPI to provide a comprehensive view of GSCs as they divide (examples are shown in Fig. 2). It is incredibly useful to co-label anti-α-Tubulin, anti-Hts, and anti-LaminC to visualize microtubules, the fusome core, and the nuclear lamina of both the GSCs and the neighboring cap cells, respectively. Labeling the cap cells provides a reliable way to identify GSCs and their dividing daughters within the context of the cellular microenvironment of the germarium [14, 22]. The challenge to this approach is that available antibodies are frequently raised in the same species, precluding clear discrimination of proteins. For example, available anti-Tubulin, anti-Hts, and anti-LaminC are all mouse monoclonal antisera (see Table 2). To work around this limitation, we altered the original protocol of Grieder and Spradling [19] to layer antibodies onto the tissue (*see* **Note 3**). Below, we describe the protocol to visualize germ cells using anti-Hts to label fusomes, anti-LaminC to label nuclear membranes of cap cells, anti-GFP to label the plasma membrane (using *Scrib-GFP*; see Table 1), and anti-α-Tubulin antisera that is directly conjugated to a stable fluorophore (AlexaFluor555) to visualize the dynamic movements of microtubules (Fig. 2c, d). We group anti-Hts and anti-LaminC (both mouse monoclonals) onto the same wavelength/channel using anti-mouse-AlexaFluor633, as their intracellular distributions are largely distinct.

This protocol could be modified to visualize fusomes and microtubules with other GFP- or RFP/mCherry-tagged proteins (endogenous or via the UASp/Gal4 system) by simply substituting secondary antibodies conjugated with different fluorophores (see Table 1 for some useful fly lines for visualizing fusomes). Alternatively, Tubulin can also be visualized in two- or three-color immunofluorescence using an endogenously tagged *Jupiter-GFP* (Table 1 and Fig. 2a), or by expressing *UASp-Tubulin-GFP* specifically in germ cells using the *nos-Gal4::VP16* driver (Table 1 and Fig. 2b).

Until the addition of primary antibody, all reagents are brought to room temperature on the day of the experiment, and blocking, rinses, and washes are performed at room temperature, unless otherwise noted. When discarding the previous solution, we always keep ~100–200 μL of liquid on the ovaries to keep the tissue from drying out. In our hands, we reduce background fluorescence by layering primary antibodies onto samples one at a time in sequence, separated by extensive washing in detergent and incubation overnight or over two nights at 4 °C. For other antibodies or tagged proteins, the exact sequence should be tested experimentally. Secondary antibodies against different species (i.e., anti-rabbit-AlexaFluor568 plus anti-chicken-AlexaFluor488) can be combined; however, the directly conjugated anti-α-Tubulin should be independently added last to prevent cross-labeling by the other anti-mouse secondary antibodies.

1. Remove all except ~100–200 μL of Grace's medium from each tube of dissected ovaries. Add 1000 μL fixative to each sample and fix on nutator at RT for exactly 10 min.
2. Quickly remove fixative to appropriate formaldehyde waste container using a 1000 μL pipettor.
3. Quickly add 1000 μL PBS-T to each sample. Invert tube 4–5 times to rinse fixative off ovaries.

4. Allow ovaries to sink to the bottom of the microfuge tubes. Remove liquid to appropriate formaldehyde waste container using a 1000 μL pipettor.
5. Add 1000 μL PBS-T to each sample and wash for at least 10 min on a nutator.
6. Allow ovaries to sink to the bottom of the microfuge tubes. Remove liquid to appropriate formaldehyde waste container using a 1000 μL pipettor.
7. Repeat **steps 5–6** twice (totaling one "rinse" and three longer "washes").
8. Add 1000 μL blocking reagent to each sample. Block for at least 30 min on a nutating mixer at RT.
9. Allow ovaries to sink to the bottom of the microfuge tubes and discard the blocking solution.
10. Add 400–500 μL of chicken anti-GFP (diluted 1:2000 in blocking solution) to each microfuge tube. Incubate in primary antibody on a nutator at 4 °C overnight.
11. The next day, allow ovaries to sink to the bottom of the microfuge tubes. Discard the first primary antibody and add 1000 μL PBS-T to each sample. Wash for at least 30 min on a nutator.
12. Discard the wash solution and repeat three more times with fresh wash solution (totaling four "washes").
13. Allow ovaries to sink to the bottom of the microfuge tubes and discard the last wash.
14. Add 400–500 μL of a primary antibody solution containing mouse anti-Hts (1:10) and mouse anti-LaminC (1:100), diluted together in blocking solution, to each microfuge tube. Incubate in primary antibody solution on a nutator at 4 °C over two nights.
15. Allow ovaries to sink to the bottom of the microfuge tubes. Discard the second primary antibody and add 1000 μL PBS-T to each sample. Wash for at least 30 min on a nutator.
16. Discard the wash solution and repeat three more times with fresh wash solution (totaling four "washes").
17. Allow ovaries to sink to the bottom of the microfuge tubes and discard the last wash.
18. Add 1000 μL blocking reagent to each sample. Block for at least 30 min on a nutating mixer at room temperature.
19. Allow ovaries to sink to the bottom of the microfuge tubes and discard the blocking solution.
20. Add 400–500 μL of a secondary antibody solution containing goat anti-chicken-AlexaFluor488 (1:200) and goat anti-

mouse-AlexaFluor633 (1:200), diluted together in blocking solution, to each microfuge tube. Incubate in the secondary antibody solution for 1–2 h at RT on a nutator. From this point forward, samples should be covered in tinfoil as much as possible to prevent fluorescence bleaching.

21. Allow ovaries to sink to the bottom of the microfuge tubes. Discard the secondary antibody solution and add 1000 μL PBS-T to each sample. Wash for at least 30 min on a nutator.
22. Discard the wash solution and repeat three more times with fresh wash solution (totaling four "washes").
23. Allow ovaries to sink to the bottom of the microfuge tubes and discard the last wash.
24. Add 400–500 μL of anti-α-Tubulin-AF555 (1:200) diluted in blocking solution. Incubate in this third primary antibody solution on a nutator at 4 °C overnight (*see* **Note 4**).
25. The next day, allow ovaries to sink to the bottom of the microfuge tubes. Discard the antibody and add 1000 μL PBS-T to each sample. Wash for at least 30 min on a nutator.
26. Discard the wash solution and repeat three more times with fresh wash solution (totaling four "washes").
27. Allow ovaries to sink to the bottom of the microfuge tubes and discard the last wash.
28. Add 400–500 μL DAPI diluted in blocking solution. Incubate at RT for 15 min on a nutator.
29. Allow ovaries to sink to the bottom of the microfuge tubes. Discard the DAPI solution and add 1000 μL PBS-T to each sample. Wash for at least 10 min on a nutator.
30. Discard the wash solution and repeat three more times with fresh wash solution (totaling four "washes").
31. Remove the last wash and add 2–3 drops of mounting medium (*see* **Note 1**). Store at 4 °C in dark until ready for mounting on slides.

3.4 Microscopy

1. Use a glass pipette to move ovaries and mounting media from tube to a glass slide. Remove enough mounting media from the slide such that ovarioles no longer float.
2. Use fine dissecting tools (needles, tungsten wire, and/or forceps) to separate ovarioles from each other and remove egg chambers larger than stage 9. These steps help to reduce the thickness of the ovariole such that germaria can be more clearly imaged (*see* **Note 5**).
3. Once the ovarioles are spread across the slide, add 1–2 drops of mounting medium to prevent air bubbles forming between ovarioles. Carefully place a #1.5 glass coverslip on top of the

media in the center of the slide and allow mounting media to spread over ovarioles by capillary action (*see* **Note 6**).

4. Flatten coverslip onto glass slide for 2–5 min. This is most easily accomplished by placing the slide onto a clean kimwipe. Then assemble a weight apparatus composed of a kimwipe, an inverted cardboard box top, and a steel weight. Carefully place the weight apparatus on top of your slide, being cautious not to let the coverslip move on the slide (*see* **Note 7**).
5. After your slide has been flattened, seal the edges of the coverslip to the slide using fingernail polish. Allow slides to dry for at least 1 h before imaging.
6. Use a confocal microscope with a 63× oil immersion lens to image samples. We use an inverted laser scanning confocal (Zeiss LSM700 or LSM800) capable of adding an optical zoom of at least 2.0× (and up to 6.0×). Images are collected as z-stacks with anywhere from 0.25 to 1.0 μm optical step size (*see* **Note 8**).

4 Notes

1. We have used both homemade and commercially available mounting media with great success. For a relatively cheap version that works well with long shelf-life, we use 20 n-propyl gallate in 90% glycerol. To create this solution, mix 1.0 g n-propyl gallate and 5 mL of PBS in a 50 mL conical and vortex to mix; add 45 mL of 100% glycerol to conical and wrap in tinfoil to keep solution in dark; nutate overnight at room temperature; store at 4 °C in dark. Alternatively, the commercially available Vectashield Vibrance is a setting formulation antifade mounting medium (Vector Laboratories, H-1700). Slides mounted with this medium can be imaged with no altered fluorescence brightness for more than one week after mounting, if kept stored at 4 °C. See manufacturer's instructions for additional details.
2. Be sure to thoroughly tease apart ovaries as no permeabilization step is present.
3. In general, for triple or quadruple labeling, we apply primary antibodies first in sequence (each incubated overnight at 4 °C), then combine species-specific secondary antibodies in the same solution. Since the fluorophore-conjugated α-Tubulin antibody is also a mouse antibody, this allows independent analysis of the Hts or α-spectrin proteins and α-tubulin in different fluorescence channels. An important note: for immunostaining with α-Spectrin antibodies, we substitute Tween-20 for Triton in all solutions.

4. Samples can be incubated in the fluorophore-conjugated α-Tubulin antibody for as little as 20 min in dark at RT, but leaving the antibody on the tissue at 4 °C overnight results in brighter staining intensity.
5. It is important to remove late-stage egg chambers (typically everything after stage 9). Removing these large egg chambers allows high-quality images to be produced, as the cells of the germarium are all in the same optical plane. Leaving the large egg chambers on the slide allows room for germaria to move around on the slide. This movement can negatively affect image quality.
6. If mounting media does not reach edges, extra can be added around the edge of the glass coverslip using a pipette and allowing capillary action to bring the mounting medium under the coverslip.
7. Place the lid of a cardboard microscope slide box on top of a folded kimwipe and place a steel weight in the middle of this box. Then set the weight/kimwipe/box apparatus on top of the coverslip. This helps to evenly distribute weight resulting in equally flattened ovarioles for optimal imaging.
8. With smaller optical sections, a more cohesive three-dimensional image can be created in post-processing, allowing more sophisticated quantitative analyses.

Acknowledgments

We are grateful to members of the *Drosophila* community for sharing reagents and protocols and to members of the Ables lab past and present for helpful comments on this manuscript. This work was supported by National Institutes of Health (NIH) grant R15-GM117502 (E.T.A.) and an East Carolina University Undergraduate Research and Creative Activity Award (A.E.W.).

References

1. Lu K, Jensen L, Lei L, Yamashita YM (2017) Stay connected: a germ cell strategy. Trends Genet 33(12):971–978. https://doi.org/10.1016/j.tig.2017.09.001
2. Matova N, Cooley L (2001) Comparative aspects of animal oogenesis. Dev Biol 231(2): 291–320. https://doi.org/10.1006/dbio.2000.0120
3. Pepling M, Lei L (2018) Germ cell nests and germline cysts. In: Skinner MK (ed) Encyclopedia of reproduction, 2nd edn. Academic Press, Oxford, pp 159–166. https://doi.org/10.1016/B978-0-12-801238-3.64710-4
4. de Cuevas M, Lilly MA, Spradling AC (1997) Germline cyst formation in Drosophila. Annu Rev Genet 31:405–428. https://doi.org/10.1146/annurev.genet.31.1.405
5. Haglund K, Nezis IP, Stenmark H (2011) Structure and functions of stable intercellular bridges formed by incomplete cytokinesis during development. Commun Integr Biol 4(1): 1–9. https://doi.org/10.4161/cib.4.1.13550

6. O'Connell JM, Pepling ME (2021) Primordial follicle formation - some assembly required. Curr Opin Endocr Metab Res 18:118–127. https://doi.org/10.1016/j.coemr.2021.03.005
7. Yamashita YM (2018) Subcellular specialization and organelle behavior in germ cells. Genetics 208(1):19–51. https://doi.org/10.1534/genetics.117.300184
8. Hinnant TD, Merkle JA, Ables ET (2020) Coordinating proliferation, polarity, and cell fate in the Drosophila female germline. Front Cell Dev Biol 8:19. https://doi.org/10.3389/fcell.2020.00019
9. Villa-Fombuena G, Lobo-Pecellín M, Marín-Menguiano M, Rojas-Ríos P, González-Reyes A (2021) Live imaging of the Drosophila ovarian niche shows spectrosome and centrosome dynamics during asymmetric germline stem cell division. Development 148(18). https://doi.org/10.1242/dev.199716
10. de Cuevas M, Lee JK, Spradling AC (1996) alpha-spectrin is required for germline cell division and differentiation in the Drosophila ovary. Development 122(12):3959–3968
11. Lighthouse DV, Buszczak M, Spradling AC (2008) New components of the Drosophila fusome suggest it plays novel roles in signaling and transport. Dev Biol 317(1):59–71. https://doi.org/10.1016/j.ydbio.2008.02.009
12. Lin H, Yue L, Spradling AC (1994) The Drosophila fusome, a germline-specific organelle, contains membrane skeletal proteins and functions in cyst formation. Development 120(4):947–956
13. Röper K (2007) Rtnl1 is enriched in a specialized germline ER that associates with ribonucleoprotein granule components. J Cell Sci 120(Pt 6):1081–1092. https://doi.org/10.1242/jcs.03407
14. de Cuevas M, Spradling AC (1998) Morphogenesis of the Drosophila fusome and its implications for oocyte specification. Development 125(15):2781–2789
15. Huynh J-R (2006) Fusome as a cell-cell communication channel of Drosophila ovarian cyst. In: Cell-cell channels. Springer, New York. https://doi.org/10.1007/978-0-387-46957-7_16
16. Koch EA, King RC (1966) The origin and early differentiation of the egg chamber of Drosophila melanogaster. J Morphol 119:283–303. https://doi.org/10.1002/jmor.1051190303
17. Ong SK, Tan C (2010) Germline cyst formation and incomplete cytokinesis during Drosophila melanogaster oogenesis. Dev Biol 337:84–98. https://doi.org/10.1016/j.ydbio.2009.10.018
18. McKearin D (1997) The Drosophila fusome, organelle biogenesis and germ cell differentiation: if you build it. BioEssays : news and reviews in molecular, cellular and developmental biology 19(2):147–152. https://doi.org/10.1002/bies.950190209
19. Grieder NC, de Cuevas M, Spradling AC (2000) The fusome organizes the microtubule network during oocyte differentiation in Drosophila. Development 127(19):4253–4264
20. Snapp EL, Iida T, Frescas D, Lippincott-Schwartz J, Lilly MA (2004) The fusome mediates intercellular endoplasmic reticulum connectivity in Drosophila ovarian cysts. Mol Biol Cell 15(10):4512–4521. https://doi.org/10.1091/mbc.e04-06-0475
21. Yue L, Spradling AC (1992) hu-li tai shao, a gene required for ring canal formation during Drosophila oogenesis, encodes a homolog of adducin. Genes Dev 6:2443–2454. https://doi.org/10.1101/gad.6.12b.2443
22. Ables ET, Drummond-Barbosa D (2013) Cyclin E controls Drosophila female germline stem cell maintenance independently of its role in proliferation by modulating responsiveness to niche signals. Development 140(3):530–540. https://doi.org/10.1242/dev.088583
23. Hinnant TD, Alvarez AA, Ables ET (2017) Temporal remodeling of the cell cycle accompanies differentiation in the Drosophila germline. Dev Biol 429(1):118–131. https://doi.org/10.1016/j.ydbio.2017.07.001
24. Kao SH, Tseng CY, Wan CL, Su YH, Hsieh CC, Pi H, Hsu HJ (2015) Aging and insulin signaling differentially control normal and tumorous germline stem cells. Aging Cell 14(1):25–34. https://doi.org/10.1111/acel.12288
25. Mathieu J, Cauvin C, Moch C, Radford SJ, Sampaio P, Perdigoto CN, Schweisguth F, Bardin AJ, Sunkel CE, McKim K, Echard A, Huynh JR (2013) Aurora B and cyclin B have opposite effects on the timing of cytokinesis abscission in Drosophila germ cells and in vertebrate somatic cells. Dev Cell 26(3):250–265. https://doi.org/10.1016/j.devcel.2013.07.005
26. Mathieu J, Huynh JR (2017) Monitoring complete and incomplete abscission in the germ line stem cell lineage of Drosophila ovaries. Methods Cell Biol 137:105–118. https://doi.org/10.1016/bs.mcb.2016.03.033
27. Matias NR, Mathieu J, Huynh JR (2015) Abscission is regulated by the ESCRT-III

protein shrub in Drosophila germline stem cells. PLoS Genet 11(2):e1004653. https://doi.org/10.1371/journal.pgen.1004653
28. Zaccai M, Lipshitz HD (1996) Role of Adducin-like (hu-li tai shao) mRNA and protein localization in regulating cytoskeletal structure and function during Drosophila Oogenesis and early embryogenesis. Dev Genet 19(3):249–257. https://doi.org/10.1002/(sici)1520-6408(1996)19:3<249::Aid-dvg8>3.0.Co;2-9
29. Pesacreta TC, Byers TJ, Dubreuil R, Kiehart DP, Branton D (1989) Drosophila spectrin: the membrane skeleton during embryogenesis. J Cell Biol 108(5):1697–1709. https://doi.org/10.1083/jcb.108.5.1697
30. Buszczak M, Paterno S, Lighthouse D, Bachman J, Planck J, Owen S, Skora AD, Nystul TG, Ohlstein B, Allen A, Wilhelm JE, Murphy TD, Levis RW, Matunis E, Srivali N, Hoskins RA, Spradling AC (2007) The carnegie protein trap library: a versatile tool for Drosophila developmental studies. Genetics 175(3):1505–1531. https://doi.org/10.1534/genetics.106.065961
31. Lye CM, Naylor HW, Sanson B (2014) Subcellular localizations of the CPTI collection of YFP-tagged proteins in Drosophila embryos. Development 141(20):4006–4017. https://doi.org/10.1242/dev.111310
32. Morin X, Daneman R, Zavortink M, Chia W (2001) A protein trap strategy to detect GFP-tagged proteins expressed from their endogenous loci in Drosophila. Proc Natl Acad Sci U S A 98(26):15050–15055. https://doi.org/10.1073/pnas.261408198
33. Nagarkar-Jaiswal S, Lee PT, Campbell ME, Chen K, Anguiano-Zarate S, Gutierrez MC, Busby T, Lin WW, He Y, Schulze KL, Booth BW, Evans-Holm M, Venken KJ, Levis RW, Spradling AC, Hoskins RA, Bellen HJ (2015) A library of MiMICs allows tagging of genes and reversible, spatial and temporal knockdown of proteins in Drosophila. elife 4. https://doi.org/10.7554/eLife.05338
34. Rørth P (1998) Gal4 in the Drosophila female germline. Mech Dev 78(1-2):113–118. https://doi.org/10.1016/s0925-4773(98)00157-9

Chapter 8

Tracking Follicle Cell Development

Adrianna Soriano, Christopher Petit, Savannah Ryan, and Jennifer C. Jemc

Abstract

Somatic follicle cells are critical support cells for *Drosophila* oogenesis, as they provide signals and molecules needed to produce a mature egg. Throughout this process, the follicle cells differentiate into multiple subpopulations and transition between three different cell cycle programs to support nurse cell and oocyte development. The follicle cells are mitotic in early egg chamber development, as they cover the germline cyst. In mid-oogenesis, follicle cells switch from mitosis to endocycling, increasing their ploidy from 2C to 16C. Finally, in late oogenesis, cells transition from endocycling to gene amplification, increasing the copy number of a small subset of genes, including the genes encoding proteins required for egg maturation. In order to explore the genetic regulation of these cell cycle switches and follicle cell development and specification, clonal analysis and the GAL4/UAS system are used frequently to reduce or increase expression of genes of interest. These genetic approaches combined with immunohistochemistry and *in situ* hybridization are powerful tools for characterizing the mechanisms regulating follicle cell development and the mitosis/endocycle and endocycle/gene amplification transitions. This chapter describes the genetic tools available to manipulate gene expression in follicle cells, as well as the methods and reagents that can be utilized to explore gene expression throughout follicle cell development.

Key words *Drosophila*, Ovary, Follicle cells, Oogenesis, *in situ* hybridization, Immunostaining

1 Introduction

The *Drosophila* ovary is made of ~15–20 units known as ovarioles, linear arrays of progressively developing egg chambers that result in egg production (Fig. 1a). Each ovariole has a germarium at its anterior end, which can be divided into four regions: 1, 2A, 2B, and 3 (Fig. 1b). Region 1 is located at the anterior end of the germarium and serves as the germline stem cell niche, which includes somatic terminal filament cells and cap cells that provide the signaling and adhesion proteins required for the maintenance of 2–3 germline stem cells [1]. Germline stem cells can asymmetrically divide and give rise to a cystoblast in region 1. In region 2A, the cystoblast undergoes four rounds of mitosis with incomplete

Michelle S. Giedt and Tina L. Tootle (eds.), *Drosophila Oogenesis: Methods and Protocols*,
Methods in Molecular Biology, vol. 2626, https://doi.org/10.1007/978-1-0716-2970-3_8,

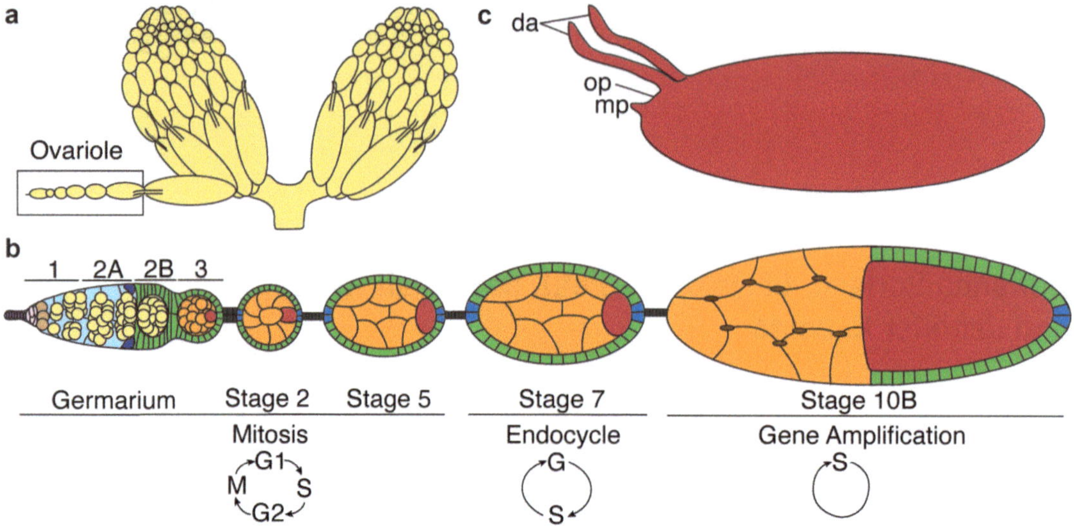

Fig. 1 Schematic of *Drosophila* egg chamber development. (**a**) Overview of the *Drosophila* ovary. (**b**) Stages of egg chamber development in a single ovariole. The germarium contains the germline stem cell niche and the follicle stem cell niche. A stage 1 egg chamber buds off the posterior of the germarium to become a stage 2 egg chamber. Early egg chamber development corresponds to stages 1–6 when follicle cells are mitotic. Mid-oogenesis includes stage 7–10A egg chamber development, when follicle cells are endocycling (undergoing G to S phase transitions without mitosis and cytokinesis). Late oogenesis includes stages 10B-14, during which a subset of genes in the follicle cells are replicated in S phase in a process known as gene amplification. Cells in the developing egg chambers are labeled as follows: terminal filament cells (purple), cap cells (light pink), germline stem cells (beige), cystoblasts and germline cysts (yellow), inner germarial sheath cells (light blue), follicle stem cells (dark blue), nurse cells (orange), oocyte (dark pink), follicle cells (green), stalk cells (dark green), polar cells (medium blue), and stretch follicle cells (brown). (**c**) Mature egg. Dorsal appendage (da), operculum (op), and micropyle (mp)

cytokinesis to give rise to a 16-cell germline cyst, in which cells are interconnected by ring canals [2]. Ring canals allow for the transfer of RNAs and proteins between the germ cells. Throughout these cell divisions, the germ cells are surrounded by a population of somatic cells known as inner germarial sheath (IGS) cells, or escort cells, which regulate germ cell differentiation [3, 4].

As the 16-cell cyst moves to the posterior of the germarium, IGS cells will be replaced by pre-follicle cells in region 2B [3]. Pre-follicle cells are derived from a population of somatic stem cells, known as follicle stem cells (FSCs) that reside at the region 2A/B boundary and divide asymmetrically (Fig. 1b, dark blue) [3, 5]. The fate of FSCs and pre-follicle cells is regulated by multiple signaling pathways, including the Wnt, Bone Morphogenetic Protein (BMP), Hedgehog (Hh), Hippo, Epidermal Growth Factor Receptor (EGFR), Notch, and Janus Kinase (JAK)/Signal Transducer and Activator of Transcription (STAT) signaling pathways [6]. By the time the germline cyst has reached region 3, each egg chamber is surrounded by a single layer of pre-follicle cells

(Fig. 1b). At this point, one germ cell is specified as the oocyte and localizes to the posterior end of the egg chamber. The other 15 germ cells are specified as nurse cells, which synthesize RNAs, proteins, and organelles for deposition in the oocyte by stage 11, after which the nurse cells undergo cell death [7]. This review focuses on methods to track the different subpopulations of follicle cells throughout the 14 stages of egg chamber development.

Pre-follicle cells have been observed to differentiate into 3 classes of follicle cells early in egg chamber development: main body follicle cells, polar cells, and stalk cells [5, 8, 9]. Main body follicle cells, or epithelial follicle cells, form a single layer of cells around the 16-cell germline cyst and proliferate through stage 6 of egg chamber development (Fig. 1, green). The main body follicle cell fate appears to be the default state, as signals specific to main body follicle cell differentiation have not yet been identified [6]. A common precursor population, dependent on Hedgehog (Hh) signaling, gives rise to polar and stalk cells [8]. A pair of polar cells are specified at the terminal ends of the egg chamber as early as region 3/stage 1 of egg chamber development--a process dependent on *fringe* expression and the presence of Notch/Delta signaling (Fig. 1b, medium blue) [8–10]. Following specification, polar cells express Unpaired (Upd), the ligand for the JAK/STAT signaling pathway, which induces the stalk cell fate in neighboring precursor cells (see Fig. 5) [11–13]. Stalk cells form a column of 6–8 cells that connect neighboring egg chambers and are hypothesized to induce polar cell fate in the anterior, younger egg chamber, thus establishing anterior-posterior polarity throughout egg chamber development (Fig. 1, dark green) [14].

Egg chambers proceed through 14 stages of development, culminating in the production of an egg [15]. Stages of egg chamber development can be distinguished based on nuclear staining, egg chamber and oocyte size, ratio of axes, and follicle cell distribution [16]. Throughout oogenesis, main body follicle cells will undergo multiple changes in their cell cycle, starting out as mitotic cells, transitioning to endocycling cells, and finally undergoing gene amplification, corresponding to early, middle, and late egg chamber development, respectively. During early egg chamber development (stages 1–6), follicle cells are mitotically active, resulting in ~650 follicle cells that cover each egg chamber (Fig. 1b). Immunostaining with the mitotic marker phospho-histone H3 suggests that stage 5 is the last stage at which mitotically active follicle cells are observed [16]. Delta signals from the germ cells to increase Notch signaling gradually in the follicle cells beginning in stage 5, inducing the mitosis/endocycle (M/E) switch at the transition between stages 6 and 7 [9, 17]. Following the switch, follicle cells undergo three rounds of endocycling, in which cells undergo G and S phases without mitosis and cytokinesis, increasing DNA content from 2C to 16C, where C equals the DNA mass in a haploid chromosome set (Fig. 1b) [18]. Broad (Br) and Hindsight (Hnt) expression are induced at this time, while expression of *cut* and the G2/M transition regulator *string* are downregulated, as is Hh signaling (see Fig. 4) [19–21].

Notch signaling remains active during endocycling and is downregulated by stage 10B for the shift from endocycle to gene amplification (E/A; Fig. 1b) [22]. Gene amplification increases copy number of a limited number of target genes needed for egg maturation, including the chorion genes [23, 24]. Recent work suggests the presence of a third follicle cell transition at stage 13/14, characterized by an increase in Hnt expression and a decrease in Cut expression, which is required for follicle cell rupture and ovulation [25].

As main body follicle cells progress through egg chamber development, they also take on new fates needed to promote egg maturation. Evidence suggests that anterior follicle cells will take on different fates based on levels of exposure to Upd from the anterior polar cells and other signaling factors [26, 27]. Follicle cells experiencing the highest levels of Upd become border cells in stage 8 and will express *slow border cells* (*slbo*) [13, 28]. JAK/STAT and ecdysone signaling promotes the migration of the border cell cluster along with the anterior polar cells toward the anterior dorsal border of the oocyte at stage 9; a process that will be complete in stage 10A [29]. Border cells will contribute to the interior of the micropyle structure, which allows for sperm entry [30]. Anterior follicle cells experiencing mid-levels of Upd, as well as Notch and BMP signaling, will become stretch follicle cells during mid-oogenesis, downregulating adhesion proteins to flatten and surround nurse cells to later promote nurse cell death [30, 31]. Anterior follicle cells exposed to the lowest levels of Upd, as well as BMP, EGFR, and Notch signaling, become centripetal follicle cells, which migrate interiorly into the egg chamber between the nurse cells and the oocyte to cover the anterior end of the oocyte [32]. They will eventually form part of the micropyle and the operculum, which is required for larval hatching (Fig. 1c) [30].

Follicle cells at the posterior end of the egg chamber also undergo further specification during egg chamber development. During early oogenesis, Gurken (Grk) from the oocyte induces EGFR signaling in the surrounding follicle cells, promoting a posterior follicle cell fate in combination with JAK/STAT signaling from posterior polar cells [33, 34]. Later in oogenesis, when the nucleus has shifted to the dorsal anterior position of the oocyte, Grk/EGFR signaling combined with BMP signaling induces a dorsal anterior fate and *mirr* expression in the overlying follicle cells [34]. These dorsal anterior cells will differentiate as roof and floor plate cells to give rise to the two dorsal appendages, which allow for respiration in the embryo (Fig. 1c) [35].

In order to better understand the genetic regulation of follicle cell development, it is necessary to examine and manipulate expression of genes of interest in this cell type. Single-cell transcriptome analysis has provided significant insight into the genes expressed in the different FSC-derived subpopulations [36, 37]. Clonal analysis and the GAL4/UAS system can be used to reduce or increase expression of genes of interest to explore gene function further [38–42]. Numerous GAL4 drivers have been demonstrated to be expressed in the FSCs and different follicle cell subpopulations in the adult ovary, as summarized in Table 1 (see examples in Fig. 2). In addition, we provide an effective procedure for the generation of follicle cell clones in part 3.1 (Fig. 3). These genetic approaches combined with immunohistochemistry (3.2) and *in situ* hybridization (3.3) are valuable tools for examining the genetic regulation of follicle cell development and the M/E and E/A switches (Figs. 4 and 5). This chapter summarizes the genetic tools available to manipulate gene expression in follicle cells, as well as the reagents that can be utilized to explore gene expression throughout follicle cell development.

2 Materials

2.1 Fly Husbandry and Clone Induction

1. Food vials and bottles with plugs.
2. Fly Food (*see* **Note 1**).
3. Fly Incubator.
4. Stereomicroscope with light source.
5. Porous plastic pad for fly pushing.
6. Paintbrush or feather for fly pushing.
7. Water bath or hybridization incubator set to 37 °C.
8. CO_2 set up with a regulator for anesthetizing flies.
9. GAL4 and UAS lines available for manipulation of gene expression in follicle cells (see Table 1 and Fig. 2).
10. FLP and FRT stocks for clonal induction.
11. Active dry yeast.

2.2 Immunostaining

1. Rocker.
2. Dumont #5 Forceps for tissue dissection.
3. Glass dissection dish (optional).
4. Slides and 22 μm Coverslips.
5. Pipettors (P2, P20, P200, P1000) and Tips.
6. 1.7 mL microcentrifuge tubes.
7. Glass pipettes with pipette bulb.

Table 1
Fly strains for follicle cell analysis

Reagent	Details	Source/Reference
109-28-Gal4	Mid/late AFs, most BCs	BDSC #7022 [45]
109-30-Gal4	FSCs; all FCs	BDSC #7023 [46]
109-53 Gal4	PFCs, PCs and SCs	BDSC #7025 [11]
109C1-Gal4	Most BCs	BDSC #7020 [45]
179Y-Gal4	FCs over oocyte	BDSC #3733 [47]
183Y-Gal4	SCs; FC over NCs	[47]
185Y-Gal4	SCs; FC over NCs	BDSC #3731 [47]
198Y-Gal4	BCs; FCs over oocyte	BDSC #3745 [47]
212Y-Gal4	Posterior pole	[47]
55b-Gal4	Some FSCs, Most FCs	BDSC #1803 [38]
645b-Gal4	FSCs, most FCs, BCs	BDSC #3748 [47]
7xEcRE-EGFP	Ecdysone sensor; FCs over NCs; BCs	V. Henrich [48]
93F-lacZ	SC enhancer trap	[49]
bab1-Gal4-2	Most FCs	[50]
boi^{NP4065}*-Gal4*	BCs; AFs	Kyoto #104588 [51]
brE-Gal4	FCs s5-9; PFCs s10A and later	[52]
brE-lacZ	FCs s5-9; PFCs s10A and later	[52]
brL-lacZ	Dorsal AFs	[52]
c2-Gal4	Some FCs over oocyte	[47]
c46-Gal4	Posterior pole	[47]
c68-Gal4	Some FCs over oocyte	[47]
c135-Gal4	Most FCs	BDSC #6978 [45]
c204-Gal4	Stage 8–14 FCs over oocyte	BDSC #3751 [47]
c208-Gal4	Most FCs	[47]
c273-Gal4	Most FCs	[47]
c306-Gal4	SC and BC	BDSC #3743 [47]
c319-Gal4	Posterior pole	[47]
c323-Gal4	Most FCs	BDSC #3732 [47]
c323a-Gal4	FSCs, Most FCs	[47]
c324a-Gal4	Some FCs over oocyte	[47]
c329b-Gal4	FC over NCs	[47]

(continued)

Table 1
(continued)

Reagent	Details	Source/Reference
c338a-Gal4	Posterior pole	[47]
c355-Gal4	FCs over oocyte	[47]
c368-Gal4	SCs	BDSC #3744 [47]
c372-Gal4	Some FCs over oocyte	[47]
c405-Gal4	Posterior pole	[47]
c409-Gal4	FCs over oocyte	[47]
c415-Gal4	FC over NCs	[47]
c418-Gal4	FCs over oocyte	[47]
c419-Gal4	FCs over oocyte	[47]
c454a-Gal4	FC over NCs	[47]
c458-Gal4	BCs, FCs at poles	BDSC #30837 [47]
c522-Gal4	BCs	BDSC #3747 [47]
c535a-Gal4	Posterior pole	[47]
c544a-Gal4	FCs over oocyte	[47]
c563b-Gal4	FCs over oocyte	[47]
c565-Gal4	FCs over oocyte	[47]
c566-Gal4	FCs over oocyte	[47]
c662a-Gal4	FCs over oocyte	[47]
c587-Gal4	FSCs/early FCs	BDSC #67747 [53]
c606-Gal4	BCs, FCs over oocyte, poles	[47]
c612a-Gal4	Posterior pole	[47]
c654-Gal4	Posterior pole	[47]
c680-Gal4	FCs over oocyte in mid/late	[47]
c685-Gal4	Some FCs over oocyte	[47]
c709-Gal4	FCs over oocyte	[47]
c734-Gal4	Posterior pole	[47]
c761-Gal4	Posterior pole	[47]
c768-Gal4	FCs over oocyte	BDSC #3742 [47]
c780-Gal4	FCs over most of oocyte	BDSC #32549 [47]
c788-Gal4	Late FCs	BDSC #32550 [47]
c804-Gal4	Posterior pole	[47]

(continued)

Table 1
(continued)

Reagent	Details	Source/Reference
c815-Gal4	BCs, some AFs, FCs over oocyte	[47]
c820-Gal4	Some FCs over oocyte	[47]
c825-Gal4	Late FCs over oocyte	BDSC #6987 [47]
c856-Gal4	BCs, AF pole	[47]
cb13-Gal4	Mid/late-AFs, PFCs	BDSC #6720 [54]
cv-2-lacZ	Stretch follicle cells	BDSC #6342 [36]
$dlp^{GMR53H05}$*-Gal4*	Some AFs, BCs	[55]
$dlp^{GMR52D02}$*-Gal4*	Some AFs, few BCs	[55]
$dlp^{GMR61C11}$*-Gal4*	Some AF, BC	[55]
EcRE-lacZ	Ecdysone sensor; FCs over NCs; BCs	BDSC #4516, 4517 [48]
E(spl)mβ-CD2	Notch activity reporter; early PCs; mid-FCs	BDSC #83353 [56]
E(spl)m7-lacZ	Notch activity reporter; early PCs; mid-FCs	[11]
e22c-Gal4	FSCs, BCs; most mid-FCs	BDSC #1973 [57]
fend-$Gal4^{NP112A}$	FSCs, BCs, AFs	[58]
fend-$Gal4^{NP5045}$	Some BCs, AFs	Kyoto #113556 [51]
fend-Gal^{NP4124}	FSCs, Most FCs, BCs, SCs	Kyoto #113397 [51]
fzr-lacZ	Mid-FCs	[59]
Gbe-Su(H)m8-lacZ	Notch activity reporter; early PCs; mid-FCs	BDSC #83352 [60]
GR1-Gal4	FSCs, most FCs, PCs	BDSC #362878 [61]
H15-lacZ	PFCs	[62]
$hh^{GMR28E04}$*-Gal4*	Some AFs, BCs	BDSC #45169 [55]
$hh^{GMR28E03}$*-Gal4*	FSCs, some AFs, PFCs, and BCs	BDSC #45546 [55]
$hh^{GMR28D09}$*-Gal4*	Some AFs	BDSC #45946 [55]
hs-Gal4-EcR	Ecdysone sensor; FCs over NCs; BCs	C. Thummel [63]
mirr-lacZ	AFs	[64]
neur-lacZ (A101)	PC enhancer trap	BDSC #4369 [65]
NRE-EGFP	Notch activity reporter; early PCs; mid-FCs	BDSC #30727, 30728 [66]
pnr^{MD237}*-Gal4*	Some AF, PF, and BC	BDSC #3039 [38]
$ptc^{559.1}$*-Gal4*	FSCs, most FCs	BDSC #2017 [67]
PZ80-lacZ	PCs	[68]
R10H05-Gal4	FSCs, Stage 1–14	BDSC #48276 [55]

(continued)

Table 1
(continued)

Reagent	Details	Source/Reference
rho-lacZ (2.2, 8.3)	FPCs	[69]
sgg^{NP0082}*-Gal4*	Some AFs, PFCs, and BCs	Kyoto #112003 [51]
sgg^{NP2253}*-Gal4*	Some AFs	Kyoto #104140 [51]
sgg^{NP4101}*-Gal4*	Some AFs, PFCs, and BCs	Kyoto #104603 [51]
sgg^{NP7167}*-Gal4*	BCs	Kyoto #114115 [51]
sgg^{NP7270}*-Gal4*	Some AFs, PFCs, and BCs	Kyoto #114191 [51]
sli^{GMR30H09}*-Gal4*	Some BCs	BDSC #54243 [55]
slbo-Gal4	BCs, PFCs	BDSC #58435 [70]
slbo^{01310}*-lacZ*	BCs, CFCs, PFCs	BDSC #12227 [28]
SNF4Aγ^{GMR31C04}	Mid-AFs	[55]
SNF4Aγ^{GMR30H03}	Mid-AFs	[55]
stg-lacZ (6.4)	Early FCs	[71]
tj-Gal4	Most FCs	Kyoto #104055 [51]
upd (E132)-Gal4	PCs	BDSC #26796 [72]
wg^{GMR16D01}*-Gal4*	AFs	BDSC #48722, 52503 [55]
wg^{GMR17D09}*-Gal4*	Some FSCs, AFs, most BCs	BDSC #48766 [55]

Abbreviations: *AFs* Anterior follicle cells, *BCs* border cells, *CFCs* centripetal follicle cells, *FPCs* floor plate cells, *FSCs* follicle stem cells, *FCs* follicle cells, *NCs* nurse cells, *PCs* polar cells, *PFCs* posterior follicle cells, and *SCs* stalk cells. In some cases, stage-specific expression is noted: early (stages 1–6), mid (7-10A), and late (stages 10B-14); more detailed information on stage-specific Gal4 expression has been described [45, 47]. Stocks number from the Bloomington *Drosophila* Stock Center (BDSC) and the Kyoto Stock Center (Kyoto) have been noted where applicable

8. 15 mL and 50 mL conical tubes.
9. Aluminum foil.
10. Blotting paper.
11. Mounting medium with DAPI (*see* **Note 2**).
12. DAPI (4', 6'-diamidino-2-phenylindole; 1 μg/mL): Dilute 2 μL of 5 mg/mL DAPI Stock solution (in deionized H_20 (d H_20)) in 10 mL 1× PBS for a final concentration of 1 μg/mL. Store at 4 °C.
13. Sylgard plates (optional): Sylgard plates are made using Sylgard 184. Tare the balance with a disposable tri-pour beaker. Weigh 10 parts of Part A liquid directly into a beaker. Add 1 part of Part B liquid by carefully pouring into the same beaker. Use a tongue depressor to mix. Weigh 1 g Norit A Charcoal Powder for every 150 g of Part A liquid and add to beaker. Mix thoroughly with a tongue depressor. Pour solution into desired

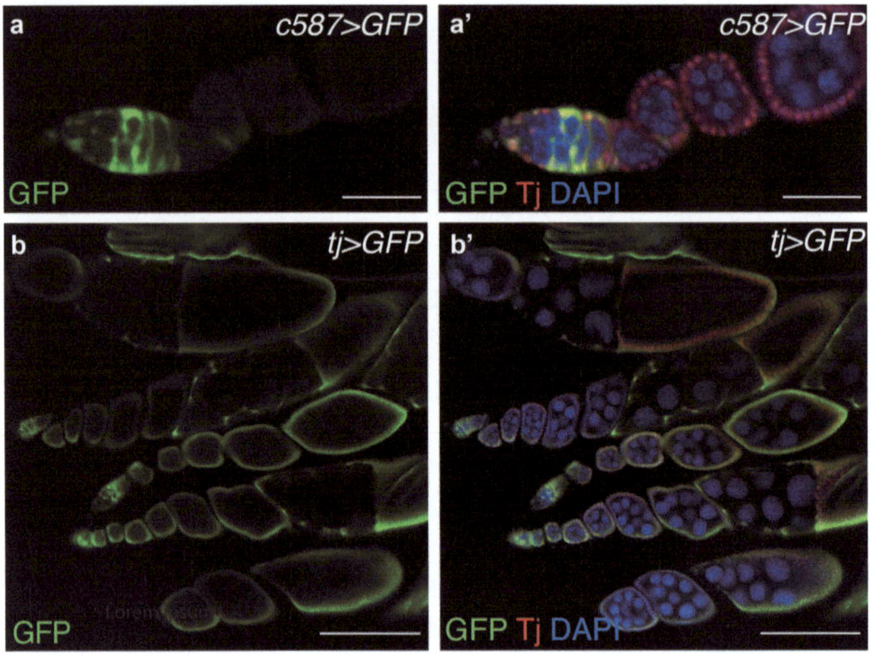

Fig. 2 Gal4 expression in follicle cells. (**a**–**a**") *c587-Gal4* > *GFP* promotes Green Fluorescent Protein (GFP) expression in somatic cells in the germarium and expression is decreased in stage 1 through stage 14 egg chambers. Scale bar is 25 μm. (**b**, **b**") *traffic jam (tj)-Gal4* > *GFP* promotes GFP expression in follicle cells throughout egg chamber development. GFP (green), Traffic jam (Tj; red), and nuclei (DAPI; blue). Scale bar is 100 μm

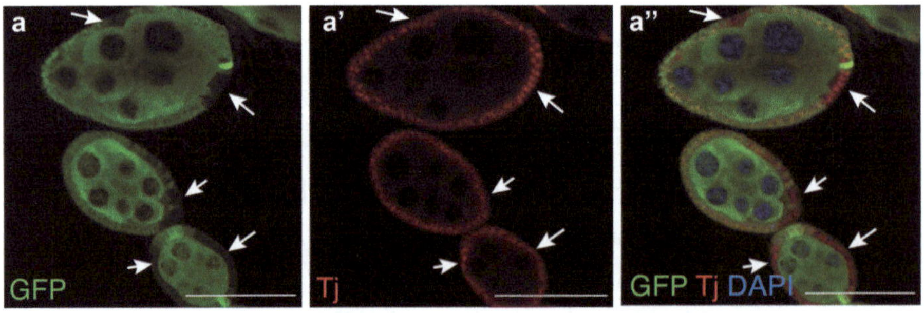

Fig. 3 Follicle cell clone generation. (**a**-**a**") Clones in stage 4–7 egg chambers of *hs-FLP*; *FRTG13/FRTG13 ubi-GFP* flies. Arrows indicate clones, which lack Green Fluorescent Protein (GFP) expression. GFP (green), Traffic jam (Tj; red), and nuclei (DAPI; blue). Scale bar is 50 μm

size of petri dishes. A Bunsen burner can be used to pop bubbles that have risen to the surface. Cover and allow to cure for at least 24 h. Allow waste to cure/harden in hood before disposal.

14. 10× Phosphate-Buffered Saline (PBS): Dissolve 160 g NaCl, 4 g KCl, 28.8 g Na_2HPO_4, and 4.8 g KH_2PO_4 in 1600 mL dH_2O. Adjust pH to 7.5 with HCl and volume to 2 L with dH_2O.

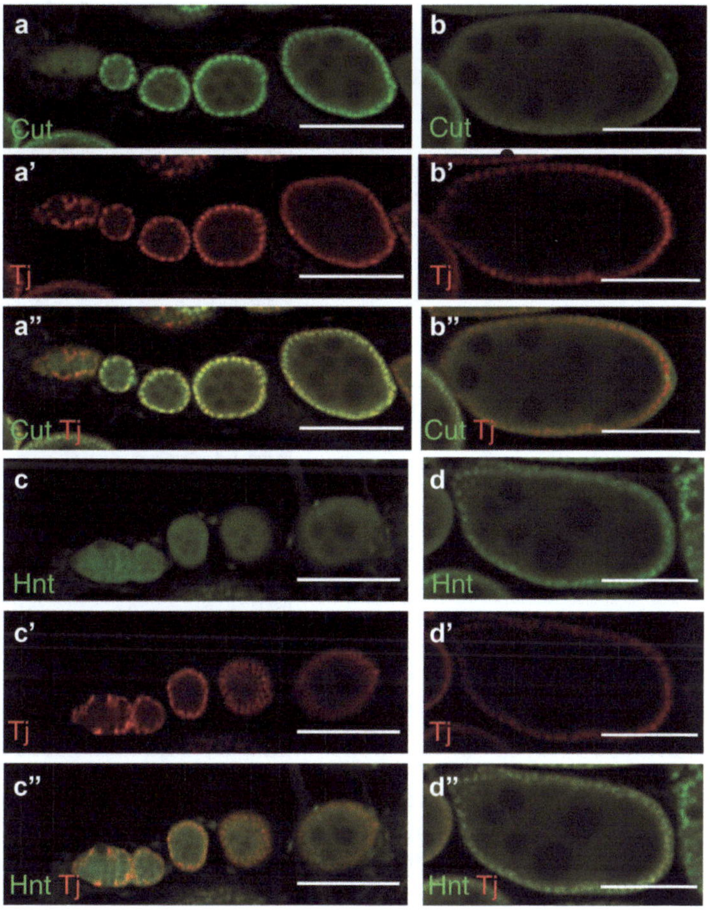

Fig. 4 Immunostaining with follicle cell markers. (**a**, **b**) Adult ovaries (w^{1118}) stained for Cut (green) and Traffic jam (Tj; red). (**a**, **a**") Germarium and stage 1–6 egg chambers. (**b**, **b**") Stage 7 egg chamber. (**c**, **d**) Adult ovaries (w^{1118}) stained for Hnt (green) and Traffic jam (Tj; red). (**c**, **c**") Germarium and stage 1–4 egg chambers. (**d**, **d**") Stage 7 egg chamber. Scale bars are 50 μm

15. 1× PBS: For 1 L of 1× PBS, dilute 100 mL 10× PBS in 900 mL dH_2O.
16. 10% Triton-X 100: Pour 5 mL Triton-X 100 into a 50 mL conical tube and add 45 mL dH_2O. Parafilm tube and place on a rocker until solution is homogeneous.
17. PBS with 0.1% Triton X-100 (PBTx): For 1 L of PBTx, combine 100 mL 10× PBS, 10 mL 10% Triton X-100, and 890 mL dH_2O into a beaker and mix.
18. 5.3% formaldehyde: Add 100 μL 37% formaldehyde to 600 μL PBTx in a microcentrifuge tube (*see* **Note 3**). *All steps requiring the addition or removal of formaldehyde should be performed in a hood.

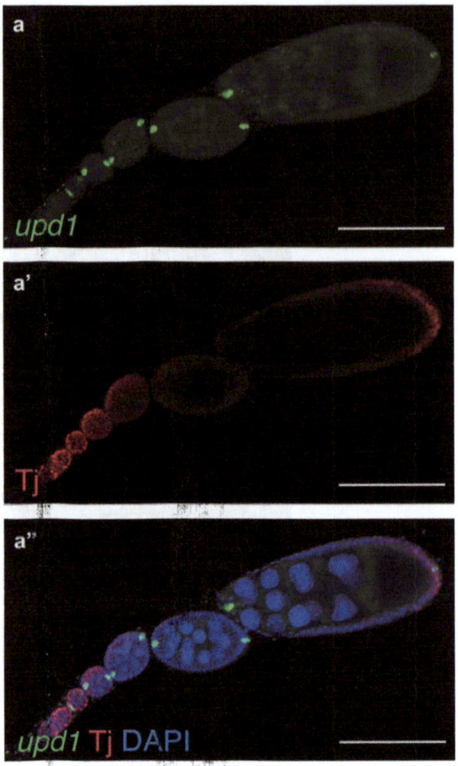

Fig. 5 Fluorescent *in situ* hybridization for *upd1*. (**a**–**a**") Adult ovaries (w^{1118}) with immunostaining and *in situ* hybridization for *unpaired1 (upd1)*. *upd1* (green), Traffic jam (Tj; red), and nuclei (DAPI; blue). Image is maximum intensity Z projection through a *Drosophila* ovariole. Scale bar is 100 μm

19. PBTx with 0.1% Bovine Serum Albumin (BBTx): For 1 L of BBTx, add 1 g Bovine Serum Albumin (BSA) to 1 L PBTx, add a stir bar, and mix on stir plate until dissolved. Filter sterilize with a 0.22 μm filter and store at 4 °C (*see* **Note 4**).
20. Blocking solution: For 500 μL, add 25 μL NGS to 475 μL BBTx.
21. Primary antibodies: Table 2.
22. Secondary antibodies (*see* **Note 5**).
23. Nail polish.
24. Laser-scanning confocal microscope for imaging (*see* **Note 6**).

2.3 Fluorescent In Situ Hybridization with Immunostaining

1. PCR machine.
2. PCR tubes.
3. RNase-free filter pipette tips.
4. Water bath or hybridization incubator.

Table 2
Antibodies for tracking follicle cell development

Antibody	Details	Concentration	Source/Reference
Mouse anti-Beta-galactosidase	lacZ-marked cells	1:500	Promega
Rabbit anti-Beta-galactosidase	lacZ-marked cells	1:2000	MP Biomedicals
Mouse anti-Broad BR-Z1 (3C11)	FCs after s5	1:30	DSHB
Mouse anti-Broad-core (25E9)	FCs after s5; restricted to roof cells late	1:30	DSHB
Mouse anti-Bromo-deoxyuridine (BrdU)	Oscillates in FCs in mitosis and endocycles; GA loci in s10B and double bars by s13	1:50	BD Biosciences
Rabbit anti-Castor	FSCs, early FCs (s1-3), PCs, and SCs	1:5000	W. Odenwald [73]
Rabbit anti-*Drosophila* cleaved caspase (Dcp-1)	Cell death marker	1:100	Cell Signaling Technology
Mouse anti-Cut (2B10)	FCs s1-6, 10B-13	1:15	DSHB
Rabbit anti-Cyclin A	Mitotic marker FCs s1-6	1:500	C. Lehner [74]
Mouse anti-Cyclin B (F2F4)	Mitotic marker FCs s1-6	1:50	DSHB
Guinea pig anti-Cyclin E	Oscillates in FCs mitosis and endocycles	1:1000	T. Orr-Weaver
Mouse anti-Cyclin E (8B10)	Oscillates in FCs in mitosis and endocycles	1:10	DSHB
Mouse monoclonal anti-Delta (Dl; C594.9B)	Enriched at germ cell surface stages 5–7	1:100	DSHB
Guinea pig anti-Double-parked (Dup)/Cdt1	GA loci in s10B and double bars by s13	1:1000	T. Orr-Weaver [75]
Guinea pig anti-dE2F1	Oscillates in FCs mitosis and endocycles; GA loci in s10B and double bars by s13	1:500	T. Orr-Weaver [76]
Rat anti-E-cadherin (DCAD2)	High in PCs; low in FCs	1:20	DSHB
Mouse anti-Eyes absent (Eya; 10H6)	FSCs, FCs	1:25	DSHB [77]

(continued)

Table 2
(continued)

Antibody	Details	Concentration	Source/Reference
Mouse anti-Fasciclin III (FasIII; 7G10)	Early FCs; restricted to PCs by s3	1:15	DSHB
Chick anti-Green Fluorescent Protein (GFP)	GFP-marked cells	1:1000	abcam
Rabbit anti-Green Fluorescent Protein (GFP)	GFP-marked cells	1:2000	Torrey Pines BioLabs, Inc.
Rabbit anti-H15	PFCs	1:2000	[78]
Mouse anti-Hindsight (Hnt; 1G9)	FCs in s7-10A, s14	1:15	DSHB
Guinea pig anti-Midline (Mid)	PFCs	1:1000	[33]
rabbit anti-Midline (Mid)	PFCs	1:1000	[79]
Mouse anti-Minichromosome maintenance (MCM)-2-7	GA loci in s10B and double bars by st13	1:200	S. Bell [80]
Mouse anti-phospho-Ser/Thr-Pro-MPM2	Oscillates in FCs mitosis and endocycles; foci in FCs at s10B	1:1000	Millipore
Mouse monoclonal anti-NICD (C17.9C6)	Highly enriched FC surface s3-5; less at surface thereafter	1:1000	DSHB
Rabbit anti-Origin recognition complex subunit 2 (ORC2)	GA loci in s10A-11	1:3000	S. Bell [81]
Rabbit anti-phospho-histone H3 (pH3)	Mitotic cells	1:100–200	Cell Signaling Technology
Rabbit anti-phospho-histone H3 (pH3)	Mitotic cells	1:200	Millipore
Rabbit anti-Proliferating cell nuclear antigen (PCNA)	GA loci in s10B and double bars by s13	1:1000	D. Henderson [82]

(continued)

Table 2 (continued)

Antibody	Details	Concentration	Source/Reference
rat anti-Slow border cells (Slbo)	BCs	1:500	P. Rørth [83]
guinea pig anti-Traffic jam (Tj)	FCs	1:1000	Generated by J. Jemc and M. Van Doren using same epitope as described in [84]
rabbit anti-Tramtrack69 (Ttk69)	FCs over oocyte s10B-13	1:200	P. Badenhorst [85]
rat anti-Tramtrack69 (Ttk69)	FCs over oocyte s10B-13	1:100	P. Badenhorst [86]

Abbreviations: *BCs* border cells, *DSHB* Developmental Studies Hybridoma Bank, *FCs* follicle cells, *GA* gene amplification, *PCs* polar cells, *PFCs* posterior follicle cells, *s* stage

5. DEPC H_2O: Add 1 mL Diethylpyrocarbonate (DEPC) per 1 L of double distilled H_2O. Mix overnight at room temperature in hood. Autoclave for 20 min to deactivate DEPC (*see* **Note 7**).
6. 10× PBS in DEPC H_2O (10× PBS-DEPC): Prepare 10× PBS as described above. Add 1 mL DEPC per 1 L of 10× PBS. Mix overnight at room temperature in hood. Autoclave for 20 min to deactivate DEPC.
7. 1× PBS in DEPC H_2O (1× PBS-DEPC): For 50 mL, dilute 5 mL 10× PBS-DEPC H_2O in 45 mL DEPC H_2O.
8. 10% Tween-20 in DEPC H_2O: Pour 2 mL Tween-20 into 15 mL conical tube and bring up to a final volume of 10 mL with DEPC H_2O. Rock until homogeneous and dilute in 50 mL conical tube with an additional 10 mL DEPC H_2O to make 10% Tween.
9. 1× PBS-DEPC with 0.1% Tween-20 (PBTw-DEPC): For 50 mL, use 5 mL 10× PBS-DEPC, 500 μL 10% Tween, and bring to 50 mL with DEPC H_2O.
10. tRNA stock solution: Dissolve 20 mg tRNA in 1 mL DEPC H_2O. Store as 500 μL aliquots at −20 °C.
11. 100 mM Dithioreitol (DTT): For 10 mL, measure 0.154 g DTT and add to 10 mL DEPC H_2O in a 15 mL conical tube and rock to mix. Aliquot 1 mL into each microcentrifuge tube and store at −20 °C. *Perform these steps in a fume hood.
12. Heparin stock solution: Dissolve 50 mg heparin in 1 mL DEPC H_2O. Store at 4 °C.
13. 1× PBS-Tween-Heparin (PBTH): Prepare only as much as you will need for one experiment (~15 mL/sample). For 15 mL,

add 15 μL heparin stock solution (final concentration is 50 μg/mL) and 187.5 μL tRNA stock solution (final concentration is 250 μg/mL) to a 15 mL conical tube and add PBTw-DEPC to 15 mL volume (*see* **Note 8**).

14. 4% formaldehyde: For 1 mL, dilute 110 μL 37% Formaldehyde in 890 μL PBTw-DEPC (*see* **Note 3**).
15. 10% formaldehyde: For 1 mL, dilute 270 μL 37% Formaldehyde, 730 μL PBTw-DEPC (*see* **Note 3**).
16. PBTH with RNase inhibitor and DTT (PBTHR): For 0.6 mL, add 0.3 μL RNase inhibitor (40 U/μL stock) and 6 μL 100 mM DTT to 594 μL PBTH. Mix well before use.
17. 20× Saline-Sodium Citrate (SSC) buffer: Dissolve 175.3 g of NaCl and 88.2 g of sodium citrate in 800 mL of DEPC H_2O. Adjust the pH to 7.0 with a few drops of a 14 N solution of HCl. Adjust the final volume to 1 L with DEPC H_2O and autoclave or filter sterilize with a 0.22 μm filter. The final concentrations are 3.0 M NaCl and 0.3 M sodium citrate.
18. 5× SSC with 0.1% Tween (5× SSCT): For 10 mL, combine 2.5 mL 20 × SSC, 100 μL 10% Tween-DEPC, and 7.4 mL DEPC H_2O in a 15 mL conical tube.
19. *In situ* probe(s). For example, the *upd1* probe set was generated by Molecular Instruments according to their guidelines (*see* **Note 9**).
20. This protocol is optimized for use of the Hybridization Chain Reaction (HCR)-RNA-FISH kit from Molecular Instruments [43]. Probe hybridization buffer, probe wash buffer, and amplification buffer were supplied with the kit and used as directed. *All steps requiring the addition or removal of probe hybridization buffer and probe wash buffer should be performed in a fume hood due to the presence of formamide.

3 Methods

Many genes that are required in the follicle cells throughout development and in the adult ovary are also required in other tissues, resulting in lethality in homozygous mutants. The GAL4/UAS system and clonal analysis are two approaches that can be used to circumvent the lethality associated with systemic mutation of genes of interest. The GAL4/UAS system utilizes the yeast GAL4 transcription factor to drive expression of a gene of interest downstream of an upstream activating sequence (UAS). This allows for the expression of cDNA to promote gene overexpression or expression of short-hairpin RNAs (shRNAs) to knockdown gene expression in specific cell types [38–40]. Numerous GAL4 drivers have been demonstrated to be expressed in the FSCs and different follicle cell subpopulations in the adult ovary, as summarized in Table 1. In addition, GAL4-driven expression of a reporter, like Green Fluorescent Protein (GFP), can be used to label specific cell types (see Fig. 2).

Clonal analysis with the FLP-FRT system has also proven an effective way to circumvent the lethality associated with mutants and to more deeply explore gene function in specific cell types [41, 42]. We provide a protocol for generating GFP-marked clones using a heat shock-inducible procedure. For this approach, one must generate flies carrying (1) flippase recombinase (FLP) under the control of a heat shock-inducible promoter (*hs*-FLP), (2) an FRT (FLP recombinase target) chromosome carrying a wild-type copy of the gene of interest and a ubiquitously expressed reporter, like *ubiquitin*-GFP, and (3) a second FRT chromosome carrying a wild-type copy of the gene of interest for the control condition or a mutant copy of the gene of interest for the experimental condition. Upon heat shock, the FLP will be expressed and will promote recombination at FRT sites. Following mitosis, cells mutant for the gene of interest will lack GFP expression, while homozygous wild-type and heterozygous cells will express GFP (Fig. 3).

3.1 Generation of GFP-Marked Follicle Cell Clones Using Heat Shock

1. Collect newly eclosed female flies carrying the desired FRT chromosomes and *hs*-FLP into a food vial with a sprinkle of active yeast and age them 1–3 d at 25 °C without males (*see* **Note 10**).
2. Set incubator to 37 °C. Verify internal temperature of the incubator with a thermometer (*see* **Note 11**).
3. To activate *flp* expression, heat shock vials in incubator at 37 °C for 30 min.
5. Allow flies to recover for 30 min at 25 °C.
4. Return flies to 37 °C for 30 min.
5. Allow flies to recover for 30 min at 25 °C.
6. Return flies to 37 °C for 30 min.
7. Incubate at 25 °C for desired time frame (1–10 d) before proceeding to analysis via immunostaining (Subheading 3.2) or *in situ* hybridization (Subheading 3.3) (*see* **Note 12**).

3.2 Immunostaining of Adult Ovaries

Previous studies have identified numerous proteins that are expressed in the different follicle cell types and stages of follicle cell development. Immunostaining for these proteins is a valuable tool for dissecting the genetic regulation of follicle cell development. In demonstrating the role of different genes in cell cycle transitions from mitosis to endocycling to gene amplification, proteins that show an OFF/ON/OFF pattern like Hindsight, or an ON/OFF/ON pattern like Cut are particularly useful in tracking follicle cell development and the ability of cells to transition from one cell cycle stage to the next (Fig. 4). Here we provide a robust protocol for immunostaining of *Drosophila* ovaries.

*All steps requiring the addition or removal of formaldehyde should be performed in a hood and solutions disposed of properly.

1. Age newly eclosed females 1–3 d in a food vial with a sprinkle of active yeast at 25 °C without males.
2. Dissect ovaries in cold 1× PBS on a Sylgard plate, glass dissection dish, or slide. Use one pair of forceps to hold the female firmly at the top of the abdomen, while using a second set of forceps to pull the posterior to expose ovaries. Splay ovarioles apart slightly with forceps to allow antibodies to penetrate ovarioles during immunostaining (*see* **Note 13**).
3. Transfer ovaries to a 1.7 mL microcentrifuge tube containing ~1.25 mL of 1× PBS on ice as they are dissected. Dissect all samples within 30–45 min and proceed to fixation.
4. Remove 1× PBS. Fix ovaries in 1 mL of 5.3% formaldehyde for 10 min at room temperature on a rocker.
5. Remove formaldehyde and replace with 1 mL PBTx. Invert to rinse and remove PBTx. Repeat once (*see* **Note 14**).
6. Remove rinse and add 1 mL PBTx. Incubate on rocker for 10 min. Repeat once.
7. Remove rinse and block samples in 500 μL blocking solution for a minimum of 45 min – 1 h on a rocker (*see* **Note 15**).
8. Remove blocking solution and incubate with 300 μL primary antibody mix (BBTx with 5% NGS and desired antibodies) overnight at 4 °C on a rocker. See Table 2 for primary antibodies.
9. Remove primary antibodies and replace with 1 mL PBTx. Invert to rinse and remove PBTx. Repeat once.
10. Remove rinse and add 1 mL PBTx. Incubate on rocker for 10 min. Repeat once.
11. Remove rinse and incubate with 300 μL secondary antibodies mix (BBTx with 5% NGS and secondary antibodies) overnight at 4 °C or 2 h at room temperature. Protect samples from light by wrapping tubes in foil from this point forward to prevent bleaching of fluorescently labeled secondary antibodies.
12. Remove secondary antibodies. *If staining with DAPI, add 300 μL DAPI stock solution and rock for 10 min. Remove DAPI stain.
13. Add 1 mL PBTx. Invert to rinse and remove PBTx. Repeat once.
14. Remove rinse and add 1 mL PBTx. Incubate on rocker for 10 min. Repeat once.

15. Remove wash, add 500 μL 1× PBS, and proceed to mounting.
16. Transfer samples to a slide using a glass Pasteur pipette in as little PBS as possible.
17. Wick away PBS with blotting paper, but do not allow samples to become completely dry.
18. Add desired mounting medium to cover samples immediately.
19. Use forceps to splay apart ovarioles in each ovary and align them in rows for ease of imaging.
20. Place a coverslip on top, seal the edges of the coverslip with nail polish, and proceed to confocal imaging (*see* **Note 6**). Slides can be stored at 4 °C protected from light before microscopy.

3.3 Fluorescent In Situ *Hybridization with Immunostaining*

While immunostaining allows for the tracking of protein localization, *in situ* hybridization allows for the tracking of mRNA and can allow for the identification of the cells expressing the gene of interest. Immunostaining combined with fluorescent *in situ* hybridization allows for more precise localization of gene expression with confocal microscopy than with previously described colorimetric methods (Fig. 5). This approach is also valuable for examining gene expression for which antibodies are not available. This protocol was adapted from [43] and [44] for use specifically with the Molecular Instruments HCR Hybridization kit. RNase-free filter tips should be used for all steps of this protocol.

*All steps requiring the addition or removal of formaldehyde, hybridization buffer (contains formamide), and probe wash buffer (contains formamide) should be performed in a fume hood and solutions disposed of properly.

1. Dissect ovaries in cold 1× PBS-DEPC as described above for immunostaining (*see* **Note 13**).
2. Transfer to a RNase-free 1.7 mL microcentrifuge tube containing ~1.25 mL cold 1× PBS-DEPC on ice as they are dissected. Dissect all samples within 30–45 min and proceed to fixation.
3. Remove PBS and add 1 mL 4% formaldehyde. Incubate 20 min at room temperature on rocker.
4. Remove formaldehdye and add 1 mL PBTH. Incubate on rocker for 5 min. Remove PBTH and repeat wash two more times (*see* **Note 14**).
5. Remove wash and add 300 μL primary antibody solution in PBTHR overnight at 4 °C on rocker. See Table 2 for primary antibodies.
6. Remove antibodies and add 1 mL PBTH. Incubate on rocker for 5 min. Remove PBTH and repeat wash two more times.
7. Remove wash and add 300 μL secondary antibody solution in PBTHR for 2 h at room temperature. Protect samples from

light by wrapping tubes in foil from this point forward to prevent bleaching of fluorescently labeled secondary antibodies. Turn on water bath/hybridization incubator to warm to 37 °C.

8. Remove antibodies and add 1 mL PBTH. Incubate on rocker for 5 min. Remove PBTH and repeat wash two more times.
9. Remove wash solution and add 1 mL 10% formaldehyde to post-fix for 20 min on rocker. While fixing, pre-warm 1 mL hybridization buffer/sample to 37 °C for 30 min before use.
10. Remove formaldehyde and add 1 mL PBTH. Incubate on rocker for 5 min. Remove PBTH and repeat wash two more times.
11. Remove PBTH from samples and add 500 μL of pre-warmed probe hybridization buffer for 30 min at 37 °C.
12. To make the probe solution, add 2 pmol of each probe set (2 μL of 1 μM stock) to 500 μL of pre-warmed probe hybridization buffer.
13. Remove pre-hybridization buffer from samples and add the probe solution to samples (*see* **Note 16**).
14. Incubate samples overnight at 37 °C (*see* **Note 17**).
15. Pre-warm 2 mL probe wash buffer/sample to 37 °C. Allow 1 mL amplification buffer/sample to come to room temperature.
16. Remove probes from samples and add 500 μL of pre-warmed probe wash buffer. Incubate at 37 °C for 15 min. Remove probe wash buffer and repeat wash three more times (*see* **Note 16**).
17. Remove probe wash buffer and add 500 μL 5× SSCT. Incubate for 5 min at room temperature on rocker. Remove 5× SSCT and repeat wash two more times.
18. Remove 5× SSCT and add 500 μL of amplification buffer for 30 min at room temperature on rocker.
19. Aliquot 30 pmol of hairpin h1 (10 μL of 3 μM stock) to one PCR tube and 30 pmol of hairpin h2 (10 μL of 3 μM stock) to a second PCR tube. Heat hairpins at 95 °C for 90 s and cool to room temperature in dark for 30 min (*see* **Note 18**).
20. Add h1 hairpin and h2 hairpin to 500 μL room temperature amplification buffer to make hairpin solution.
21. Remove amplification buffer from samples and add hairpin solution to samples.
22. Incubate overnight at room temperature on rocker (*see* **Note 19**).

23. Remove hairpin solution. Add 500 μL 5× SSCT at room temperature. Incubate on rocker for 5 min. Remove 5× SSCT and repeat wash.
24. Remove 5× SSCT and add 500 μL 5× SSCT at room temperature. Incubate on rocker for 30 min. Remove 5× SSCT and repeat wash.
25. Remove 5× SSCT and add 500 μL 5× SSCT at room temperature. Incubate on rocker for 5 min (*see* **Note 20**).
26. Remove 5× SSCT and add 500 μL DEPC-PBS and proceed to mounting.
27. Mount as described above for immunostaining and proceed to confocal imaging. Slides can be stored at 4 °C protected from light before microscopy.

4 Notes

1. Bloomington *Drosophila* stock center standard cornmeal food recipe was used for raising flies in these studies.
2. DAPI-Fluoromount-G® from Southern BioTech was used for mounting in these studies.
3. Fixative should be diluted fresh at time of use to avoid degradation.
4. BSA-containing solutions should be filter-sterilized rather than autoclaved to avoid protein denaturation.
5. Alexa Fluor 488, 555, and 633 conjugated secondary antibodies were used at a dilution of 1:500 in this study, but dilutions may vary based on secondary antibodies used.
6. All imaging was performed with a Zeiss LSM880 laser-scanning confocal microscope.
7. An alternative to preparing DEPC-treated water is purchasing RNase-free water.
8. PBTH should be prepared as needed, as the recommended storage temperatures for tRNA and heparin are −20 °C and 4 °C, respectively. This maintains activity of reagents.
9. The *in situ* probe set used in these studies was prepared by Molecular Instruments according to their guidelines. The *upd* RNA target sequence was provided to Molecular Instruments and probes were generated to recognize multiple sites in the target RNA [43].
10. The Bloomington *Drosophila* Stock Center has many stocks available for clone induction using the FLP-FRT system. This study used females flies of the genotype hs-FLP; *FRT G13/FRT G13 ubi-GFPnls* generated from Bloomington stocks # 1929, 5752, and 5826.

11. For follicle cell clone induction, a Thermo Scientific Heratherm Incubator was used. However, a water bath can also be an effective means for clone induction. If using a water bath incubator, it is critical to use a weight to keep vials submerged in the water bath to the same depth as the plug in the vial. This will ensure all flies are being effectively heat-shocked. Times for heat shock may require adjustments based on the success of clone induction.
12. For clonal analysis, it is often necessary to examine clones at multiple time points following heat shock to examine the effects of gene mutation at different stages in follicle cell development. The best time points will need to be empirically determined for each experiment and may differ based on the gene being studied and the function of that gene in the follicle cells.
13. When dissecting ovaries, occasionally the ovaries are retained in the abdomen after pulling at the posterior. Pushing the forceps holding the anterior end of the abdomen gently into the Sylgard as you pull at the posterior can help eliminate this issue. If ovaries continue to be retained in the abdomen, gently pushing on the anterior end of the abdomen with forceps can help to push them out.
14. For immunostaining, it is helpful to allow the samples to settle for ~1 min after removing from the rocker before aspiration to prevent sample loss.
15. The incubation time for blocking may need to be optimized for the antibodies used.
16. When processing *in situ* hybridization samples in hybridization buffer, the ovaries have a tendency to remain suspended in the hybridization buffer and float. When removing the buffer, use a small-bore pipet tip to avoid sucking up samples and start from the bottom of the microcentrifuge tube. Remove small amounts of buffer until you have removed as much as possible. The final volume of sample in hybridization buffer should be about 100 μL.
17. For hybridization, 12–16 h worked well for the probe used here. It may be necessary to increase or decrease hybridization based on the probe used and the transcript being studied.
18. The *in situ* method described here utilizes fluorescently labeled hairpins, generated by Molecular Instruments that will recognize the *in situ* probes and amplify the signal from hybridization by forming polymers [43]. For heating of hairpins for amplification, a PCR machine was used to warm hairpin solutions at 95 °C for 90 s. Do not add the PCR tube until the machine has reached 95 °C.

19. Overnight amplification incubations were performed for 12–18 h. It may be necessary to increase or decrease amplification based on the probe used and the transcript being studied.
20. Additional washes with 5× SSCT may be necessary if background for the *in situ* probe sets is high.

Acknowledgments

This work was supported by NSF MRI grant # 1828164 to J.J. and Loyola University Chicago to A.S., C. P., S. R., and J.J.

References

1. Morrison SJ, Spradling AC (2008) Stem cells and niches: mechanisms that promote stem cell maintenance throughout life. Cell 132(4): 598–611. https://doi.org/10.1016/j.cell.2008.01.038
2. Koch EA, King RC (1966) The origin and early differentiation of the egg chamber of Drosophila melanogaster. J Morphol 119(3):283–303. https://doi.org/10.1002/jmor.1051190303
3. Decotto E, Spradling AC (2005) The Drosophila ovarian and testis stem cell niches: similar somatic stem cells and signals. Dev Cell 9(4):501–510. https://doi.org/10.1016/j.devcel.2005.08.012
4. Kirilly D, Wang S, Xie T (2011) Self-maintained escort cells form a germline stem cell differentiation niche. Development 138(23):5087–5097. https://doi.org/10.1242/dev.067850. Epub 2011 Oct 26
5. Margolis J, Spradling A (1995) Identification and behavior of epithelial stem cells in the Drosophila ovary. Development 121(11): 3797–3807
6. Rust K, Nystul T (2020) Signal transduction in the early Drosophila follicle stem cell lineage. Curr Opin Insect Sci 37:39–48. https://doi.org/10.1016/j.cois.2019.11.005
7. Foley K, Cooley L (1998) Apoptosis in late stage Drosophila nurse cells does not require genes within the H99 deficiency. Development 125(6):1075–1082
8. Tworoger M, Larkin MK, Bryant Z, Ruohola-Baker H (1999) Mosaic analysis in the drosophila ovary reveals a common hedgehog-inducible precursor stage for stalk and polar cells. Genetics 151(2):739–748. https://doi.org/10.1093/genetics/151.2.739
9. Lopez-Schier H, St Johnston D (2001) Delta signaling from the germ line controls the proliferation and differentiation of the somatic follicle cells during Drosophila oogenesis. Genes Dev 15(11):1393–1405. https://doi.org/10.1101/gad.200901
10. Grammont M, Irvine KD (2001) fringe and Notch specify polar cell fate during Drosophila oogenesis. Development 128(12):2243–2253
11. Assa-Kunik E, Torres IL, Schejter ED, Johnston DS, Shilo BZ (2007) Drosophila follicle cells are patterned by multiple levels of Notch signaling and antagonism between the Notch and JAK/STAT pathways. Development 134(6):1161–1169. https://doi.org/10.1242/dev.02800
12. McGregor JR, Xi R, Harrison DA (2002) JAK signaling is somatically required for follicle cell differentiation in Drosophila. Development 129(3):705–717
13. Silver DL, Montell DJ (2001) Paracrine signaling through the JAK/STAT pathway activates invasive behavior of ovarian epithelial cells in Drosophila. Cell 107(7):831–841. https://doi.org/10.1016/s0092-8674(01)00607-9
14. Torres IL, López-Schier H, St Johnston D (2003) A Notch/Delta-dependent relay mechanism establishes anterior-posterior polarity in Drosophila. Dev Cell 5(4):547–558. https://doi.org/10.1016/s1534-5807(03)00272-7
15. Spradling AC (1993) Drosophila genetics of oogenesis. In: Arias MB, A M (eds) The development of Drosophila melanogaster, vol 1. Cold Spring Harbor Laboratory Press, Plainview, pp 1–70
16. Jia D, Xu Q, Xie Q, Mio W, Deng WM (2016) Automatic stage identification of Drosophila egg chamber based on DAPI images. Sci Rep 6:18850. https://doi.org/10.1038/srep18850
17. Deng WM, Althauser C, Ruohola-Baker H (2001) Notch-Delta signaling induces a transition from mitotic cell cycle to endocycle in

Drosophila follicle cells. Development 128(23):4737–4746

18. Lilly MA, Spradling AC (1996) The Drosophila endocycle is controlled by Cyclin E and lacks a checkpoint ensuring S-phase completion. Genes Dev 10(19):2514–2526. https://doi.org/10.1101/gad.10.19.2514
19. Jia D, Tamori Y, Pyrowolakis G, Deng WM (2014) Regulation of broad by the Notch pathway affects timing of follicle cell development. Dev Biol 392(1):52–61. https://doi.org/10.1016/j.ydbio.2014.04.024
20. Sun J, Deng WM (2005) Notch-dependent downregulation of the homeodomain gene cut is required for the mitotic cycle/endocycle switch and cell differentiation in Drosophila follicle cells. Development 132(19): 4299–4308. https://doi.org/10.1242/dev.02015
21. Sun J, Deng WM (2007) Hindsight mediates the role of notch in suppressing hedgehog signaling and cell proliferation. Dev Cell 12(3): 431–442. https://doi.org/10.1016/j.devcel.2007.02.003
22. Sun J, Smith L, Armento A, Deng WM (2008) Regulation of the endocycle/gene amplification switch by Notch and ecdysone signaling. J Cell Biol 182(5):885–896. https://doi.org/10.1083/jcb.200802084
23. Calvi BR, Lilly MA, Spradling AC (1998) Cell cycle control of chorion gene amplification. Genes Dev 12(5):734–744. https://doi.org/10.1101/gad.12.5.734
24. Claycomb JM, Benasutti M, Bosco G, Fenger DD, Orr-Weaver TL (2004) Gene amplification as a developmental strategy: isolation of two developmental amplicons in Drosophila. Dev Cell 6(1):145–155. https://doi.org/10.1016/s1534-5807(03)00398-8
25. Knapp EM, Li W, Sun J (2019) Downregulation of homeodomain protein Cut is essential for. Development 146(18). https://doi.org/10.1242/dev.179002
26. Xi R, McGregor JR, Harrison DA (2003) A gradient of JAK pathway activity patterns the anterior-posterior axis of the follicular epithelium. Dev Cell 4(2):167–177. https://doi.org/10.1016/s1534-5807(02)00412-4
27. Denef N, Schüpbach T (2003) Patterning: JAK-STAT signalling in the Drosophila follicular epithelium. Curr Biol 13(10):R388–R390. https://doi.org/10.1016/s0960-9822(03)00317-8
28. Montell DJ, Rorth P, Spradling AC (1992) Slow border cells, a locus required for a developmentally regulated cell migration during oogenesis, encodes Drosophila C/EBP. Cell 71(1):51–62. https://doi.org/10.1016/0092-8674(92)90265-e
29. Montell DJ, Yoon WH, Starz-Gaiano M (2012) Group choreography: mechanisms orchestrating the collective movement of border cells. Nat Rev Mol Cell Biol 13(10): 631–645. https://doi.org/10.1038/nrm3433
30. Duhart JC, Parsons TT, Raftery LA (2017) The repertoire of epithelial morphogenesis on display: progressive elaboration of Drosophila egg structure. Mech Dev 148:18–39. https://doi.org/10.1016/j.mod.2017.04.002
31. Timmons AK, Mondragon AA, Schenkel CE, Yalonetskaya A, Taylor JD, Moynihan KE, Etchegaray JI, Meehan TL, McCall K (2016) Phagocytosis genes nonautonomously promote developmental cell death in the Drosophila ovary. Proc Natl Acad Sci U S A 113(9): E1246–E1255. https://doi.org/10.1073/pnas.1522830113. Epub 2016 Feb 16
32. Wu X, Tanwar PS, Raftery LA (2008) Drosophila follicle cells: morphogenesis in an eggshell. Semin Cell Dev Biol 19(3):271–282. https://doi.org/10.1016/j.semcdb.2008.01.004
33. Fregoso Lomas M, Hails F, Lachance JF, Nilson LA (2013) Response to the dorsal anterior gradient of EGFR signaling in Drosophila oogenesis is prepatterned by earlier posterior EGFR activation. Cell Rep 4(4):791–802. https://doi.org/10.1016/j.celrep.2013.07.038
34. Fregoso Lomas M, De Vito S, Boisclair Lachance JF, Houde J, Nilson LA (2016) Determination of EGFR signaling output by opposing gradients of BMP and JAK/STAT activity. Curr Biol 26(19):2572–2582. https://doi.org/10.1016/j.cub.2016.07.073
35. Osterfield M, Berg CA, Shvartsman SY (2017) Epithelial patterning, morphogenesis, and evolution: drosophila eggshell as a model. Dev Cell 41(4):337–348. https://doi.org/10.1016/j.devcel.2017.02.018
36. Rust K, Byrnes LE, Yu KS, Park JS, Sneddon JB, Tward AD, Nystul TG (2020) A single-cell atlas and lineage analysis of the adult Drosophila ovary. Nat Commun 11(1):5628. https://doi.org/10.1038/s41467-020-19361-0
37. Jevitt A, Chatterjee D, Xie G, Wang XF, Otwell T, Huang YC, Deng WM (2020) A single-cell atlas of adult Drosophila ovary identifies transcriptional programs and somatic cell lineage regulating oogenesis. PLoS Biol 18(4): e3000538. https://doi.org/10.1371/journal.pbio.3000538

38. Brand AH, Perrimon N (1993) Targeted gene expression as a means of altering cell fates and generating dominant phenotypes. Development 118(2):401–415
39. Dietzl G, Chen D, Schnorrer F, Su KC, Barinova Y, Fellner M, Gasser B, Kinsey K, Oppel S, Scheiblauer S, Couto A, Marra V, Keleman K, Dickson BJ (2007) A genome-wide transgenic RNAi library for conditional gene inactivation in Drosophila. Nature 448(7150):151–156
40. Ni JQ, Zhou R, Czech B, Liu LP, Holderbaum L, Yang-Zhou D, Shim HS, Tao R, Handler D, Karpowicz P, Binari R, Booker M, Brennecke J, Perkins LA, Hannon GJ, Perrimon N (2011) A genome-scale shRNA resource for transgenic RNAi in Drosophila. Nat Methods 8(5):405–407. https://doi.org/10.1038/nmeth.1592. Epub 2011 Apr 3
41. Golic KG, Lindquist S (1989) The FLP recombinase of yeast catalyzes site-specific recombination in the Drosophila genome. Cell 59(3): 499–509
42. Lee T, Luo L (1999) Mosaic analysis with a repressible cell marker for studies of gene function in neuronal morphogenesis. Neuron 22(3):451–461
43. Choi HMT, Schwarzkopf M, Fornace ME, Acharya A, Artavanis G, Stegmaier J, Cunha A, Pierce NA (2018) Third-generation *in situ* hybridization chain reaction: multiplexed, quantitative, sensitive, versatile, robust. Development 145(12). https://doi.org/10.1242/dev.165753
44. Toledano H, D'Alterio C, Loza-Coll M, Jones DL (2012) Dual fluorescence detection of protein and RNA in Drosophila tissues. Nat Protoc 7(10):1808–1817. https://doi.org/10.1038/nprot.2012.105
45. Hartman TR, Ventresca EM, Hopkins A, Zinshteyn D, Singh T, O'Brien JA, Neubert BC, Hartman MG, Schofield HK, Stavrides KP, Talbot DE, Riggs DJ, Pritchard C, O'Reilly AM (2015) Novel tools for genetic manipulation of follicle stem cells in the Drosophila ovary reveal an integrin-dependent transition from quiescence to proliferation. Genetics 199(4):935–957. https://doi.org/10.1534/genetics.114.173617. Epub 2015 Feb 12.
46. Hartman TR, Zinshteyn D, Schofield HK, Nicolas E, Okada A, O'Reilly AM (2010) Drosophila Boi limits Hedgehog levels to suppress follicle stem cell proliferation. J Cell Biol 191(5):943–952. https://doi.org/10.1083/jcb.201007142. Epub 2010 Nov 22.
47. Manseau L, Baradaran A, Brower D, Budhu A, Elefant F, Phan H, Philp AV, Yang M, Glover D, Kaiser K, Palter K, Selleck S (1997) GAL4 enhancer traps expressed in the embryo, larval brain, imaginal discs, and ovary of Drosophila. Dev Dyn 209(3):310–322. https://doi.org/10.1002/(sici)1097-0177(199707)209:3<310::aid-aja6>3.0.co;2-l
48. Koelle MR, Talbot WS, Segraves WA, Bender MT, Cherbas P, Hogness DS (1991) The Drosophila EcR gene encodes an ecdysone receptor, a new member of the steroid receptor superfamily. Cell 67(1):59–77. https://doi.org/10.1016/0092-8674(91)90572-g
49. Ruohola H, Bremer KA, Baker D, Swedlow JR, Jan LY, Jan YN (1991) Role of neurogenic genes in establishment of follicle cell fate and oocyte polarity during oogenesis in Drosophila. Cell 66(3):433–449. https://doi.org/10.1016/0092-8674(81)90008-8
50. Bolívar J, Pearson J, López-Onieva L, González-Reyes A (2006) Genetic dissection of a stem cell niche: the case of the Drosophila ovary. Dev Dyn 235(11):2969–2979. https://doi.org/10.1002/dvdy.20967
51. Hayashi S, Ito K, Sado Y, Taniguchi M, Akimoto A, Takeuchi H, Aigaki T, Matsuzaki F, Nakagoshi H, Tanimura T, Ueda R, Uemura T, Yoshihara M, Goto S (2002) GETDB, a database compiling expression patterns and molecular locations of a collection of Gal4 enhancer traps. Genesis 34(1-2):58–61
52. Fuchs A, Cheung LS, Charbonnier E, Shvartsman SY, Pyrowolakis G (2012) Transcriptional interpretation of the EGF receptor signaling gradient. Proc Natl Acad Sci U S A 109(5): 1572–1577. https://doi.org/10.1073/pnas.1115190109
53. Song X, Wong MD, Kawase E, Xi R, Ding BC, McCarthy JJ, Xie T (2004) Bmp signals from niche cells directly repress transcription of a differentiation-promoting gene, bag of marbles, in germline stem cells in the Drosophila ovary. Development 131(6):1353–1364
54. Ward EJ, Thaipisuttikul I, Terayama M, French RL, Jackson SM, Cosand KA, Tobler KJ, Dorman JB, Berg CA (2002) GAL4 enhancer trap patterns during Drosophila development. Genesis 34(1-2):46–50. https://doi.org/10.1002/gene.10138
55. Jenett A, Rubin GM, Ngo TT, Shepherd D, Murphy C, Dionne H, Pfeiffer BD, Cavallaro A, Hall D, Jeter J, Iyer N, Fetter D, Hausenfluck JH, Peng H, Trautman ET, Svirskas RR, Myers EW, Iwinski ZR, Aso Y, DePasquale GM, Enos A, Hulamm P, Lam SC, Li

HH, Laverty TR, Long F, Qu L, Murphy SD, Rokicki K, Safford T, Shaw K, Simpson JH, Sowell A, Tae S, Yu Y, Zugates CT (2012) A GAL4-driver line resource for Drosophila neurobiology. Cell Rep 2(4):991–1001. https://doi.org/10.1016/j.celrep.2012.09.011

56. de Celis JF, Tyler DM, de Celis J, Bray SJ (1998) Notch signalling mediates segmentation of the Drosophila leg. Development 125(23):4617–4626
57. Duffy JB, Harrison DA, Perrimon N (1998) Identifying loci required for follicular patterning using directed mosaics. Development 125(12):2263–2271
58. Larkin MK, Holder K, Yost C, Giniger E, Ruohola-Baker H (1996) Expression of constitutively active Notch arrests follicle cells at a precursor stage during Drosophila oogenesis and disrupts the anterior-posterior axis of the oocyte. Development 122(11):3639–3650
59. Schaeffer V, Althauser C, Shcherbata HR, Deng WM, Ruohola-Baker H (2004) Notch-dependent Fizzy-related/Hec1/Cdh1 expression is required for the mitotic-to-endocycle transition in Drosophila follicle cells. Curr Biol 14(7):630–636. https://doi.org/10.1016/j.cub.2004.03.040
60. Furriols M, Bray S (2001) A model Notch response element detects Suppressor of Hairless-dependent molecular switch. Curr Biol 11(1):60–64. https://doi.org/10.1016/s0960-9822(00)00044-0
61. Gupta T, Schüpbach T (2003) Cct1, a phosphatidylcholine biosynthesis enzyme, is required for Drosophila oogenesis and ovarian morphogenesis. Development 130(24):6075–6087. https://doi.org/10.1242/dev.00817
62. Brook WJ, Ostafichuk LM, Piorecky J, Wilkinson MD, Hodgetts DJ, Russell MA (1993) Gene expression during imaginal disc regeneration detected using enhancer-sensitive P-elements. Development 117(4):1287–1297
63. Kozlova T, Thummel CS (2002) Spatial patterns of ecdysteroid receptor activation during the onset of Drosophila metamorphosis. Development 129(7):1739–1750
64. Ruohola-Baker H, Grell E, Chou TB, Baker D, Jan LY, Jan YN (1993) Spatially localized rhomboid is required for establishment of the dorsal-ventral axis in Drosophila oogenesis. Cell 73(5):953–965. https://doi.org/10.1016/0092-8674(93)90273-s
65. Bier E, Vaessin H, Shepherd S, Lee K, McCall K, Barbel S, Ackerman L, Carretto R, Uemura T, Grell E (1989) Searching for pattern and mutation in the Drosophila genome with a P-lacZ vector. Genes Dev 3(9):1273–1287. https://doi.org/10.1101/gad.3.9.1273
66. Saj A, Arziman Z, Stempfle D, van Belle W, Sauder U, Horn T, Dürrenberger M, Paro R, Boutros M, Merdes G (2010) A combined ex vivo and in vivo RNAi screen for notch regulators in Drosophila reveals an extensive notch interaction network. Dev Cell 18(5):862–876. https://doi.org/10.1016/j.devcel.2010.03.013
67. König A, Yatsenko AS, Weiss M, Shcherbata HR (2011) Ecdysteroids affect Drosophila ovarian stem cell niche formation and early germline differentiation. EMBO J 30(8):1549–1562. https://doi.org/10.1038/emboj.2011.73
68. Karpen GH, Spradling AC (1992) Analysis of subtelomeric heterochromatin in the Drosophila minichromosome Dp1187 by single P element insertional mutagenesis. Genetics 132(3):737–753. https://doi.org/10.1093/genetics/132.3.737
69. Ip YT, Park RE, Kosman D, Bier E, Levine M (1992) The dorsal gradient morphogen regulates stripes of rhomboid expression in the presumptive neuroectoderm of the Drosophila embryo. Genes Dev 6(9):1728–1739. https://doi.org/10.1101/gad.6.9.1728
70. Rorth P (1998) Gal4 in the Drosophila female germline. Mech Dev 78(1-2):113–118
71. Lehman DA, Patterson B, Johnston LA, Balzer T, Britton JS, Saint R, Edgar BA (1999) Cis-regulatory elements of the mitotic regulator, string/Cdc25. Development 126(9):1793–1803
72. Halder G, Callaerts P, Gehring WJ (1995) Induction of ectopic eyes by targeted expression of the eyeless gene in Drosophila. Science 267(5205):1788–1792. https://doi.org/10.1126/science.7892602
73. Kambadur R, Koizumi K, Stivers C, Nagle J, Poole SJ, Odenwald WF (1998) Regulation of POU genes by castor and hunchback establishes layered compartments in the Drosophila CNS. Genes Dev 12(2):246–260. https://doi.org/10.1101/gad.12.2.246
74. Lehner CF, O'Farrell PH (1989) Expression and function of Drosophila cyclin A during embryonic cell cycle progression. Cell 56(6):957–968. https://doi.org/10.1016/0092-8674(89)90629-6
75. Whittaker AJ, Royzman I, Orr-Weaver TL (2000) Drosophila double parked: a conserved, essential replication protein that colocalizes with the origin recognition complex and links DNA replication with mitosis and the

down-regulation of S phase transcripts. Genes Dev 14(14):1765–1776

76. Bosco G, Du W, Orr-Weaver TL (2001) DNA replication control through interaction of E2F-RB and the origin recognition complex. Nat Cell Biol 3(3):289–295. https://doi.org/10.1038/35060086
77. Bai J, Montell D (2002) Eyes absent, a key repressor of polar cell fate during Drosophila oogenesis. Development 129(23):5377–5388. https://doi.org/10.1242/dev.00115
78. Leal SM, Qian L, Lacin H, Bodmer R, Skeath JB (2009) Neuromancer1 and Neuromancer2 regulate cell fate specification in the developing embryonic CNS of Drosophila melanogaster. Dev Biol 325(1):138–150. https://doi.org/10.1016/j.ydbio.2008.10.006
79. Liu QX, Hiramoto M, Ueda H, Gojobori T, Hiromi Y, Hirose S (2009) Midline governs axon pathfinding by coordinating expression of two major guidance systems. Genes Dev 23(10):1165–1170. https://doi.org/10.1101/gad.1774209
80. Claycomb JM, MacAlpine DM, Evans JG, Bell SP, Orr-Weaver TL (2002) Visualization of replication initiation and elongation in Drosophila. J Cell Biol 159(2):225–236. https://doi.org/10.1083/jcb.200207046
81. Royzman I, Austin RJ, Bosco G, Bell SP, Orr-Weaver TL (1999) ORC localization in Drosophila follicle cells and the effects of mutations in dE2F and dDP. Genes Dev 13(7):827–840. https://doi.org/10.1101/gad.13.7.827
82. Henderson DS, Wiegand UK, Norman DG, Glover DM (2000) Mutual correction of faulty PCNA subunits in temperature-sensitive lethal mus209 mutants of Drosophila melanogaster. Genetics 154(4):1721–1733. https://doi.org/10.1093/genetics/154.4.1721
83. Borghese L, Fletcher G, Mathieu J, Atzberger A, Eades WC, Cagan RL, Rorth P (2006) Systematic analysis of the transcriptional switch inducing migration of border cells. Dev Cell 10(4):497–508
84. Li MA, Alls JD, Avancini RM, Koo K, Godt D (2003) The large Maf factor Traffic Jam controls gonad morphogenesis in Drosophila. Nat Cell Biol 5(11):994–1000
85. Badenhorst P (2001) Tramtrack controls glial number and identity in the Drosophila embryonic CNS. Development 128(20):4093–4101
86. Lehembre F, Badenhorst P, Müller S, Travers A, Schweisguth F, Dejean A (2000) Covalent modification of the transcriptional repressor tramtrack by the ubiquitin-related protein Smt3 in Drosophila flies. Mol Cell Biol 20(3):1072–1082. https://doi.org/10.1128/mcb.20.3.1072-1082.2000

Chapter 9

Optimized Fixation and Phalloidin Staining of Basally Localized F-Actin Networks in Collectively Migrating Follicle Cells

Mitchell T. Anderson, Kristin Sherrard, and Sally Horne-Badovinac

Abstract

The follicular epithelial cells of the *Drosophila* egg chamber have become a premier model to study how cells globally orient their actin-based machinery for collective migration. The basal surface of each follicle cell has lamellipodial and filopodial protrusions that extend from its leading edge and an array of stress fibers that mediate its adhesion to the extracellular matrix; these migratory structures are all globally aligned in the direction of tissue movement. To understand how this global alignment is achieved, one must be able to reliably visualize the underlying F-actin; however, dynamic F-actin networks can be difficult to preserve in fixed tissues. Here, we describe an optimized protocol for the fixation and phalloidin staining of the follicular epithelium. We also provide a brief primer on relevant aspects of the image acquisition process to ensure high quality data are collected.

Key words *Drosophila*, Egg chamber, Follicle, Collective cell migration, Morphogenesis, Fixed imaging, Staining, Phalloidin, Protrusions, Stress fibers, Actin

1 Introduction

The collective migration of epithelial cells is a critical process underlying tissue remodeling during morphogenesis, wound healing, and cancer metastasis [1, 2]. For epithelial cells to move in a coherent direction, the actin-based machinery that powers each cell's motility must be aligned across the tissue. The polarization of these actin structures is particularly striking in the follicular epithelium of *Drosophila*. At the basal surface of each cell are leading edge protrusions, composed of lamellipodia and filopodia, that push the plasma membrane forward, as well as a parallel array of stress fibers that organize the cell's integrin-based adhesions to the extracellular matrix (Fig. 1a–c) [3–9]. To study the alignment of these actin structures and how they power collective migration, one must be able to reliably visualize them. This is not a trivial task,

Michelle S. Giedt and Tina L. Tootle (eds.), *Drosophila Oogenesis: Methods and Protocols*,
Methods in Molecular Biology, vol. 2626, https://doi.org/10.1007/978-1-0716-2970-3_9,

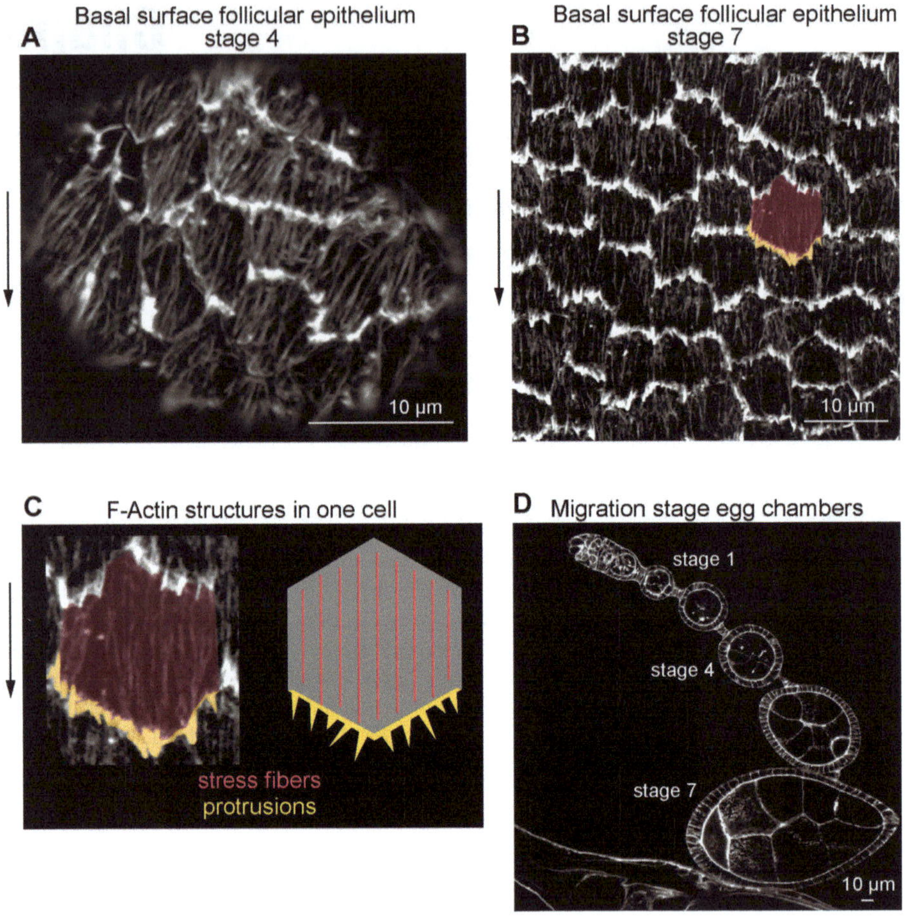

Fig. 1 Phalloidin-stained F-actin structures at the basal surface of the follicular epithelium. (**a**) Image of the basal surface of the follicular epithelium at stage 4. (**b**) Image of the basal surface of the follicular epithelium at stage 7. (**c**) Image of one follicle cell from (**b**), highlighting leading-edge protrusions (gold) and stress fibers fibers (red). (**d**) Image of a transverse section through a developmental array of egg chambers at stages when the follicular epithelium is migrating

however, as it has been noted for several decades that preserving these dynamic F-actin networks by chemical fixation can be difficult [6, 9]. Live F-actin labels can also be problematic as they often disrupt the native organization of these structures [7, 10]. Here, we present a protocol for optimal fixation and phalloidin staining of basal F-actin networks in the migratory follicular epithelium and provide tips on how to best image them.

The follicular epithelium is part of a larger multicellular structure called the egg chamber, which is the precursor to the egg [11]. Each egg chamber is composed of an inner germ cell cyst, consisting of 15 nurse cells and 1 oocyte, which is surrounded by the somatic follicular epithelium, also called "follicle cells". The follicle cells are arranged with their apical surfaces contacting the germ cells and their basal surfaces facing outwards. This entire structure is encapsulated by a basement membrane extracellular matrix. Egg chambers progress through 14 developmental stages, starting as a spherical shape and ending as a highly elongated shape (Fig. 1d). From stage 1 or 2 through stage 8, the follicle cells collectively migrate on the basement membrane, and this motion is required to lengthen the egg chamber [3, 12, 13]. Because the basal surfaces of the follicle cells face outward, the F-actin structures powering their migration are located close to the coverslip in slide-mounted samples, facilitating high-resolution imaging.

Although this chapter focuses on the preservation and phalloidin staining of F-actin structures in migration-stage egg chambers, the protocol is also compatible with other experimental goals. First, with minor adjustments to the tissue dissection technique, the protocol is also useful for visualizing follicle cell stress fibers in post-migratory-stage egg chambers. Second, co-visualization of basal F-actin networks and fluorescently tagged proteins is readily achievable, and we provide corresponding guidance on the uses of different phalloidin conjugates, which will vary based on the fluorescent tags present. However, to obtain the highest quality images of the basal F-actin networks, we do not recommend co-staining with antibodies, as this weakens the phalloidin signal. If you do combine this protocol with immunofluorescence, performing antibody staining followed by phalloidin staining yields the best results.

We have previously published a chapter in Methods in Molecular Biology on live imaging of follicle cell migration [14]. The protocol for dissection is shared between these chapters and the chapter by Zajac et al., on flow chambers in the current edition, and sections are duplicated here for convenience. We refer the reader to the live imaging chapter for detailed images and movies of the egg chamber dissection process, which are not included here [14].

2 Materials

2.1 Preparing Female Flies for Dissection

1. Vials with fly food.
2. Yeast powder, dry active yeast ground to a fine powder in coffee grinder.

2.2 Egg Chamber Dissection

1. Pen/Strep: penicillin G-sodium 10,000 U/mL, streptomycin sulfate 10,000 μg/mL in 0.85% saline.
2. Acidified water: 1 μL concentrated (12.1 N) HCl in 1 mL deionized water.
3. Insulin: 1 mg dissolved in 100 μL acidified water (*see* **Note 1**).
4. Live imaging medium [15]: Schneider's S2 medium, 0.6× Pen/Strep, 15% vol/vol fetal bovine serum, 0.1 mg/mL insulin (*see* **Note 2**).

5. Pyrex 9-Cavity Spot Plate.
6. Dumont forceps: #5, 0.1 × 0.06 mm tip, and #55, 0.05 × 0.02 mm tip (*see* **Note 3**).
7. Eyelash tool: insert an eyelash into a slightly melted p1000 pipettor tip.
8. Disposable needle, 27G × ½ in.
9. Glass Pasteur pipets, 5 ¾ in.
10. 5 mL pipet pump.
11. Stereomicroscope with magnification of at least 10×.
12. CO_2 pad to anesthetize flies.

2.3 Fixation, Staining, and Mounting

1. Phosphate buffered saline (PBS): 137 mM NaCl, 2.7 mM KCl, 10 mM Na_2HPO_4, 1.8 mM KH_2PO_4.
2. PBS + Triton X-100 (PBT): PBS, 0.1% vol/vol Triton X-100.
3. 16% EM grade formaldehyde (*see* **Note 4**).
4. 1.5 mL microfuge tubes.
5. p10 pipettor and tips.
6. Glass Pasteur pipets, 5 ¾ in.
7. 5 mL pipet pump.
8. Alexa Fluor™ 488-, Alexa Fluor™ 555-, or Alexa Fluor™ 647-phalloidin (*see* Fig. 2 and caption for a comparison of their relative staining quality).
9. Nutating mixer.
10. Liquid mounting medium (*see* **Note 5**).
11. Glass microscope slide, 3″ × 1″ × 1 mm.
12. Coverslip, 22 mm × 50 mm #1.5.
13. Nail polish.

2.4 Imaging

1. Confocal microscope with high-resolution objectives.

3 Methods

3.1 Preparing Female Flies for Dissection

1. Sprinkle yeast powder on fly food in a vial, covering about one half of the surface. Add up to six 1–2 day old females and at least 1–2 young males to the vial. Yeast is essential for proper egg chamber production. Incomplete nutrition will slow egg chamber production by inducing cell death in the germarium and in stage 8 egg chambers [16–18].
2. Age females for 1–3 days. Move flies to a new vial with fresh yeast every 2 days before dissection (*see* **Note 6**).

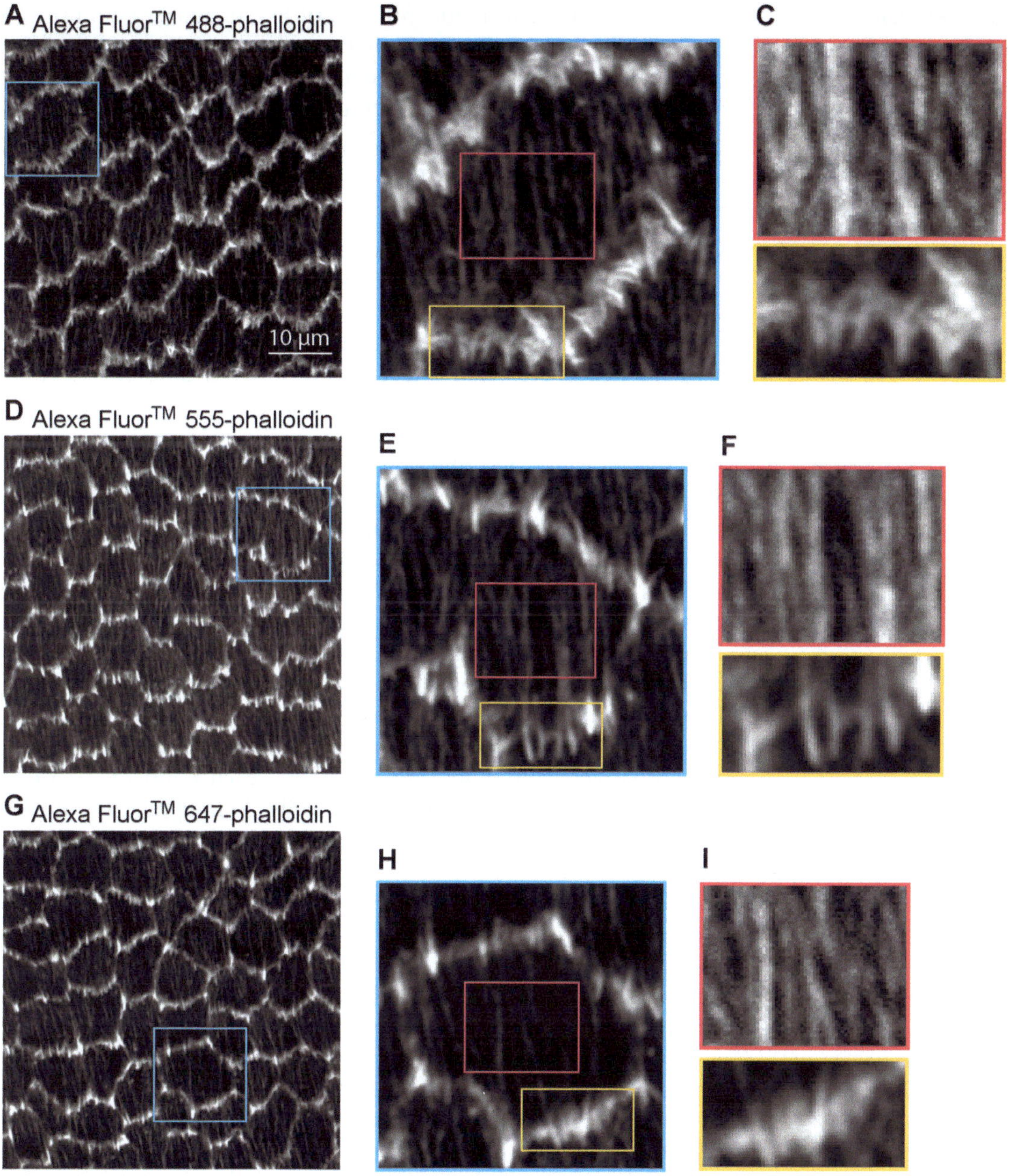

Fig. 2 Comparison of different fluorophore conjugates of phalloidin. Images of (**a**–**c**) Alexa Fluor™ 488-phalloidin, (**d**–**f**) Alexa Fluor™ 555-phalloidin, and (**g**–**i**) Alexa Fluor™ 647-phalloidin staining at the basal surface of the follicular epithelium of a stage 8 egg chamber. (**a**) Image of Alexa Fluor™ 488-phalloidin staining of a field of follicle cells. (**b**) Zoom-in of an individual cell in the blue boxed region of **a**. (**c**) Gold and red rectangles show a zoom-in of leading edge protrusions and stress fibers, respectively, in the boxed regions of b. (**d**) Image of Alexa Fluor™ 555-phalloidin staining of a field of follicle cells. (**e**) Zoom-in of an individual cell in the blue boxed region of **d**. (**f**) Gold and red rectangles show a zoom-in of leading edge protrusions and stress fibers, respectively, in the boxed regions of **e**. (**g**) Image of Alexa Fluor™ 647-phalloidin staining of a field of follicle cells. (**h**) Zoom-in of an individual cell in the blue boxed region of **g**. (**i**) Gold and red rectangles show a zoom-in of leading-edge protrusions and stress fibers, respectively, in the boxed regions of **h**. Brightness is increased specific to each image in red boxes to better show stress fibers (**c**, **f**, **i**). All images

3.2 Reagent Preparation

1. Prior to dissection, add 500 μL of live imaging medium to a well of a Pyrex 9-Cavity Spot Plate. Use a separate well of live imaging medium for each genotype you are dissecting. Allow medium to come to room temperature prior to dissection.
2. Prepare 300 μL of fixative in a microfuge tube by diluting EM grade formaldehyde in PBT to 4% vol/vol. Allow to come to room temperature.
3. Prepare 25–150 μL of staining solution in a microfuge tube by diluting phalloidin in PBT to the following specified concentrations: Alexa Fluor™ 488-phalloidin at 1:200, Alexa Fluor™ 555-phalloidin at 1:300, and Alexa Fluor™ 647-phalloidin at 1:50 (*see* **Note** 7). Protect the tube from ambient light by wrapping in foil or placing in a drawer.

3.3 Egg Chamber Dissection

In this section, we describe how to isolate egg chambers from the ovary, which requires a basic understanding of the ovary's structure. We also provide some guidance on potential reasons for isolating egg chambers of a given stage. We refer the reader to our previous chapter in Methods in Molecular Biology on live imaging of follicle cell migration for detailed images and movies of the egg chamber dissection process [14].

Each ovary is composed of 15–18 developmental arrays of egg chambers called ovarioles. At the anterior of each ovariole is the germarium, the site of egg chamber production. Moving posteriorly are successively older egg chambers, each connected to one another like beads on a string by stalk cells. At the posterior of each ovariole are the oldest egg chambers or mature eggs waiting to be laid. Each ovariole, as well as the entire ovary, is surrounded by a muscle sheath that will need to be removed during dissection. Egg chambers can be staged by eye based on both their overall size and shape, as well as the relative size and shape of the oocyte. Older egg chambers are also distinguished by their partly opaque appearance, due to the onset of vitellogenesis at stage 8.

Based on your experimental goals, you may choose egg chambers of various stages to dissect. Although stage 7 and 8 egg chambers have the most prominent and easily identifiable protrusions (Fig. 1b), protrusions are present in younger egg chambers as well if you wish to focus on earlier stages of follicle cell migration (Fig. 1a). Dissection technique varies based on which stage egg chambers you wish to collect, and dissection will damage egg chambers of other stages, so do not attempt to collect egg chambers of stages for which your dissection is not optimized.

Fig. 2 (continued) were taken with a 63 × 1.4 NA Plan Apochromat oil lens, a digital zoom of 1×, and the pinhole set to 1 Airy Unit. Detailed images in colored boxes are shown scaled by pixel number to illustrate how fluorophore wavelength affects the maximum resolution that can be achieved. Resolution of images in left-most column: (**a**) 13.4493 pixels/μm. (**d**) 12.1642 pix/μm. (**g**) 11.4424 pix/μm

Finally, we caution that the entire dissection process described below should not last longer than 10 minutes, as the preservation of dynamic F-actin structures is improved if egg chambers are quickly moved into fixative following their removal from the female.

1. Position a spot plate well containing live imaging medium underneath microscope. Adjust magnification to 10×–16× and focus to the bottom of the well.
2. Dissect ovaries from female flies. First, anesthetize the flies with CO_2. Using the #5 forceps in your nondominant hand, pinch a female fly at the anterior-most part of her abdomen (*see* **Note 8**). While still pinching, transfer the fly to the live imaging medium and completely submerge. Using the #55 forceps in your dominant hand, pinch the abdomen between the two posterior-most segments and pull posteriorly. The ovaries should release from the abdomen (if ovaries do not release, *see* **Note 9**). Remove all non-ovary tissue from the well. To obtain stage 1–5 egg chambers, *see* **step 3**. To obtain stage 5–8 egg chambers, *see* **step 4**.
3. To obtain stage 1–5 egg chambers, pinch the posterior of an ovary with the #5 forceps and pin it to the bottom of the well. Use the #55 forceps to pinch a single ovariole just anterior to where the ovary is pinned. Pull the ovariole orthogonally and then anteriorly to remove it from the muscle sheath and isolate it from the ovary. Ovarioles that have been removed from their muscle sheaths will look and move like beads on a string, rather than like a single packaged unit. Repeat this process until ovarioles do not easily separate from the ovary, as repeated pulling risks damaging egg chambers. Repeat for the second ovary.
4. To obtain stage 5–8 egg chambers, pinch the posterior of an ovary with the #5 forceps and pin it to the bottom of the well. Use the #55 forceps to pinch single ovarioles at the anterior tip of the ovary and pull anteriorly to remove ovarioles from their muscle sheaths and isolate them from the ovary. Ovarioles that have been removed from their muscle sheaths will look and move like beads on a string, rather than like a single packaged unit. Repeat this process until ovarioles do not easily separate from the ovary, as repeated pulling risks damaging egg chambers. Repeat for the second ovary.
5. Repeat **steps 2–5** as necessary to achieve a desired number of egg chambers for a given stage, typically 10–20 ovarioles. Fixing too many ovarioles in a single microfuge tube could limit penetration of the phalloidin stain. Be aware during dissection of any damaged egg chambers or other tissues and

isolate them in another part of the dish away from healthy egg chambers. Damaged tissues in the vicinity of undamaged egg chambers can lower staining quality.

6. Separate healthy ovarioles from unwanted material using an eyelash tool. Do not take ovarioles that retained their muscle sheaths, as this actin-rich tissue will prevent visualization of the F-actin structures at the basal epithelial surface. If you are attempting to gather pre-stage 5 egg chambers, trim ovarioles containing egg chambers older than stage 5 by placing the needle on the stalk after the stage 5 egg chamber and pressing down using a slicing motion. Older egg chambers will prevent the compression of younger egg chambers under the coverslip and thereby reduce the area of the basal surface available for imaging. If you are imaging older egg chambers, trimming is usually unnecessary.
7. Gather the trimmed ovarioles in the center of the well with the eyelash tool.

3.4 Fixation, Staining, and Mounting

1. Using a p10 pipettor, transfer egg chambers from the well to fixative in as little live imaging medium as possible to avoid diluting the fixative (*see* **Note 10**).
2. Fix egg chambers for 15 minutes at room temperature without rocking. Avoid fixing for any longer, as over-fixation can reduce penetration of the phalloidin stain.
3. Using a 5 mL pipet pump fitted with a glass Pasteur pipet, remove as much fixative as possible from the microfuge tube without removing the egg chambers. Wash egg chambers 3× with 500 μL PBT for 5 minutes each at room temperature without rocking, removing and replacing each wash (*see* **Note 11**).
4. After the final wash, remove PBT with the glass Pasteur pipet and pipet pump (*see* **Note 12**). Add the prepared phalloidin staining solution and allow the egg chambers to stain for the following times based on your conjugated fluorophore: Alexa Fluor™ 488-phalloidin or Alexa Fluor™ 555-phalloidin, 15 minutes at room temperature without rocking; Alexa Fluor™ 647-phalloidin, 2 hours at room temperature, rocking on nutator, covered in foil to prevent bleaching from exposure to ambient light.
5. Using the glass Pasteur pipet and pipet pump, remove the phalloidin solution and replace with 500 μL PBT. Wash 3× with 500 μL PBT for 5 minutes each at room temperature without rocking, removing and replacing each wash.
6. After removing the final wash, add 35 μL of mounting medium to the microfuge tube (*see* **Note 13**).

7. Using the glass Pasteur pipet and pipet pump, remove mounting medium and egg chambers from microfuge tube and deposit on the glass slide, moving horizontally to create a single line of medium. Ensure the line of medium is shorter than the cover slip. Apply the coverslip to the slide (*see* **Note 14**).
8. Apply a small amount of nail polish to each corner of the coverslip and allow to dry for a few minutes to stabilize the coverslip position. Apply a small amount of nail polish to each edge of the coverslip without gaps and allow to dry. Store the slide horizontally at 4 °C until ready for imaging.

3.5 Imaging

In this section, we provide brief tips on image acquisition of basal actin structures using confocal microscopy. This is not intended to be a comprehensive protocol, but to describe aspects of the process that are important for capturing high quality images. The main issue faced in imaging middle to late migration stages (stages 5–8) is that the protrusive F-actin structures are extremely bright compared to the stress fibers. Thus, the challenge is to image the brightest structures accurately, while still collecting adequate information from the dimmer structures. Although it is not necessary to optimize all the settings discussed below for every image, it is advantageous to understand the elements of image collection, where introduction of noise can best be minimized, and wherein the limits of achieving maximal resolution lie. We use a Zeiss LSM 800 upright scanning confocal microscope controlled with Zen 2.3 Blue acquisition software (Zeiss) for imaging, though the steps described below are broadly applicable to other confocal systems.

- Choose the objective with the best resolving power available on your microscope, as determined by numerical aperture (NA) written on the lens. Higher NA equals higher resolving power. If you are imaging two or more colors, apochromat lenses ensure the wavelengths are properly aligned in the same z-plane. If imaging a single color, fluor or achromat lenses are also fine.
- Scan at a fast speed to get the area of interest in approximate focus, then use a slow scan speed to fine-tune the focus.
- Select the excitation spectra in your software that best matches the excitation and emission wavelengths of your fluorophore. For example, 488 can differ from EGFP on many systems.
- Set data collection to 16-bit; this will give a greatly expanded dynamic range that preserves information from the dimmer structures while maintaining reasonable pixel intensity of the brightest structures (*see* **Note 15**). If data are collected at 8 bits, much information about the dimmer structures will be lost.

- Set the gain as low as possible to reduce the noise in the image.
- Adjust the laser power so that it is just high enough that most of the dynamic range is used, without oversaturating any pixels. Most microscopes have software tools to check for range and oversaturation.
- If you wish to obtain the maximum possible z-resolution for the lens and fluorophore you are using, set the pinhole to 1 Airy unit (software option available on most systems). This will be smaller for higher NA lenses and shorter fluorophore wavelengths, meaning for the same 63× lens a z-section of 0.7 μm can be resolved for Alexa Fluor™ 488-phalloidin, but this expands to 0.9 μm for Alexa Fluor™ 647-phalloidin. Note that (within reasonable parameters) the pinhole size does not affect the x-y resolution. Setting the pinhole to a larger diameter than 1 Airy unit can sometimes be advantageous, as it lets in more light and includes structures from a slightly thicker plane. In contrast, setting the pinhole smaller than 1 Airy unit presents no advantages.
- To obtain the maximum x-y resolution possible with a given combination of fluorophore, lens, and digital zoom, choose the appropriate number of pixels per μm to sample. Most microscopes contain software settings to select the maximum number of pixels for a given set of conditions. Over-collecting pixels does no harm other than using more microscope time and requiring more storage space for the resulting images, as phalloidin is not prone to rapid bleaching. Under-collecting pixels, however, will needlessly reduce your image quality. *See* Fig. 2 for illustration of how wavelength limits maximum resolution for three different fluorophore conjugates of phalloidin. Shorter wavelengths give higher pixel resolution; however, a very nice stain can be obtained with Alexa Fluor™ 647-phalloidin.
- Center the area of interest and increase the digital zoom to the desired magnification.
- Orient the egg chamber such that its anterior-posterior axis is horizontal in the field of view by rotating the stage either manually or virtually. This prevents having to rotate the image later by an increment other than 90 degrees, which requires either resampling the pixels (considered undesirable in many contexts) or reducing the resolution substantially.
- Scan at a sufficiently slow speed that noise is not introduced at this step (e.g., for a 743 × 743 pixel image, a pixel dwell time of 10 μs or longer is preferable; *see* **Note 16**).

4 Notes

1. Insulin solution can be stored at 4 °C for up to 1 week. Insulin solution should be added to the live imaging medium immediately before use.
2. Schneider's S2 medium and Pen/Strep can be combined and stored at 4 °C. Fetal bovine serum can also be combined with the Schneider's S2 medium and Pen/Strep mixture and stored at 4 °C for up to 1 week. We make 10 mL of the S2, Pen/Strep, and fetal bovine serum mixture at a time.
3. To minimize tissue damage, maintain forceps with care and sharpen as needed.
4. We dilute methanol-free EM grade formaldehyde in PBT. Formaldehyde should be stored at 4 °C for no more than 1 week after an ampule is opened and should be added to PBT immediately before use.
5. Mounting medium that hardens will cause cracking of the egg chamber, so it is best to use a mounting medium that remains liquid.
6. Depending on the genotype, age, and temperature, the rate of oogenesis may vary, and flies may need to be aged for different numbers of days to achieve an optimal number of egg chambers of your desired stage for dissection. Younger females produce more eggs than older females, colder temperatures slow oogenesis, and warmer temperatures speed oogenesis. Flies must be moved to fresh yeast regularly or females will cease laying eggs and retain them in their ovaries, resulting in unhealthier egg chambers and fewer younger egg chambers. Certain genetic conditions can result in round eggs which will block the oviduct, which results in unhealthy egg chambers; dissecting younger flies prior to oviduct blockage is often useful in these backgrounds.
7. To minimize the amount of phalloidin used, we typically prepare a volume of staining solution such that 0.5 μL of phalloidin will be diluted to the correct concentration.
8. Proper pinching of the fly will ensure the rest of your dissection proceeds smoothly. Pinching the thorax increases the risk that attempting to open the abdomen to remove the ovaries will instead remove the entire abdomen without opening it. Pinching a more posterior part of the abdomen increases the risk that you will damage young egg chambers.
9. If the ovaries do not initially release from the abdomen, there are several methods to remove them. If the oviduct is visible, first attempt to tug posteriorly on the oviduct with your #55

forceps to remove the ovaries, as this is the method least likely to damage egg chambers. If the oviduct is not visible or tugging did not work, there are two other options. If late-stage egg chambers are desired, the abdomen can be gently squeezed like a tube of toothpaste in an anterior-to-posterior direction to extrude the ovaries. If younger egg chambers are desired, the posterior of the ovaries can be pulled on directly. We recommend practicing dissections as more maneuvering to remove the ovaries increases the likelihood of damage.

10. A p20 pipettor could also be used in lieu of a p10 pipettor.

11. Ovarioles will sink to the bottom of the tube; removing most but not all of the liquid will ensure that ovarioles are not also removed. Ensure the first wash does not proceed for longer than 5 minutes, as some residual formaldehyde will remain after removal of the fixative and could result in over-fixation.

12. If antibody staining is being combined with phalloidin staining, antibody staining should be performed at this step and followed with phalloidin staining as described by the subsequent steps of this protocol.

13. Remove as much wash as possible without removing the ovarioles. Remaining wash will dilute the mounting medium.

14. To apply the coverslip, our preferred method to avoid introducing air bubbles is to first place the short end of the coverslip on the glass slide. Then, using a p10 pipet tip, gently lower the other end of the coverslip to the glass slide, and pull the pipet tip out from under the cover slip once the tip is in contact with the glass slide. Other methods for applying the coverslip are acceptable, provided that air bubbles are kept to a minimum. If any mounting medium spills out from under the coverslip, it can be removed by gently dabbing with the edge of a delicate task wipe. Excess liquid can make it difficult to seal the coverslip with nail polish. Be cautious not to remove too much medium, however, as insufficient medium can result in cracking of egg chambers due to pressure from the cover slip.

15. Dimmer structures can be highlighted in post-processing using various means such as alternate look-up tables or inversion in Fiji/ImageJ before the image is set to 8-bit for use in figures.

16. If you are co-imaging another label like GFP, it is often preferable to use Alexa Fluor™ 647-phalloidin, as this allows good separation of excitation and emission spectra to avoid issues of bleed-through. Alexa Fluor™ 647-phalloidin is also inherently dimmer than Alexa Fluor™ 488- or Alexa Fluor™ 555-phalloidin, which can help avoid bleed-through from the shorter wavelengths.

Acknowledgments

We thank members of the Horne-Badovinac lab for helpful comments on the protocol. M.T.A. was supported by NIH T32 GM007183 and work in the Horne-Badovinac lab is supported by NIH R01s GM126047 and GM136961.

References

1. Friedl P, Gilmour D (2009) Collective cell migration in morphogenesis, regeneration and cancer. Nat Rev Mol Cell Biol 10(7):445–457. https://doi.org/10.1038/nrm2720
2. Mayor R, Etienne-Manneville S (2016) The front and rear of collective cell migration. Nat Rev Mol Cell Biol 17(2):97–109. https://doi.org/10.1038/nrm.2015.14
3. Cetera M, Ramirez-San Juan GR, Oakes PW, Lewellyn L, Fairchild MJ, Tanentzapf G, Gardel MG, Horne-Badovinac S (2014) Epithelial rotation promotes the global alignment of contractile actin bundles during Drosophila egg chamber elongation. Nat Commun 5:5511. https://doi.org/10.1038/ncomms6511
4. Gutzeit HO (1991) Organization and in vitro activity of microfilament bundles associated with the basement membrane of Drosophila follicles. Acta Histochem Suppl 41:201–210
5. Gutzeit HO, Eberhardt W (1991) Laminin and basement membrane-associated microfilaments in wild-type and mutant Drosophila ovarian follicles. J Cell Sci 100(4):781–788. https://doi.org/10.1242/jcs.100.4.781
6. Gutzeit HO (1990) The microfilament pattern in the somatic follicle cells of mid-vitellogenic ovarian follicles of Drosophila. Eur J Cell Biol 53(2):349–356
7. Sherrard KM, Cetera M, Horne-Badovinac S (2021) DAAM mediates the assembly of long-lived, treadmilling stress fibers in collectively migrating epithelial cells in Drosophila. eLife 10:e72881. https://doi.org/10.7554/eLife.72881
8. Bateman J, Reddy RS, Saito H, Van Vactor D (2001) The receptor tyrosine phosphatase Dlar and integrins organize actin filaments in the Drosophila follicular epithelium. Curr Biol 11(17):1317–1327. https://doi.org/10.1016/S0960-9822(01)00420-1
9. Delon I, Brown NH (2009) The integrin adhesion complex changes its composition and function during morphogenesis of an epithelium. J Cell Sci 122(23):4363–4374. https://doi.org/10.1242/jcs.055996
10. Spracklen J, Fagan TN, Lovander KE, Tootle T (2014) The pros and cons of common actin labeling tools for visualizing actin dynamics during Drosophila oogenesis. Dev Biol 393(2):209–226. https://doi.org/10.1016/j.ydbio.2014.06.022
11. Horne-Badovinac S, Bilder D (2005) Mass transit: epithelial morphogenesis in the Drosophila egg chamber. Dev Dyn 232(3):559–574. https://doi.org/10.1002/dvdy.20286
12. Haigo SL, Bilder D (2011) Global tissue revolutions in a morphogenetic movement controlling elongation. Science 331(6020):1071–1074. https://doi.org/10.1126/science.1199424
13. Horne-Badovinac S (2014) The drosophila egg chamber—a new spin on how tissues elongate. Integr Comp Biol 54(4):667–676. https://doi.org/10.1093/icb/icu067
14. Cetera M, Lewellyn L, Horne-Badovinac S (2016) Cultivation and live imaging of drosophila ovaries. In: Dahmann C (ed) Drosophila. Methods in molecular biology, vol 1478. Humana Press, New York, NY, pp 215–226
15. Prasad M, Jang AC-C, Starz-Gaiano M, Melani M, Montell DJ (2007) A protocol for culturing Drosophila melanogaster stage 9 egg chambers for live imaging. Nat Protoc 2(10):2467–2473. https://doi.org/10.1038/nprot.2007.363
16. Drummond-Barbosa D, Spradling AC (2001) Stem cells and their progeny respond to nutritional changes during Drosophila oogenesis. Dev Biol 231(1):265–278. https://doi.org/10.1006/dbio.2000.0135
17. Mazzalupo S, Cooley L (2006) Illuminating the role of caspases during Drosophila oogenesis. Cell Death Differ 13(11):1950–1959. https://doi.org/10.1038/sj.cdd.4401892
18. Pritchett TL, Tanner EA, McCall K (2009) Cracking open cell death in the Drosophila ovary. Apoptosis 14(8):969–979. https://doi.org/10.1007/s10495-009-0369-z

Chapter 10

Quantitative Image Analysis of Dynamic Cell Behaviors During Border Cell Migration

Yujun Chen, Nirupama Kotian, and Jocelyn A. McDonald

Abstract

Drosophila border cells have emerged as a genetically tractable model to investigate dynamic collective cell migration within the context of a developing organ. Studies of live border cell cluster migration have revealed similarities with other migrating collectives, including formation and restriction of cellular protrusions to the front of the cluster, supracellular actomyosin contractility of the entire collective, and intra-collective cell motility. Here, we describe protocols to prepare ex vivo cultures of stage 9 egg chambers followed by live time-lapse imaging of fluorescently labeled border cells to image dynamic cell behaviors. We provide options to perform live imaging using either a widefield epifluorescent microscope or a confocal microscope. We further outline steps to quantify various cellular behaviors and protein dynamics of live migrating border cells using the Fiji image processing package of ImageJ. These methods can be adapted to other migrating cell collectives in cultured tissues and organs.

Key words Collective cell migration, Live imaging, Border cell, Cell protrusions, Cell shape, Fluorescent microscopy, Confocal microscopy, Protein dynamics, Egg chamber, Ex vivo culturing

1 Introduction

Small to large groups of cells coordinate their movement in a process termed collective cell migration. Migrating cell collectives are essential for the proper development of many tissues and organs and contribute to pathological processes such as tumor invasion and metastasis [1, 2]. To better understand the conserved developmental, cellular, and molecular mechanisms that underlie collective cell migration, a number of in vivo models have been used, including the zebrafish lateral line, *Xenopus* neural crest cells, *Drosophila* embryonic tracheal branching and dorsal closure, and *Drosophila* ovarian border cells [2, 3]. Border cells are a small group of epithelial-derived cells that migrate collectively during mid-to-late oogenesis [4]. Importantly, border cells can be genetically manipulated and can be imaged over several hours in live ex vivo cultured egg chambers [4, 5]. Studies using border cells have revealed

Michelle S. Giedt and Tina L. Tootle (eds.), *Drosophila Oogenesis: Methods and Protocols*,
Methods in Molecular Biology, vol. 2626, https://doi.org/10.1007/978-1-0716-2970-3_10,

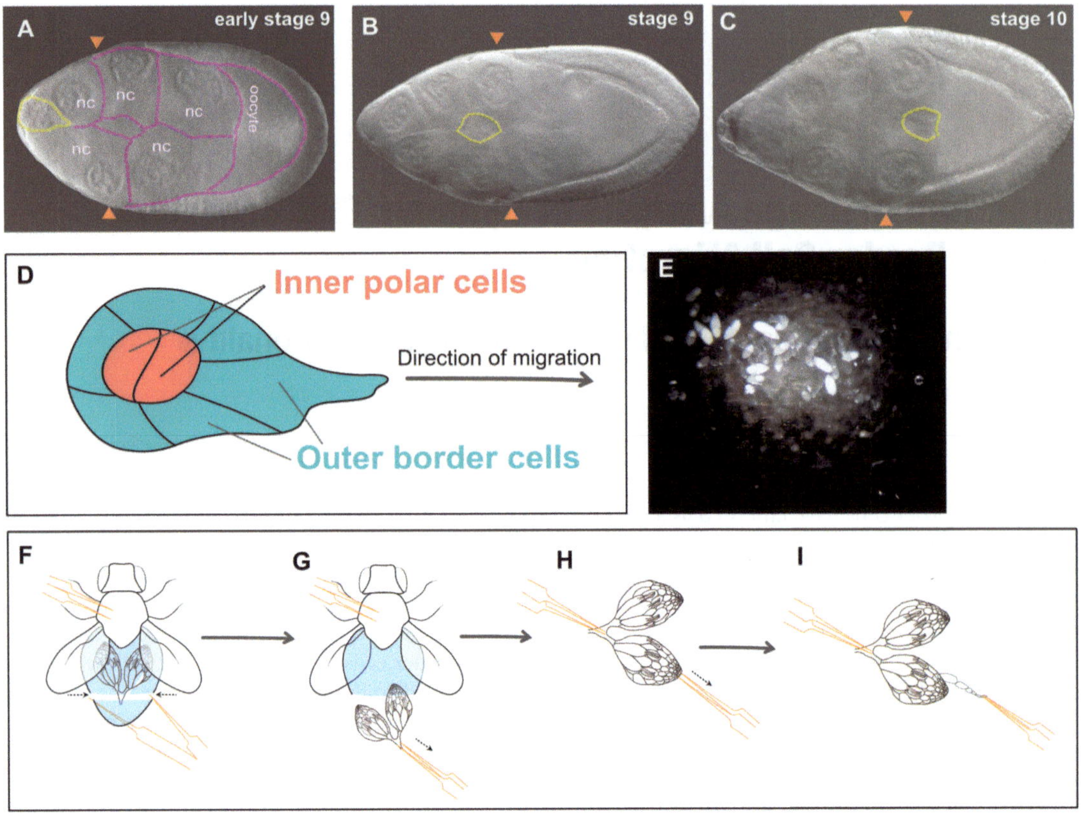

Fig. 1 Illustration of ovary dissection and staging of egg chambers for imaging border cell migration. (**a–c**) Differential interference contrast (DIC) images of *Drosophila* egg chambers from early stage 9 (**a**), mid-migration (**b**), and stage 10 (**c**) with yellow line outlining the border cell cluster in each egg chamber and orange arrowheads showing the movement of outer follicle cells as egg chamber maturation progresses. In (**a**), the nurse cells are labeled "nc"; both the oocyte and nurse cells are outlined in magenta. (**d**) Schematic of a border cell cluster. Shown in the drawing are the inner polar cells (pink) and outer border cells (blue). The arrow points towards the direction of migration. Anterior is to the left in (**a–d**). (**e**) Field of view on a stereo microscope showing a typical number of dissected egg chambers. The stage 9 egg chambers will be used for live imaging; older and younger staged egg chambers are also present. Tissue debris and older egg chambers will be discarded prior to mounting the sample. (**f–i**) Illustrated stepwise instructions for ovary dissection from flies and further ovariole dissection from a pair of whole ovaries. In all drawings, the dominant forceps are on the right and the non-dominant forceps are on the left

mechanisms that control collective cell polarization, cell-cell communication and adhesion for cooperative movement, and interactions with the surrounding tissue microenvironment [4, 6–12].

Border cells migrate as a cohesive cluster of 6–10 cells, navigating the surrounding germline-derived nurse cells to reach the oocyte at the anterior end of the egg chamber, the functional unit of the ovary (Fig. 1a–c). At late stage 8, a pair of non-motile polar cells at the anterior end of the egg chamber specifies and recruits surrounding follicle cells to become border cells. The outer border cells physically surround the inner polar cell pair at the center to jointly form the migratory border cell cluster (Fig. 1d). The border cell cluster then delaminates from the follicular epithelium at early stage 9 (Fig. 1a) and migrates between the 15 large nurse cells (Fig. 1b). By stage 10, border cells complete their migration when they arrive at the anterior border of the oocyte (Fig. 1c). Border cells further align with the dorsal-anterior corner of the oocyte. Later, border cells contribute to the formation of the micropyle, the sperm entry pore into the oocyte [13]. Once border cells delaminate, the migration phase takes ~3–4 hours, though the entire process can take 4–6 hours [5].

The development of robust methods of ex vivo live culturing of stage 9 egg chambers along with time-lapse imaging has revealed complex dynamic cellular behaviors of border cells [5, 14, 15]. The border cell cluster initially undergoes fast polarized migration with one or two cellular protrusions that extend and retract from the leading border cells [16, 17]. This is followed by a slower migration phase in which the cluster "tumbles" or "rotates" and extends multiple protrusions prior to reaching the oocyte [16–18]. Protrusions help border cells crawl between nurse cells and sense signals in the environment [15, 16]. During migration, border cells maintain adhesions to the polar cells and to each other yet border cells are able to exchange positions within the cluster (Fig. 1d) [12, 18]. Individual border cells, and the cluster itself, have specific cellular shapes, which help the group navigate the egg chamber [8, 11, 19]. Visualization of fluorescently labeled proteins and biosensors has revealed the dynamics of non-muscle myosin II (myosin II), F-actin, and E-cadherin proteins and the activity of the small GTPase Rac in live migrating border cell clusters [8, 11, 20, 21]. This includes a supracellular organization of actomyosin in the border cell cluster [8, 10, 11].

Here, we describe a protocol to culture stage 9 egg chambers and perform live time-lapse imaging of fluorescently labeled border cells followed by quantification of cellular behaviors or protein dynamics. We describe flexible imaging options using either an epifluorescent microscope or a confocal microscope. We include a recently described strategy to immobilize egg chambers during time-lapse imaging using a fibrinogen-thrombin clot [22], which is useful in imaging and analyzing live border cell migration. Depending on which cellular or protein behaviors are of interest, imaging can be done on timescales of minutes to hours (e.g., 15 minutes for protein dynamics; 4–6 hours for protrusions). We outline steps to process images from time-lapse imaging using the open-source Fiji software [23]. We further describe quantification methods to measure and analyze border cell cluster migration speed, cluster and cell shapes, protrusion dynamics, and protein localization and dynamics using Fiji. While for simplicity we focus on wild-type border cell migration, many options exist to genetically manipulate border cells in the developing egg chamber. The

GAL4-UAS system can be used to knock down genes using RNA interference (RNAi), to overexpress proteins, or to manipulate protein levels or activity with various available optogenetic tools (e.g., [8, 19, 21, 24, 25]). Alternatively, researchers can perform classical genetic analyses using homozygous mutant alleles or mosaic mutant analyses [25, 26].

2 Materials

2.1 Preparation of Fly Stocks for Time Lapse Imaging

1. Select an appropriate fluorescent reporter fly stock for live imaging. Table 1 lists commonly used reporters and GAL4/UAS lines used for live time-lapse imaging of border cell migration (*see* **Note 1**). As needed, set up a genetic cross to obtain flies of the correct genotype to express the fluorescent reporter in the desired cell types.
2. Wet yeast paste: Dissolve dry (active) yeast with water and mix to produce a spreadable paste that is not too runny but also will not dry out too quickly. Feeding the females wet yeast paste is essential for promoting healthy oogenesis prior to dissection. The yeast paste can be stored at 4 °C for up to a week.
3. Vials containing fly food.
4. Incubators, set at 25 °C and 29 °C.

2.2 Ovary Dissection

1. Two pairs of dissecting forceps, e.g., Dumont #5, 0.1 × 0.06 mm tips, Inox.
2. One two-well glass concavity ("depression") slide, e.g., thickness 3.12–3.22 mm.
3. Stereo microscope with good illumination (e.g., a ring light; a polarizer can be added to reduce glare).
4. CO_2 and CO_2 pad to anesthetize flies for dissection.
5. 70% ethanol to clean the work surfaces.
6. Kimwipes.
7. Dissection medium: 1× Schneider's *Drosophila* medium, 20% FBS, 1× antibacterial-antimycotic solution. To make, aliquot 39.5 mL of Schneider's medium into a 50 mL tube. To the same tube, add 10 mL of FBS and 0.5 mL of the 100 × antibiotic-antimycotic solution. Bring the medium to room temperature, calibrate the pH meter and measure the pH of the solution. Adjust the pH to 6.95 (*see* **Note 2**). Filter sterilize using a 50 mL tube top vacuum filter. Aliquot into 1.5 mL microcentrifuge tubes. Store at 4 °C for up to 2 months. Bring to room temperature before dissection.

Table 1
Commonly used fluorescent reporters and other fly stocks useful for live imaging of border cells

Stock genotype	Source (stock number)	Advantages	Disadvantages	Live imaging references
UAS-PLCδPH::EGFP	BDSC (39693)	Marks cell membranes with almost no cytoplasmic signal; bright with very good signal to noise; use to label cell shapes, cluster shape, and protrusions		[8]
slbo-LifeAct::GFP [2 M]	BDSC (58364)	Labels F-actin; LifeAct-GFP driven directly by the *slbo* enhancer; specific to border cells and a few follicle cells; very good signal to noise; can be used to measure protrusions and actin dynamics; independent of GAL4/UAS system	Strong expression of LifeAct can cause severe F-actin defects in some cells and tissues, for example the germline [35]; live imaging conditions should be optimized to avoid artifacts induced by LifeAct overexpression	[8, 12]
UAS-LifeAct::GFP	BDSC (35544)	Labels F-actin; UAS-driven LifAct-GFP; excellent signal to noise; can be used to measure protrusions and actin dynamics	Strong expression of LifeAct can cause severe F-actin defects in some cells and tissues, for example the germline [35]; choice of GAL4 is crucial; live imaging conditions should be optimized to avoid artifacts induced by LifeAct overexpression	[19]
sqh fTRG library ($sqh^{fTRG00600.sfGFP\text{-}TVPTBF}$)	VDRC (318484)	Labels non-muscle myosin II; bright with very good signal to noise; tagged FlyFos transgene; can be used as a heterozygote		[8]
sqh^{AX3}; sqh::GFP	BDSC (57144)	Labels myosin II; bright with good signal to noise; rescue transgene in mutant background so is the only source of MRLC		[30, 36]

(continued)

Table 1
(continued)

Stock genotype	Source (stock number)	Advantages	Disadvantages	Live imaging references
c306-GAL4; sqh::GFP	J. McDonald (derived from BDSC 57144)	Labels myosin II; bright with good signal to noise; rescue transgene, can be used as a heterozygote		[30]
sqh-mCherry	BDSC (59024)	Labels myosin II; good signal; rescue transgene	mCherry can sometimes aggregate into abnormal punctae; brighter at central polar cells	[20, 37, 38]
UAS-mCherry::Jupiter	C.Q. Doe	Labels microtubules and cytoplasm; bright with good signal to noise	Mostly cytoplasmic in border cells; does not visibly label distinct microtubules in live border cells; occasional aggregated mCherry punctae are observed	[39, 40]
c306-GAL4;UAS-mCherry::Jupiter	J.A. McDonald	Stock carrying mCherry-Jupiter to express in border cells	Best to outcross to wild-type chromosome and image as a heterozygote	
UAS-mCD8::GFP or UAS-mCD8::ChRFP	Various stocks, BDSC (e.g., 5136, 27392)	General membrane marker; bright with good signal to noise; labels protrusions and general cell and cluster shape	Can appear mostly cytoplasmic in border cells with less enrichment at the cell membrane	
UAS-nuclear::GFP or UAS-nuclear::ChRFP	Various stocks, BDSC (e.g., 4776, 38425)	Use to label nuclei; especially good for tracking positions of border cells within the cluster		[20]
E-cad::GFP (TI{TI}shg[GFP])	BDSC (60584)	Labels endogenous E-cadherin (Shg) in border cells; GFP engineered into the *shg* gene locus		[20, 41]
UAS-GMA	Various stocks, BDSC (e.g., 31775)	Moesin actin-binding domain tagged with GFP; enriched at cell membranes; good signal to noise		[15, 42]

(continued)

Table 1 (continued)

Stock genotype	Source (stock number)	Advantages	Disadvantages	Live imaging references
***c306*-GAL4**	BDSC (3743)	GAL4 driver that turns on early in border cells and maintains expression; especially useful for effective RNAi-mediated knockdown in border cells	Occasional slow border cell delamination and/or mild cluster splitting; somewhat heterogeneous expression in border cells; use outcrossed to UAS or to a wild-type chromosome; expressed in other follicle cells; expressed at earlier stages of development so is often used in combination with ts-GAL80	
slbo-GAL4	D. Montell (insertion on II without UAS-GFP); BDSC (58435; insertion on III)	GAL4 driver that turns on in border cells just prior to delamination; strong expression in border cells; not expressed during earlier stages of oogenesis	Not expressed in the central polar cells; expressed in a few anterior and posterior follicle cells in addition to the border cells	
ts-GAL80	BDSC various stocks (e.g., 7019)	Used to suppress GAL4/UAS expression at earlier stages of development	Flies carrying ts-GAL80 need to be incubated at 29 °C for at least 14 hours prior to ovary dissection to ensure GAL80 is off and GAL4/UAS is on	
w^{1118} or *Oregon R*	BDSC various stocks (e.g., 3605, 5)	Stocks used to outcross GAL4-driven fluorescent reporter stocks; many P-element stocks are made in a *w* background making *w*^{1118} a good control		

Bolded genotypes: stocks used in this chapter

8. Transfer pipettes to move egg chambers and dissection or live imaging medium (e.g., 1 mL ultra-fine tip) or a P200 pipettor with the pipette tip cut off.
9. 0.6 mL microcentrifuge tubes.

2.3 Mounting Egg Chambers for Live Imaging

1. 10 mg/mL insulin solution: To prepare a stock solution of insulin, first prepare acidified water. Add 1.2 mL 1 N HCl to 100 mL of dH_2O; store at room temperature. Then add 5 mg insulin powder to 500 μL acidified water and aliquot 10 μL into separate 0.6 mL microcentrifuge tubes (to a final concentration of 10 mg/mL). The insulin stock solution can be stored at − 80 °C for at least 6 months.
2. Dissection medium.
3. Live imaging medium: 1× Schneider's *Drosophila* medium, 20% FBS, 0.2 mg/mL insulin, 1 × antibacterial-antimycotic solution. To prepare, add 490 μL dissection medium to a microcentrifuge tube containing 10 μL of 10 mg/mL insulin. Bring to room temperature before mounting egg chambers. We will need 100 μL × 2 per sample.
4. Transfer pipettes or P200 pipettor with tips.
5. Lumox dish or other types of culture dish (*see* **Note 3**).
6. 22 × 22 mm^2 coverslip (no. 1 thickness).
7. 22 × 40 mm^2 coverslip (no. 1 or no. 1.5 thickness).
8. Diamond tip glass cutter (optional).
9. Kimwipes.
10. P20 pipettor with tips.
11. Halocarbon oil 27 (*see* **Note 4**).
12. Fibrinogen, Bovine Plasma (10 mg/mL; *see* **Note 5**; optional).
13. Thrombin (10 U/mL; *see* **Note 5**; optional).

2.4 Live Time-Lapse Imaging and Quantitative Analysis of Live Migrating Border Cells

1. Motorized microscope with an upright or inverted setup and either widefield epifluorescence or confocal imaging modality (*see* **Note 6**). Microscope software, e.g., Zen software (Zeiss) or similar.
2. Objective: e.g., Plan-Apochromat 20× 0.75 Numerical Aperture (NA) air or 40× 1.2 NA water immersion.
3. Fiji (https://fiji.sc/) open-source image processing package based on ImageJ2 [23], or other imaging processing software.

3 Methods

3.1 Prepare Flies for Dissection

1. Obtain fly stocks with fluorescent markers and/or set up a genetic cross to express the fluorescent marker using the GAL4/UAS system (Table 1; *see* **Note 1**)
2. Collect 8–10 newly eclosed females (3–5 days old) of the correct genotype, along with 2–3 males, to ensure the females have mated.
3. Transfer flies to a fresh food vial with wet yeast paste for 18–20 hours at 25 °C or 29 °C prior to ovary dissection. For maximal expression of GAL4/UAS, and inactivation of temperature sensitive-GAL80 (ts-GAL80) as needed, incubation at 29 °C is often required.
4. Verify that the female has a swollen abdomen that is filled by an enlarged ovary pair, which indicates that the female is properly mated and fattened. The ovaries should contain ovarioles with most stages of oogenesis represented.

3.2 Dissection of Ovaries for Live Imaging

3.2.1 Pull Out Whole Ovaries

1. Allow the dissection medium to warm up to room temperature.
2. Prepare the dissection station by cleaning with 70% ethanol. Prepare a Kimwipe for disposal of fly tissue debris, and to wipe off forceps, during dissection.
3. Transfer a small amount of dissection medium to both wells of the glass concavity slide using a transfer pipette or P200 pipettor and focus on the slide under the stereomicroscope.
4. Anesthetize the flies and place onto a CO_2 pad (males can be discarded). Using the non-dominant hand, pick up one female using one forceps. Place the fly in one well of the depression slide while simultaneously pressing down gently on the thorax and the top of the abdomen to make sure the fly is fully submerged under the medium. With the forceps in the dominant hand, grasp the dorsal posterior end of the abdominal cuticle and pierce it (*see* **Note 7**). Pull the cuticle gently away, which will cause the pair of ovaries to be released (Fig. 1f, g). Pull the ovary pair completely away from the rest of the fly carcass (Fig. 1g).
5. Remove the fly carcass by cleaning the forceps on a wet Kimwipe. Remove any other tissue debris that may be attached to the ovary using the forceps. Transfer the whole ovary to another well with fresh dissecting media using a transfer pipette or P200 pipettor (*see* **Note 8**). Repeat these steps until you obtain three to four pairs of whole ovaries.

3.2.2 Dissect Ovaries into Ovarioles

1. Remove ovarioles from the whole ovary. This is essential to image the specific egg chamber stages for border cell migration as well as to remove the overlying muscle sheath (*see* **Note 7**). To dissect the ovary into ovarioles, grasp one ovary from the pair with your non-dominant forceps at the larger posterior end (Fig. 1h). Push down to anchor the ovary at the bottom of the dissecting well. While continuing to hold a gentle grip on the older egg chambers of the whole ovary, use the other forceps (dominant hand) to grasp the early egg chambers at the anterior end. Slowly and carefully pull out 1–2 ovariole chains at a time (Fig. 1i). Repeat, until you are confident you have obtained at least 5–10 stage 9 egg chambers from approximately 10–15 dissected ovarioles (Fig. 1e, i).
2. Once the ovarioles are pulled out, remove late-staged egg chambers (anything older than stage 9, typically egg chambers with the oocyte filling >50% of the length of the egg chamber) and other unwanted ovarian tissue using the forceps (*see* **Note 7**). Clean the forceps on a wet Kimwipe.
3. Carefully aspirate excess media from the well that has the dissected egg chambers by looking through the stereo microscope. Leave behind enough media to prevent the ovarioles from drying up. Add 100 μL of live imaging medium to the same well. Aspirate the egg chambers along with the medium using a fine transfer pipette or a P200 pipettor with the tip cut off. To ensure that fresh live imaging media is used for mounting the sample, first transfer the egg chambers to a 0.6 mL tube and wait for the egg chambers to settle to the bottom. Next, remove the existing media from the tube with a fine pipette tip or transfer pipette, leaving behind a few microliters of media. Add 100 μL of fresh live imaging medium (*see* **Note 9**).

3.3 Mounting Egg Chambers for Live Imaging

The mounting of samples for live imaging will depend on the type of microscope being used. We provide protocols for both upright and inverted microscope platforms (Fig. 2a, d–g; *see* **Note 6** and Subheading 3.4). In the case of upright microscope imaging, we often mount the sample on the inside of the dish (Subheading 3.3.1), then place the sample on the microscope stage with the coverslip oriented towards the objective (Fig. 2a, d, e). For inverted microscope imaging (or upright imaging), the sample can be mounted on the outside of the dish (Subheading 3.3.2), then placed on the microscope with the coverslip facing the objective (Fig. 2a, f, g). In both cases, a fibrinogen-thrombin clot can be used to immobilize egg chambers during imaging [22, 27], which is especially helpful when using a 40× water-immersion objective or any objective with short working distance (Fig. 2a; *see* **Note 5**).

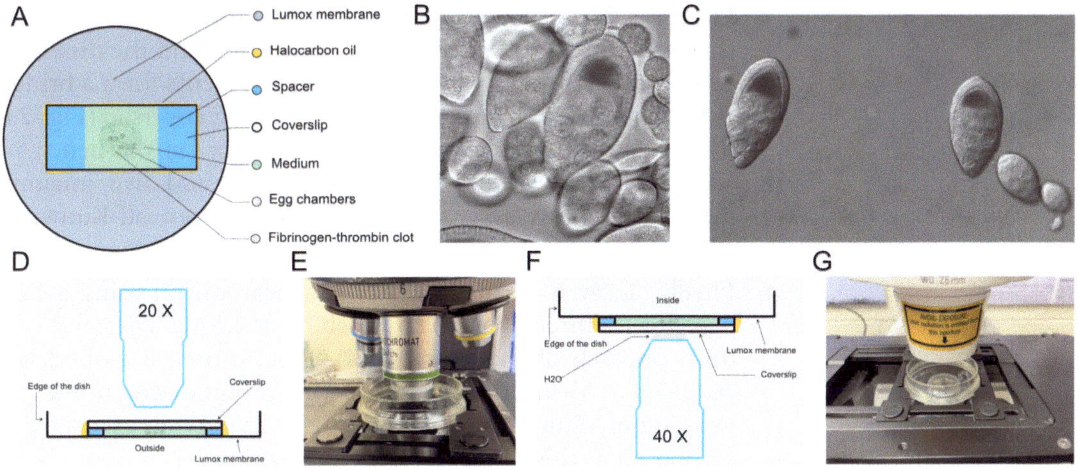

Fig. 2 Sample mounting and positioning on the microscope. (**a**) Illustration of the mounted egg chambers ready for imaging. Shown is the sample prep with the optional fibrinogen-thrombin clot. (**b**) DIC image of an overcrowded field of view of a sample with egg chambers too close to each other. (**c**) DIC image of a sample with egg chambers at an ideal density for imaging. (**d**) Illustration of the imaging setup for an upright widefield microscope with the sample mounted on the inside of the dish. (**e**) Image of the setup on an upright microscope, the Zeiss Axio Imager. (**f**) Illustration of the sample setup for an inverted confocal microscope with the sample mounted on the outside of the dish. (**g**) Image of the setup on an inverted confocal microscope, the Zeiss LSM-800

3.3.1 Mount Egg Chambers for Upright Microscope Live Imaging

1. Prepare a Lumox culture dish (new or reused; *see* **Note 10**).
   (a) Use the blunt end of forceps (or a diamond tip glass cutter) to break a 22 × 22 mm^2 coverslip in half in a petri dish; these coverslip shard spacers will be used to support the overlying coverslip as a bridge to prevent crushing of egg chambers (Fig. 2a, d).
   (b) Pipet two 6 μL drops of fresh live imaging medium onto the inner membrane ("inside") of the Lumox dish (Fig. 2d). Place the drops about 18 mm apart.
   (c) Place the broken coverslip halves onto each drop, with the smooth edge facing the center (Fig. 2a).
2. Take the microcentrifuge tube containing the dissected egg chambers and gently aspirate to mix the egg chambers with the live imaging media, being careful to not introduce bubbles into the media.
3. Pipette 88 μL of the egg chambers with live imaging media and dispense the mixture to the Lumox dish in the center between the coverslip bridge spacers (Fig. 2a, d). Use a circular motion to avoid overcrowding of egg chambers (Fig. 2b). Only a few egg chambers should be in the field of view (Fig. 2c). Avoid any bubbles while pipetting the mixture onto the dish (*see* **Note 11**).

4. Pick up a 22 × 40 mm^2 coverslip using a forceps. Gently place the coverslip on top of the broken coverslips and the dissected egg chambers, such that the coverslip fragments form a bridge between the egg chambers and the overlying coverslip (Fig. 2a, d). Gently tap the dish using your hand or on the bench to allow even spread of the egg chambers and live imaging medium. Remove any excess media using a torn off Kimwipe, wicking away liquid near the broken coverslip bridges.
5. Finally, apply 6–9 small drops of halocarbon oil using a P20 pipettor (~ three drops of oil each on the long sides and one drop on each of the short sides). Pipette the oil around the coverslip to form a thin layer that will prevent drying of the live imaging medium during imaging (Fig. 2a). Avoid excess halocarbon oil as this can interfere with imaging.

3.3.2 Mount Egg Chambers for Inverted Microscope Live Imaging and Immobilize Egg Chambers with a Fibrinogen-Thrombin Clot

1. Prepare a Lumox culture dish (new or reused; *see* **Note 10**) as described in Subheading 3.3.1 of **step 1**, except prepare to mount on the outside membrane of the dish (Fig. 2f).
2. Rinse the egg chambers one time with 100 μL of live imaging medium. Remove as much live imaging medium as possible.
3. If desired, make a fibrinogen-thrombin clot by adding 10 μL fibrinogen (10 mg/mL) to the egg chambers in the tube (Fig. 2a; *see* **Note 5**). Next, transfer the egg chambers with fibrinogen to the outside of the Lumox dish onto the membrane between the broken coverslips (Fig. 2f). Quickly add 1 μL thrombin (10 U/mL), then wait 5–10 minutes to allow the fibrinogen-thrombin clot to form.
4. Add 78 μL of live imaging medium to the egg chambers with the fibrinogen-thrombin clot.
5. Add the coverslip and halocarbon oil, as described in Subheading 3.3.1 of **steps 4** and **5**.

3.4 Live Time-Lapse Imaging

The choice between widefield epifluorescence and confocal imaging will depend on the goal of the experiment and the availability of microscopes. We often use widefield epifluorescence imaging to perform longer time-lapse imaging (3–4 hours), which allows analysis of migration speed, protrusion dynamics, and general protein localization [8, 28, 29]. To avoid phototoxicity, an LED illuminator can be used (*see* **Note 6**), although other light sources are compatible with live imaging. Confocal microscopy offers significant advantages by reducing out-of-focus illumination. Acquisition of images on the confocal occurs at high speed, allowing for shorter time-lapse intervals and faster z-stack acquisition, both of which are needed for detailed analysis of cellular and protein dynamics. We typically use confocal imaging to perform shorter time-lapse imaging (30 minutes to 2 hours), which is useful for analysis of faster protein dynamics and cellular behaviors [8]. In all cases, care must be made to avoid photobleaching of fluorescent proteins and phototoxicity of the egg changers (*see* **Note 6**).

3.4.1 Live Imaging Using a Widefield Epifluorescent Microscope

1. Place the mounted egg chambers on the Lumox dish onto the stage of a widefield epifluorescent microscope. If using an upright microscope, your sample will typically be mounted on the inside of the dish (Fig. 2d, e); if using an inverted microscope, your sample will be on the outside of the dish (Fig. 2f, g).
2. Place the Lumox dish on the stage of a widefield epifluorescent microscope (Fig. 2e, g). Move the objective into place and bring the egg chambers on the dish into focus (*see* **Note 12**).
3. Using either brightfield, or the fluorescence channel of interest, scan the dish to find an egg chamber at the appropriate stage (Figs. 1a and 2c). Avoid egg chambers that are visibly damaged, look abnormal in overall shape, or show a migration defect when not expected (*see* **Note 13**). Additionally, avoid egg chambers that are too early (e.g., border cells not yet rounded up or are not ready to delaminate) or are too late (Fig. 1c; border cells finished migrating). Selection of mid-stage 9 egg chambers (Fig. 1b; border cells in mid-migration) is sometimes appropriate when focusing on specific cellular phenomena such as protrusions, cell exchange, and/or cell shape. *See* ref. [13] for guidance on staging of egg chambers.
4. After finalizing the egg chamber to be imaged, switch to the final desired objective to be used for time-lapse imaging and, as needed, manually refocus.

3.4.2 Set Up the Widefield Epifluorescent Time-Lapse Experiment

1. Using the available microscope software (e.g., Zeiss Zen), set up the time-lapse experiment. Set the time interval and duration of the experiment as preferred. Check the exposure for your channel of interest (*see* **Note 14**).
2. Set up a z-stack for the sample to capture different slices of the border cell cluster as it migrates (*see* **Note 15**).
3. Start the time-lapse imaging experiment (*see* **Note 14**). Check the status of the experiment every 15–30 minutes. Check if the border cell cluster is still in focus and that the egg chamber continues to develop. Pause the experiment in the software, if needed, readjust the focus and/or exposure, then continue with the experiment. If the egg chambers are not developing correctly (e.g., no growth of oocyte or no movement of the outer follicle cells towards the oocyte) or begin to appear abnormal as the experiment proceeds, either find a new egg chamber or start over with a new live imaging preparation (*see* **Note 13**).

3.4.3 Live Imaging Using a Confocal Microscope

1. After turning on the confocal microscope, open the available microscope software (e.g., Zeiss Zen software), load the Lumox dish with the mounted egg chambers onto the stage (Fig. 2f, g). *See* **step 1** of Subheading 3.4.1 if using an upright microscope.
2. If desired, locate multiple imaging positions using a scanning x-y stage: Switch the objective lens to 10× for ease of finding egg chambers to image. Use brightfield to locate egg chambers and mark their location using the microscope software (*see* **Note 12**). Use the desired fluorescent channel to check the fluorescence intensity in the border cell cluster to help choose the sample(s) to image. Scan the slide and repeat to select multiple egg chambers to image (*see* **Note 16**). *See* ref. [13] for guidance on staging of egg chambers.
3. Switch to the 20× objective and make fine focus adjustments. Add water to the 40× lens (if using a water-immersion objective) and move the 40× objective into place on the sample.
4. Set the Scan Area: Change the scan area, frame size, and scan line speed to optimize for faster scanning time. For imaging the fast dynamics of proteins, increase the scanning speed accordingly. It is best to adjust the frame size and scan speed to meet the purpose of the specific experiment (*see* **Note 17**).
5. Set Channels and Intensity: Adjust the laser intensity (usually ranging from 0.1 ~ 3%), pinhole, and the gain (master) with the range indicator to the appropriate level (bright enough but not oversaturated). As needed, crop the image, modify the scan area, and rotate the orientation of the egg chamber (anterior to the left and posterior to the right). Including unnecessary scanning areas in the field of view will increase acquisition time.
6. Set z-Stack: For cell shape and protrusion analysis, find the top and bottom of the cluster, then set the interval of the z-stack as 1 μm to optimize for speed of acquisition (*see* **Notes 15** and **16**). For analyses of protein dynamics, we typically image a single focal plane.
7. Set Time Series: Set the interval time and duration of the experiment. Stop the imaging early if needed. For cell shape and protrusion analysis, a time interval of 3 minutes and duration time of 4–6 hours is sufficient. For protein dynamics, a time interval of 15 seconds and duration time of 10–15 minutes is sufficient (*see* **Note 14**).
8. Start the time-lapse movie experiment. Once finished, save the file in the native microscope software file format to preserve metadata information. As time permits, move to the next egg chamber position, then repeat **steps 3–6** to finish time-lapse imaging of all egg chambers (e.g., when performing shorter duration movies for protein dynamics; *see* **Note 16**).

3.5 Quantitative Analysis of Live Migrating Border Cells

In this section, we describe how to process and edit the movie using the open-source imaging platform Fiji (Subheading 3.5.1), although other microscope imaging software packages can be used. We then describe how to analyze cellular behaviors such as migration speed, cluster and/or cell shapes, and protrusion characteristics such as size, length, direction and lifetime (Subheading 3.5.2). Finally, we provide guidance on how to measure protein localization and dynamics in live migrating border cells (Subheading 3.5.3). Depending on the goal of the experiment, one or more parameters can be measured to reveal the cellular and molecular dynamics of both wild-type and mutant border cells.

3.5.1 General Image Processing

1. Open the movie file (typically the native file format; e.g., ".czi" movie from Zeiss) in Fiji with "Bio-Formats Import Options" (Fig. 3a; *see* **Note 18**).
2. As needed, create a maximum z-stack projection. Select "Image › Stacks › Z Project" (Fig. 3b). For tracking protein dynamics, in which a single focal plane is imaged, skip to **step 3**.
3. As appropriate, crop the region of interest using "Image › Crop" (Fig. 3c). Be sure to duplicate the image; this will ensure that the image can be re-edited as needed.
4. Change the visualization to invert the fluorescence image (*see* **Note 19**). Use "Image › Lookup Tables › Invert LUT" (Fig. 3d). As needed, adjust the image to best display the range of the signals using "Image › Adjust › Brightness/Contrast."
5. Add a scale bar using "Analyze › Tools › Scale Bar" (Fig. 3e).
6. Add a timestamp using "Image › Stacks › Time Stamper" (Fig. 3f).
7. Convert the image to RGB mode. Save the file in a common video format, such as ".avi," ".mov," or ".mp4," or in a no compression storage image format like ".tif." Still frames from several successful live time-lapse border cell movies, with dynamic protrusion extension and retraction, are shown in Fig. 4 (widefield epifluorescence long time-lapse, Fig. 4a, b'''''; confocal short time-lapse, Fig. 4c, c''''').

3.5.2 Measure Cellular Behaviors in Fiji

1. Quantify the speed of the migrating border cell cluster. Although border cells undergo an initial faster migration followed by a slower tumbling migration phase, here we describe how to measure the total speed for the entire migration process (speed = migrated distance/overall time). *See* Refs. [17, 18, 20] for discussion of methods to measure the different phases of motility. Select the "Straight line" tool from the main toolbar and draw a line along the migration pathway, from the anterior end of the egg chamber to the anterior end of the

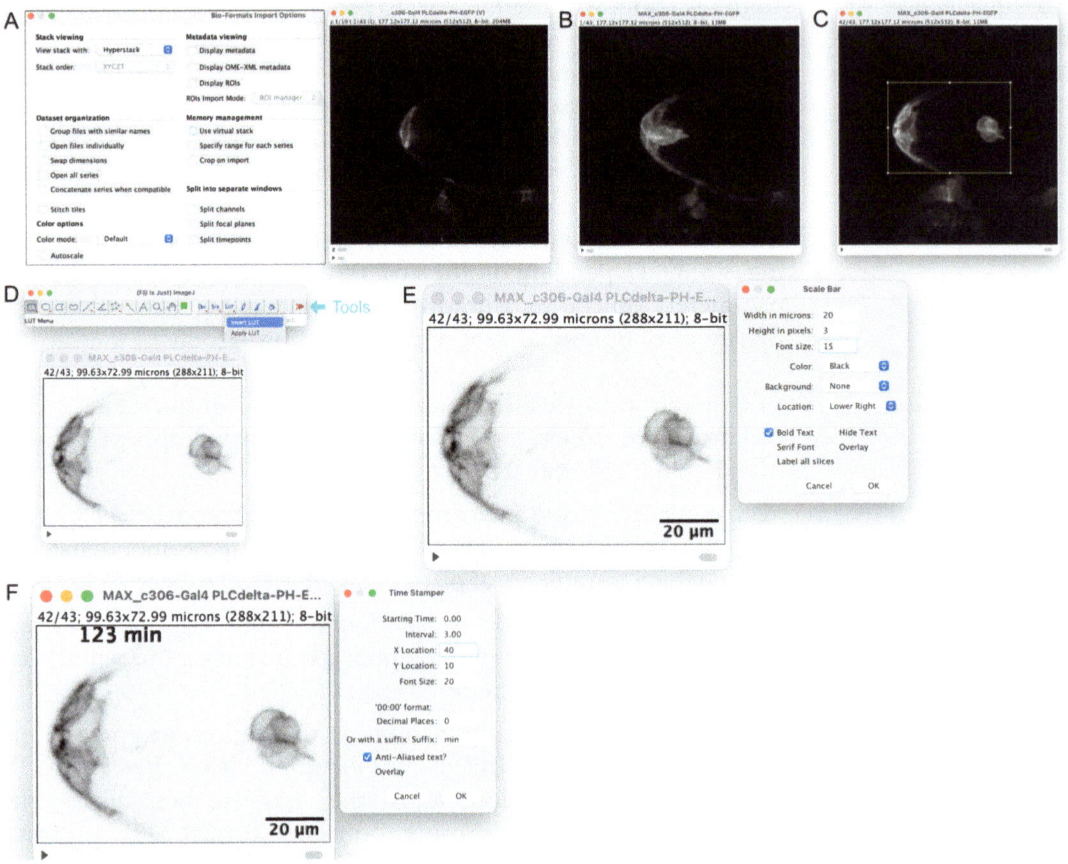

Fig. 3 Overview of steps in Fiji to process a time-lapse movie. Screen captures to illustrate how to process time-lapse imaging files in Fiji. (**a**) Open the movie in Fiji with the Bio-Formats plugin; shown here is the default setting. (**b**) Create a maximum z-stack projection. (**c**) Crop the image as needed. (**d**) Invert the LUT as desired; the tool bar is shown. (**e**) Add a scale bar. (**f**) Add the time using the time stamper tool

oocyte (green dashed line in Fig. 4a'''''; Fig. 5a). Next, select the "Analyze › Measure" to obtain the measurement of the distance migrated. Divide this distance by the overall time of the movie to obtain the overall speed.

2. To assess cluster and cell shape, use the "Freehand selections" tool to outline and select the main body of the cluster or a single cell (*see* Fig. 5a–c). Next, use "Analyze › Measure" to measure the area, circularity, and aspect ratio of the selected region. Before measuring, ensure that the "Area" and "Shape descriptors" are selected under "Analyze › Set Measurements."
3. Quantify the size, length, direction, and lifetime of protrusions.
   (a) Measure the area and length of protrusions. First, hold the shift key and use the "Oval" function to draw a circle surrounding the cell cluster (Fig. 5a, d). Any cellular

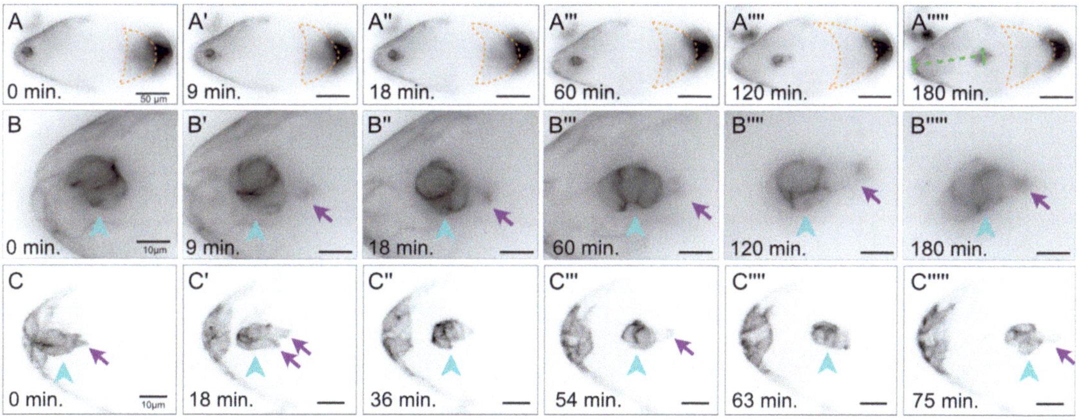

Fig. 4 Time-lapse images showing key stages of border cell migration. (**a**-**a**''''') Still images from a 240 minute (imaged at 3 minute intervals) time-lapse movie of a *c306*-GAL4/+; UAS- PLCδ-PH-EGFP/+ egg chamber generated using a widefield epifluorescent microscope. Six time points, from 0 to 180 minutes, show the border cell cluster as it migrates to the oocyte anterior border. (**b**-**b**''''') Magnified views of the same border cell cluster (arrowhead) from (**a**-**a**'''''), showing a protrusion (arrow) at various time points. (**c**-**c**''''') Still images of a 129 minute (imaged at 3 minute intervals) time-lapse movie of a *c306*-GAL4/+; UAS- PLCδ--PH-EGFP/+ generated using a confocal microscope. Six time points are shown. The border cell cluster (arrowhead) delaminated from the epithelium (**c**; 0 min) and migrated (**c**'-**c**'''''; 18–75 minutes), with protrusions (arrows) extending and retracting during this time. Anterior is to the left in all panels and scale bars indicate the image magnification

extensions greater than 4 μm are defined as major (also termed 'prominent') protrusions [8, 19, 28] (*see* **Note 20**). Use the "Freehand selections"tool to select the protrusion and use "Analyze › Measure" to measure the area. Before measuring, ensure that the "Area" box under "Analyze › Set Measurements" is selected. Use the "Straight Line" function and "Analyze › Measure" tools to measure the overall length of the protrusion (Fig. 5e).

(b) To assess the directionality of protrusions, select the "Angle Tool" (found on the main toolbar; Fig. 5a), which will be used to measure the angle of the protrusion (Fig. 5f). With the anterior-posterior axis of the egg chamber aligned with the anterior at the left, measure the protrusion angles off a center line that cuts through that axis. Protrusions can be classified into front (0° to 45° and 0° to 315°), side (45° to 135°and 225° to 315°), and back (135° to 225°) directions based on their measured position around the cluster (Fig. 5g).

(c) To calculate protrusion lifetime, subtract the time of the frame in which the protrusion begins to extend from the cluster from the time at which the protrusion fully retracts back inside the cluster.

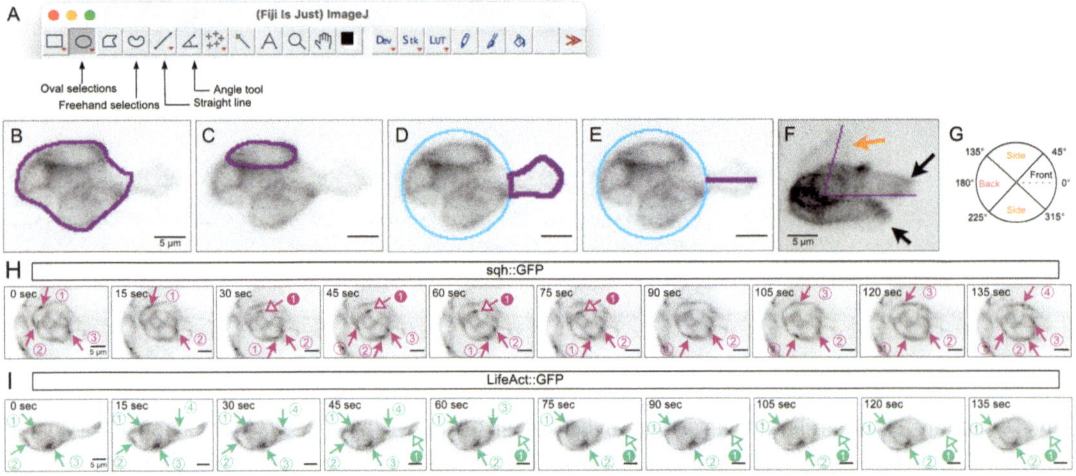

Fig. 5 Analysis of protrusion and protein dynamics in live border cells. (**a**–**g**) Measurement and selection of live border cell cellular features using Fiji: (**a**) screen capture of the FIJI toolbar; (**b**) cluster shape; (**c**) single border cell shape; (**d**) area of the major protrusion; (**e**) length of the major protrusion; and (**f**, **g**) the direction of the major protrusion. (**b**–**e**) A still image of the same *c306*-GAL4/+; UAS- PLCδ-PH-EGFP/+ border cell cluster shown at high magnification. (**f**) A still image of a different c306-GAL4/+; UAS-PLCδ-PH-EGFP/+ border cell cluster shown at high magnification. (**h**, **i**) Measurement of enriched accumulations of proteins in live border cells using fluorescent markers of myosin II (**h**; sqh::GFP) and F-actin (**i**; LifeAct::GFP). (**h**) Accumulation of sqh::GFP (fTRG Line; Table 1) at the cluster perimeter (solid pink arrows) and inside the cluster (hollow pink arrows) at the indicated times (in seconds). Numbers indicate focal accumulations of sqh::GFP per frame; open numbers indicate foci at cluster perimeter and closed numbers indicate foci inside the cluster. (**i**) Accumulation of LifeAct::GFP (driven directly by the *slbo* enhancer; Table 1) at the cluster perimeter (solid teal arrows) and in protrusions (hollow teal arrows) at the indicated times (in seconds); open numbers indicate focal accumulation of F-actin at the cluster perimeter and closed numbers indicate accumulation in the protrusion. Anterior is to the left in all panels and scale bars indicate the image magnification

3.5.3 Measure Protein Localization and Dynamics in Fiji

Protein localization and dynamics can be assessed in live control and mutant border cells. Here, we focus on myosin II and F-actin, for which we have extensive experience imaging and quantifying [8, 11, 30]. GFP-tagged Spaghetti Squash (Sqh:GFP) labels the *Drosophila* non-muscle myosin II regulatory subunit (MRLC) and thus visualizes the dynamics of myosin II, particularly at cell membranes. LifeAct:GFP detects F-actin and has been used extensively in border cells (Table 1; [8, 19, 20]). Other proteins can be measured (e.g., E-cadherin, Moesin; [19, 20]). Measurements and analyses for each specific fluorescently labeled protein needs to be empirically determined. Moreover, experimental validation needs to be performed for each fluorescently tagged-protein to ensure accurate reporting of the protein localization and dynamics in vivo.

During border cell migration, Sqh:GFP accumulation occurs mostly at the outer periphery of the border cell cluster (supracellular localization), with occasional accumulation observed inside the cluster at contacts between border cells in some movie frames (Fig. 5h). Accumulation of LifeAct:GFP can be found at the protrusion and the periphery of the border cell cluster (Fig. 5i). Both Sqh-GFP and LifeAct-GFP foci at the periphery should be dynamic at cell membranes while the cluster is migrating. The average intensity of fluorescent signals ("foci" or spots of accumulated protein) can be compared between control clusters and genetically manipulated clusters. These measurements can provide useful insights into the mechanical forces and other cellular behaviors that regulate border cell migration.

1. To count the number of Sqh:GFP or LifeAct:GFP foci in each frame for the entire movie, use the "Freehand selections"tool to manually select identified foci (Fig. 5a). Example fluorescent protein foci are shown in Fig. 5h, i. It is helpful to play the movie backward and forward to accurately distinguish these dynamic foci.
2. To measure the intensity of each focal accumulation of the fluorescently labeled protein, select the "Analyze › Set Measurements" tool. Next, select the "Mean gray value" and use the "Analyze › Measure" function (Fig. 5a).
3. Quantify the average number of fluorescent protein foci by dividing the total number of fluorescent focal accumulations by the total number of movie frames.
4. Alternatively, or in addition, calculate the number and average intensity of foci at the periphery of the cluster versus inside the cluster (internal cell membranes) as described above in **steps 2** and **3** and shown in Fig. 5h, i.

4 Notes

1. Selecting an appropriate fluorescent reporter (and GAL4 driver, if required) is crucial for setting up a successful time-lapse imaging experiment (Table 1). *slbo*-GAL4 and *c306*-GAL4 are the two most commonly used GAL4 lines for time-lapse imaging of border cell migration. If driving RNAi, *c306*-GAL4 is often used; however, because *c306*-GAL4 drives expression during early oogenesis [11, 31], ts-Gal80 should be employed to block early GAL4/UAS expression. All genotypes should be tested prior to scaling up experiments to make sure no unexpected defects in border cell migration occur. For example, *c306*-GAL4, tsGal80; UAS-PLCδPH:GFP (Table 1; Fig. 4) is typically outcrossed to a control fly stock (e.g., w^{1118}) to ensure normal migration; two copies of *c306*-GAL4 in this stock can cause developmental defects, whereas outcrossing eliminates these defects. *See* Table 1 for additional known

advantages and disadvantages for specific fluorescent reporters and GAL4/UAS combinations that have been used to image migrating border cells. When performing genetic manipulations (e.g., using RNAi, mutant alleles, overexpression), be sure to have controls for live imaging. This will ensure that any defect observed in border cells is a real phenotype and not due to any extraneous issues such as reagents that have gone bad, issues with dissecting or mounting the samples, or phototoxicity during imaging.

2. The pH of the live imaging medium is one of the most crucial factors for time-lapse imaging of border cells [5]. Ideally, the pH should be between 6.85 and 6.95. Egg chambers tend to show developmental defects and/or border cell migration defects in media with pH outside this range. Before using the dissection or live imaging medium, check for any signs of contamination or growth, such as cloudiness. It is best to prepare fresh batches of media or enough to last 1–2 months.
3. Oxygen is very important for successful live imaging of border cells. We have successfully used the Lumox dish (Sarstedt Lumox 50 tissue culture dish, 50 mm, adherent) with its oxygen-permeable membrane for both widefield and confocal imaging. Other types of dishes or slides can be used [30, 32, 33], including use of MatTek glass bottom dishes. However, special care with these other types of preparations must be made to prevent drying out of the imaging medium, and thus the egg chambers, during imaging.
4. We have used Halocarbon 27 from MilliporeSigma (cat. no. H8773) with success.
5. Mounting in a fibrinogen-thrombin clot is an optional step but is useful when using objectives with small working distances, such as the 40× water immersion objective. Fibrinogen and thrombin together form a clot that immobilizes the egg chambers [22].
6. We have successfully time-lapse imaged border cells using a Zeiss Axio Imager Z1 upright epifluorescence microscope, which has a motorized z-focus drive, and a Zeiss 880 confocal with an inverted microscope and motorized scanning x-y stage. If using an upright microscope, care must be made to ensure that the objective does not hit the sides of the dish. While different fluorescent light sources can be used for epifluorescence illumination, LED illuminators (e.g., Zeiss Colibri) offer significant benefits for time-lapse imaging. LED light sources provide gentle light for live imaging due to the precise control of light intensity and wavelength and ability to be precisely switched on and off. When using a confocal microscope, care

must be made to reduce phototoxicity and potential photobleaching of fluorescent proteins even with lower laser intensity.

7. Handling the ovaries too much increases the risk of tissue damage and potential failure of egg chamber development and/or impairment of border cell migration during live imaging. Likewise, leaving behind excess tissue can result in unwanted movement of egg chambers during imaging or depletion of nutrients in the medium by older egg chambers. It is important to very slowly and gently pull out only one or two ovarioles to allow full removal of the ovarioles from the overlying muscle sheath. Failure to remove the muscle sheath will potentially cause unwanted movement of the egg chamber due to the muscle contracting during imaging.
8. It is important to transfer dissected whole ovaries to a well with fresh dissecting media. This is because the initial dissection of ovaries from the fly will be accompanied by other organs and tissues that can contaminate the dissecting medium.
9. To ensure proper egg chamber health, minimize the time it takes to dissect the ovarioles and mount the egg chambers. We try to limit dissection and mounting to ~30 minutes from start to finish.
10. The Lumox dish can be reused multiple times but discard the dish if the membrane tears. To reuse the Lumox dish, carefully remove the cover glass and cover glass shards after imaging. Immerse and soak the dish in "Windex Glass Cleaner" (or similar cleaning solution) overnight while shaking to remove the halocarbon oil. While wearing gloves, gently rub the membrane with your fingers under running tap water to ensure no residual halocarbon oil remains. Rinse the dish with deionized water and air dry. Store the cleaned Lumox dish in a closable container to avoid dust or disturbance.
11. There should be no air bubbles between the dish and coverslip. As needed, gently nudge the coverslip to push any bubbles out.
12. When scanning the dish to find a desired egg chamber to image, it is best to start from the center of the coverslip instead of the sides. We often use a 10× objective to find a sample, then switch to a 20× high NA objective for the time-lapse imaging experiment. Avoid egg chambers that are near the edge of the coverslip if you are imaging with a 20× objective, or another objective with a short working distance, so that the halocarbon oil does not come into contact with the objective lens.
13. Sometimes egg chambers only show defects during imaging. Some egg chambers are either developmentally behind, show migration defects when not expected, or the egg chamber simply moves out of the frame of view during imaging. Normal

development of the egg chamber includes the growth of the oocyte, movement of the follicle cell layer towards the oocyte, and cytoplasmic flow within the oocyte (visualized by yolk granule movement in the brightfield channel). If defects are observed, the best solution is either to find another egg chamber within the dish, being sure to avoid any egg chambers that might be even slightly damaged, or to start over and prepare a fresh sample. If you encounter too many problems during imaging, it is a good idea to recheck the pH of the live imaging media.

14. The mounted live imaging preparation is generally good for up to 6 hours. Generally, one preparation is used to perform one longer length movie to analyze cell shapes and protrusions. For tracking protein dynamics, one preparation can be used for multiple movies. Time and phototoxicity are correlated when it comes to setting up time intervals for the experiment. An optimal combination of time intervals and speed of acquisition is key to this process. Shorter time intervals can lead to defects caused by phototoxicity due to continuous exposure while capturing the slices. However, longer time intervals allow the entire cluster to be imaged using z-stacks while also decreasing the chance for phototoxicity to occur. Short intervals (15-- 30 seconds) allow the capture of more cellular details and dynamics but increase the chances of photobleaching and phototoxicity. Long intervals (e.g., 2–3 minutes), on the other hand, capture fewer cellular details and dynamics but with less chance of phototoxicity. Using a widefield epifluorescent microscope, an LED light source, and 2–3 minute intervals, we typically image one egg chamber for 2–4 hours. Depending on the time interval, live imaging on a confocal can be done for 1–4 hours. Signs of phototoxicity include shrinking of the egg chamber and/or loss of yolk granule movement in the oocyte. If this is observed, move to a new sample and adjust the light levels and/or time interval.

15. The border cell cluster can sometimes go out of focus during migration, either because of sample drift or the border cell migration path between nurse cells. Therefore, it is important to consider the thickness of the border cell cluster when setting up z-stacks. We recommend setting a z-stack of at least 20 μm with 10 slices (or 1 μm interval between z-slices) to ensure that most of the entire cluster height (estimated to be ~30–50 μm) is captured. Alternatively, manually set the top and the bottom of the cluster *z*-positions.

16. If using a scanning stage, it is useful to pre-select multiple egg chambers to image (e.g., we typically choose 4–10 egg chambers). Having pre-selected egg chambers saves time finding samples during the imaging experiment because not all egg

chambers will be at the right stage or develop properly. Although we typically do not perform imaging on multiple samples in the same experiment, this can be done. In this case, it is useful to limit imaging to a few egg chambers at a time, as this will depend on the time intervals of the experiment. If performing imaging on multiple samples in the same experiment, most microscope software packages only allow a single z-stack range and image frame size to be set up. Therefore, adjust the z-stack range and image frame size accordingly.

17. Using an image with a frame size of "512 × 512" is generally sufficient for publication quality imaging.
18. When opening a large movie file, in "Bio-Formats Import Options" we often select "Data Browser" from the pull-down menu ("View stack with," select "Data Browser" rather than "Hyperstack") to view the z-stack and then select "Use Virtual Stack" in "Memory Management" (Fig. 3a).
19. We often find it easier to invert the fluorescence image, which allows for better contrast when visualizing and measuring cellular behaviors and protein localization. As preferred, leave the image as acquired. Alternatively, use "Image › Lookup Tables" to change the color and/or use a color range such as "Rainbow RGB," or other appropriate LUT, to visualize pixel intensity of fluorescent proteins.
20. We define major protrusions as those cellular extensions that project away from the cell body and are greater than 4 μm in length from the base of the protrusion to the tip. Other studies use the same method to identify where in the cluster protrusions form but use different criteria to identify major protrusions. These criteria include any cellular extension greater than 2 μm [21], any protrusion greater than 3 μm long (base to tip) and 3 μm wide at the base of the protrusion [34], or any cellular extension that is thinner in diameter than the cell body [18].

Acknowledgments

This work was supported by the National Science Foundation, awards OIA-1738757 and IOS-2027617 to J.A.M. The Confocal Core, funded by the Kansas State University College of Veterinary Medicine, provided the use of the Zeiss LSM 880 confocal microscope. The authors thank Drs. Joseph Campanale and Denise Montell for previous advice on live cell imaging and quantification methods. Thank you to Emily Burghardt and C. Luke Messer for helpful comments on the manuscript.

References

1. Nagai T, Ishikawa T, Minami Y, Nishita M (2020) Tactics of cancer invasion: solitary and collective invasion. J Biochem 167:347–355
2. Scarpa E, Mayor R (2016) Collective cell migration in development. J Cell Biol 212: 143–155
3. Friedl P, Gilmour D (2009) Collective cell migration in morphogenesis, regeneration and cancer. Nat Rev Mol Cell Biol 10:445–457
4. Saadin A, Starz-Gaiano M (2016) Circuitous genetic regulation governs a straightforward cell migration. Trends Genet 32:660–673
5. Prasad M, Jang AC-C, Starz-Gaiano M et al (2007) A protocol for culturing Drosophila melanogaster stage 9 egg chambers for live imaging. Nat Protoc 2:2467–2473
6. Dai W, Guo X, Cao Y et al (2020) Tissue topography steers migrating Drosophila border cells. Science 370:987–990
7. Zeledon C, Sun X, Plutoni C et al (2019) The ArfGAP drongo promotes actomyosin contractility during collective cell migration by releasing myosin phosphatase from the trailing edge. Cell Rep 28:3238–3248
8. Chen Y, Kotian N, Aranjuez G et al (2020) Protein phosphatase 1 activity controls a balance between collective and single cell modes of migration. elife 9:e52979
9. Ramel D, Wang X, Laflamme C et al (2013) Rab11 regulates cell-cell communication during collective cell movements. Nat Cell Biol 15: 317–324
10. Lamb MC, Kaluarachchi CP, Lansakara TI et al (2021) Fascin limits Myosin activity within *Drosophila* border cells to control substrate stiffness and promote migration. elife 10: e69836
11. Aranjuez G, Burtscher A, Sawant K et al (2016) Dynamic myosin activation promotes collective morphology and migration by locally balancing oppositional forces from surrounding tissue. Mol Biol Cell 27:1898–1910
12. Cai D, Chen S-C, Prasad M et al (2014) Mechanical feedback through E-cadherin promotes direction sensing during collective cell migration. Cell 157:1146–1159
13. Spradling AC (1993) Developmental genetics of oogenesis. In: Bate, Martinez-Arias (eds) The development of Drosophila melanogaster. Cold Spring Harbor Laboratory Press, Cold Spring Harbor
14. Bianco A, Poukkula M, Cliffe A et al (2007) Two distinct modes of guidance signalling during collective migration of border cells. Nature 448:362–365
15. Prasad M, Montell DJ (2007) Cellular and molecular mechanisms of border cell migration analyzed using time-lapse live-cell imaging. Dev Cell 12:997–1005
16. Poukkula M, Cliffe A, Changede R, Rorth P (2011) Cell behaviors regulated by guidance cues in collective migration of border cells. J Cell Biol 192:513–524
17. Combedazou A, Choesmel-Cadamuro V, Gay G et al (2017) Myosin II governs collective cell migration behaviour downstream of guidance receptor signalling. J Cell Sci 130:97–103
18. Cliffe A, Doupé DP, Sung H et al (2017) Quantitative 3D analysis of complex single border cell behaviors in coordinated collective cell migration. Nat Commun 8:14905
19. Plutoni C, Keil S, Zeledon C et al (2019) Misshapen coordinates protrusion restriction and actomyosin contractility during collective cell migration. Nat Commun 10:3940
20. Mishra AK, Mondo JA, Campanale JP, Montell DJ (2019) Coordination of protrusion dynamics within and between collectively migrating border cells by myosin II. MBoC 30:2490–2502
21. Wang X, He L, Wu YI et al (2010) Light-mediated activation reveals a key role for Rac in collective guidance of cell movement in vivo. Nat Cell Biol 12:591–597
22. Wilcockson SG, Ashe HL (2021) Live imaging of the Drosophila ovarian germline stem cell niche. STAR Protoc 2:100371
23. Schindelin J, Arganda-Carreras I, Frise E et al (2012) Fiji: an open-source platform for biological-image analysis. Nat Methods 9: 676–682
24. Wang H, Guo X, Wang X et al (2020) Supracellular actomyosin mediates cell-cell communication and shapes collective migratory morphology. iScience 23:101204
25. Lamb MC, Anliker KK, Tootle TL (2020) Fascin regulates protrusions and delamination to mediate invasive, collective cell migration in vivo. Dev Dyn 249:961–982
26. Badmos H, Cobbe N, Campbell A et al (2021) Drosophila USP22/nonstop polarizes the actin cytoskeleton during collective border cell migration. J Cell Biol 220:e202007005
27. Wilcockson SG, Ashe HL (2019) Drosophila ovarian germline stem cell cytocensor projections dynamically receive and attenuate BMP signaling. Dev Cell 50:296–312.e5
28. Sawant K, Chen Y, Kotian N et al (2018) Rap1 GTPase promotes coordinated collective cell

migration in vivo. Mol Biol Cell 29:2656–2673
29. Starz-Gaiano M, Melani M, Wang X et al (2008) Feedback inhibition of JAK/STAT signaling by apontic is required to limit an invasive cell population. Dev Cell 15:330
30. Majumder P, Aranjuez G, Amick J, McDonald JA (2012) Par-1 controls myosin-II activity through myosin phosphatase to regulate border cell migration. Curr Biol 22:363–372
31. Murphy AM, Montell DJ (1996) Cell type-specific roles for Cdc42, Rac, and RhoL in Drosophila oogenesis. J Cell Biol 133:617–630
32. Manning L, Starz-Gaiano M (2015) Culturing drosophila egg chambers and investigating developmental processes through live imaging. Methods Mol Biol 1328:73–88
33. Cetera M, Lewellyn L, Horne-Badovinac S (2016) Cultivation and live imaging of drosophila ovaries. Methods Mol Biol 1478:215–226
34. Wang H, Qiu Z, Xu Z et al (2018) aPKC is a key polarity determinant in coordinating the function of three distinct cell polarities during collective migration. Development 145: dev158444
35. Spracklen AJ, Fagan TN, Lovander KE et al (2014) The pros and cons of common actin labeling tools for visualizing actin dynamics during Drosophila oogenesis. Dev Biol 393: 209–226
36. Royou A, Field C, Sisson JC et al (2004) Reassessing the role and dynamics of nonmuscle myosin II during furrow formation in early Drosophila embryos. Mol Biol Cell 15:838–850
37. Chang Y-C, Wu J-W, Hsieh Y-C et al (2018) Rap1 negatively regulates the hippo pathway to polarize directional protrusions in collective cell migration. Cell Rep 22:2160–2175
38. Rauzi M, Lenne P-F, Lecuit T (2010) Planar polarized actomyosin contractile flows control epithelial junction remodelling. Nature 468: 1110–1114
39. Cabernard C, Doe CQ (2009) Apical/basal spindle orientation is required for neuroblast homeostasis and neuronal differentiation in Drosophila. Dev Cell 17:134–141
40. McDonald JA, Khodyakova A, Aranjuez G et al (2008) PAR-1 kinase regulates epithelial detachment and directional protrusion of migrating border cells. Curr Biol 18:1659–1667
41. Huang J, Zhou W, Dong W et al (2009) From the cover: directed, efficient, and versatile modifications of the Drosophila genome by genomic engineering. Proc Natl Acad Sci U S A 106:8284–8289
42. Sharma A, Halder S, Felix M et al (2018) Insulin signaling modulates border cell movement in Drosophila oogenesis. Development 145: dev166165-43

Chapter 11

Live Imaging of Nurse Cell Behavior in Late Stages of *Drosophila* Oogenesis

Jonathan A. Jackson, Jasmin Imran Alsous, and Adam C. Martin

Abstract

Drosophila oogenesis is a powerful and tractable model for studies of cell and developmental biology due to the multitude of well-characterized events in both germline and somatic cells, the ease of genetic manipulation in fruit flies, and the large number of egg chambers produced by each fly. Recent improvements in live imaging and ex vivo culturing protocols have enabled researchers to conduct more detailed, longer-term studies of egg chamber development, enabling insights into fundamental biological processes. Here, we present a protocol for dissection, culturing, and imaging of late-stage egg chambers to study intercellular and directional cytoplasmic flow during "nurse cell dumping." This critical developmental process towards the latter stages of oogenesis (stages 10b/11) results in rapid growth of the oocyte and shrinkage of the nurse cells and is accompanied by dynamic changes in cell shape. We also describe a procedure to record high-time-resolution movies of the flow of unlabeled cytoplasmic contents within nurse cells and through cytoplasmic bridges in the nurse cell cluster using reflection microscopy, and we describe two ways to analyze data from nurse cell dumping.

Key words *Drosophila* egg chamber, Oogenesis, Intercellular flow, Live Imaging, Nurse cell dumping, Reflection microscopy

1 Introduction

Oogenesis in the fruit fly *Drosophila melanogaster* is a process that begins in the germarium, at the anterior tip of the ovary. Germline and somatic stem cells give rise to cystoblasts and follicle cells, respectively, after which each cystoblast undergoes four rounds of incomplete cell division to produce the interconnected germline cyst. The 16-cell germline cyst is then encapsulated by somatic cells before pinching off from the germarium; at this point, the collection of germline and somatic cells is known as an "egg chamber." *Drosophila* oocytes develop and increase dramatically in volume while remaining connected to the fifteen supportive "nurse cells" in a stereotyped manner through cytoplasmic bridges called ring canals. During the roughly 96 h course of oogenesis following

Michelle S. Giedt and Tina L. Tootle (eds.), *Drosophila Oogenesis: Methods and Protocols*,
Methods in Molecular Biology, vol. 2626, https://doi.org/10.1007/978-1-0716-2970-3_11,

encapsulation, an egg chamber progresses through 14 stages of development while moving toward the posterior of the ovary (representative images of a germarium and egg chambers in stages 2–14 are shown in Fig. 1), finally becoming a mature egg and entering the oviduct [1, 2]. Oogenesis in the fruit fly is highly parallelized: egg chambers develop in strings known as ovarioles, with germaria at their anterior tip and mature eggs at the posterior, and 15–20 ovarioles are bound together to form one ovary.

Because of the reproducibility of egg chamber structure, the high number of chambers per fly, the abundance of genetic tools available for *Drosophila*, and the amenability of egg chambers to ex vivo live imaging, studies of fruit fly oogenesis have provided insight into processes including stem cell regulation, tissue-level shape changes, cell migration, and three-dimensional morphogenesis in epithelia [3–6]. One behavior for which *Drosophila* oogenesis provides an exciting model is cytoplasmic transport, both within cells, as with cytoplasmic streaming in the oocyte [7, 8], and between cells, as has also been observed during mouse oocyte development [9]. Around stages 11 and 12 of *Drosophila* oogenesis, nurse cells transfer upwards of 75 percent of their cytoplasmic contents through ring canals into the oocyte over approximately 90 min, accompanied by actomyosin-driven cell shape changes. This process, known as "nurse cell dumping," results in the rapid growth of the oocyte as it receives the materials necessary to eventually support early embryonic development [10–12]. Fluid transport is largely unidirectional, proceeding towards the oocyte at the posterior of the chamber, and it is hierarchical, with fluid from more anterior nurse cells passing through up to three more posterior nurse cells before entering the oocyte. Nurse cell dumping thus presents an opportunity to study the interplay of cytoplasmic pressure, contractility, and cell geometry during fluid transport within a multicellular network.

In recent years, improvements in imaging modalities and ex vivo culturing techniques have opened the door to longer-term, more quantitative studies of developmental processes during oogenesis [3, 6, 11, 13]. Several excellent protocols for live imaging of egg chambers have been previously published, using various designs for the imaging apparatus and focusing on different stages or developmental processes [13–16]. Here, we present a relatively simple protocol for long-term (~2–4 h) live imaging of late-stage egg chambers (stages 10–14), originally developed for capturing the dynamics of cytoplasmic flow during nurse cell dumping in stage 11/12 chambers [11] but readily adaptable to younger egg chambers. Additionally, we describe the use of confocal reflection microscopy [17, 18] to image intra- and intercellular flow of unlabeled cellular contents. Finally, we provide a brief description of two possible types of analysis that can be performed on data collected using this protocol.

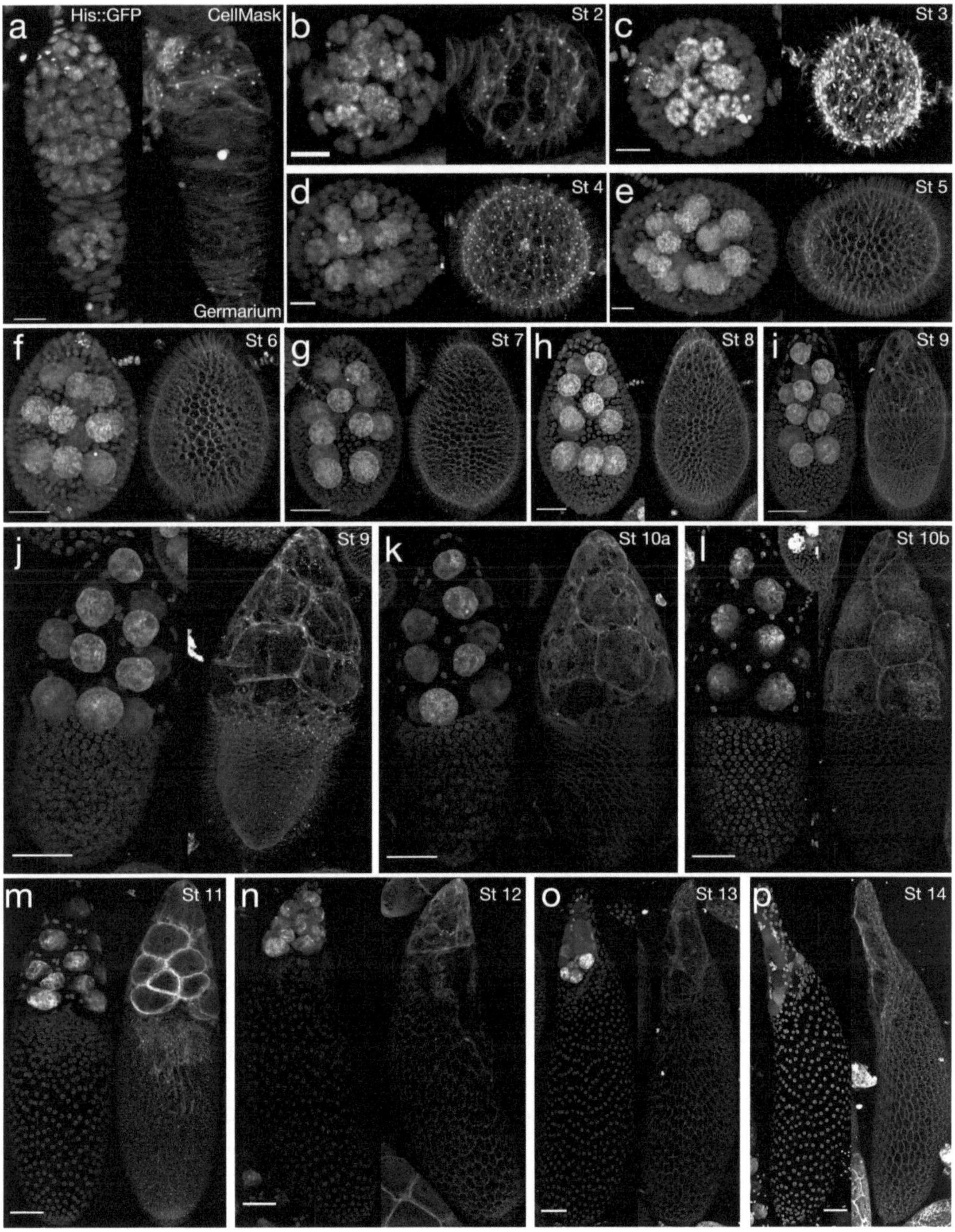

Fig. 1 The 14 stages of *Drosophila melanogaster* oogenesis. Histone H2A::GFP (left) and CellMask signal (right) for egg chambers from youngest to oldest, with stage (St) for each. (**a**) germarium; (**b**) stage 2; (**c**) stage 3; (**d**) stage 4; (**e**) stage 5; (**f**) stage 6; (**g**) stage 7; (**h**) stage 8; (**i**) early stage 9, just prior to border cell migration; (**j**) later in stage 9, when the border cells are partway to the oocyte; (**k**) stage 10a; (**l**) stage 10b; (**m**) stage 11, relatively early into the nurse cell dumping process; (**n**) stage 12, roughly at the end of nurse cell dumping; (**o**) stage 13; (**p**) stage 14. Note: inhomogeneity in follicle cell layer CellMask signal in (**m**, **n**) is partly a result of uneven staining where the egg chamber was in contact with the glass dish (slight spatial discontinuities in (**m**–

2 Materials

2.1 Materials for Dissection of Individual Follicles

1. Fly stocks (exact stocks used will depend on the goals of the experiment).
2. Fly vials with food.
3. Dry yeast.
4. Schneider's insect medium (*see* **Notes 1** and **2**).
   (a) Fetal bovine serum (FBS), added to Schneider medium to a final concentration of 15% (vol/vol) (optional for stage 10+ egg chambers, but necessary for stage 9 or younger; *see* **Note 3**).
   (b) Insulin, added to Schneider medium to a final concentration of 250 μg/mL (optional for stage 10+ egg chambers, but necessary for stage 9 or younger).
5. 1.5 mL microcentrifuge tubes to hold dissection medium.
6. 100 μL pipette with tips cut off about ¼ inch from the end to avoid shearing egg chambers during transfer.
7. CO_2 source and diffusing pad.
8. Fine brush and/or feather tool for sorting flies.
9. Straight forceps (110 mm length × 8 mm width, tip diameter ~0.1 mm).
10. Stereomicroscope with a black stage and adjustable light source.
11. Glass dissecting slide with concave wells.
12. Fine tungsten needles (tip diameter 0.001 mm).
13. Metal needle holders.
14. 35 mm, no. 1.5 glass-bottomed dish.
15. Kimwipes.
16. 70% isopropanol for cleaning forceps and the dissecting slide.

2.2 Imaging

1. (Optional) Membrane stain (e.g., CellMask Deep Red Plasma membrane stain, Invitrogen).
2. Laser-scanning confocal (or similar) microscope able to perform both fluorescence and reflection microscopy (*see* **Note 4**).

Fig. 1 (continued) p) result from imperfect merging of tile scans). Scale bars: 10 μm (**a**–**e**), 20 μm (**f**), 30 μm (**g**,**h**), 50 μm (**i**–**p**). Stages are determined from a combination of egg chamber dimensions, nuclear size, and other morphological considerations detailed previously [1, 20]. Anterior is at the top in all panels except (**b**–**e**), in which anterior is to the left. Other objects in the corners of images are portions of nearby egg chambers that are partly in the field of view

2.3 Image Analysis

1. Computer with image analysis software such as FIJI [19] installed. More specialized software intended for 3D/4D viewing and volume measurements, such as Imaris (Bitplane), is also useful.

3 Methods

All procedures are carried out at room temperature unless otherwise specified.

3.1 Dissection of Individual Follicles

1. The day before dissection, transfer females 3–5 days post-eclosion, along with a roughly equal number of males of any age, to a new vial with a thin layer of dry yeast added. Flies fed with yeast produce more eggs, making it easier to remove ovaries (*see* **Note 5**).
2. Transfer about 500 μL of medium to an Eppendorf tube. Allow this tube to come to room temperature.
3. Anesthetize flies on the CO_2 pad. Separate out 2–5 females to dissect and return the remaining flies to the vial.
4. Place a dissecting slide on the stage of the stereomicroscope. Add 80–100 μL of medium to one well of the slide.
5. Using one pair of forceps, pick up a fly by its legs, transferring it to the filled well of the glass slide, wings down (*see* **Note 6**).
6. Grasp the fly around the anterior end of the abdomen using the forceps in the non-dominant hand and apply gentle pressure to distend the abdomen slightly. With the pair of forceps in the dominant hand, pinch the abdomen near the ovipositor and pull directly away from the first pair of forceps. The ovaries will usually separate from the fly (Fig. 2a) (*see* **Notes 7–9**).
7. Immediately remove the fly carcass and crush it using a Kimwipe.
8. If necessary, use the tungsten needles or forceps to separate the two ovaries by severing the oviduct connecting them.
9. Either pin one ovary down with gentle pressure from a tungsten needle or grasp the ovary carefully with one pair of forceps (*see* **Notes 10** and **11**).
10. While keeping the ovary immobilized with one hand, use a tungsten needle to slide between ovarioles from the posterior to the anterior tip, disrupting the peritoneal muscle sheath to separate ovarioles (Fig. 2b). Ideally, the ovarioles will splay out, allowing easy access to the egg chambers (*see* **Note 12**).
11. Identify stage 10b-11 egg chambers, shortly before or during nurse cell dumping (Fig. 2c). For a more detailed discussion of stage determination, *see* [20].

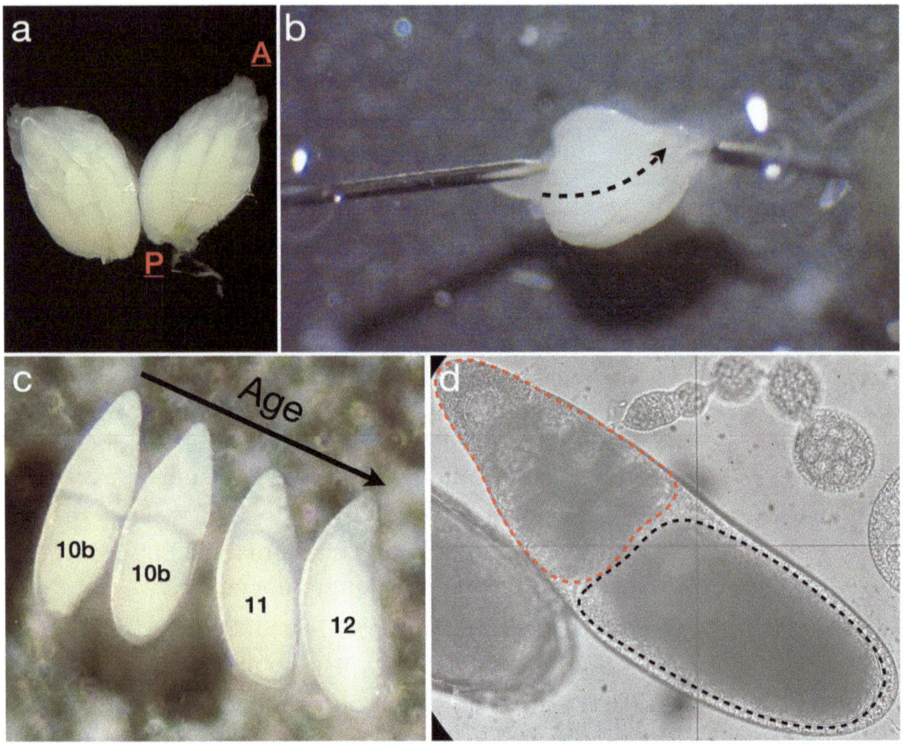

Fig. 2 Ovary dissection and identification of egg chambers undergoing nurse cell dumping. (**a**) One pair of ovaries removed from a fruit fly; A is the anterior end, where the germaria are located, and P is the posterior end, where mature stage 14 eggs reside. (**b**) Disruption of the peritoneal muscle sheath. The needle on the left is used to immobilize the ovary, while the needle on the right is moved gently along the dashed arrow, between ovarioles, to remove the muscle sheath holding the ovarioles together. (**c**) Four egg chambers in stages 10b-12, with the youngest (stage 10b) just beginning the nurse cell dumping process at left, and the oldest (stage 12) shortly after dumping is completed at right. (**d**) Brightfield image of an egg chamber approximately 10–20 min into the dumping process with the oocyte (black dashed outline) and the nurse cell cluster (red dashed outline) highlighted

(a) These egg chambers are typically near the posterior end of ovarioles, often just anterior to fully mature eggs.

(b) At stage 10b, the oocyte is ~50% the length of the germline cluster. If it is slightly larger, but the nurse cells are still individually visible, dumping has begun relatively recently (Fig. 2d).

12. Stage 10–12 egg chambers will often be ejected from the epithelial muscle sheath surrounding the ovariole due to contractions of the sheath (*see* **Note 13**). If they are not, use both needles in a scissoring motion to cut through the stalk cells between egg chambers and remove egg chambers from either end of the desired one. Ensure there is no remaining epithelial

muscle sheath around the egg chamber of interest or any egg chamber still connected to it, as its contractions will cause the egg chamber to move sporadically while imaging.

13. Using the side of one needle, move all egg chambers to be imaged to one side of the well and all remaining tissue as far away from them as possible.
14. Pipette 100–200 μL of medium into the well of the glass-bottomed dish.
17. Place the cut-off tip of a pipette near the group of stage 10b-11 chambers, on the side away from the remaining unwanted chambers. Gently pipette up the egg chambers, avoiding the other chambers, and transfer them to a 35 mm, no. 1.5 glass-bottomed dish (*see* **Note 14**). The total volume of medium in the dish should be at least 200 μL now; if not, add more. Having excess medium is better than having too little.
15. Clean the dissecting slide and forceps with 70% isopropanol and dry with a Kimwipe.

3.2 Imaging

1. (Optional) If desired, add CellMask to the dish to a final concentration of 5 μg/mL, or use a different membrane dye (*see* **Note 15**). Swirl the dish gently to mix and allow egg chambers to incubate in CellMask 5–10 min before imaging (*see* **Note 16**).
2. Place the dish, with its lid on to prevent drying of the medium and sample, on an inverted microscope.
3. For fluorescence microscopy, image z-stacks as usual.
4. To adapt a fluorescence microscopy setup to a reflection microscopy setup (*see* **Notes 17** and **18**):
   (a) Replace the dichroic beamsplitter normally used in the beam path for fluorescence microscopy with a partial mirror. This will allow some incident light to reach the sample and some reflected light of the same wavelength to reach the detector (*see* **Note 19**).
   (b) Set the detection wavelength equal to the excitation wavelength (*see* **Note 20**).
5. Set up time-lapse z-stack imaging. Egg chambers can be cultured and imaged for ~2–4 h with no noticeable ill effects depending on intensity and frequency of irradiation (*see* **Notes 21** and **22**).
6. When finished imaging, dispose of the glass-bottomed dish in the sharps disposal.

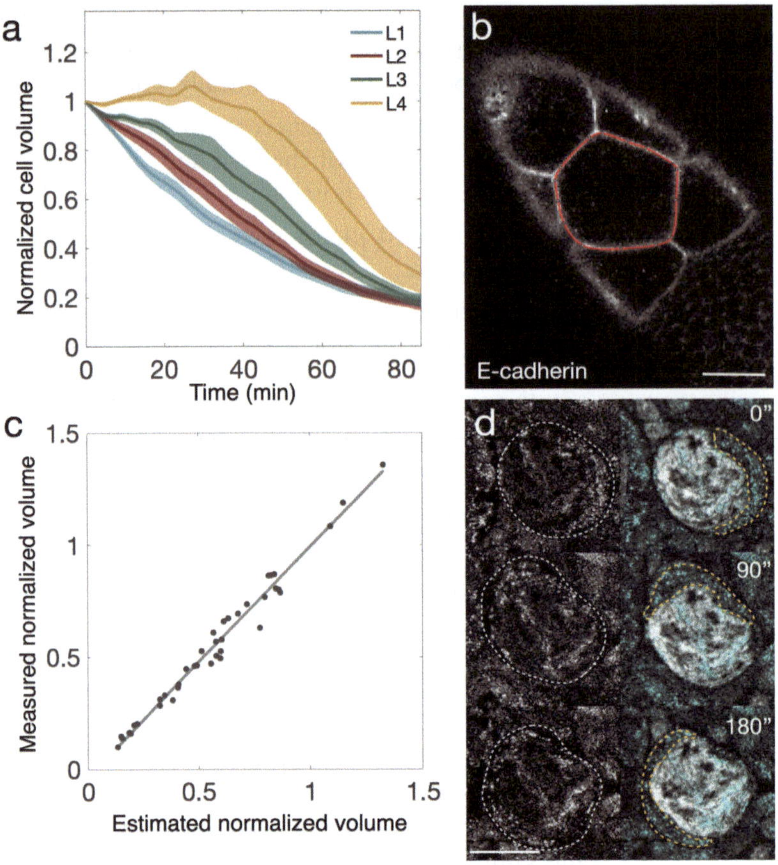

Fig. 3 Analysis of cell volume changes and reflection microscopy images of intracellular flow during nurse cell dumping. (**a**) Averaged, normalized cell volume trajectories from 41 nurse cells. $t = 0$ is the onset of nurse cell dumping. Lines show averages, envelopes standard error; colors correspond to rows of cells at different distances from the oocyte. Layer 1 (L1) cells are connected directly to the oocyte, while each successive layer is one ring canal farther away from the oocyte ($n = 15$, 12, 9, and 5 cells for layers 1, 2, 3, and 4, respectively). (**b**) Example of one midplane outline (red) in an E-cadherin::GFP-expressing egg chamber shortly after dumping onset. Scale bar: 40 μm. (**c**) Plot showing correlation between normalized volume measured from Bitplane's Imaris and normalized volume estimated from midplane area measurements in FIJI. Solid line shows the best fit ($R^2 = 0.98$, with slope 1.03). (**d**) Left column: reflection microscopy signal from a histone H2A::GFP-expressing egg chamber. Right column: merge of the same reflection signal (cyan) and H2A signal (gray), showing a pocket of cytoplasm (orange outline) flowing counterclockwise around the nucleus. Scale bar: 20 μm. (**a**–**c** reproduced from [11])

3.3 Image Analysis

This section describes two examples of the type of analysis possible with images acquired using the protocol above. The first involves determining nurse cell size trajectories from fluorescence images (*see* Fig. 3a), and the second describes more rapid cytoplasmic flow behaviors observed using reflection microscopy [11].

3.3.1 Size Trajectories

1. Open a 4D (x, y, z, t) movie in FIJI, preferably one in which at least one channel includes a cortical or cell body marker. If the

file size is larger than the memory available to FIJI, use the Bio-Formats importer to open the movie as a virtual stack instead.

2. Determine the approximate frame around which dumping begins. This is most easily accomplished by observing the boundary between the oocyte anterior and the posterior side of the nurse cell cluster, as the frame of dumping onset corresponds to the sudden increase in the rate at which this boundary moves towards the anterior (*see* **Note 23**).
3. For each nurse cell of interest, at each time point to use for analysis, step through the z-dimension to locate the plane with the maximal cross-sectional area for that nurse cell (*see* **Notes 24** and **25**).
4. Use the selection tool to trace the cell membrane (Fig. 3b), then measure the cell's cross-sectional area. Assuming the cells are roughly spherical, volume can be approximated by $V = \frac{4}{3\sqrt{\pi}} A^{3/2}$ (*see* **Note 26** and Fig. 3c regarding validation of this estimate using the full 3D data).
5. Repeat for each cell and time point to use for analysis. Ensure that one time point includes the initial frame corresponding to dumping onset ($t = 0$).
6. For each nurse cell, normalize the size measurements to the size at dumping onset (thus the normalized volume at time t is $V(t)/V_0 = (A(t)/A_0)^{3/2}$), where $A(t)$ is the measured cross-sectional area at time t and A_0 is the cross-sectional area at dumping onset.
7. Size trajectories can now be plotted for each cell if desired, showing the fractional change in volume over time (*see* **Note 27**).

3.3.2 Cytoplasmic Flow Behaviors

1. Open a reflection-mode microscopy movie in FIJI. Cell boundaries should be discernible from the reflection signal, as should a darker region in the cell interior corresponding to the nucleus (Fig. 3d).
2. In wild-type egg chambers, during the first ~45–60 min of dumping, flow of bright spots through ring canals should be visible, manifesting as bright puncta moving in a straight line between cells.
3. Once cell shape contractions begin, large-scale cytoplasmic movements can be tracked by searching for coordinated movement of large groups of puncta.
4. Flow of reflective cytoplasmic contents is in theory amenable to measurement via methods such as particle image velocimetry (PIV, *see* **Note 28**).

4 Notes

1. Egg chambers stage 10 or older can be cultured in Schneider's medium without supplementation, but younger egg chambers require Schneider's medium supplemented with insulin and fetal bovine serum.
2. Store Schneider's medium and insulin at 4 °C and FBS in aliquots at −20 °C. If supplementing with insulin and FBS, add Schneider's medium and insulin to the FBS aliquot the day of the experiment and keep at 4 °C when not in use.
3. Adding FBS to the dissection medium can also reduce the chance of egg chambers sticking to the pipette tip or the glass dish during transfer.
4. Many commercial systems can perform both reflection and fluorescence microscopy. For those that cannot, the dichroic beamsplitter used for fluorescence microscopy must be replaced by a mirror.
5. For egg chamber studies, flies are kept at room temperature (22–23 °C). Higher temperatures can also be used, e.g., to increase the efficiency of GAL4/UAS-driven RNA interference [21]. In that case, flies are allowed to eclose and are then transferred to the desired temperature to develop for at least 4 days.
6. Surface tension in the pool of dissecting medium in the well can pull the fly to the side of the pool and occasionally turn it on its side. **Step 6** can still be performed in this case, or the fly can be turned using the other pair of forceps and held in place.
7. If the ovaries remain inside the abdomen after pulling, use one pair of forceps to make a slit in the ventral side of the abdomen from the anterior end of the ovaries to the tear made in **step 6**. Carefully use the forceps to then remove the ovaries. This may require using the forceps to pull away connective tissue just anterior to the tip of the ovary.
8. Other tissues, especially the part of the posterior cuticle and intestines, usually come with the ovaries when pulling with the forceps. In this case, use the tungsten needles to separate the ovaries from all remaining tissue, then remove everything else from the dissecting medium using forceps.
9. Avoid rupturing the fat body or any part of the intestinal tract when performing **step 6**. Doing so will fill the dissection medium pool with debris; in this case, it is often best to start over with fresh medium in another well on the slide to avoid any possible effects of the released contents on the remaining egg chambers.

10. To avoid damaging stage 10/11 egg chambers, grasp or pin down the ovary as close to its posterior end as possible. If flies are well-fed, there should be several mature eggs posterior to the stage 10/11 egg chambers, which will allow pinning the ovary down without damaging the egg chambers of interest.
11. If at any point an egg chamber is ruptured, move it as far away from the remaining ones as possible to prevent its contents from leaking and surrounding the intact chambers, which can potentially affect the development of the intact chambers.
12. If the method in **steps 9** and **10** is difficult or if the egg chambers of interest are too close to the posterior end, anchor the ovary by inserting one needle through the egg chambers at the anterior end. Slide the other needle from anterior to posterior to disrupt the peritoneal sheath or place both needles at the anterior end and pull them apart in the direction perpendicular to the anterior–posterior axis.
13. Occasionally an egg chamber will slide only partly out of the epithelial muscle sheath, and the sheath will contract around the middle of the egg chamber, deforming it for a prolonged period. To be safe, avoid using these egg chambers as this contraction can cause subtle damage that leads to a failure of nurse cell dumping.
14. Avoid adding too many stage 10b or later egg chambers to the medium in the dish. Doing so can deplete nutrients from the medium too quickly and reduce the time egg chambers remain healthy in the culture medium, and it will also crowd the dish and cause chambers to contact one another.
15. A dye such as CellMask can be used to assess the integrity of egg chambers and to non-specifically label membranes. If a cell is damaged such that membrane integrity is compromised, the cell interior will appear bright, in contrast to the usual pattern of a dark interior with a few bright puncta and a bright membrane.
16. CellMask, if added directly to the imaging medium, will continue to bind the cell membranes, which will increase in brightness over time. Alternatively, the egg chambers can be incubated in a medium supplemented with CellMask before being washed and imaged in a fresh medium without CellMask.
17. Reflection microscopy can be used to image the flow of unlabeled cytoplasmic contents through ring canals. To do so, adjust the focal plane until it bisects a ring canal, visible as a small gap between cells through which bright puncta move. Set the acquisition time to ~0.5–2 s per stack, and flow should be visible when playing the acquired movie. A fluorescent marker for membranes or ring canals can be used to locate ring canals before switching to reflection mode.

18. Depending on the software, it is possible to perform both reflection and fluorescence microscopy in the same movie acquisition. However, doing so requires switching between a beamsplitter and a mirror during the acquisition of each z-stack, which in our experience greatly decreases the temporal resolution.
19. We use the Zen Black software, which has a preset box that can be checked to switch to reflection mode.
20. Reflection microscopy is not wavelength-specific. In our experience, 488 nm excitation works well, as it provides slightly better resolution than longer wavelengths, and shorter wavelengths often cause tissue damage more rapidly.
21. Because the egg chamber is approximately 150–175 μm in width at this point in oogenesis, imaging through the entire tissue is not generally possible at laser powers that are safe for long-term imaging of the egg chamber.
22. Reflection microscopy is the simplest modality for imaging flow in this stage. Some fluorescently labeled proteins useful for imaging the changes in cell size and shape accompanying cytoplasmic flow include:
    (a) Gap43::mCherry [22], E-cadherin::GFP [23], or CellMask for membranes.
    (b) Clip170::GFP [24], a microtubule-binding protein, which in these tissues labels cell bodies fairly uniformly when imaged at low laser power. This label can also be used to image cytoplasmic streaming in the oocyte in stages 10 and 11.
    (c) Pavarotti::GFP [25], a kinesin-like-protein. This label localizes to ring canals, although they only appear clearly in the latter part of dumping.
    (d) PCNA::GFP [26], which changes location from nucleus to cytoplasm during nurse cell dumping shortly before the onset of cell-shape contractions.
23. If it is difficult to identify the onset of nurse cell dumping by eye, an alternative approach involves a kymograph. Using a marker with a clear distinction between the oocyte and nurse cell cluster, create a kymograph showing intensity along the anterior–posterior axis of the egg chamber. There should be a clear timepoint in the kymograph at which the feature corresponding to the posterior edge of the nurse cells abruptly begins moving quickly; this is the time at which dumping begins.
24. If there is not a slice with a local maximum of cross-sectional area for a cell, then the cell's midplane is out of the field of view. Such a cell can no longer be analyzed using this method from this time point onward.

25. Because the nurse cells are quite large and there is a physical limit on imaging depth, most cells will have at least a portion of their volume outside the field of view. For this reason, volume measurements are often not possible.
26. Volume measurements have been performed using Bitplane's Imaris to validate the use of cross-sectional area in volume estimates (Fig. 3c) [11]. Overall, changes in normalized volume measured directly correlated well with changes in normalized, estimated volume from midplane measurements; however, cells must be roughly spherical and not change shape significantly for the correlation to hold.
27. Nurse cells show a size gradient along the anterior-posterior axis, with cells nearest the oocyte being the largest [27]. Because of this, normalizing to the initial volume allows comparisons between different nurse cells. For applications in which knowledge of the absolute cell size is important, the normalization step can be skipped.
28. There are several free software packages available for performing PIV analysis. Two such examples are PIVlab (for MATLAB) [28] and a PIV plugin for ImageJ [29].

Acknowledgments

This work was supported by NIH grant R01GM125646 to A.C.M. We would like to thank members of the Martin lab for helpful discussions regarding this project.

References

1. Spradling AC (1993) The developmental genetics of oogenesis. Cold Spring Harbor Laboratory Press, Plainview N.Y, pp 3–6
2. McLaughlin JM, Bratu D (2015) Drosophila melanogaster Oogenesis: An Overview. Methods Mol Biol 1328:1–20
3. Morris LX, Spradling AC (2011) Long-term live imaging provides new insight into stem cell regulation and germline-soma coordination in the *Drosophila* ovary. Development 138(11):2207–2215
4. Haigo SL, Bilder D (2011) Global tissue revolutions in a morphogenetic movement controlling elongation. Science 331:1071–1074
5. Cai D, Chen S-C, Prasad M et al (2014) Mechanical feedback through E-Cadherin promotes direction sensing during collective cell migration. Cell 157:1146–1159
6. Osterfield M, Du X, Schüpbach T et al (2013) Three-dimensional epithelial morphogenesis in the developing *Drosophila* egg. Dev Cell 24(4): 400–410
7. Ganguly G, Williams LS, Palacios IM, Goldstein RE (2012) Cytoplasmic streaming in *Drosophila* oocytes varies with kinesin activity and correlates with the microtubule cytoskeleton architecture. Proc Nat Acad Sci 109(38): 15109–15114
8. Theurkauf WE (1994) Premature microtubule-dependent cytoplasmic streaming in cappuccino and spire mutant oocytes. Science 265(5181):2093–2096
9. Lei L, Spradling AC (2016) Mouse oocytes differentiate through organelle enrichment from sister cyst germ cells. Science 352(6281):95–99

10. Gutzeit HO, Koppa R (1982) Time-lapse film analysis of cytoplasmic streaming during late oogenesis of *Drosophila*. J Embryol Exp Morph 67:101–111
11. Imran Alsous J, Romeo N, Jackson JA et al (2021) Dynamics of hydraulic and contractile wave-mediated fluid transport during *Drosophila* oogenesis. Proc Nat Acad Sci 118(10): e2019749118
12. Mahajan-Miklos S, Cooley L (1994) Intercellular cytoplasm transport during *Drosophila* oogenesis. Dev Biol 165(2):336–351
13. Peters NC, Berg C (2016) In vitro culturing and live imaging of *Drosophila* egg chambers: a history and adaptable method. Methods Mol Biol 1457:35–68
14. Prasad M, Jang AC-C, Starz-Gaiano M et al (2007) A protocol for culturing *Drosophila melanogaster* stage 9 egg chambers for live imaging. Nat Protocols 2(10):2467–2473
15. Wilcockson SG, Ashe HL (2021) Live imaging of the *Drosophila* ovarian germline stem cell niche. STAR Protocols 2(1):100371
16. Cetera M, Lewellyn L, Horne-Badovinac S (2016) Cultivation and live imaging of *Drosophila* ovaries. Methods Mol Biol 1478:215–226
17. Gáspár I, Szabad J (2009) In vivo analysis of MT-based vesicle transport by confocal reflection microscopy. Cell Motil Cytoskeleton 66(2):68–79
18. Guggenheim EJ, Lynch I, Rappoport JZ (2017) Imaging in focus: reflected light imaging: techniques and applications. Int J Biochem Cell Biol 83:65–70
19. Schindelin J, Arganda-Carreras I, Frise E et al (2012) Fiji: an open-source platform for biological-image analysis. Nat Methods 9(7): 676–682
20. Jia D, Xu Q, Xie Q et al (2016) Automatic stage identification of Drosophila egg chamber based on DAPI images. Sci Reports 6:18850
21. Duffy JB (2002) GAL4 system in Drosophila: a fly geneticist's Swiss army knife. Genesis 34(1–2):1–15
22. Martin AC, Gelbart M, Fernandez-Gonzalez R, Kaschube M, Wieschaus EF (2010) Integration of contractile forces during tissue invagination. J Cell Biol 188(5):735–749
23. Oda H, Tsukita S (2001) Real-time imaging of cell–cell adherens junctions reveals that Drosophila mesoderm invagination begins with two phases of apical constriction of cells. J Cell Sci 114:493–501
24. Beaven R, Dzhindzhev NS, Qu Y et al (2015) *Drosophila* CLIP-190 and mammalian CLIP-170 display reduced microtubule plus end association in the nervous system. Mol Biol Cell 26(8):1491–1508
25. Minestrini G, Máthé E, Glover DM (2002) Domains of the Pavarotti kinesin-like protein that direct its subcellular distribution: effects of mislocalisation on the tubulin and actin cytoskeleton during Drosophila oogenesis. J Cell Sci 115:725–736
26. Kisielewska J, Lu P, Whitaker M (2012) GFP-PCNA as an S-phase marker in embryos during the first and subsequent cell cycles. Biol Cell 97(3):221–229
27. Imran Alsous J, Villoutreix P, Berezhkovskii AM, Shvartsman SY (2017) Collective growth in a small cell network. Curr Biol 27(17): 2670–2676
28. Thielicke W, Sonntag R (2021) Particle image velocimetry for MATLAB: accuracy and enhanced algorithms in PIVlab. J Open Res Softw 9:12
29. Tseng Q https://sites.google.com/site/qingzongtseng/piv. Accessed 25 Feb 2022

Chapter 12

Visualizing Lipid Droplets in *Drosophila* Oogenesis

Roger P. White and Michael A. Welte

Abstract

Lipid droplets (LDs) are fat storage organelles highly abundant in oocytes and eggs of many vertebrates and invertebrates. They have roles both during oogenesis and in provisioning the developing embryo. In *Drosophila*, large numbers of LDs are generated in nurse cells during mid-oogenesis and then transferred to oocytes. Their number and spatial distribution changes developmentally and in response to various experimental manipulations. This chapter demonstrates how to visualize LDs in *Drosophila* follicles, both in fixed tissues and living samples. For fixed samples, the protocol explains how to prepare female flies, dissect ovaries, isolate follicles, fix, apply stains, mount the tissue, and perform imaging. For live samples, the protocol shows how to dissect ovaries, apply a fluorescent LD dye, and culture follicles such that they remain alive and healthy during imaging. Finally, a method is provided that employs in vivo centrifugation to assess colocalization of markers with LDs.

Key words *Drosophila*, Oogenesis, Follicles, Lipid droplet, Neutral lipids, Dissection, Label free

1 Introduction

Lipid droplets (LDs) are cellular organelles with critical roles in lipid metabolism, energy homeostasis, and the handling of specific proteins [1–4]. They have a unique structure, with a core of neutral lipids surrounded by a monolayer of amphipathic lipids and proteins. This core stores fatty acids and sterols in the form of triglycerides and sterol esters as well as other lipids, depending on the cell type. Such storage prevents lipotoxicity by removing toxic lipids from other cellular compartments and provides cells with lipids for the production of energy, membrane components, and signaling molecules. The proteins at the LD surface mediate various aspects of lipid metabolism, promote contact to other organelles, as well as traffic LDs along the cytoskeleton [5–7]. They also facilitate the handling of certain proteins from other cellular compartments, including their maturation, storage, and turnover [8]. As such, LDs play fundamental roles in cell biology and physiology [9, 10]. They are particularly abundant in the oocytes and embryos

Michelle S. Giedt and Tina L. Tootle (eds.), *Drosophila Oogenesis: Methods and Protocols*,
Methods in Molecular Biology, vol. 2626, https://doi.org/10.1007/978-1-0716-2970-3_12,

of many species, including *Drosophila*, where they play important roles during oogenesis, from lipid signaling to protein storage [11, 12], and provision the developing embryos with energy, lipids, and proteins [13–15].

Drosophila females bear a pair of ovaries within the abdomen [16]. Each ovary consists of 15–20 ovarioles, strings of follicles of increasing stages of maturity. Follicles arise from germline stem cells in the germarium, develop through 14 stages, and consist of somatic follicle cells and 16 germline cells (one oocyte and 15 nurse cells). In initial stages, LDs are sparse, but during mid-oogenesis (Stages 8–10), the nurse cells accumulate massive amounts of LDs [13, 15, 17]. In contrast to many other tissues, these LDs are uniform in size and relatively small, with diameters around 0.5 μm. In Stage 11, nurse cells contract and transfer most of their cytoplasmic contents, including LDs, to the oocyte. LDs are also present in follicle cells, but their dynamics and functions in this cell type remain to be documented.

Given the myriad of biological roles of LDs in other cells, it appears likely that they also have critical functions during *Drosophila* oogenesis. They have already been shown to provide precursors for prostaglandin signaling [11] and to sequester certain histones to protect them from degradation [12]. They are highly motile in nurse cells and oocytes [18, 19], and they have been employed to monitor transport from nurse cells to oocytes [19] as well as ooplasmic streaming [20]. Finally, their abundance, size, and distribution are altered in mutant conditions [13, 17, 19–23]. It is therefore critical to reliably visualize LDs in follicles.

The protocols in this chapter describe multiple strategies to visualize LDs by confocal microscopy in both fixed and live preparations (*see* **Note 1**). We also describe an approach for centrifuging entire flies or dissected ovaries that separates organelles by density in nurse cell and oocytes; such separation is useful for evaluating colocalization of markers with LDs.

2 Materials

2.1 Dissecting Ovaries/Follicles

1. Standard fly food.
2. Dry yeast.
3. CO_2 pad for anaesthetization.
4. Depression spot plate.
5. P1000 micropipette and tips (*see* **Note 2**).
6. Pasteur pipettes (*see* **Note 2**).
7. Dumont tweezers #5.

8. For dissections to generate fixed samples for imaging: Grace's insect medium supplemented with L-Glutamine (0.6 g/L) and sodium bicarbonate (0.35 g/L).
9. For dissections to generate live samples for imaging: 1× Maturation Medium (MM): 4.12 mL Schneider's *Drosophila* medium with L-Glutamine, pH 6.6–6.95; 750 μL 15% fetal bovine serum, heat inactivated; 150 μL insulin from bovine pancreas, 10 mg/mL in 25 mM HEPES, pH 8.2; 30 μL Penicillin–Streptomycin (10,000 U/mL); 5 mL total volume.
10. 1× Phosphate buffered saline (PBS): 37 mM NaCl, 2.7 mM KCl, 10 mM Na_2HPO_4, 1.8 mM KH_2PO_4.
11. PBT: PBS with 0.1% TritonX-100 (*see* **Note 3**).

2.2 Detecting LDs in Fixed Samples

1. Pasteur pipette (*see* **Note 2**).
2. Dumont tweezers #5.
3. Scintillation vial (*see* **Note 4**).
4. 1× Phosphate buffered saline (PBS): 37 mM NaCl, 2.7 mM KCl, 10 mM Na_2HPO_4, 1.8 mM KH_2PO_4.
5. PBT: PBS with 0.1% TritonX-100 (*see* **Note 3**).
6. Ovary block: PBT with 10% bovine serum albumin (BSA) and 0.02% sodium azide (*see* **Note 5**).
7. Stock solutions for LD dyes (*see* Table 1 for details).
8. 37% Formaldehyde.
9. P1000, P200, and P10 micropipettes and tips.
12. Nutator.
13. Aluminum foil.
14. Microscope slides (25 × 75 × 1.0 mm).

Table 1
Dyes for detecting LDs in follicles

LD dye	Excitation max (nm)	Emission max (nm)	Buffer for stock solution	Dye concentration for stock solution	Recommended incubation time
Nile red (*see* **Note 7**)	515	585	Acetone or Methanol or Ethanol	1 mg/mL	1 h
BODIPY-FL	503	512	DMSO	2 mM	20 min
LipidSpot (488)	427	585	DMSO	1000×	20 min
LipidSpot (610)	592	638	DMSO	1000×	20 min
Monodansylpentane (MDH)	405	480	DMSO	0.1 M	1 h

15. Disposable wipes.
16. 22 × 22 mm coverslips (*see* **Note 6**).
17. Nail polish.
18. Aqua-Poly/Mount or another mounting medium.

2.3 Detecting LDs by Live Imaging

1. 35 mm petri dish with a coverslip in the bottom.
2. Low melt agarose.
3. Schneider's medium.
4. 1× Maturation Medium (MM): 4.12 mL Schneider's Drosophila medium with L-Glutamine, pH 6.6–6.95; 750 μL 15% fetal bovine serum, heat inactivated; 150 μL insulin from bovine pancreas, 10 mg/mL in 25 mM HEPES, pH 8.2; 30 μL Penicillin–Streptomycin (10,000 U/mL); 5 mL total volume.
5. 2× Maturation Medium (MM): 162 μL Schneider's medium; 75 μL heat inactivated fetal bovine serum; 10 μL 10 mg/mL insulin from bovine pancreas; 3 μL penicillin/streptomycin; 250 μL total volume.
6. P1000 Micropipette and tips.
7. Pasteur pipettes.
8. Two microfuge tubes (1.5 mL).
9. Two heat blocks.
10. BODIPY-FL stock solution (*see* Table 1 details).
11. Aluminum foil.
12. For living samples for short-term imaging, we mount in Voltalef Oil 10S (*see* **Note 8**). For long-term imaging, we generate a mounting medium out of low melt agarose, Schneider's medium, and 2× MM (*see* Subheadings 3.3.1 and 3.3.3).
13. Confocal microscope, inverted.

2.4 Enriching LDs by In Vivo Centrifugation for Colocalization Studies

1. Refrigerated microcentrifuge.
2. CO_2 pad.
3. Low melt agarose.
4. P1000 micropipette and tips.
5. Dumont tweezers #5.
6. 1× PBS.
7. PBT.
8. 1× MM.
9. Microcentrifuge tubes (1.5 mL).
10. Scintillation vial.
11. Confocal microscope, upright, or inverted.

3 Methods

3.1 Dissecting Ovaries or Follicles Out of Adult Females

Here, we describe how to dissect ovaries and follicles. This procedure is used in all the remaining protocols for detecting LDs in fixed tissues (Subheading 3.2) or in living samples (Subheading 3.3) or for in vivo centrifugation (Subheading 3.4).

3.1.1 Preparation

1. In the days preceding the dissection: Transfer 5–10 females (preferably a few days post eclosion) of the desired genotype along with a similar number of males into a new fly vial with fly food and dry yeast sprinkled on top. This feeding step is important to fatten up the ovaries. Feeding duration will affect which stages of oogenesis are present. To enrich for follicles in mid and late stages, feed overnight at 25 °C or for two nights at room temperature.
2. On the day of dissection, warm the desired amount of dissection medium to room temperature.
   - For dissections to generate fixed samples for imaging, use Grace's medium as the dissection medium.
   - For dissections to generate live samples for imaging, use 1× MM as the dissection medium.

3.1.2 Dissection

1. Anesthetize ~10 flies on a CO_2 pad.
2. Place 200 μL of room-temperature dissection medium each into three separate wells of a nine-well depression spot plate.
3. Transfer a single fly into the first well, wing-side facing down to expose abdomen.
4. Hold the fly in place, with one tweezer gently grabbing the thorax. A second tweezer is used to pinch the cuticle of the abdomen near the connection to the thorax (Fig. 1a). Move the second tweezer away from the thorax (arrow in Fig. 1a), which will rupture the cuticle and peel it back, exposing the internal organs. Remove any guts or other extraneous tissues.
5. Using the second tweezer, grab the ovaries at the oviduct (Fig. 1b) and place them into a second well.
6. To isolate follicles of specific stages, hold the posterior of the ovary (where the Stage 14 follicles are located) with one tweezer. With a second tweezer, grab the germarium end of the ovary (Fig. 1c) and gently pull until a single ovariole is pulled out of the muscle sheath that surrounds the ovary (Fig. 1d). Typically, the ovariole breaks, with Stages 13 and 14 remaining in the ovary. Often early stages through Stage 9 remain together in a row, with some later stages individualized. It is best to not further break up ovarioles, as that could lead to damage to follicles. To obtain Stage 13 and 14 follicles, a

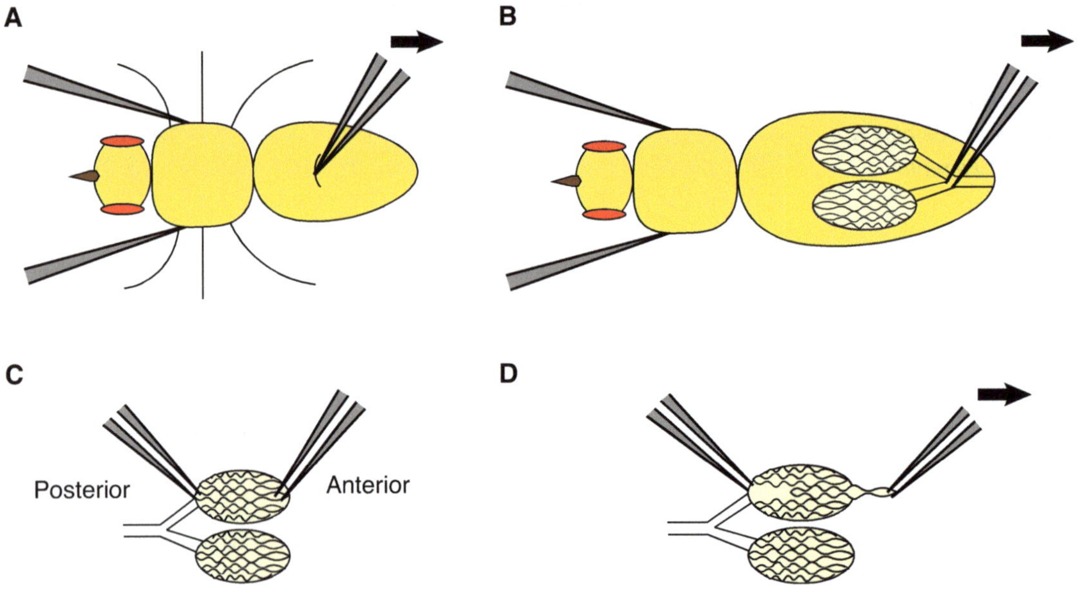

Fig. 1 Schematic of ovary dissection and follicle isolation. (**a**) An anesthetized adult female is placed in one well of a depression spot plate, wing side down. The left tweezer lightly grabs the thorax to hold the fly in place; the right tweezer pinches the cuticle of the abdomen and is moved in the direction of the arrow. The cuticle breaks, allowing the right tweezer to pull it back and expose the ovary. (**b**) While the left tweezer keeps holding the thorax, the right tweezer grips the oviduct and removes the ovary from the abdomen. (**c**) After the ovary is placed into a separate well, one tweezer (left) immobilizes it while the other (right) is used to pinch an ovariole at the germarium end. (**d**) The right tweezer is gently moved in the direction of the arrow to pull the ovariole out from its muscle sheath

different approach is used (*see* **Note 9**). For fixed imaging of a few (~10) intact ovaries, a quick alternative dissection method can be used (*see* **Note 10**).

7. Transfer follicles into a third well using a lubricated P1000 micropipette (*see* **Note 2**). We typically collect ~5–10 follicles/ovarioles per preparation.
8. Use a lubricated Pasteur pipette to transfer follicles in dissection medium from the dissection dish into a microfuge tube.

3.2 Detecting LDs in Fixed Tissue

Here, we describe how to fix dissected follicles, stain them with LD-specific dyes, mount, and image. The protocol can be adapted to visualize LDs by antibody staining against endogenous LD proteins (*see* **Note 11**) or targeting fluorescent proteins to LDs (*see* **Note 12**). Protein detection and use of LD dyes can be combined (*see* **Note 13**) [12, 24].

3.2.1 Preparation and Dissection

1. Same as in Subheading 3.1.1.
2. In addition, prepare a scintillation vial with 3.6 mL 1× PBS; this vessel will be used to fix the follicles.
3. Follow the procedure in Subheading 3.1.2 to dissect the ovaries.

3.2.2 Fixation and Staining with LD Dyes

1. Aspirate the Grace's medium from the microfuge tube using a P1000 micropipette, and if necessary, remove the remaining medium with a P200.
2. Add ~200 μL of PBS (from the scintillation vial) to the microfuge tube.
3. Using a lubricated P1000 micropipette (*see* **Note 2**), transfer the follicles in PBS from the microfuge to the scintillation vial.
4. Add 400 μL of 37% formaldehyde to the scintillation vial with the sample (for a final concentration of 3.7% formaldehyde) and gently mix by inversion (*see* **Note 14**).
5. Incubate at room temperature for 15 min with slight agitation, e.g., on a nutator.
6. Using a lubricated Pasteur pipette (*see* **Note 2**), remove follicles from the vial and transfer into a 1.5 mL microfuge tube.
7. Replace the formaldehyde mixture with PBT (1 mL) to wash. Wash three times with PBT for 5–10 min each on a nutator.
8. Follicles can be stored at 4 °C in PBT for up to several days before proceeding with staining or stained immediately. If samples are stored, allow to come to room temperature before proceeding.
9. Replace PBT with 980 μL of ovary block and 20 μL of Nile red stock solution or with 100 μL of ovary block and 1 μL of stock solution for BODIPY-FL/LipidSpot/MDH (*see* Table 1 for details).
10. Cover tube with foil to protect from light and incubate at room temperature on a nutator for 20 min to 1 h. Staining intensity will increase over time. *See* Table 1 for recommended times for each dye.
11. Wash three times for 5 min with PBT at room temperature.
12. If desired, samples can be stored at 4 °C, protected from light, for a few days or can be mounted immediately.

3.2.3 Mounting

1. Using a lubricated Pasteur pipette, transfer follicles in buffer onto a glass slide.
2. Remove most excess buffer by aspirating it with a P200 micropipette, leaving the follicles in a tiny drop.
3. Often, the buffer has spread over the slide and is still present as a thin film (too thin to be pipetted up) or as multiple, dispersed droplets in addition to the drop with the follicles. If so, gently wipe away this excess buffer with a disposable wipe.
4. Place a drop of Aqua-Poly/Mount on top of the follicles (*see* **Note 15**).

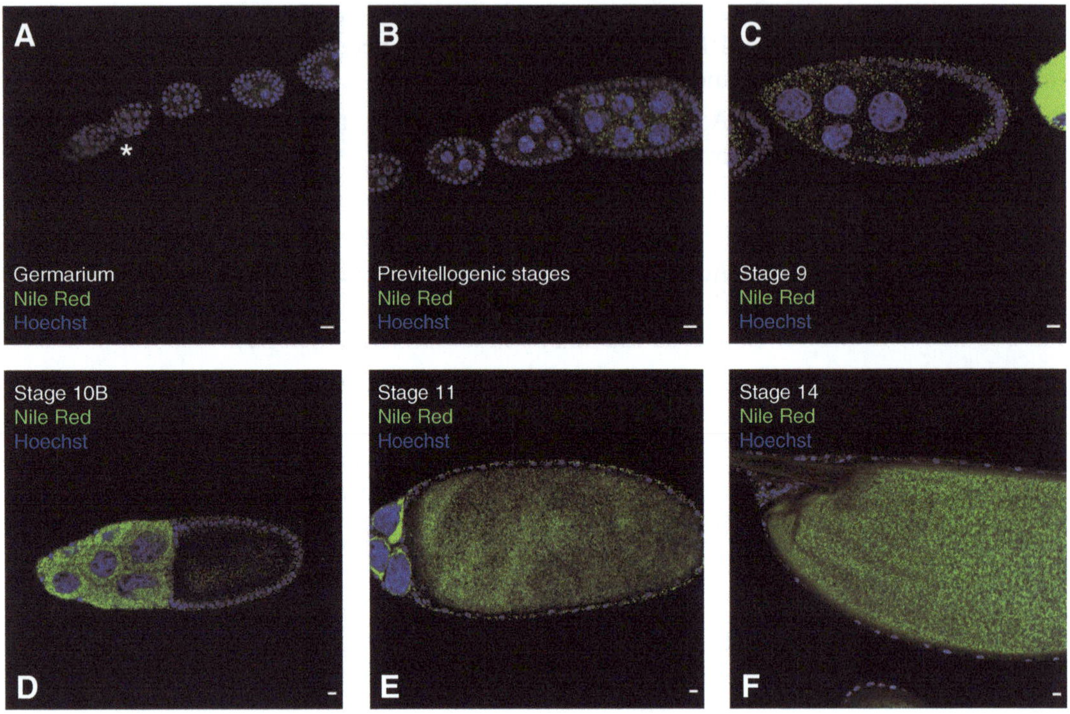

Fig. 2 LD staining of follicles at different stages of oogenesis. Follicles were dissected from well-nourished females, fixed in formaldehyde, and stained at room temperature for LDs (green, Nile red for 1 h) and DNA (blue, Hoechst 33342 for 20 min). *indicates location of germarium. (**a**, **b**) Germarium and early-stage follicles within the same ovariole; LDs are relatively sparse. (**c**) Stage 9; LDs start to accumulate in nurse cells and follicle cells. (**d**) Stage 10B; LDs are very dense in nurse cells and less abundant in oocyte and follicles cells. (**e**) Stage 11 follicle in the process of dumping LDs into the oocyte. (**f**) Stage 14 follicle with abundant LDs in the oocyte. Images were captured with a 40× (**d**, **f**) or 63× (**a**, **b**, **c**, **e**) objective on a Leica Sp5 confocal microscope. Scale bars = 10 μm

5. Place a 22 × 22 mm coverslip on top of the follicles and allow the Aqua-Poly/Mount to spread so that it fills the entire space between coverslip and glass slide.
6. Seal with nail polish and store at 4 °C for up to a week (*see* **Note 16**).

3.2.4 Imaging

1. Analyze the preparations by confocal microscopy. The excitation/emission maxima of the various dyes are shown in Table 1.
2. Examples of successful LD staining are shown in Figs. 2 and 3.

3.3 Detecting LDs in Living Follicles

Staining with LD-specific dyes also works for living samples, allowing the motion of LDs and transient interactions of LDs with other cellular structures to be recorded. Here, we describe how to stain follicles with an LD dye, mount them, and image them on an inverted confocal microscope. Slightly modified approaches can

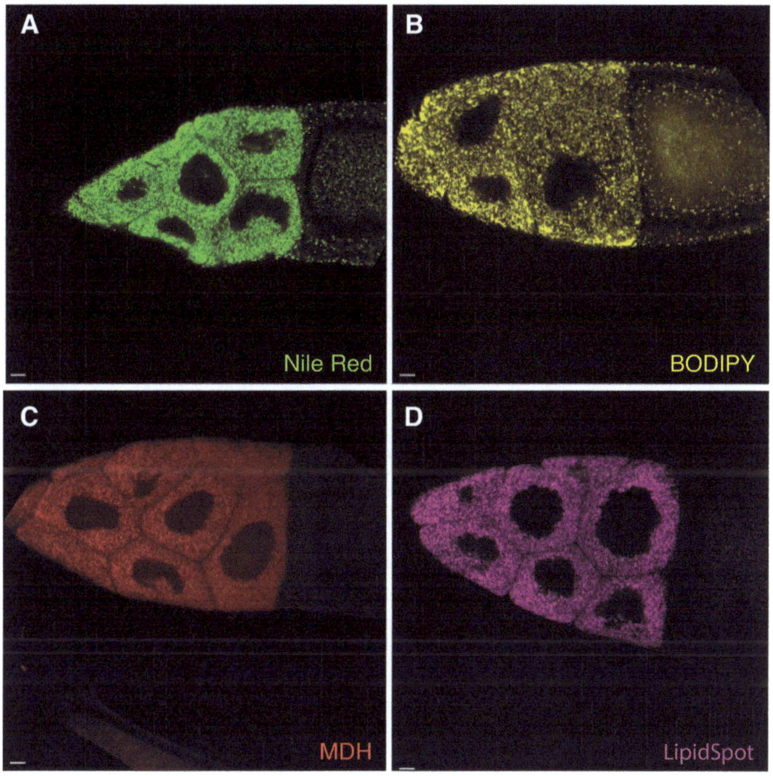

Fig. 3 Confocal images of fixed Stage 10 follicles stained with neutral lipid dyes. (**a**) Nile red for 1 h. (**b**) BODIPY-FL for 20 min. (**c**) MDH for 1 h. (**d**) LipidSpot 610 nm for 20 min. Images were captured using a 63× objective on a Leica Sp5 confocal microscope. Scale bars = 10 μm. The stronger signal in **a** and **b** allows for visualization of the sparser LDs in follicle cells and the oocyte

be used to detect LDs with other dyes (*see* **Note 17**), fluorescent proteins targeted to LDs (*see* **Note 18**), or even label-free (*see* **Note 19**). The described mounting protocol (Subheading 3.3.3) makes it possible to observe LD motion in healthy follicles for well over 30 min. For shorter observations, Subheading 3.3.3 can be replaced with mounting in Voltalef oil (*see* **Note 8**).

3.3.1 Preparation and Dissection

1. Prepare flies as in Subheading 3.1.1, **step 1**.
2. Prepare 5 mL of 1× MM and 250 μL of 2× MM and bring to room temperature.
3. Prepare 2.5% low melt agarose (LMA) in Schneider's medium. Mix 125 mg of agarose and 5 mL of Schneider's medium, and briefly (for a few seconds) heat in microwave to a boil. Confirm that LMA is fully molten and dispersed (*see* **Note 20**).

4. Transfer 250 μL of LMA to a 1.5 mL microfuge tube and put in a heat block at 50 °C.
5. Put another 1.5 mL microfuge tube with 250 μL of 2× MM on a second heat block at 42 °C until ready to mix with LMA in Subheading 3.3.3 **step 3**.
6. Follow the procedure in Subheading 3.1.2 to dissect the ovaries.

3.3.2 Staining with BODIPY-FL

1. In the following, the 1.5 mL microfuge tube with ovaries/follicles in 1× MM is referred to as Tube 1.
2. In another tube (Tube 2), mix 1 mL of 1× MM with 2.5 μL of BODIPY-FL stock solution (*see* Table 1 for details).
3. Aspirate 1× MM from Tube 1 and replace with the BODIPY-FL/MM mixture from Tube 2.
4. Cover Tube 1 with foil to protect from light and incubate on a nutator for 20 min at room temperature.
5. Remove BODIPY-FL/MM and replace with 1× MM and proceed immediately to mounting.

3.3.3 Mounting

1. Lubricate a micropipette tip (P1000) with PBT to prevent follicles from sticking (*see* **Note 2**) and use it to transfer follicles to a coverslip dish (Fig. 4), ideally in the center of the coverslip.
2. Remove all excess medium using a P200 micropipette but leave the ovaries in a small drop of liquid so that they do not dry out.
3. Prepare embedding medium: mix 2× MM and LMA 1:1 by pipetting 2× MM into LMA. Pipette up and down until the mixture is homogenous.
4. Using a P200 micropipette, pipette 200 μL of embedding medium onto the coverslip of the coverslip dish. This step is time-sensitive (*see* **Note 21**). Point the pipette tip close to the coverslip, at a location between the follicles and the edge of the indentation in the petri dish, and then release the embedding medium slowly while turning the dish 360°. The goal is to have the medium push all the follicles into the center of the coverslip for ease of imaging (*see* **Note 22**).
5. If one plans to follow LDs for 30 min or less, a simpler mounting technique can be used instead (*see* **Note 8**).

3.3.4 Imaging

1. Analyze by confocal microscopy.
2. Place coverslip dish on microscope stage and add ~2.5 mL of 1× MM to cover both the solidified embedding medium and the entire bottom of the dish.
3. Excite sample with 488 nm laser.

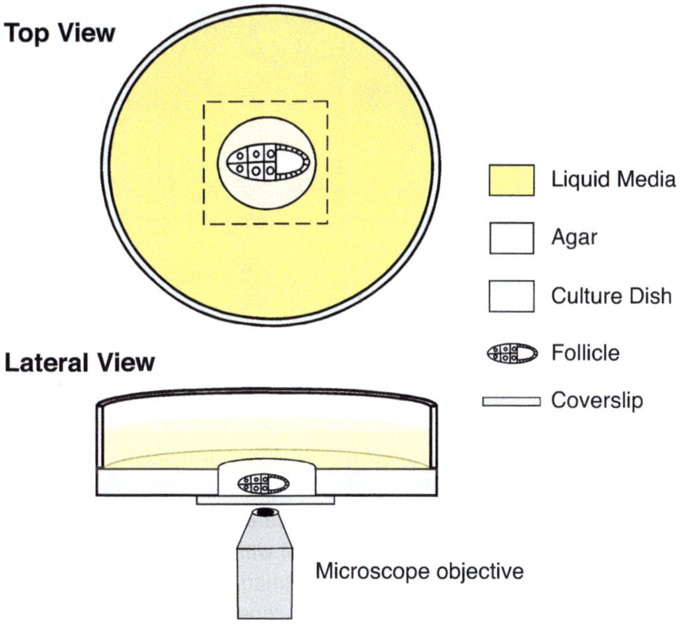

Fig. 4 Using a coverslip dish to embed follicles in supplemented agar for live imaging. Top view (top) and lateral view (bottom). The coverslip dish consists of a plastic petri dish with a round hole in its bottom; the hole is covered on the underside with a coverslip. The hole contains agar (supplemented with culture medium) and the follicles. Ideally, the follicles are as close to the coverslip as possible. The rest of the petri dish contains liquid medium to prevent the agar from drying out. The follicles are imaged from below. The dashed line in the top view represents the outline of the coverslip below the dish

4. Collect 490–515 nm emission.
5. Imaging duration depends in part on the stage of the follicle and exact imaging conditions (*see* **Note 23**).

3.4 Detecting Colocalization with LDs Using In Vivo Centrifugation

As LDs are abundant in the cytoplasm of nurse cells and oocytes, it can be challenging to determine if a particular marker colocalizes with LDs or is merely present throughout the cytoplasm or on another abundant organelle. A simple way to distinguish between these possibilities is to separate the major organelles by density via in vivo centrifugation. Because of the low density of LDs, they "float up" during centrifugation, i.e., in the direction towards the center of rotation; in both oocyte and nurse cells, they therefore accumulate in the region of each cell that was closest to the axis of the centrifuge. Examples of stained centrifuged follicles are shown in Fig. 5, with distinct LD layers in the nurse cells (the corresponding LD layers of the oocytes are outside the field of view). If a marker colocalizes with LDs, it will also be enriched in this layer. The protocol below has been modified from an approach used for *Drosophila* embryos [25] and explains how either whole

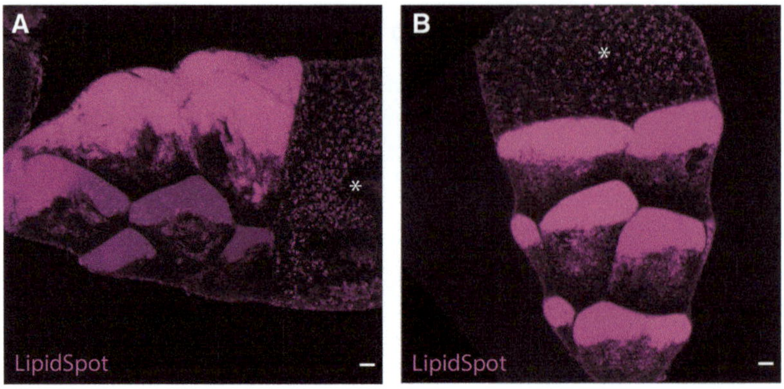

Fig. 5 LD staining after in vivo centrifugation. Flies were anesthetized and beheaded prior to centrifugation. After centrifugation, ovaries were dissected and fixed with formaldehyde, then stained with LipidSpot 610 nm (shown in magenta). * indicates the position of the oocyte. LDs can accumulate laterally (**a**) or posteriorly (**b**) within each nurse cell, depending on how the fly settles during centrifugation. Images were captured using a 63× objective on a Leica Sp5 confocal microscope. Scale bars = 10 μm

flies or isolated ovaries can be centrifuged. Centrifugation of whole flies (Subheading 3.4.2) is easier and should be tried first. If ovaries need to be treated ex vivo first before testing for colocalization, the centrifugation protocol Subheading 3.4.3 provides a suitable alternative.

3.4.1 Preparing a Vessel to Keep Flies/Ovaries Oriented during Centrifugation

1. Mix 2.5 g of LMA with 100 mL of dH_2O and dissolve by microwaving agarose until it reaches a boil.
2. Allow to cool until the solution is just warm to the touch and use a P1000 micropipette to fill 1.5 mL microfuge tubes with the agarose solution until the solution is just under the rim of the tube.
3. Allow agarose to completely solidify and cool before use. Microfuge tubes can be used immediately once cooled or stored at 4 °C.
4. In order to hold a fly or an ovary in a fixed orientation during centrifugation, a small cavity needs to be present in the solidified agarose in the microfuge tube. This is accomplished by scooping out agarose using tweezers to create a fly-sized hole. The hole should be wide enough to fit the entire fly and to a depth of about ¾ of the fly (*see* **Note 24**). Proceed to either Subheading 3.4.2 or 3.4.3.

3.4.2 In Vivo Centrifugation of Entire Flies

1. Prepare ~8 flies and dissection medium as in Subheading 3.1.1, and scintillation vial as in Subheading 3.2.1.
2. Anesthetize flies on a CO_2 pad and behead them (*see* **Note 25**).

3. Using a tweezer, place them thorax down (upside down) into the holes (*see* **Note 26**).
4. In a microcentrifuge, centrifuge tubes at ~9500 g at 4 °C for 10 min.
5. Pull flies out of the agar with a tweezer.
6. Dissect ovaries as in Subheading 3.1.2 (starting at **step 2**).
7. Then follow method Subheading 3.2 to detect LDs in fixed tissues.

3.4.3 *In Vivo* Centrifugation of Ovaries

1. Prepare flies and 1× MM as in Subheading 3.1.1.
2. Perform dissection as in Subheading 3.1.2, through **step 5**. It is best to try to keep ovaries intact.
3. Gently place ovaries into the hole in the agarose and add 1× MM medium (around 10 μL) to prevent desiccation during centrifugation (*see* **Note 27**).
4. Centrifuge tubes at ~6000 g at 4 °C for 10 min.
5. Using a tweezer, remove centrifuged ovaries from microfuge tubes and drop them into a scintillation vial with 3.6 mL 1× PBS.
6. Proceed with ovary fixation and staining as in method Subheading 3.2 (starting at Subheading 3.2.2 **step 4**).

4 Notes

1. We prefer these fluorescence-based approaches over staining methods that detect LDs by bright-light microscopy [21, 26], as the latter are less suited for detecting individual LDs or for quantitation.
2. Ovaries and follicles tend to stick to glass and plastic and thus can easily get lost when being transferred using a Pasteur pipette or micropipette tip. To minimize sticking, "lubricate" the pipette first with PBT: aspirate a small volume of PBT into the pipette to wet the surfaces with which the tissue will come into contact and then expel the liquid from the pipette. Immediately use the pipette to transfer the tissue.
3. Because TritonX-100 is extremely viscous, pipetting it accurately is challenging. Rather than trying to accurately pipette small volumes of pure TritonX-100, it is better to make a 10% stock solution first.
4. Many other fixation vessels can be used instead of scintillation vials, e.g., microfuge tubes. If so, adjust the total fixation volume accordingly. We prefer scintillation vials because the

clear glass makes it easy to spot the tissues, and the fixation solution is in large excess even with abundant numbers of ovaries.

5. BSA reduces tissue sticking and binds lipophilic dyes, enhancing their solubility. Azide acts as antibacterial agent.
6. Choose a coverslip thickness appropriate for the microscope objective to be used, typically #1.5 = 0.17 mm.
7. Nile red also labels membranes. However, when incorporated into a phospholipid bilayer, its excitation and emission maxima are 554 nm/638 nm.
8. Mounting for short-term imaging involves placing a drop of Voltalef oil on a 20 × 60 mm coverslip and then transferring follicles with a fine-tip transfer pipette to the oil droplet. The coverslip can then be imaged directly.
9. To isolate Stage 13 and 14 follicles, hold the posterior of the ovary (where the Stage 13/14 follicles are located) with one tweezer. Use a second tweezer to grab the muscle sheath near the posterior of the ovary to rupture the muscle sheath. Once a noticeable rupture is made, use the second tweezer to gently push Stage 13/14 follicles out of the ovary. The follicles can also be removed from the ovary by gently grabbing the dorsal appendages; be careful to not poke the follicles.
10. Place a small drop of 1× PBS onto a microscope slide and then transfer a fly from the CO_2 pad into the drop, wing-side facing down to expose abdomen. Dissect as in Subheading 3.1.2 **step 4**. Then grab the ovaries at the oviduct (Fig. 1b) and place them into the scintillation vial. Repeat until all ovaries are dissected; replace coverslip if it becomes too messy with tissue remnants.
11. Table 2 summarizes antibodies that have been used to detect LDs in *Drosophila* ovaries. Protocol Subheading 3.2 is followed, with Subheading 3.2.2 **steps 9–11** modified as described in [12].
12. Table 3 summarizes the fluorescently labeled LD proteins whose LD localization in *Drosophila* ovaries has been verified by a publication at the time of this writing. When using them, the protocol in Subheading 3.2.2 can be followed as is, but **steps 8–11** are omitted.
13. Because Nile red has a larger Stokes shift than GFP, it is possible to simultaneously excite Nile red and GFP with a 488 nm laser and collect the corresponding emissions separately. In this scenario, we typically collect GFP emission from 500 to 530 nm, and Nile red emission from 550 to 650 nm. To assure the specificity of signal, we image samples with GFP only or Nile red only under the same conditions.

Table 2
Published antibodies used to detect LDs in follicles

LD protein targeted	Description	Reference for ovary staining	Comment
Jabba	Jabba is an LD protein that recruits histones to LDs	[12]	*See* **Note 28**
H2Av	Histone H2Av is sequestered on LDs	[12]	*See* **Note 28**
PLIN2 (also known as LSD-2)	PLIN2 is an LD protein regulating lipid metabolism and LD motility	[22, 23]	*See* **Note 29**

Table 3
Fluorescent proteins to label LDs in follicles

Fluorescent construct	Description	Expression	Reference for use in oogenesis	Comment
GFP-LD	In this context, LD refers to a C-terminal domain of the Klarsicht protein that targets to LDs [24, 33]. Klarsicht controls the activity of microtubule motors on LDs [32, 33].	Gal4 dependent [24]	[19, 20, 24]	*See* **Note 30**
H2Av-GFP, H2Av-RFP, H2Av-Dendra	In nurse cells and ovaries, histone H2Av localizes to both LDs and the nucleus.	Transgene under control of endogenous promoter [34–36]	[12, 37]	*See* **Note 28**
H2B-mEOS	In nurse cells and ovaries, histone H2B localizes to both LDs and the nucleus.	Transgene under control of endogenous promoter [38]	[12]	*See* **Note 28**
JabbaB-mCherry	Jabba is an LD protein that recruits certain histones to LDs.	Gal4 dependent [12]	[12]	

14. Many other types of fixations preserve LDs well. For example, we have successfully used 4% paraformaldehyde (final concentration) and heat fixation (boiling in Triton salt solution [0.03% Triton, 70 mM NaCl] for 1–2 min). However, buffers containing appreciable amounts of organic solvents (such as methanol, heptane) should be avoided since they extract the neutral lipids from the core of LDs. Even though stock solutions of 37% formaldehyde contain methanol as stabilizer, at the dilution used in Subheading 3.2.2 **step 4**, we have not noticed detrimental effects on LDs.

15. Aqua-Poly/Mount is very viscous, and it is easy for air bubbles to get trapped in the spout of the dispensing bottle. When medium is dispensed from such a bottle, the air bubbles end up in the mounting medium around the follicles, potentially obscuring parts of the sample. To avoid air from getting trapped, we store the bottle of Aqua-Poly/Mount upside down (with the spout/cap facing downward) and keep it always in this inverted state, including when dispensing mounting medium. Occasionally, air bubbles get trapped in the spout anyway; to remove them, we dispense a small amount of medium (from an inverted bottle) into a disposable wipe until the bubbles are cleared from the spout.
16. Older slides are susceptible to artificial fusion of LDs.
17. All the dyes in Table 1 have been used to label LDs for live imaging e.g., [27–30], but for follicles we have not yet assessed best dye concentrations and staining times, other than for BODIPY.
18. If the flies used express a fluorescently tagged protein targeted to LDs (Table 3), Subheading 3.3.2 (staining) can be omitted, but mounting and imaging (using appropriate excitation and emission) are performed as described.
19. LDs can also be detected label free using confocal reflection microscopy [18], while other cellular structures can be simultaneously visualized using conventional confocal imaging. Using a conventional confocal, samples are illuminated with a 633 nm He/Ne laser beam, and the reflected signal of similar wavelengths is collected. This approach predominately reveals LDs, with a minor signal from yolk vesicles [18]. The only difference to the protocol described is that method Subheading 3.3.2 is omitted and that in Subheading 3.3.4 the sample is excited with a 633 nm laser and emission is collected from 623 to 643 nm. This method works most reliably in live samples; we do not recommend it for fixed samples.
20. A gentler method of preparing 2.5% LMA in Schneider's medium is to mix the ingredients, transfer them to microfuge tubes, and put the tubes in a 65 °C heat block for the agarose to melt. However, this takes much longer, and we have noticed no ill effects from the boiling method described in the main protocol.
21. On the one hand, the medium needs to have cooled enough not to harm the follicle; on the other hand, it must remain warm enough to not solidify. *See* **Note 23** for how to recognize damaged follicles. This step requires practice. Our rule of thumb is to wait 10–15 s after mixing before starting to pipette the embedding medium into the coverslip dish.

22. The goal is to have the follicles settled near the center of the coverslip and close to the coverslip. Excessive turbulence while adding the medium will lift the follicles up, away from the coverslip, and possibly out of the working distance of the objective, resulting in blurry images. Follicles displaced to the edge of the indentation in the petri dish are also hard to image.
23. Long or frequent laser exposure causes phototoxicity. As imaging set ups differ, safe conditions for acquisition that do not damage the follicle have to be determined empirically. We use the following criteria to judge damage to follicles: the follicle as a whole or individual cells shrink or shrivel; follicle appears misshapen; follicular epithelium does not follow a smooth line; nuclei condense; LD motion noticeably slows or stops altogether; abnormal clustering of organelles such as LDs, mitochondria, ER, etc. If we observe any of these phenotypes, we discard the follicle.
24. When scooping out the agar, it is important not to crack the surface of the agar elsewhere. Cracks will cause the agar to split during centrifugation, potentially resulting in the fly/ovary becoming dislodged and losing its orientation.
25. To behead the fly, put it in a small drop of PBT on a coverslip, wing-side down. Holding the thorax gently with one tweezer, grip the neck of the fly with another tweezer and push the head away from the body. Remove any tissue that has spilled out of the thorax.
26. When flies are mounted into the holes as described, the posterior of most follicles will be oriented towards the center of rotation, resulting in LD accumulation in the posterior regions of each nurse cell, those facing the oocyte (as in Fig. 5b). However, some flies/follicles tilt during centrifugation, so other patterns (like the sidewise accumulation in Fig. 5a) are also observed.
27. We prefer placing the ovary with the germarium end pointing up. But because ovaries are fragile, it is critical to transfer the ovary into the hole as quickly as possible. Thus, it is more important to get the ovary into the hole at all than to achieve a particular orientation.
28. Only labels LDs in nurse cells and the oocyte, not in follicle cells.
29. Antibody used [31] is temperamental in our hands and requires extensive pre-absorption against ovaries or embryos from *PLIN2* null mutants. Specificity of signal should be confirmed with *PLIN2* null mutant ovaries.
30. Only a subset of all LDs per cell are labeled; may depend on the GAL4 driver used.

Acknowledgments

We thank Jonathon Thomalla, Marcus Kilwein, Pakinee Phromsiri, Jinghong (James) Tang, Alicia Shipley, and the editors for comments on the manuscript. Research in the Welte laboratory on LDs in oogenesis is supported by NIH grant GM102155 to M.A.W.

References

1. Hashemi HF, Goodman JM (2015) The life cycle of lipid droplets. Curr Opin Cell Biol 33:119–124. https://doi.org/10.1016/j.ceb.2015.02.002
2. Walther TC, Farese RV Jr (2012) Lipid droplets and cellular lipid metabolism. Annu Rev Biochem 81:687–714. https://doi.org/10.1146/annurev-biochem-061009-102430
3. Welte MA (2015) Expanding roles for lipid droplets. Curr Biol 25(11):R470–R481. https://doi.org/10.1016/j.cub.2015.04.004
4. Zechner R, Zimmermann R, Eichmann TO, Kohlwein SD, Haemmerle G, Lass A, Madeo F (2012) FAT SIGNALS–lipases and lipolysis in lipid metabolism and signaling. Cell Metab 15(3):279–291. https://doi.org/10.1016/j.cmet.2011.12.018
5. Schuldiner M, Bohnert M (2017) A different kind of love – lipid droplet contact sites. Biochim Biophys Acta Mol Cell Biol Lipids 1862(10 Pt B):1188–1196. https://doi.org/10.1016/j.bbalip.2017.06.005
6. Kilwein MD, Welte MA (2019) Lipid droplet motility and organelle contacts. Contact (Thousand Oaks) 2:2515256419895688. https://doi.org/10.1177/2515256419895688
7. Gross DA, Silver DL (2014) Cytosolic lipid droplets: from mechanisms of fat storage to disease. Crit Rev Biochem Mol Biol 49(4):304–326. https://doi.org/10.3109/10409238.2014.931337
8. Welte MA, Gould AP (2017) Lipid droplet functions beyond energy storage. Biochim Biophys Acta Mol Cell Biol Lipids 1862(10 Pt B):1260–1272. https://doi.org/10.1016/j.bbalip.2017.07.006
9. Kuhnlein RP (2012) Thematic review series: lipid droplet synthesis and metabolism: from yeast to man. Lipid droplet-based storage fat metabolism in Drosophila. J Lipid Res 53(8):1430–1436. https://doi.org/10.1194/jlr.R024299
10. Greenberg AS, Coleman RA, Kraemer FB, McManaman JL, Obin MS, Puri V, Yan QW, Miyoshi H, Mashek DG (2011) The role of lipid droplets in metabolic disease in rodents and humans. J Clin Invest 121(6):2102–2110. https://doi.org/10.1172/JCI46069
11. Giedt MS, Thomalla JM, Johnson MR, Lai ZW, Tootle TL, Welte MA (2022) Adipose triglyceride lipase promotes prostaglandin-dependent actin remodeling by regulating substrate release from lipid droplets. bioRxiv. https://doi.org/10.1101/2021.08.02.454724
12. Stephenson RA, Thomalla JM, Chen L, Kolkhof P, White RP, Beller M, Welte MA (2021) Sequestration to lipid droplets promotes histone availability by preventing turnover of excess histones. Development 148(15):dev199381. https://doi.org/10.1242/dev.199381
13. Teixeira L, Rabouille C, Rorth P, Ephrussi A, Vanzo NF (2003) Drosophila Perilipin/ADRP homologue Lsd2 regulates lipid metabolism. Mech Dev 120(9):1071–1081. https://doi.org/10.1016/s0925-4773(03)00158-8
14. Li Z, Thiel K, Thul PJ, Beller M, Kuhnlein RP, Welte MA (2012) Lipid droplets control the maternal histone supply of Drosophila embryos. Curr Biol 22(22):2104–2113. https://doi.org/10.1016/j.cub.2012.09.018
15. Welte MA (2015) As the fat flies: the dynamic lipid droplets of Drosophila embryos. Biochim Biophys Acta 1851(9):1156–1185. https://doi.org/10.1016/j.bbalip.2015.04.002
16. Prasad M, Jang AC, Starz-Gaiano M, Melani M, Montell DJ (2007) A protocol for culturing Drosophila melanogaster stage 9 egg chambers for live imaging. Nat Protoc 2(10):2467–2473. https://doi.org/10.1038/nprot.2007.363
17. Buszczak M, Lu X, Segraves WA, Chang TY, Cooley L (2002) Mutations in the midway gene disrupt a Drosophila acyl coenzyme A: diacylglycerol acyltransferase. Genetics 160(4):1511–1518
18. Gaspar I, Szabad J (2009) In vivo analysis of MT-based vesicle transport by confocal reflection microscopy. Cell Motil Cytoskeleton

66(2):68–79. https://doi.org/10.1002/cm.20334

19. Lu W, Lakonishok M, Gelfand VI (2021) Gatekeeper function for short stop at the ring canals of the Drosophila ovary. Curr Biol 31(15):3207–3220 e3204. https://doi.org/10.1016/j.cub.2021.05.010
20. Monteith CE, Brunner ME, Djagaeva I, Bielecki AM, Deutsch JM, Saxton WM (2016) A mechanism for cytoplasmic streaming: kinesin-driven alignment of microtubules and fast fluid flows. Biophys J 110(9):2053–2065. https://doi.org/10.1016/j.bpj.2016.03.036
21. Parra-Peralbo E, Culi J (2011) Drosophila lipophorin receptors mediate the uptake of neutral lipids in oocytes and imaginal disc cells by an endocytosis-independent mechanism. PLoS Genet 7(2):e1001297. https://doi.org/10.1371/journal.pgen.1001297
22. Vereshchagina N, Ramel MC, Bitoun E, Wilson C (2008) The protein phosphatase PP2A-B' subunit Widerborst is a negative regulator of cytoplasmic activated Akt and lipid metabolism in Drosophila. J Cell Sci 121(Pt 20):3383–3392. https://doi.org/10.1242/jcs.035220
23. Vereshchagina N, Wilson C (2006) Cytoplasmic activated protein kinase Akt regulates lipid-droplet accumulation in Drosophila nurse cells. Development 133(23):4731–4735. https://doi.org/10.1242/dev.02659
24. Yu YV, Li Z, Rizzo NP, Einstein J, Welte MA (2011) Targeting the motor regulator Klar to lipid droplets. BMC Cell Biol 12:9. https://doi.org/10.1186/1471-2121-12-9
25. Tran SL, Welte MA (2010) In-vivo centrifugation of Drosophila embryos. J Vis Exp 40:2005. https://doi.org/10.3791/2005
26. McMillan EA, Longo SM, Smith MD, Broskin S, Lin B, Singh NK, Strochlic TI (2018) The protein kinase CK2 substrate Jabba modulates lipid metabolism during Drosophila oogenesis. J Biol Chem 293(8):2990–3002. https://doi.org/10.1074/jbc.M117.814657
27. Targett-Adams P, Chambers D, Gledhill S, Hope RG, Coy JF, Girod A, McLauchlan J (2003) Live cell analysis and targeting of the lipid droplet-binding adipocyte differentiation-related protein. J Biol Chem 278(18):15998–16007. https://doi.org/10.1074/jbc.M211289200
28. Greenspan P, Mayer EP, Fowler SD (1985) Nile red: a selective fluorescent stain for intracellular lipid droplets. J Cell Biol 100(3):965–973. https://doi.org/10.1083/jcb.100.3.965
29. Wang S, Li W, Hui H, Tiwari SK, Zhang Q, Croker BA, Rawlings S, Smith D, Carlin AF, Rana TM (2020) Cholesterol 25-Hydroxylase inhibits SARS-CoV-2 and other coronaviruses by depleting membrane cholesterol. EMBO J 39(21):e106057. https://doi.org/10.15252/embj.2020106057
30. Chen BH, Yang HJ, Chou HY, Chen GC, Yang WY (2017) Staining of lipid droplets with monodansylpentane. Methods Mol Biol 1560:231–236. https://doi.org/10.1007/978-1-4939-6788-9_17
31. Welte MA, Cermelli S, Griner J, Viera A, Guo Y, Kim DH, Gindhart JG, Gross SP (2005) Regulation of lipid-droplet transport by the perilipin homolog LSD2. Curr Biol 15(14):1266–1275. https://doi.org/10.1016/j.cub.2005.06.062
32. Welte MA, Gross SP, Postner M, Block SM, Wieschaus EF (1998) Developmental regulation of vesicle transport in Drosophila embryos: forces and kinetics. Cell 92(4):547–557. https://doi.org/10.1016/s0092-8674(00)80947-2
33. Guo Y, Jangi S, Welte MA (2005) Organelle-specific control of intracellular transport: distinctly targeted isoforms of the regulator Klar. Mol Biol Cell 16(3):1406–1416. https://doi.org/10.1091/mbc.e04-10-0920
34. Clarkson M, Saint R (1999) A His2AvDGFP fusion gene complements a lethal His2AvD mutant allele and provides an in vivo marker for Drosophila chromosome behavior. DNA Cell Biol 18(6):457–462. https://doi.org/10.1089/104454999315178
35. Johnson MR, Stephenson RA, Ghaemmaghami S, Welte MA (2018) Developmentally regulated H2Av buffering via dynamic sequestration to lipid droplets in Drosophila embryos. eLife 7:e36021. https://doi.org/10.7554/eLife.36021
36. Pandey R, Heidmann S, Lehner CF (2005) Epithelial re-organization and dynamics of progression through mitosis in Drosophila separase complex mutants. J Cell Sci 118(Pt 4):733–742. https://doi.org/10.1242/jcs.01663
37. Cermelli S, Guo Y, Gross SP, Welte MA (2006) The lipid-droplet proteome reveals that droplets are a protein-storage depot. Curr Biol 16(18):1783–1795. https://doi.org/10.1016/j.cub.2006.07.062
38. Mir M, Stadler MR, Ortiz SA, Hannon CE, Harrison MM, Darzacq X, Eisen MB (2018) Dynamic multifactor hubs interact transiently with sites of active transcription in Drosophila embryos. eLife 7:e40497. https://doi.org/10.7554/eLife.40497

Chapter 13

Assessing Ovulation in *Drosophila melanogaster*

Andrew Beard, Rebecca Oramas, and Jianjun Sun

Abstract

Ovulation is a critical reproductive process by which a mature oocyte is released from the ovary for fertilization. This process requires the coordination of multiple cellular and molecular events including the spatiotemporal breakdown of the follicle wall. Recent work has shown that ovulation in *Drosophila* utilizes conserved cellular processes and molecular pathways as in mammals. Thus, *Drosophila* ovulation can serve as a good model to decipher the fundamental mechanisms of ovulation utilized across species. In past decades, several methods have been developed to study *Drosophila* ovulation, but all of them have drawbacks. This chapter offers a strategy and detailed protocols for performing and analyzing the necessary assays to evaluate the ovulation process in *Drosophila*.

Key words *Drosophila*, Ovulation, Follicle rupture, Oviposition

1 Introduction

Ovulation is a crucial step for successful reproduction and involves the proteolytic breakdown of mature follicles to release fertilizable oocytes from the ovary. In mammals, ovulation is triggered by the surge of luteinizing hormone (LH) that binds to the LH receptor in the outer granulosa cells of the preovulatory follicles [1]. It triggers multifaceted cellular signaling pathways, including epidermal growth factor, progesterone, and prostaglandin signaling pathways, which ultimately lead to the resumption of meiosis, the expansion of cumulus granulosa cells, and the activation of proteolytic enzymes that breaks down the follicle wall [2–4]. As a model organism, *Drosophila* has played an eminent role in many aspects of modern biology. Its role in deciphering ovulation mechanisms has not been obvious until recently. With the powerful genetic tools available in *Drosophila*, we characterized *Drosophila* ovulation at the

Supplementary Information The online version contains supplementary material available at https://doi.org/10.1007/978-1-0716-2970-3_13.

Michelle S. Giedt and Tina L. Tootle (eds.), *Drosophila Oogenesis: Methods and Protocols*,
Methods in Molecular Biology, vol. 2626, https://doi.org/10.1007/978-1-0716-2970-3_13,

cellular and molecular levels and uncovered that *Drosophila* ovulation shows multiple similarities to ovulation in mammals and other vertebrates. For example, we found that posterior follicle cells in a stage 14 egg chamber (or mature follicle) are degraded before the encased mature oocyte can be released into the oviduct [5]. After ovulation, the remaining follicle cells are retained at the end of the ovariole to form a corpus luteum-like structure as in mammals [5]. We demonstrated that octopamine (OA) functions as the signal to induce ovulation via activation of its receptor Oamb in mature follicle cells, which leads to the activation of matrix metalloproteinase 2 (MMP2) and NADPH oxidase (NOX) [6, 7]. We also showed that the breakdown of the follicle wall requires MMP2 and ecdysteroid signaling [8]. With these advancements, *Drosophila* ovulation not only serves as a model to understand conserved molecular mechanisms for ovulation but also can be used as a platform for screening non-steroidal contraceptive compounds [9].

Over the years, several different methods have been developed to evaluate the ovulation capabilities of *Drosophila*. The first method is to push on the female's abdomen and to record if there is an egg ejecting from the uterus (*see* Fig. 1 for the female reproductive system) [10]. This method provides a rough estimate of whether there is an ovulated egg in the uterus. A slightly more accurate assay derived from the one above is to examine the presence and/or the location of an egg in the lower reproductive tract (LRT; including the bilateral and common oviducts and the uterus) through dissection after mating [11–13]. A reduction of ovulated eggs in the lower reproductive tract could indicate an ovulation defect; however, this reduction could also be caused by an

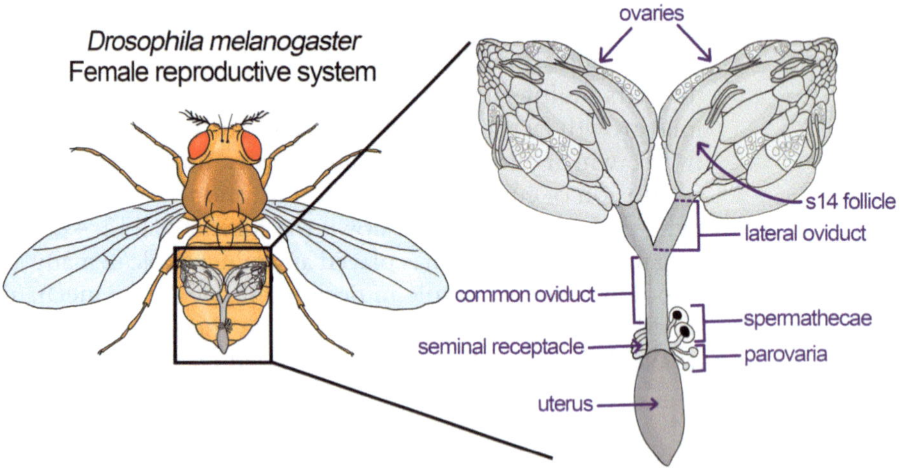

Fig. 1 An illustration of the female reproductive system in *Drosophila melanogaster*. The female reproductive system is located in the abdomen and consists of two ovaries, bilateral oviducts, the common oviduct, and the uterus. In addition, there are ventrally localized seminal receptacle and dorsally localized spermathecae and parovaria connecting the uterus. A stage 14 (s14) egg chamber is indicated

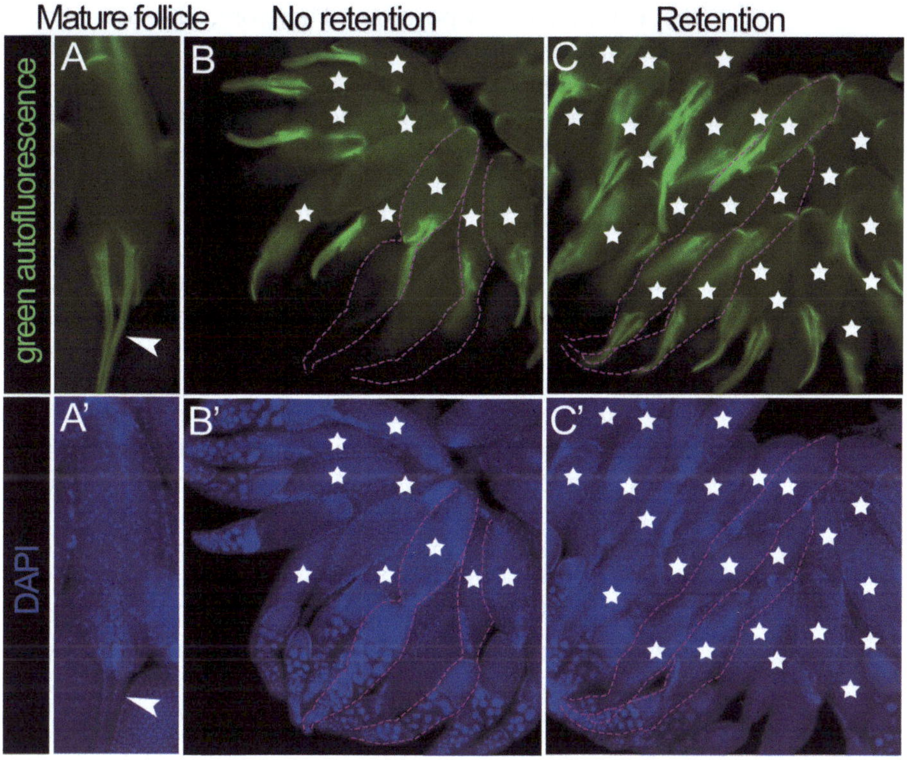

Fig. 2 Mature follicle retention phenotype. (**a**–**a′**) A picture of a mature follicle with green autofluorescence (**a**) and stained with DAPI (blue in **a′**). The arrowhead points to the long dorsal appendage. (**b**, **b′**) an example of the normal ovary without egg retention phenotype. The stars denote mature follicles, and the dashed lines outline the ovariole. (**c**, **c′**) An example of the ovary with an egg-retention phenotype. The stars denote the mature follicles, and the dashed lines outline the ovariole

oogenesis defect or influenced by the rate of oviposition (the release of the egg from the inside uterus to the outside environment). In our experience, we seldom observed two eggs in the lower reproductive tract of wild-type females even when oviposition was prevented, indicating that the presence of an egg in the uterus likely sends a feedback signal to inhibit ovulation. Another commonly used method to assess ovulation is to examine the morphology of the ovary including counting stage 14 egg chambers [14–17]. The accumulation of more than two stage 14 egg chambers in each ovariole, called a "backed-up" or "egg-retention" phenotype, could indicate an ovulation defect (*see* Fig. 2a–c); however, any blockage in the egg-laying process (including ovulation, transportation via the oviduct, and oviposition) could secondarily induce the "backed-up" phenotype. In other instances, measuring egg-laying capacity or fecundity has been used to assess the ovulation defect [18]. It is obvious that a reduction in fecundity could be influenced by the rate of oogenesis, ovulation, transportation via the oviduct, or oviposition. Therefore, the use of any above-mentioned methods individually could run the risk of making an inaccurate conclusion.

In order to account for the drawbacks of each assay, our laboratory has developed a series of assays to quantitatively assess the ovulation capacities in *Drosophila* [6, 8, 19]. We first evaluate the egg-laying capacity using the *egg-laying assay*. If there's a reduction in the number of eggs laid, we then proceed to examine whether there is an oogenesis defect by counting mature follicles inside the ovaries of the females after the egg-laying assay (*mature follicle counting*). If there is no reduction in mature follicles, we then examine whether there is an ovulation defect using the *egg-laying time assay*, in which we estimate the time required for each step in the egg-laying process including ovulation, transportation through the oviduct, and oviposition. A significant increase in time spent in ovulation indicates an ovulation defect. Finally, we examine whether the ovulation defect is due to mature follicles' incompetence for OA-induced follicle rupture using our *ex vivo follicle rupture assay*. Using this strategy, our lab gets consistent and repeatable results regardless of the researcher and time of experiments. Thus, these are robust assays for investigating the underlying mechanisms of ovulation. In this chapter, we provide detailed protocols for these assays.

2 Materials

2.1 Common Equipment and Materials

1. Refrigerator (4 °C).
2. Fly incubator.
3. Fly anesthetizing equipment.
4. Stereomicroscope for dissection.
5. Compound fluorescent microscope with camera.
6. Fine dissection forceps.
7. Nine-well dissection glass plates.
8. Pipet set.
9. Plastic transfer pipets, disposable, 5.8 mL.
10. 1.5 mL microcentrifuge tubes.
11. Fly vials with fresh standard cornmeal and molasses food.
12. Active dry yeast.
13. Grace's medium, with L-glutamine.
14. Image processing software such as ImageJ or Photoshop.
15. Assessing Ovulation Template (*see* Supplemental 1).

2.2 Egg-Laying Assay and Counting Mature Follicles

1. High-resolution scanner.
2. Hand tally counter.
3. Peristaltic pump.
4. Microwave.
5. Bunsen burner.
6. Syringe needle (18 gauge).
7. Assorted colored tape (0.75 inches wide).
8. Microscope slides.
9. Cover slip 24 × 55.
10. Delicate task wipes.
11. Clear nail polish.
12. 1 L flask.
13. Egg-laying bottles with holes.
14. 35 × 10 mm Petri Dishes.
15. Plastic cling wrap.
16. Agar.
17. Molasses.
18. 5% Tegosept.
19. 4% paraformaldehyde.
20. Phosphate-buffered saline (PBS) with Triton X-100 (PBT): 1× PBS with 0.2% Triton X-100.
21. 4′,6-diamidino-2-phenylindole (DAPI) stock solution (0.1 mg/mL).
22. Mounting medium.

2.3 Egg-Laying Time Assay

1. Freezer (−80 °C).
2. 25 mm tube rack.
3. 1× phosphate-buffered solution (PBS).

2.4 Materials for Ex Vivo Follicle Rupture Assay

1. 500 mL utility boxes.
2. Stainless steel needles, 0.25 mm (D), 36 mm (L).
3. Aluminum foil.
4. Paper towels.
5. Culture medium: Add 1 mL of fetal bovine serum and 100 μL of penicillin–streptomycin (10,000 U/mL) to 9 mL of Grace's medium.
6. Octopamine stock solution: Dissolve octopamine hydrochloride in distilled water to 10 mM. Aliquots can be stored at −20 °C for up to 6 months.
7. Mature follicle reporter: *47A04-Gal4/UAS-RFP, 44E10-Gal4/UAS-RFP, or Oamb-RFP.*

3 Methods

3.1 The Egg-Laying Assay

The first assay used to evaluate the ovulation capacity of *Drosophila* females is the egg-laying assay. A defect in egg-laying does not necessarily indicate an ovulation defect, but a defect in ovulation will certainly lead to an egg-laying defect. Therefore, this assay serves as a good entry point to identify/screen potential genes involved in ovulation. Once an egg-laying defect is observed, it is important to dissect the ovaries and count the mature follicles from the females to exclude the possibility that the egg-laying defect is due to the lack of mature follicles (oogenesis defect). In contrast, an increase in mature follicles (similar to the backed-up/egg-retention phenotype; Fig. 2b, c) serves as strong evidence of an ovulation defect and prompts the subsequent analyses.

3.1.1 Preparation of Reagents

1. *Fresh Wet Yeast*
   - (a) Mix 6 g of active dry yeast with 13.5 mL distilled water until the paste is homogeneous, and grains of dry yeast are no longer visible.
   - (b) Wait approximately 30 min after mixing for the wet yeast to reach its final consistency (*see* **Note 1**).
2. *Molasses Plates*
   - (a) Add 11 g of agar, 45 mL of molasses, and 500 mL of cold water to a 1 L flask and swirl the flask to mix all the contents together.
   - (b) Microwave the flask for 30 s to 1 min, swirl the flask to mix the ingredients and repeat this process until the agar is completely dissolved and the mixture appears completely translucent.
   - (c) Add 9 mL of 5% Tegosept once the mixture has cooled to approximately 60 °C.
   - (d) Dispense 4.3 mL of molasses mixture into clear petri dish lids using a liquid dispenser (i.e., peristaltic pump). Allow plates to cool and solidify. The plates can then be used for the experiment. Alternatively, the plates can be sealed with plastic cling wrap and stored in a 4 °C refrigerator for up to a week (*see* **Note 2**).
3. *Egg-Laying Bottles*
   - (a) Heat an 18-gauge syringe needle using a Bunsen burner and gently push the heated needle through the bottle. Make nine holes per side, and make sure each hole is not too big to prevent flies from escaping.

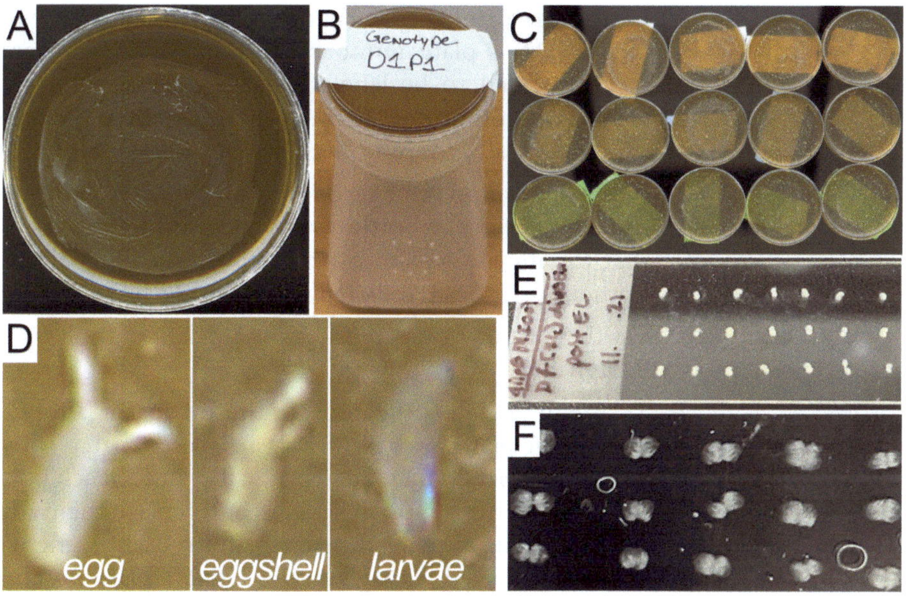

Fig. 3 Egg-laying and mature follicle counting. (**a**) A picture of a properly wet yeasted egg-laying plate. (**b**) An egg-laying bottle with the egg-laying plate taped down and labeled. (**c**) An example of a high-resolution scanned image of organized egg-laying plates. Multiple genotypes can be scanned at once.(**d**) Examples of an egg, an eggshell, and a larva. (**e**, **f**) Dissected ovaries spread out across a slide during the mounting process before (**e**) and after (**f**) the addition of the mounting media and coverslip

3.1.2 Egg Laying

1. Select 25–30 virgin females from each desired genotype and raise them in a vial for 5 days before the egg-laying assay (*see* **Note 3**). On the fourth day (1 day before the egg-laying assay begins), feed females with ~1 teaspoon of fresh wet yeast paste (*see* **Note 1**) applied to the wall of the fresh food vial.
2. Isolate 50–60 young wild-type males (e.g., Oregon-R) for each genotype 2 days before the egg-laying assay and keep in a fresh food vial at room temperature (*see* **Note 4**).
3. On the day of the egg-laying assay, prepare molasses plates by applying a thin layer of fresh wet yeast paste on the plate surface. Make sure to apply similar amounts of yeast paste to each plate and cover two-thirds of the surface area, avoiding the edges of the plate (*see* Fig. 3a; *see* **Note 5**).
4. Anesthetize the 25–30 prepared virgin females on the fly sorting pad using CO_2, and divide them into five groups, each containing five females. Discard any females with obvious morphological defects (*see* **Note 6**).
5. Anesthetize the prepared males on the same fly sorting pad using CO_2, and divide them into five groups, each containing 10 males. Discard any leftover males that were not used.
6. Add five females and ten males into each egg-laying bottle and top the bottle with the molasses plate prepared in **step 3**.

Secure the plate to the bottle using tape labeled with genotype + D1 + plate number (*see* Fig. 3b). Each genotype is recommended to be labeled with a unique color of tape to facilitate identification after scanning (*see* **Step 9**). Repeat this step to sort all groups of flies into bottles (*see* **Note** 7).

7. Finish setting up the egg-laying bottles for all genotypes. Wait until all flies in the bottles wake up, and then invert bottles so that the molasses plate is on the bottom. Keep all egg-laying bottles in the desired fly incubator for 22 h (*see* **Note 8**).
8. On the next day, prepare the same amount of molasses plates as in **step 3**.
9. Take egg-laying bottles out of the incubator after 22 h and replace the plates with new molasses plates coated with the wet yeast paste. To do that, anesthetize flies inside each bottle using CO_2, remove old molasses plates, and replace them with a new molasses plate. Secure the new plate to each bottle with rainbow tape labeled with genotype + D2 + plate number. Make sure to prevent any flies from escaping during plate replacement. Once you finish setting up the bottles, wait until all flies wake up, and then invert the bottles so that the molasses plate is on the bottom. Again, keep all egg-laying bottles in the desired incubator for another 22 h (*see* **Note 8**).
10. Cool the molasses plates with laid eggs (from D1 bottles) to room temperature, scan the plates facing down with a high-resolution document scanner, and save as a tiff file for counting the number of eggs laid on day 1 (*see* Fig. 3c). Flip the plates and scan the labeled side of the egg laying plates on the scanner as a record of the order, so you can repeat the same order for day 2 (*see* **Note 9**).
11. After the second 22 h period, retrieve the egg-laying bottles from the incubator. Anesthetize the flies in each bottle and transfer all females of the same genotype to a fresh food vial with the same color tape on the vial. Store the vials in a refrigerator (*see* **Note 10**) for subsequent dissection (*see* Subheading 3.1.3).
12. Cool the molasses plates with laid eggs (from D2 bottles) to room temperature, scan with a high-resolution document scanner, and save as a tiff file for counting the number of eggs laid on day 2 (*see* **Note 9**).
13. *Egg-laying analysis*:
    (a) Drag the scanned egg-laying high-resolution tiff file into the image processing software of your choice.
    (b) Once the image is loaded into the software, it is suggested to overlay a grid on the image or manually draw lines to split the plate up into sections to ensure an accurate count (*see* **Note 11**).

(c) Start counting by either using the image processing software's built-in count feature or using a mechanical/digital counter. Only count eggs and eggshells, not hatched larvae (*see* Fig. 3d).

(d) Record the total number of eggs and eggshells on each plate and repeat this step for all genotypes.

(e) After all the plates on both days 1 and 2 have been counted, calculate the number of laid eggs/female/day for each bottle, from which, the average number of laid eggs/female/day for each genotype, the standard deviation, and the *P* value can be calculated to determine the significant difference between genotypes.

(f) If the Assessing Ovulation Template (*see* Supplemental 1) is being used, record the number of eggs in the corresponding cells (*see* Fig. 4). Repeat this step for all plates and genotypes. Once all the data have been entered, the graph reporting the number of eggs/female/day will generate in the "Egg-Laying Summary" tab (*see* Fig. 5a).

3.1.3 Mature Follicle Count

1. Dissect the female flies stored in the refrigerator within an hour (*see* **Note 10**). Retrieve the ovary pairs in Grace's medium using forceps, fix with 4% paraformaldehyde for 15 min, wash with PBT 3 × 15 min, stained with DAPI (1:200) for 15 min, and mount on slides (*see* Fig. 3e).

2. To mount, use forceps to gently organize the ovary pairs in lines. Make sure the ovaries are far enough from each other to ensure that when the coverslip is added they do not overlap. Using a delicate task wipe, remove all excess liquid around each ovary pair. Add ~200 μL of mounting medium to the slide and make sure each ovary pair is submerged in medium (*see* **Note 12**). After adding the mounting medium, place the 24 × 55 cover slip over the tissue and press down with forceps, spreading out the mounting medium. The ovaries will now look two-dimensional and acquire a flower-like shape (*see* Fig. 3f). Seal the cover slip to the slide using nail polish and allow it to dry. Label the slide with the genotype and the date.

3. *Mature follicle counting*:

   (a) Using a fluorescent microscope, count the number of mature follicles in each ovary pair. Mature follicles have a long dorsal appendage (*see* Fig. 2a) that when observed using the fluorescent filter for green fluorescent protein (GFP), displays green autofluorescence (*see* Fig. 2a). Mature follicles can also be identified by the DAPI staining, as they lack nurse cell nuclei in the anterior region of the follicle (*see* Fig. 2a′).

   (b) Record the number of mature follicles in the top and bottom ovaries for each female.

Full Genotype	Simple Genotype		22h	22h				
UAS-dcr2/OregonR ; 44E10,RG6 / +	44E10> OregonR		D1 eggs	D2 eggs		D1 eggs /female	D2 eggs /female	Eggs /female/ day
		1	320	249		64.0	49.8	56.9
		2	435	230		87.0	46.0	66.5
		3	381	189		76.2	37.8	57.0
		4	449	333		89.8	66.6	78.2
		5	267	270		53.4	54.0	53.7
	# Groups		5	5	Average	74.08	50.84	62.46
	Total Eggs		1852	1271	StDev	15.39	10.64	10.02
					SEM	6.88	4.76	4.48

UAS-dcr2/hnt[v3788] ; 44E10,RG6 / +	44E10> hnt[v3788]		Day 1 eggs	Day 2 eggs		D1 eggs /female	D2 eggs /female	Eggs /female/ day
		1	232	166		46.4	33.2	39.8
		2	143	106		28.6	21.2	24.9
		3	207	230		41.4	46.0	43.7
		4	254	202		50.8	40.4	45.6
		5	212	205		42.4	41.0	41.7
	# Groups		5	5	Average	41.92	36.36	39.14
	Total Eggs		1048	909	StDev	8.32	9.63	8.25
					SEM	3.72	4.30	3.69

UAS-dcr2/hnt[v101325] ; 44E10,RG6 / +	44E10> hnt[v101325]		Day 1 eggs	Day 2 eggs		D1 eggs /female	D2 eggs /female	Eggs /female/ day
		1	171	167		34.2	33.4	33.8
		2	211	196		42.2	39.2	40.7
		3	221	170		44.2	34.0	39.1
		4	189	187		37.8	37.4	37.6
		5	220	123		44.0	24.6	34.3
	# Groups		5	5	Average	40.48	33.72	37.10
	Total Eggs		1012	843	StDev	4.35	5.63	3.00
					SEM	1.95	2.52	1.34

Fig. 4 An example of a completed "egg laying" sheet using the Assessing Ovulation Template. The first box of the table contains the genotype information. The second box is where the number of eggs is recorded. Days 1 and 2 egg-laying number is recorded under "D1 eggs," and "D2 eggs," respectively. The number of eggs per female is calculated in the third box, and the average, standard deviation, and standard error of the mean are calculated at the bottom

(c) Once all the genotypes have been counted, calculate the total number of mature follicles for each female, the average number of mature follicles/female for each genotype, the standard deviation, and the *P* value.

(d) If the Assessing Ovulation Template is being used, record the number of mature follicles in the top and bottom ovaries in the corresponding cells of the "mature follicles" tab. The average number of mature follicles/female, standard deviation, and *P* value will be calculated at the bottom of the sheet (*see* Fig. 6) and the graph will be plotted (*see* Fig. 5b).

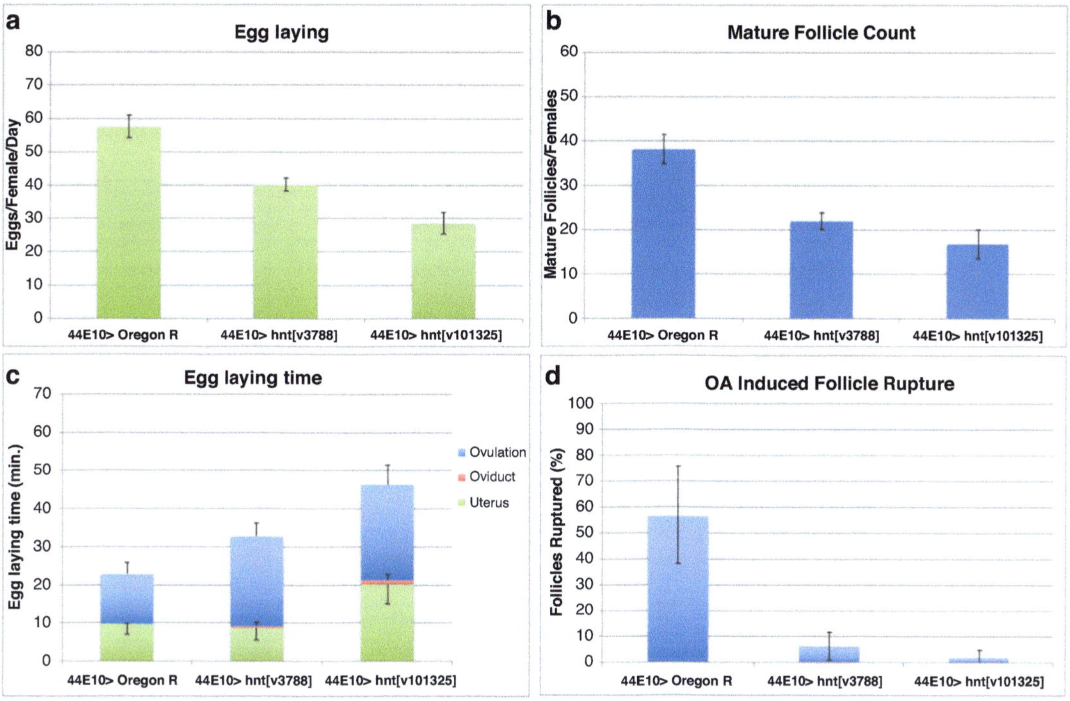

Fig. 5 Sample data graphs. (**a**) An example graph of how the egg-laying data can be reported. (**b**) An example graph of how the mature follicle data can be reported. (**c**) An example graph of how the egg laying time data can be reported. (**d**) An example graph of how the ex vivo follicle rupture assay data can be reported

3.2 Egg-Laying Time Assay

If the observed egg-laying defect is not due to an oogenesis defect, it is important to determine whether this defect is caused by an ovulation defect or another step in the egg-laying processes, such as egg transportation through the oviduct and oviposition. Using the data from the egg-laying assay, it is possible to calculate the average time required to lay an egg. The average time required to lay an egg can be partitioned into three parts: ovulation time (the time needed for ovulation), oviduct time (the time needed to transport eggs through the oviduct), and uterus time (the time an egg spends in the uterus including the time of oviposition). If there is an ovulation defect, the egg will spend a significantly longer time during ovulation compared to controls. In order to determine the partition factor, the frequency of eggs located in each part of the reproductive tract during egg laying must be determined. The more frequently an egg is in a particular part of the reproductive tract, the more time an egg is spending in this part of the reproductive tract. To measure the partition factor, use another mating assay to determine the location of eggs in the female reproductive tract 6 h after mating.

Full Genotype	UAS-dcr2/OregonR ; 44E10,RG6 / +			UAS-dcr2/hnt[v3788] ; 44E10,RG6 / +			UAS-dcr2/hnt[v101325] ; 44E10,RG6 / +		
Simple Genotype	**44E10>OregonR**			**44E10>hnt[v3788]**			**44E10>hnt[v101325]**		
	s14 (top ovary)	s14 (bottom ovary)	total s14	s14 (top ovary)	s14 (bottom ovary)	total s14	s14 (top ovary)	s14 (bottom ovary)	total s14
1	16	17	33	8	8	16	1	3	4
2	20	21	41	8	9	17	10	11	21
3	25	24	49	7	8	15	15	13	28
4	16	17	33	11	11	22	6	5	11
5	11	14	25	5	10	15	4	2	6
6	21	24	45	12	13	25	6	20	26
7	21	21	42	10	13	23	4	5	9
8	16	11	27	10	8	18	2	4	6
9	18	19	37	13	8	21	4	11	15
10	49	16	65	13	15	28	5	13	18
11	61	7	68	2	7	9	7	8	15
12	25	29	54	5	8	13	13	11	24
13	23	22	45	9	8	17	0	0	0
14	8	3	11	13	10	23	15	15	30
15	21	17	38	10	4	14	12	7	19
16	24	18	42	7	7	14	6	5	11
17	25	21	46	12	5	17	5	3	8
18	71	5	76	12	10	22	17	12	29
19	22	16	38	10	11	21	3	3	6
20	46	10	56	9	13	22	12	6	18
21									
Average			43.55			18.60			15.20
StDev			15.27			4.71			9.08
# of N			20			20			20
Ttest			1.00			0.00			0.00

Fig. 6 An example of a completed "mature follicles" sheet using the Assessing Ovulation Template. For each genotype, the first two columns record the number of mature follicles in the top and bottom ovaries of each female. The total mature follicles for each female will be calculated in the third column. The average number of follicles/female, standard deviation, and *P* value is calculated at the bottom

1. Prepare 30–35 virgin females and 50 young wild-type males for each desired genotype following **steps 1** and **2** in egg laying assay (*see* Subheading 3.1.2).
2. On the day of the egg-laying time assay, anesthetize the 30–35 prepared virgin females on the fly sorting pad using CO_2. Divide the virgin females into three groups, each containing ten females. Discard any females with obvious morphological defects, such as distorted wings or a significantly smaller body than the rest of your flies (*see* **Note 6**).
3. Anesthetize 50 prepared males on the same fly sorting pad using CO_2, and divide them into three groups, each containing 15 males.

4. Add 10 females and 15 males into a vial of fresh food with active dry yeast added to it and label each vial with the correct genotype (*see* **Note 13**). Repeat this step until all groups of flies are added to fresh food vials. Make sure to finish **steps 3–5** as soon as possible to prevent CO_2 overexposure (*see* **Note 7**).
5. Finish setting up the vials for all genotypes and wait until all flies in the vials wake up; then store all vials in the desired incubator for 6 h.
6. After 6 h, invert the vials (flugs facing down) and move them from the incubator to a −80 °C freezer for 5 min to freeze any eggs in situ in the female reproductive tract. After 5 min, store vials in a refrigerator for subsequent dissection (*see* **Note 14**).
7. For dissection, remove the cuticle of the ventral abdominal region to expose the female reproductive tract. Record the location of the egg in the reproductive tract. If there is no egg present in the reproductive tract, that female is in the process of ovulation and is scored as "*N*" for none (*see* Fig. 7a). The assumption is that the absence of an egg in the reproductive tract means that the egg is currently in the ovary undergoing ovulation. If an egg is more than 75% in the lateral oviduct (*see* Fig. 7b), the female has an egg in the bilateral oviduct, and is scored as "*B*." If an egg is completely within the common oviduct (*see* Fig. 7c), the female is scored as "*C*." The females with eggs in the uterus and ejected out of the uterus are scored as "*U*" and "*E*", respectively (*see* Fig. 7d, e). See an example of the scoring/labeling process in Fig. 8.
8. *Egg distribution analysis*: Once all the females for each genotype have been dissected and recorded, calculate the egg distribution for each genotype.
   (a) To calculate the egg distribution, calculate the percentage of eggs in the process of ovulation (females scored "*N*"), in the oviduct (females scored "*B*" or "*C*"), and in the uterus/oviposition (females scored "*U*" or "*E*") for each genotype. The egg distribution data can be reported in a 100% stacked column graph and will be used as partition factor for egg-laying time analysis.
   (b) If the assessing ovulation template (*see* Supplemental 1) is used, input the scores of each genotype in the corresponding cells under the "egg distribution" tab. The template will generate the percentage of eggs in the process of ovulation ($\%N$), in the oviduct ($\%B + \%C$), or in the uterus/oviposition ($\%U + \%E$), representing the frequency of an egg in each of these processes (*see* Fig. 8).

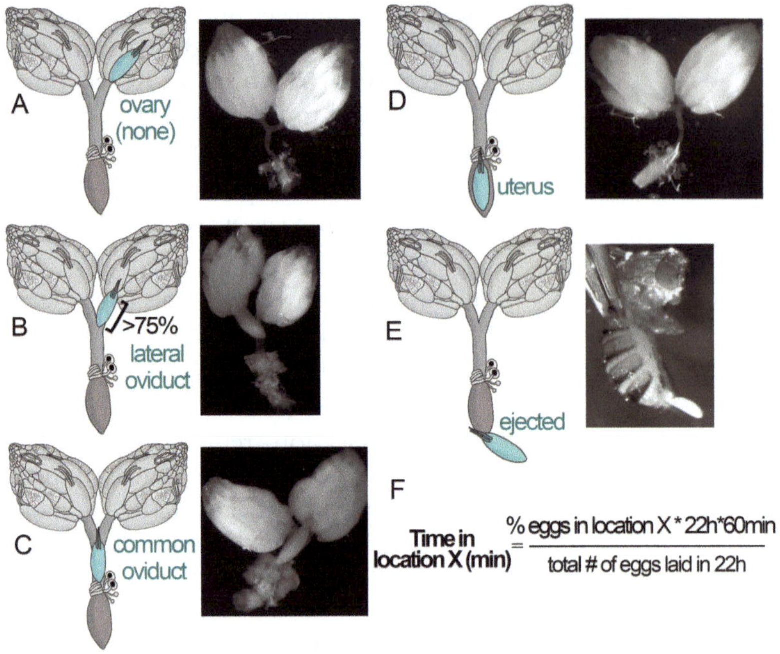

Fig. 7 Egg distribution in the female reproductive tract. (**a**–**e**) Examples of female reproductive tracts without any eggs (**a**) or with an egg in the lateral oviduct (**b**), the common oviduct (**c**), the uterus (**d**), or ejected (**e**). The left panel is a cartoon illustration, and the right panel is the actual reproductive tract. The egg is considered inside the lateral oviduct when more than 75% of the egg is out from the ovary. (**f**) The equation for estimating the time in each part of the egg-laying processes including ovulation (**a**), oviduct transportation (**b**, **c**), and uterus/oviposition (**d**, **e**)

9. *Egg-laying time analysis:*
   (a) Using the number of eggs/female/day (22 h, *see* **Note 8**) from the egg-laying assay, calculate the average time required for laying an egg (egg-laying time), which is equal to 22 × 60 min divided by the number of eggs/female/day. To calculate the ovulation time, oviduct time, and uterus time, multiply the egg-laying time by the percentage of eggs in ovulation (% N), oviduct (%B + % C), and uterus (%U + %E), respectively (*see* Fig. 7f).
   (b) To perform statistical analysis, we propagate the error as a 95% confidence interval from egg-laying data and egg distribution data to calculate the Z score and P value for ovulation time, oviduct time, and uterus time.
   (c) If the Assessing Ovulation Template is used, the calculated results can be found in the "egg-laying time" tab (*see* Fig. 9) and the graph will be plotted (see an example in Fig. 5c) once the data are entered for egg-laying assay and egg distribution.

Full Genotype	UAS-dcr2/OregonR ; 44E10,RG6 / +	UAS-dcr2/hnt[v3788] ; 44E10,RG6 / +	UAS-dcr2/hnt[v101325] ; 44E10,RG6 / +
Simple Genotype	**44E10> OregonR**	**44E10> hnt[v3788]**	**44E10> hnt[v101325]**
1	N	N	N
2	U	U	N
3	N	U	U
4	N	N	U
5	N	N	E
6	U	N	N
7	U	N	N
8	U	N	N
9	U	U	C
10	N	N	N
11	U	N	N
12	U	N	N
13	N	N	U
14	N	N	U
15	U	N	N
16	N	N	N
17	N	N	U
18	U	N	U
19	U	U	U
20	N	U	U
Mating pairs (n)	20	20	20
Number of N	10	15	10
Total oviduct (B+C	0	0	1
Total U (U+E)	10	5	9
Ovulation (N%)	50	75	50
Oviduct (B+C%)	0	0	5
Uterus (U+E%)	50	25	45

Fig. 8 An example of a completed "egg distribution" sheet using the Assessing Ovulation Template. The egg location in the female reproductive tract (N = None, B = Bilateral oviducts, C = Common oviduct, U = Uterus, E = Ejecting) is recorded under each genotype in the blue area. The percentage of eggs at each location (ovulation, oviduct, and uterus) is calculated at the bottom of the sheet

3.3 Ex Vivo Follicle Rupture Assay

The above-mentioned assays will help to determine whether females of a specific genotype have an ovulation defect in comparison to control females. Keep in mind, all these assays are indirect measurements of ovulation. As mentioned in the introduction, ovulation is a complicated process involving not only OA-induced follicle rupture but also coordinated ovarian and oviduct muscle contraction [17, 20, 21]. To further characterize the ovulation defect, the ex vivo follicle rupture assay was developed to determine

		Egg Laying in 48h				Egg Distribution in LRT							Egg Laying Time (min)											
Genotype	Genotype	# of Groups	Egg Laying	SEM	95% CI	N	Ovulation (%)	95% CI	Oviduct (%)	95% CI	Uterus (%)	95% CI	Ovulation (min)	95% CI	Oviduct (min)	95% CI	Uterus (min)	95% CI	Z score for Ovulation	Z score for Oviduct	Z score for Uterus	P values for Ovulation	P values for Oviduct	P values for Uterus
UAS-dcr2/OregonR ; 44E10,RG6 / +	44E10> OregonR	10	57.54	3.35	2.07	61	57.38	12.41	0.00	0.00	42.62	12.41	13.16	2.89	0.00	#DIV/0!	9.78	2.87						
UAS-dcr2/hnt[v3788] ; 44E10,RG6 / +	44E10> hnt[v3788]	10	40.27	1.90	1.18	75	72.00	10.16	1.33	2.60	26.67	10.01	23.60	3.40	0.44	0.85	8.74	3.29	2.33983	#DIV/0!	-0.24	0.02583	#DIV/0!	0.388
UAS-dcr2/hnt[v101325] ; 44E10,RG6 / +	44E10> hnt[v101325]	10	28.46	3.23	2.00	85	54.12	10.59	2.35	3.22	43.53	10.54	25.10	5.22	1.09	1.50	20.19	5.09	2.00097	#DIV/0!	1.782	0.05389	#DIV/0!	0.082

Fig. 9 An example of a completed "egg laying time" sheet using the Accessing Ovulation Template. The entire sheet is split up into three sections: egg laying in 48 h, egg distribution in the lower reproductive tract (LRT), and egg laying time. Data from "egg laying" and "egg distribution" sheets will be populated to egg laying in 48 h and egg distribution in LRT sections, respectively. The egg-laying time section will calculate the time spent in ovulation, oviduct, and uterus. The errors for egg laying time are propagated according to error propagation in multiplication and the Z scores and *P* values are further calculated based on the errors to determine the significant difference

whether the ovulation defect is due to mature follicles' incompetency for OA-induced follicle rupture, but not muscle contraction. The ex vivo follicle rupture assay tests an individual follicle's ability to rupture upon OA stimulation. Therefore, this assay is a direct measurement of the ovulation capacities of mature follicles and is used to support the in vivo ovulation defect. The key element for this assay to work is to select fully matured and intact egg chambers. To guide the selection of mature follicles, we recommend using *47A04-Gal4/UAS-RFP*, *44E10-Gal4/UAS-RFP*, or *Oamb-RFP* reporters [6, 19, 22] that label all follicle cells of mature egg chambers (*see* **Note 15**). The ex vivo follicle rupture assay is only briefly described here, and we refer readers to a detailed protocol described in Knapp et al. 2018 [23].

1. Collect 10–15 virgin females from each desired genotype and raise them in a vial of fresh food for 3 days (*see* **Note 3**). On the third day, transfer females to a fresh food vial with ~1 teaspoon of fresh wet yeast paste (*see* **Note 1**). Raise the females in the same vial for an additional 2–3 days before the ex vivo follicle rupture assay.
2. On the day of the ex vivo follicle rupture, warm Grace's medium to room temperature and prepare the culture medium.
3. Anesthetize prepared virgin females with CO_2 and place them on a fly sorting pad.
4. Dissect two ovary pairs at a time in the "dissection well" (one well of a nine-well glass dissection dish filled with Grace's medium). To liberate stage 14 egg chambers from each ovary, use one pair of forceps to hold the anterior end of the ovary and the other forceps to carefully peel off the muscle sheath at the posterior end of the ovary (*see* Fig. 10b). Then use the latter forceps to gently squeeze the middle portion of the ovary toward the posterior end (*see* Fig. 10c).

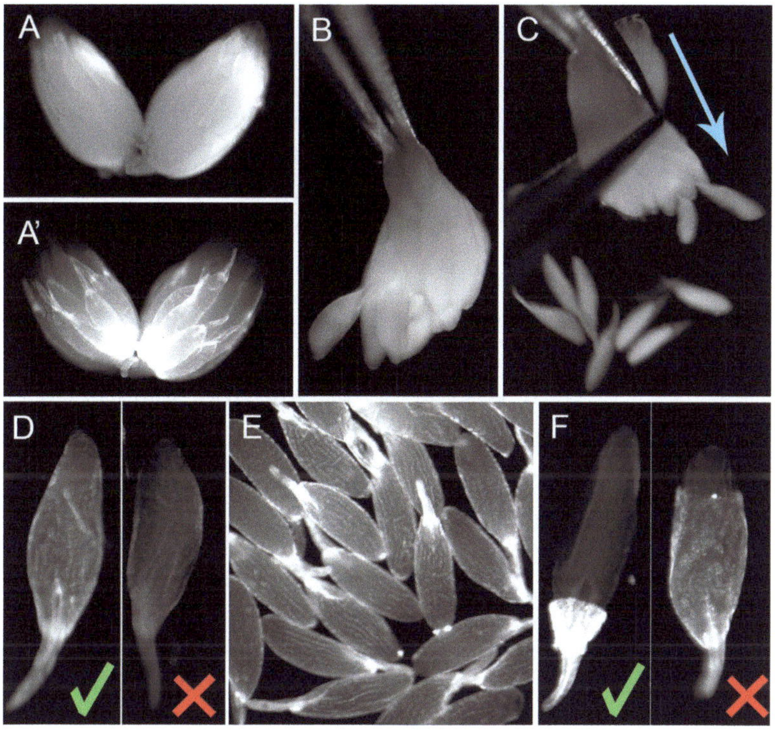

Fig. 10 Ex vivo follicle rupture assay. (**a**, **a′**) The dissected ovaries in the bright field (**a**) and under the fluorescent field (**a′**) showing reporter expression (*Oamb-RFP*) in follicle cells of mature egg chambers but not in younger ones. (**b**) The ovary after peeling off the muscle sheath at the posterior end. (**c**) The process of squeezing the middle portion of the ovary toward the posterior end (arrow points towards posterior) to liberate mature follicles. (**d**) The intact mature follicles with bright fluorescence (left one) will be selected for ex vivo follicle rupture, while the one with low fluorescent signal (right one) will not be selected. (**e**) The pool of selected intact and fully matured follicles in holding wells. (**f**) A follicle with more than 75% of the oocyte not covering with follicle cells (left one) is considered ruptured. The follicle on the right is not considered ruptured because more than 75% of the oocyte is covered with follicle cells

5. Select intact (no damage to the follicle cell layer) and fully matured follicles (with high reporter expression; *see* Fig. 10d, e) and pile them using a needle under a fluorescent stereomicroscope. Immediately move the pile of intact mature follicles to the "holding well" (a new well-containing culture medium) using a P20 pipet (*see* **Note 16**). Repeat until you have enough follicles or until ~40 min have passed since the initial dissection (*see* **Note 17**).
6. Remove any mature follicles in the "holding well" with younger egg chambers attached to them using bright field illumination or having a broken posterior follicle-cell layer using

fluorescent illumination (*see* **Note 18**). The leftover follicles will be good follicles for ex vivo follicle rupture assay (*see* Fig. 10f).

7. Pile 25–35 intact mature follicles together in the "holding well." Transfer the pile using a P20 pipet (set at 20 μL) into the "experimental wells" (new empty wells). Repeat this step until you have as many groups of 25–35 follicles as desired (*see* **Note 19**).
8. Add 1 mL of culture medium and 2 μL of octopamine stock solution (10 mM) to each "experimental well" containing 25–35 follicles. Make sure to mix immediately after adding the octopamine stock solution (*see* **Note 20**).
9. Put the nine-well glass dissection dish in a utility box with a moist paper towel and aluminum foil lining to control the humidity and prevent light exposure. Store the utility box in a 29 °C incubator for 3 h (*see* **Note 21**).
10. Remove the box from the 29 °C incubator and bring the plate to a fluorescent stereomicroscope with a camera for imaging. Use a needle to group the follicles in the center of the well but avoid having them overlap. Image with both fluorescence and bright field illumination.
11. Ex vivo *follicle rupture analysis:*
    (a) Using the image processing software of your choice, open the images and count the number of follicles using the bright field image and the ruptured follicles using the fluorescent image. A follicle is considered ruptured when more than 75% of the area of the oocyte has no follicle cell covering (*see* Fig. 10f).
    (b) Once the number of ruptured follicles has been counted, calculate the percentage of ruptured follicles for each well. The average percentage of ruptured follicles is calculated for each genotype, and the standard deviation and the *P* value can be calculated to determine the significance between genotypes.
    (c) If the Assessing Ovulation Template is being used, add the numbers to the appropriate cells in the Assessing Ovulation Template under the "OA culture" tab (*see* Fig. 11a). The bottom section of the assessing ovulation template generates the average percentage of follicles rupture and statistical analysis (*see* Figs. 5d and 11b; *see* **Note 22**).

A

Genotype	Genotype	OA Concentration	Well1-Total	Well1-Ruptured	Well2-total	Well2-Ruptured	Well3-total	Well3-Ruptured	Well4-total	Well4-Ruptured
UAS-dcr2/+;44E10,RG6/ OregonR	44E10> OregonR	20 μM	26	17	23	17	26	18		
UAS-dcr2/hnt[v3788]; 44E10,RG6/+	44E10> hnt[v3788]	20 μM	28	1	35	5	30	0	25	2
UAS-dcr2/hnt[v101325]; 44E10,RG6/+	44E10> hnt[v101325]	20 μM	29	0	27	1	30	0		

B

Full Genotype	Simple Genotype	OA Concentration	Well1-%	Well2-%	Well3-%	Well4-%	# of Groups	Total # of Eggs	Average %	StDev %	Ttest
UAS-dcr2/+; 44E10,RG6 /OregonR	44E10> OregonR	20 μM	65.4	73.9	69.2		3	75	69.51	4.27	
UAS-dcr2/hnt[v3788] ; 44E10,RG6 / +	44E10> hnt[v3788]	20 μM	3.6	14.3	0.0	8.0	4	118	6.46	6.16	2.33664E-05
UAS-dcr2/hnt[v101325] ; 44E10,RG6 / +	44E10> hnt[v101325]	20 μM	0.0	3.7	0.0		3	115	1.23	2.14	1.57968E-05

Fig. 11 An example of a completed "follicle rupture" sheet using the Accessing Ovulation Template. (**a**) The experimental information table where the experimental conditions, total numbers of follicles, and the number of ruptured follicles per well are entered. (**b**) The results table where the data from the experimental information table is used to calculate the percentage of ruptured follicles per well, which is then used to calculate the average, standard deviation, and *P* value for each genotype

4 Notes

1. The consistency of the wet yeast paste is critical and should be a smooth and thick batter. Upon lifting with a spatula, a short and thin string should form. Do not make too much wet yeast at a time, since it can become fermented in a couple of days. Old wet yeast looks darker and has an unpleasant fermented smell. Nutrition is important for proper egg development; poor nutrition will result in less mature follicle production.
2. The amount of molasses mixture in the petri dish lids should be implicitly tested if you are not using the exact size of the petri dish listed. The goal is to have enough agar in the plate that it is thick enough for the scanner to clearly visualize the laid eggs. It is also important that agar does not come all the way to the top of the plate to prevent crushing eggs during scanning. In our case, 4.3 mL of mixture works well for 35 × 10 mm petri dish lids. We typically store these plates for no more than 1 week in the refrigerator to avoid water evaporation.
3. Virgin females are collected within one or 2 days to minimize the age variation and are stored in a vial of good condition without males. If the vial starts to have bacteria/fungus growth

on the bottom, flies should be transferred into a new food vial. Mating induces ovulation, therefore, using virgin females will ensure they have retention of mature follicles, which will prime the females for egg laying and other assays.

4. For both the egg-laying assay and the egg-laying time assay, young healthy wild-type males should be used to ensure optimal mating conditions. Make sure to store them in vials of good condition. Males should be kept isolated for a couple of days so that they do not engage in mating and can produce enough sperm and seminal fluid for the subsequent mating experiments.
5. We typically set up mating for egg laying assay between 3 and 6 pm to prevent any interference with the fly's circadian rhythm and to reduce variability. The wet yeasted molasses plate should be prepared right before the mating. To make counting egg-laying plates easier, wet yeast should be applied over two-thirds of the molasses plates so that eggs would not be laid on the edge of the plates. In addition, evenly spread the wet yeast to prevent piling eggs in a clump of yeast. The amount of yeast should be similar across plates and should be enough so that there is still tiny wet yeast remaining on the plates after egg laying. In this way, flies will get enough nutrients during egg laying. Avoid using molasses plates in which the agar has cracks, dried edges, or bubbles (also *see* **Note 2**).
6. We discard any females with obvious morphological defects, such as distorted wing(s) and significantly small body size, which is why we typically select a few more virgins for each genotype. Females with obvious morphological defects may have developmental defects that could affect the results of the assays.
7. When sorting flies on a CO_2 pad to set up egg-laying bottles, make sure to finish **steps 4–6** as soon as possible to prevent CO_2 overexposure, which can lead to adverse effects or death.
8. We only incubate flies on egg-laying plates for 22 h, so that fewer embryos/eggs hatch into larvae. Identifying and counting eggs is easier than eggshells and will improve the accuracy of the counting.
9. Make sure to check the scanned image of your egg plates before discarding them. If the plates have not cooled enough, the plates will fog over, making egg visualization difficult and causing counting inaccuracy.
10. Ovary dissection for mature follicle counting should be performed right after egg-laying to reflect the ovary status during egg-laying. In addition, flies should be kept at 4 °C before dissection to avoid the accumulation of mature follicles due to lack of oviposition substrate and to prevent any mature follicles from leaving the ovary.

11. In our lab, we use either ImageJ or Abode Photoshop to count our egg-laying images. Photoshop will handle larger egg-laying tagged image file formats (TIFFs) better than ImageJ. To split up an egg-laying plate in ImageJ, either overlay a grid by going to "Analyze" > "Tools" > "Grid" or draw lines across the plate using "segmented line" in the control panel. For the latter, you should try to separate regions where a lot of eggs have been laid. In Photoshop, overlay a grid by going to "View" > "Show" > "Grid" so the plates are split up evenly. After the egg-laying plate has been split up, zoom in to the desired amount (we recommend 200%) for counting.
12. Any type of mounting medium can be used for this assay. To reduce the cost, our lab makes its own mounting medium using the following recipe: Mix 40 mL of glycerol, 5 mL of H_2O, 5 mL of 10× PBS, and 1 g of N-propyl gallate until homogeneous and store at 4 °C covered in aluminum foil.
13. According to our experience, almost all females will finish the mating within an hour when you add 10 females and 15 males into the fresh food vials. It is less variable compared to the 10 females/10 males or 10 females/20 males mating condition. To reduce variability, we always begin our egg-laying time assay between 8 am and 10 am.
14. It is critical not to use CO_2 to anesthetize flies for dissection. We have seen eggs ejecting out of the uterus when CO_2 is used to anesthetize flies, which indicates that eggs still move in the reproductive tract and will affect the accuracy of the analysis. Instead, a −80 °C freezer can efficiently freeze the eggs in situ in the reproductive tract. Flies should be dissected as soon as possible after freezing and stored in the refrigerator so that tissue morphology and location of the eggs can be easily observed.
15. For ex vivo follicle rupture, different mature follicle reporters will result in a slightly different rupture rate. For example, *44E10-Gal4/UAS-RFP*, *47A04-Gal4/UAS-RFP*, and *Oamb-RFP* reporters typically show ~50%, >80%, and 60–70% rupture, respectively. The difference is due to an inaccurate selection of slightly immature follicles as different fluorescent reporters start to be expressed at slightly different time points in stage 14 egg chambers.
16. It is most critical to transfer intact mature follicles to holding wells immediately after they are liberated from the ovary. In this way, it will prevent them from exposure to endogenous octopamine released from broken nerve terminals that innervate the ovary. For new learners, it is recommended to practice **steps 4** and **5** to shorten the time to finish these two steps

before performing real experiments. New learners also need to demonstrate that their rupture rate for negative control groups (without OA) is less than 15%, while their positive controls (with OA) should be close to the typical rupture rate for different reporter lines.

17. For setting up the ex vivo follicle rupture assay, each replicate needs about 25–35 mature follicles per culture well. We typically have three to five replicates for each genotype. Therefore, you will need 75–175 mature follicles for each genotype.
18. Using a higher magnification when checking the posterior ends of the follicles helps prevent the selection of any unwanted follicles, such as follicles that have begun to rupture or follicles that have the corpus luteum attached to the posterior end since they may not rupture normally.
19. After allocating the mature follicles into new empty wells, make sure to add the culture medium as soon as possible so that the mature follicles won't dry out. Alternatively, you can add the culture medium to the well first before allocating the mature follicles into the well.
20. Mixing the octopamine stock solution in a culture medium immediately improves the rupture rate. This is likely related to the slow diffusion of 2 μL into 1 mL solution if not mixed immediately.
21. Handle the utility box with ex vivo culture plates carefully to avoid spilling over liquid from one well to another during transportation. As a practice, we always have aluminum foil in the utility box to prevent light exposure. However, we never tested whether this is necessary. We always use a 29 °C incubator for our experiments since we want to increase the efficiency of RNA interference. Other temperatures could also be used to assess for follicle rupture, however, the dynamics and rupture rate may be slightly different. In our experimental setting, we have demonstrated that the follicle rupture rate reaches the plateau after 3 h culture [6]. Therefore, we typically culture the follicles for 3 h and then examine the rupture rate.
22. It should be noted that the statistical analysis used to determine the significance of the mean in the Assessing Ovulation Template is all based on the Student's t-test and for estimation purposes. Users should use more appropriate statistical analysis depending on their experimental design. For example, one-way analysis of variance should be used to compare the means of more than two groups.

Acknowledgments

We thank members of Sun Lab for their careful reading and suggestions. This work is supported by the National Institute of Child Health and Human Development grants (R01-HD086175 and R01-HD097206) to J.S.

References

1. Espey LL (2006) Comprehensive analysis of ovarian gene expression during ovulation using differential display. Methods Mol Biol 317:219–241. https://doi.org/10.1385/1-59259-968-0:219
2. Duffy DM, Ko C, Jo M, Brannstrom M, Curry TE (2019) Ovulation: parallels with inflammatory processes. Endocr Rev 40(2):369–416. https://doi.org/10.1210/er.2018-00075
3. Fan HY, Liu Z, Mullany LK, Richards JS (2012) Consequences of RAS and MAPK activation in the ovary: the good, the bad and the ugly. Mol Cell Endocrinol 356(1–2):74–79. https://doi.org/10.1016/j.mce.2011.12.005
4. Robker RL, Hennebold JD, Russell DL (2018) Coordination of ovulation and oocyte maturation: a good egg at the right time. Endocrinology 159(9):3209–3218. https://doi.org/10.1210/en.2018-00485
5. Deady LD, Shen W, Mosure SA, Spradling AC, Sun J (2015) Matrix metalloproteinase 2 is required for ovulation and corpus luteum formation in *Drosophila*. PLoS Genet 11(2):e1004989. https://doi.org/10.1371/journal.pgen.1004989
6. Deady LD, Sun J (2015) A follicle rupture assay reveals an essential role for follicular adrenergic signaling in *Drosophila* ovulation. PLoS Genet 11(10):e1005604. https://doi.org/10.1371/journal.pgen.1005604
7. Li W, Young JF, Sun J (2018) NADPH oxidase-generated reactive oxygen species in mature follicles are essential for *Drosophila* ovulation. Proc Natl Acad Sci U S A 115(30):7765–7770. https://doi.org/10.1073/pnas.1800115115
8. Knapp E, Sun J (2017) Steroid signaling in mature follicles is important for *Drosophila* ovulation. Proc Natl Acad Sci U S A 114(4):699–704. https://doi.org/10.1073/pnas.1614383114
9. Jiang K, Zhang J, Huang Y, Wang Y, Xiao S, Hadden MK, Woodruff TK, Sun J (2021) A platform utilizing *Drosophila* ovulation for nonhormonal contraceptive screening. Proc Natl Acad Sci U S A 118(28):e2026403118. https://doi.org/10.1073/pnas.2026403118
10. Aigaki T, Fleischmann I, Chen PS, Kubli E (1991) Ectopic expression of sex peptide alters reproductive behavior of female *D. melanogaster*. Neuron 7(4):557–563. https://doi.org/10.1016/0896-6273(91)90368-a
11. Heifetz Y, Lung O, Frongillo EA Jr, Wolfner MF (2000) The *Drosophila* seminal fluid protein Acp26Aa stimulates release of oocytes by the ovary. Curr Biol 10(2):99–102. https://doi.org/10.1016/s0960-9822(00)00288-8
12. Lee HG, Rohila S, Han KA (2009) The octopamine receptor OAMB mediates ovulation via Ca2+/calmodulin-dependent protein kinase II in the *Drosophila* oviduct epithelium. PLoS One 4(3):e4716. https://doi.org/10.1371/journal.pone.0004716
13. Lim J, Sabandal PR, Fernandez A, Sabandal JM, Lee HG, Evans P, Han KA (2014) The octopamine receptor Octbeta2R regulates ovulation in *Drosophila melanogaster*. PLoS One 9(8):e104441. https://doi.org/10.1371/journal.pone.0104441
14. Cole SH, Carney GE, McClung CA, Willard SS, Taylor BJ, Hirsh J (2005) Two functional but noncomplementing *Drosophila* tyrosine decarboxylase genes: distinct roles for neural tyramine and octopamine in female fertility. J Biol Chem 280(15):14948–14955. https://doi.org/10.1074/jbc.M414197200
15. Monastirioti M (2003) Distinct octopamine cell population residing in the CNS abdominal ganglion controls ovulation in *Drosophila melanogaster*. Dev Biol 264(1):38–49. https://doi.org/10.1016/j.ydbio.2003.07.019
16. Monastirioti M, Linn CE Jr, White K (1996) Characterization of *Drosophila* tyramine beta-hydroxylase gene and isolation of mutant flies lacking octopamine. J Neurosci 16(12):3900–3911
17. Ritsick DR, Edens WA, Finnerty V, Lambeth JD (2007) Nox regulation of smooth muscle contraction. Free Radic Biol Med 43(1):

31–38. https://doi.org/10.1016/j.freeradbiomed.2007.03.006

18. Liao S, Nassel DR (2020) *Drosophila* insulin-like peptide 8 (DILP8) in ovarian follicle cells regulates ovulation and metabolism. Front Endocrinol (Lausanne) 11:461. https://doi.org/10.3389/fendo.2020.00461
19. Deady LD, Li W, Sun J (2017) The zinc-finger transcription factor Hindsight regulates ovulation competency of *Drosophila* follicles. Elife 6: e29887. https://doi.org/10.7554/eLife.29887
20. Luo W, Liu S, Zhang W, Yang L, Huang J, Zhou S, Feng Q, Palli SR, Wang J, Roth S, Li S (2021) Juvenile hormone signaling promotes ovulation and maintains egg shape by inducing expression of extracellular matrix genes. Proc Natl Acad Sci U S A 118(39):e2104461118. https://doi.org/10.1073/pnas.2104461118
21. Rubinstein CD, Wolfner MF (2013) *Drosophila* seminal protein ovulin mediates ovulation through female octopamine neuronal signaling. Proc Natl Acad Sci U S A 110(43): 17420–17425. https://doi.org/10.1073/pnas.1220018110
22. Knapp EM, Li W, Singh V, Sun J (2020) Nuclear receptor Ftz-f1 promotes follicle maturation and ovulation partly via bHLH/PAS transcription factor Sim. Elife 9:e54568. https://doi.org/10.7554/eLife.54568
23. Knapp EM, Deady LD, Sun J (2018) Ex vivo follicle rupture and in situ Zymography in *Drosophila*. Bio Protoc 8(10):e2846. https://doi.org/10.21769/BioProtoc.2846

Chapter 14

A Low-Tech Flow Chamber for Live Imaging of *Drosophila* Egg Chambers During Drug Treatments

Allison L. Zajac, Audrey Miller Williams, and Sally Horne-Badovinac

Abstract

The *Drosophila* egg chamber is a powerful system to study epithelial cell collective migration and polarized basement membrane secretion. A strength of this system is the ability to capture these dynamic processes in ex vivo organ culture using high-resolution live imaging. Ex vivo culture also allows acute pharmacological or labeling treatments, extending the versatility of the system. However, many current ex vivo egg chamber culture setups do not permit easy medium exchange, preventing researchers from following individual egg chambers through multiple treatments. Here we present a method to immobilize egg chambers in an easy-to-construct flow chamber that permits imaging of the same egg chamber through repeated solution exchanges. This will allow researchers to take greater advantage of the wide variety of available pharmacological perturbations and other treatments like dyes to study dynamic processes in the egg chamber.

Key words *Drosophila*, Egg chamber, Follicle, Collective cell migration, Secretion, Morphogenesis, Live imaging, Flow chamber, Drug treatment

1 Introduction

The follicular epithelium of the *Drosophila* egg chamber has emerged as a powerful system to investigate the dynamics of collective cell migration and extracellular matrix (ECM) secretion and assembly. The egg chamber consists of a central cluster of germ cells surrounded by a single layer of somatic follicular epithelial cells (follicle cells, *see* Fig. 1a) [1]. The follicle cells are polarized such that their apical surfaces contact the germ cells and their basal surfaces contact a sheet of basement membrane (BM) ECM that wraps the tissue (*see* Fig. 1b). During the early stages of development (up to stage 8), the follicle cells undergo a collective migration along the BM [2, 3]. During this time, they are also actively secreting new BM proteins and depositing new layers of BM as they

Allison L. Zajac and Audrey Miller Williams contributed equally to this work.

Michelle S. Giedt and Tina L. Tootle (eds.), *Drosophila Oogenesis: Methods and Protocols*,
Methods in Molecular Biology, vol. 2626, https://doi.org/10.1007/978-1-0716-2970-3_14,

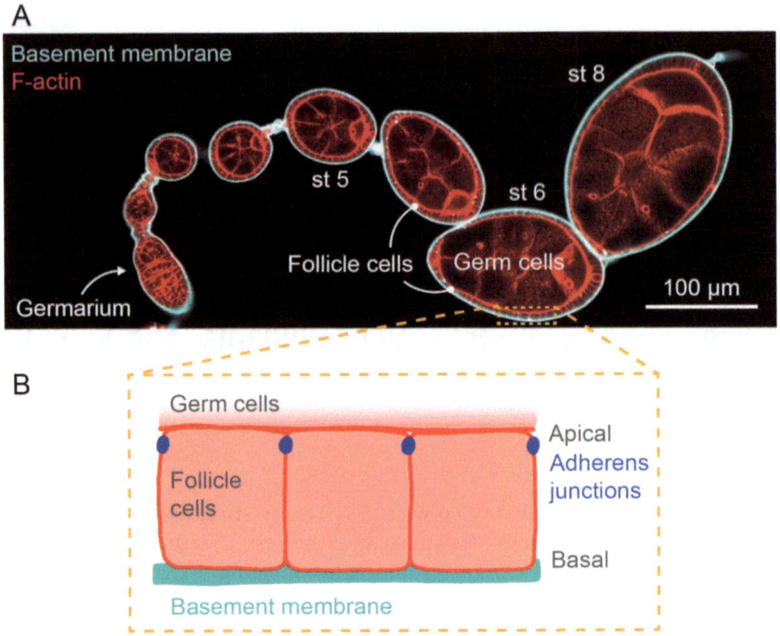

Fig. 1 *Drosophila* egg chamber organization. (**a**) Confocal cross-section image through an ovariole stained for F-actin (phalloidin) and expressing GFP-Collagen IV to label the basement membrane. Note that different sizes of egg chambers require different levels of compression to hold them in place in the flow chamber. (**b**) Illustration of the organization of the follicular epithelial cells that surround the egg chamber. The apical surfaces of follicle cells contact the germline cells and the basal surfaces contact the basement membrane on the exterior of the tissue

migrate [2, 4]. The geometry of the egg chamber places much of the machinery involved in collective cell migration and BM secretion near the exterior of the tissue. This useful property has allowed cutting-edge imaging techniques to be applied to study these dynamic processes in intact, developing tissue [2–17].

Genetic screens have been useful for identifying the machinery involved in collective cell migration and BM secretion. However, genetic perturbations often result in the cells completely failing to migrate or secrete BM proteins and leave open many questions about the normal dynamics of these processes. In dynamic systems, being able to make acute perturbations using pharmacological inhibitors, and to watch recovery from these perturbations after washing out the treatment, can provide important mechanistic information. Since a variety of pharmacological inhibitors are available that target machinery involved in cell migration and secretion, this approach is likely to provide new insights into the dynamics of these processes.

Although conditions for ex vivo culture are established [2, 18–20], it has been technically challenging to mount egg chambers in a way that allows imaging of a single egg chamber through rounds of solution exchange. Attachment of egg chambers to glass-bottomed dishes would allow easy exchange. However, past studies have found that, although using an adhesive coating like Concavalin A can immobilize egg chambers, this coating also stops the migration of cells [9]. An additional imaging challenge is the egg chamber's curved surface. To view a larger surface area of an egg chamber, we gently compress it between two pieces of glass, which makes solution exchange challenging. In past work, our lab used small fragments of glass to compress egg chambers in a dish, which does allow new solutions to be added to the bulk medium [3, 20]. However, this approach still has the drawbacks of not allowing rapid mixing of the treatment into the bulk medium, and not allowing for easy wash-out of the drug-treated medium.

Here, we report a simple modification to our current mounting protocol that creates a flow chamber and allows multiple rounds of solution exchange while imaging the same egg chamber. Many single-molecule in vitro studies use hand-made flow chambers built from a slide and a coverslip that have a channel with walls created by tape or vacuum grease [21]. We have combined this approach with our standard mounting of egg chambers between a slide and a coverslip that is supported by spacer beads [13, 20]. This is an easy-to-make flow chamber that uses widely available materials and is compatible with both upright and inverted microscopes. This flow chamber allows multiple rounds of solution exchange for drug treatment or pulse-labeling of endocytic tracers or other labels.

Finally, we want to bring the reader's attention to two alternative approaches to immobilize egg chambers in either gels [22] or fibrin clots [23, 24]. These methods also permit solution exchanges and may be preferable depending on the experimental goals.

In this chapter, we describe how to dissect and prepare ovariole strands containing stage 1–8 egg chambers and immobilize them in a flow chamber suitable for repeated media exchanges. We have previously published a chapter in Methods in Molecular Biology on live imaging of follicle cell migration [20]. The protocol for dissection is shared between these chapters, and the chapter by Anderson et al., on F-actin staining in the current edition, and sections are duplicated here for convenience. We refer the reader to the chapter on live imaging of migration [20] for detailed images and movies of the egg chamber dissection process, which are not included here.

2 Materials

2.1 Preparing Female Flies for Dissection

1. Vial with fly food.
2. Yeast powder: dry active yeast ground to a fine powder in a coffee grinder.

2.2 Flow Chamber Preparation

1. Glass microscope slide, 3″ × 1″ × 1 mm.
2. 10 mL syringe.
3. High-vacuum silicone grease.
4. 100 or 200 μL pipette tip.

2.3 Egg Chamber Dissection

1. Acidified water: 1 μL concentrated (12.1 N) HCl in 1 mL deionized water.
2. Insulin: 1 mg dissolved in 100 μL acidified water (*see* **Note 1**).
3. Live imaging medium: Schneider's S2 medium, 0.6× Pen/-Strep (Pen/Strep: penicillin G-sodium 10,000 U/ml, streptomycin sulfate 10,000 μg/ml in 0.85% saline), 15% vol/vol fetal bovine serum, 0.1 mg/ml insulin (*see* **Note 2**).
4. Pyrex 9-Cavity Spot Plate.
5. Dumont forceps: #5, 0.1 × 0.06 mm tip, and #55, 0.05 × 0.02 mm tip (*see* **Note 3**).
6. Eyelash tool: insert an eyelash into a slightly melted p1000 pipettor tip.
7. Disposable needle, 27G × ½ in.
8. Glass pasteur pipettes, 5 ¾ in.
9. 5 mL pipette pump.
10. Stereomicroscope with a magnification of at least 10×.
11. Dye to visualize plasma membranes, for example, CellMask (*see* **Note 4**).
12. 10 μL pipette and tips.
13. Glass microscope slide, 3″ × 1″ × 1 mm.
14. CO_2 pad to anesthetize flies.

2.4 Egg Chamber Mounting in Flow Chamber

1. Polystyrene or glass beads: ~30 μm diameter for stages 4–5 and ~50 μm diameter for stages 6–8 (*see* **Notes 5–6**).
2. 10 or 20 μL pipette and tips.
3. Coverslip, 22 mm × 30 mm #1.5 (*see* **Note 7**).
4. Filter paper or delicate task wiper.

2.5 Exchanging Media and Imaging

1. Live imaging medium.
2. Desired treatments: pharmacological inhibitors, dyes, etc.
3. 100 or 200 μL pipette and tips.
4. Filter paper or delicate task wiper.
5. Vaseline.
6. Heat block set to 65 °C.
7. Paintbrush.

3 Methods

3.1 Prepare Female Flies for Dissection

1. Sprinkle yeast powder on fly food in a vial, covering about one-half of the surface. Add up to six 1- to 2-day-old females and at least one to two males to the vial. Yeast is essential for proper egg chamber production. Incomplete nutrition will slow egg chamber production by inducing cell death in the germarium and in stage 8 egg chambers [25–27].
2. Age females for 1–3 days. Move flies to a new vial with fresh yeast every two days before dissection (*see* **Note 8**).

3.2 Prepare Flow Chamber

1. Make a vacuum grease-filled 10 mL syringe by removing the plunger and squeezing the vacuum grease tube into the barrel (save for repeated use indefinitely). Use the plunger to push the vacuum grease into the syringe (with no attached needle). Attach a 100 or 200 μL pipette tip to the Luer lock of the syringe in place of a needle (*see* Fig. 2a).
2. Make a vacuum grease channel on the slide. The pipette tip can detach from the Luer lock of the syringe while you are depressing the plunger, so hold it in place while drawing vacuum grease lines. The channel opening needs to be accessible to exchange media with the slide mounted on the coverslip. Therefore, the geometry of the flow chamber will be different depending on whether you are using an inverted or an upright microscope (*see* Fig. 2b–e).
   - *For an upright microscope*, take a standard glass slide and use your vacuum grease-filled syringe to draw two thin lines, spaced ~1 cm apart, parallel to the long axis of the slide (*see* Fig. 2b, d). Make sure the lines are slightly longer than the 30 mm coverslips that will be used to cover this chamber.
   - *For an inverted microscope*, instead orient the two thin lines of vacuum grease perpendicular to the slide's long axis (*see* Fig. 2c, e). These lines should extend the full width of the slide.
3. Smooth the vacuum grease lines by running a finger over each line to partially flatten it and make sure there are no gaps in the line, as gaps will cause leaks (*see* Fig. 2f; **Note 9**). Set aside for use after the dissection.

3.3 Egg Chamber Dissection

In this section, we describe how to isolate egg chambers from the ovary, which requires a basic understanding of the ovary's structure. Each ovary is composed of 15–18 developmental arrays of egg chambers called ovarioles. At the anterior of each ovariole is the germarium, the site of egg chamber production. Moving posteriorly are successively older egg chambers, connected to one another like beads on a string by stalk cells. At the posterior of each ovariole

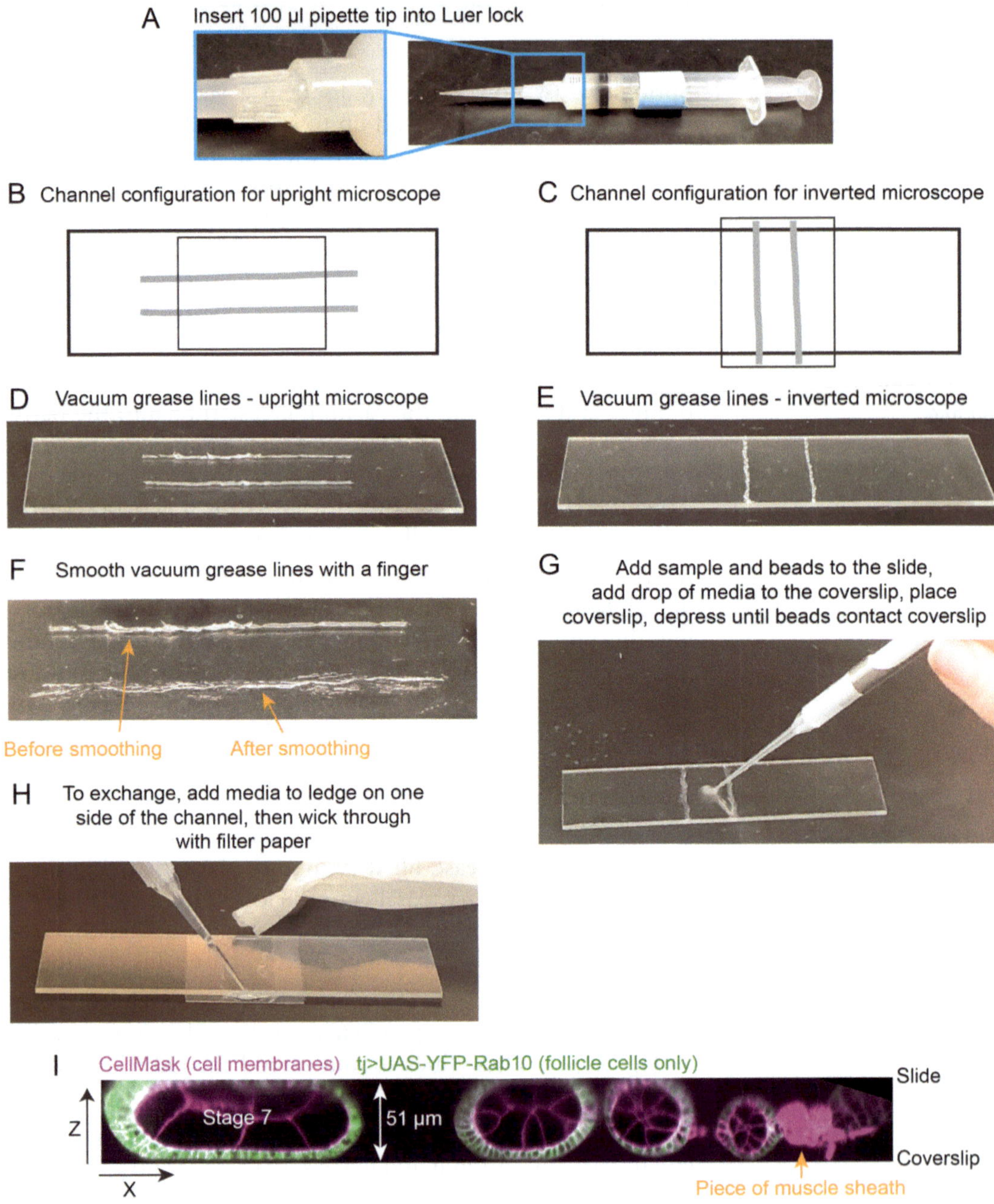

Fig. 2 How to construct flow chamber. (**a**) Vacuum grease-filled 10 mL syringe with a 100 μL pipette tip attached to the Luer lock in place of a needle. (**b**, **c**) Diagram of flow chamber construction for an inverted or upright microscope. Light grey thick lines indicate where vacuum grease lines should be drawn on a slide. Smaller rectangle indicates the position of the 20 × 30 mm coverslip. (**d**, **e**) Image of the desired spacing and thickness of vacuum grease lines. Note that these should be thin and you will likely have some gaps in the original lines. (**f**) Image of the same vacuum grease line after smoothing it with a finger (bottom). Now it is flatter and will require less manual compression when you add the coverslip. You should make sure there are no gaps that will allow leaks in the final line. (**g**) Image showing where to initially add egg chambers and beads to slide. (**h**) Image of the assembled and compressed flow chamber, showing where to add media to the

are the oldest egg chambers or mature eggs waiting to be laid. Each ovariole, as well as the entire ovary, is surrounded by a muscle sheath that needs to be removed during dissection. Egg chambers can be staged by eye based on both their overall size and shape, as well as the relative size and shape of the oocyte. Older egg chambers are also distinguished by their partly opaque appearance, caused by the onset of vitellogenesis at stage 8.

Based on your experimental goals, you may choose egg chambers of various stages to dissect. Dissection techniques vary based on which stage egg chambers you wish to collect, and dissection will damage egg chambers of other stages, so it is important to tailor your dissection approach to your stage of interest. Finally, we caution that the entire dissection process described below not last for longer than ~10 min to ensure that you can stain, mount, and image your egg chambers with minimal time in ex vivo culture. We usually limit the ex vivo culture of egg chambers to <1 h and recommend dissecting only one to two females at a time. In genotypes with healthy ovaries, this will yield plenty of ovarioles for an hour of imaging. In addition to the methods below, we refer the reader to a previously published chapter in Methods in Molecular Biology on live imaging of follicle cell migration [20].

1. If staining cell membranes for live imaging is desired, dilute CellMask 1:2000 or alternative dyes in a live imaging medium. This staining step is also helpful to assess damage to egg chambers, as damaged regions will accumulate dye.
2. Add live imaging medium to the Pyrex 9-Cavity Spot Plate and allow the medium to come to room temperature prior to dissection. If you are not staining cell membranes, you will need 1 well with 500 μL and 1 well with 250 μL of live imaging medium per dissection. If you are staining with CellMask, you will need one well with 500 μL live imaging medium for dissection, one well with 250 μL live imaging medium with CellMask, and two wells with 250 μL live imaging medium for washes.
3. Position a Spot Plate well containing live imaging medium underneath a dissecting microscope. Adjust magnification to ~16× if the microscope allows, and focus on the bottom of the well.

Fig. 2 (continued) coverslip ledge that extends past one edge of the channel, and image of torn filter paper being used to wick media through flow chamber. Note that the coverslip is on the bottom. (**i**) Zeiss Lattice Light Sheet volume of an ovariole compressed between slide and coverslip, supported by 51 μm diameter beads. An x-z projection is shown to highlight the compression of the stage 7 egg chamber that will hold it in place during medium exchanges

4. Dissect ovaries. First, anesthetize the flies. We normally use CO_2. Using the #5 forceps in your non-dominant hand, pinch a female fly at the anterior-most part of the abdomen (*see* **Note 10**). While still pinching, transfer the fly to the live imaging media and completely submerge it. Using the #55 forceps in your dominant hand, pinch the abdomen between the two posterior-most segments and pull posteriorly. The ovaries should release from the abdomen (if ovaries do not release, *see* **Note 11**). Remove all non-ovary tissue from the well. To obtain stage 1–5 egg chambers, *see* **step 5**. To obtain stage 5–8 egg chambers, *see* **step 6**.
5. To obtain stage 1–5 egg chambers, pinch the posterior of an ovary with the #5 forceps and pin it to the bottom of the well. Use the #55 forceps to pinch a single ovariole just anterior to where the ovary is pinned. Pull the ovariole orthogonally and then anteriorly to remove it from the muscle sheath and isolate it from the ovary. Ovarioles that have been removed from their muscle sheaths will look and move like beads on a string, rather than like a single packaged unit. Repeat this process until ovarioles do not easily separate from the ovary, as repeated pulling risks damaging egg chambers. Repeat for the second ovary.
6. To obtain stage 5–8 egg chambers, pinch the posterior of an ovary with the #5 forceps and pin it to the bottom of the well. Use the #55 forceps to pinch single ovarioles at the anterior tip of the ovary and pull anteriorly to remove ovarioles from their muscle sheaths and isolate them from the ovary. Ovarioles that have been removed from their muscle sheaths will look and move like beads on a string, rather than like a single packaged unit. Repeat this process until ovarioles do not easily separate from the ovary, as repeated pulling risks damaging egg chambers. Repeat for the second ovary.
7. Separate healthy ovarioles from any other unwanted material using an eyelash tool or 27-gauge needle. Do not take ovarioles that retained their muscle sheaths, as this sheath will contract in culture and move the egg chambers.
8. Trim ovarioles containing egg chambers older than the stage you plan to image by placing the needle on the stalk between two egg chambers and pressing down using a slicing motion. Older egg chambers will prop up the coverslip and prevent the needed compression of younger egg chambers under the coverslip. This will both reduce the area of the basal surface available for single-plane imaging and cause egg chambers to move when fluids are exchanged. This step takes practice (*see* **Note 12**).

9. Gather trimmed ovarioles in a cluster using the eyelash tool. Use a 10–20 μL pipette to transfer ovarioles to a well of fresh media (either with CellMask for staining or fresh medium if not staining), and cover with a slide to slow evaporation. It is important to transfer the desired, healthy-looking ovarioles to a new medium as quickly as you can. We notice even undamaged egg chambers can have failed cell migration if they spend extended time in a medium containing debris from damaged egg chambers. If staining with CellMask, incubate for ~10 min in a medium with stain.
10. To wash egg chambers from CellMask, collect egg chambers in the center of the Spot Plate by swirling. Use a 10–20 μL pipette to transfer ovarioles to the first well with 250 μL of the live imaging medium. Minimize the amount of stain-containing medium transferred with the egg chambers. Swirl the dish to facilitate rinsing. Use a fresh 10–20 μL tip to transfer ovarioles to the second 250 μL well of live imaging media. Cover with a slide to slow evaporation until ready to mount (*see* **Note 13**).

3.4 Flow Chamber Construction and Egg Chamber Mounting

1. Transfer ovarioles to the flow chamber slide. Use a 10 μL pipette to transfer 1–5 ovarioles in 10 μL of the medium, creating a drop in the center of the two lines of vacuum grease. Add a small drop of spacer beads from your bead-media stock (*see* **Note 6**)—the goal is ~20–100 beads per slide. Choose bead size based on the desired compression and stage/size of the egg chambers you are imaging (*see* **Note 5**). Use the eyelash tool to arrange ovarioles in the center of the slide surrounded by beads that will support the weight of the coverslip (*see* Fig. 2g).
2. Add a coverslip to complete the flow chamber. Add a 10 μL drop of live imaging medium to the center of a 22 × 30 mm coverslip. This will reduce bubbles when you lower it onto the sample. For an inverted microscope, the coverslip needs to extend past the slide to create a platform to allow you to exchange the fluid in the vacuum grease channel; *see* Fig. 2c for desired geometry. For an upright microscope, the coverslip will lie parallel to the long axis of the slide, but it is important that the vacuum grease lines extend past the coverslip on both sides to create a full channel (*see* Fig. 2b).
3. Once you have lowered the coverslip into place, cover it with a delicate task wiper to prevent fingerprints, and press gently to compress the vacuum grease to the same height as the spacer beads. Wick away excess medium with filter paper or a delicate task wiper. Check under a dissecting scope that your egg chambers are sufficiently compressed to stop egg chamber movement due to fluid flow (*see* Fig. 2h). This step requires practice. A sufficiently compressed stage 7 egg chamber is shown in Fig. 2i.

3.5 Exchange Fluids in Flow Chamber

1. To dislodge egg chambers that are not compressed well enough, you can exchange media in the chamber for fresh, untreated media before beginning your imaging (optional). Have torn pieces of filter paper or delicate task wipers ready to use. Add a 20 μL drop of fresh live imaging medium to one side of the flow chamber (*see* Fig. 2h). The drop must contact the liquid in the channel. Put the piece of filter paper/delicate task wiper on the opposite side and wick the medium through (*see* Fig. 2h). Make sure the channel remains filled with media so the sample does not dry out during pre-treatment imaging.
2. Put the slide onto the microscope. Be sure to anchor the slide against 1 corner of the slide holder so that if it is bumped during liquid exchange, or if you choose to exchange media off the microscope, you can quickly move the slide back to its initial position and find the same egg chamber. Perform "pre-treated" imaging of egg chambers compressed in the flow chamber.
3. Exchange medium to add the experimental treatment (drug, etc.). To make it easier to access the slide, you may want to pull back the microscope condenser head. Add a 20 μL drop of experimental medium and wick to the other side (*see* Fig. 2h). Repeat ~2×. There will be a short (1–2 min) time gap where the slide is out of focus while you exchange the medium. Restart imaging. Repeat exchanges with media containing different treatments as desired. If you are imaging for more than a few minutes after the last medium exchange, seal the two ends of the flow chamber by using a paintbrush to place a drop of Vaseline (melted in a 65 °C heat block) over the openings on both sides of the channel. Be careful not to damage the microscope with the medium or Vaseline. If you cannot do this easily with the slide on the microscope, remove the slide, exchange the medium on the bench and seal it with Vaseline, and replace the slide on the microscope.

4 Notes

1. Insulin solution can be stored at 4 °C for up to 1 week and the solution should be added to the live imaging medium immediately before use.
2. Schneider's S2 medium and Pen/Strep can be combined and stored at 4 °C. Fetal bovine serum can also be combined with Schneider's S2 medium and Pen/Strep mixture and stored at 4 °C for up to 1 week. We make 10 mL of the S2, Pen/Strep, fetal bovine serum mixture at a time.

3. To minimize tissue damage, maintain forceps with care and sharpen as needed.
4. We regularly use CellMask dyes to label the plasma membrane. An alternative plasma membrane label is FM 4-64 [20].
5. Both glass and polystyrene beads work well. Experiment with different sizes depending on the amount of compression you need.
6. We resuspend beads in a live imaging medium (without insulin) to make it easy to add them to the imaging chamber. Add dry beads to an Eppendorf tube. Wash beads with live imaging medium before use: add 500 μL of live imaging media, spin briefly to sediment, exchange to new media, and repeat 2×. You can store beads in the medium at 4 °C until you notice contamination.
7. Our microscope slide holder can only accommodate a 30 mm-long coverslip. If you have more space and are using an inverted microscope, using a longer coverslip will be easier because it will create a longer platform for medium exchanges.
8. Depending on the genotype, age, and temperature, the rate of oogenesis may vary, and flies may need to be aged for different numbers of days to achieve an optimal number of egg chambers of your desired stage for dissection. Younger females produce more eggs than older females, colder temperatures slow oogenesis, and warmer temperatures speed oogenesis. Flies must be moved to fresh yeast regularly or females will cease laying eggs and retain them in their ovaries, resulting in unhealthy egg chambers and fewer younger egg chambers. Certain genetic conditions can result in round eggs that will block the oviduct, which results in unhealthy egg chambers; dissecting younger flies prior to oviduct blockage is often useful in these backgrounds.
9. You will have to experiment with making these vacuum grease lines. The goal is to have a thin, unbroken line that can be compressed to the height of the beads you are using. The most common mistakes made are using lines of vacuum grease that are too thick or having gaps in the lines.
10. Proper pinching of the fly will ensure the rest of your dissection proceeds smoothly. Pinching the thorax increases the risk that attempting to open the abdomen to remove the ovaries will instead remove the entire abdomen without opening it. Pinching a more posterior part of the abdomen increases the risk that you will damage young egg chambers.
11. If the ovaries do not initially release from the abdomen, there are several methods to remove them. If the oviduct is visible, first attempt to tug posteriorly on the oviduct with your #55

forceps to remove the ovaries, as this is the least likely method to damage egg chambers. If the oviduct is not visible or tugging did not work, there are two other options. If late-stage egg chambers are desired, the abdomen can be gently squeezed like a tube of toothpaste in an anterior-to-posterior direction to extrude the ovaries. If younger egg chambers are desired, the posterior of the ovaries can be pulled on directly.

12. Cutting the stalks between egg chambers takes practice and can result in damage to egg chambers even with experience. If you are using a plasma membrane dye, you will see dye uptake in damaged regions of egg chambers. If you are not using a dye, check egg chambers for other signs of tissue damage or abnormal cell migration, and, if present, exclude them from analysis. An alternative to using a needle is a thinned tungsten wire tool [20].
13. It is important to wash out excess dye to prevent it from sticking to the coverslip and creating high background fluorescence. Fingerprints or dust on the coverslip will pick up excess dye so be careful to not leave coverslips uncovered or touch them.

Acknowledgments

American Heart Association 16POST2726018, American Cancer Society 132123-PF-18-025-01-CSM, and Chicago Biomedical Consortium FP064171-01-PR postdoctoral fellowships to A.L. Z., NIH T32 HD055164 to A.M.W., and work in the Horne-Badovinac Lab is supported by NIH R01s GM126047 and GM136961 to S.H-B.

References

1. Horne-Badovinac S, Bilder D (2005) Mass transit: epithelial morphogenesis in the *Drosophila* egg chamber. Dev Dyn 232:559–574. https://doi.org/10.1002/dvdy.20286
2. Haigo SL, Bilder D (2011) Global tissue revolutions in a morphogenetic movement controlling elongation. Science 331:1071–1074. https://doi.org/10.1126/science.1199424
3. Cetera M, Juan GRR-S, Oakes PW, Lewellyn L, Fairchild MJ, Tanentzapf G, Gardel ML, Horne-Badovinac S (2014) Epithelial rotation promotes the global alignment of contractile actin bundles during *Drosophila* egg chamber elongation. Nat Commun 5:5511. https://doi.org/10.1038/ncomms6511
4. Isabella AJ, Horne-Badovinac S (2016) Rab10-mediated secretion synergizes with tissue movement to build a polarized basement membrane architecture for organ morphogenesis. Dev Cell 38:47–60. https://doi.org/10.1016/j.devcel.2016.06.009
5. Lerner DW, McCoy D, Isabella AJ, Mahowald AP, Gerlach GF, Chaudhry TA, Horne-Badovinac S (2013) A Rab10-dependent mechanism for polarized basement membrane secretion during organ morphogenesis. Dev Cell 24:159–168. https://doi.org/10.1016/j.devcel.2012.12.005
6. Lewellyn L, Cetera M, Horne-Badovinac S (2013) Misshapen decreases integrin levels to promote epithelial motility and planar polarity in *Drosophila*. J Cell Biol 200:721–729. https://doi.org/10.1083/jcb.201209129
7. Viktorinová I, Dahmann C (2013) Microtubule polarity predicts direction of egg chamber

rotation in *Drosophila*. Curr Biol 23:1472–1477. https://doi.org/10.1016/j.cub.2013.06.014

8. Chen D-Y, Lipari KR, Dehghan Y, Streichan SJ, Bilder D (2016) Symmetry breaking in an edgeless epithelium by Fat2-regulated microtubule polarity. Cell Rep 15:1125–1133. https://doi.org/10.1016/j.celrep.2016.04.014
9. Squarr AJ, Brinkmann K, Chen B, Steinbacher T, Ebnet K, Rosen MK, Bogdan S (2016) Fat2 acts through the WAVE regulatory complex to drive collective cell migration during tissue rotation. J Cell Biol 212:591–603. https://doi.org/10.1083/jcb.201508081
10. Viktorinová I, Henry I, Tomancak P (2017) Epithelial rotation is preceded by planar symmetry breaking of actomyosin and protects epithelial tissue from cell deformations. PLoS Genet 13:e1007107. https://doi.org/10.1371/journal.pgen.1007107
11. Alégot H, Pouchin P, Bardot O, Mirouse V (2018) Jak-Stat pathway induces *Drosophila* follicle elongation by a gradient of apical contractility. elife 7:e32943. https://doi.org/10.7554/elife.32943
12. Dent LG, Manning SA, Kroeger B, Williams AM, Hilmi AJS, Crea L, Kondo S, Horne-Badovinac S, Harvey KF (2019) The dPix-Git complex is essential to coordinate epithelial morphogenesis and regulate myosin during *Drosophila* egg chamber development. PLoS Genet 15:e1008083. https://doi.org/10.1371/journal.pgen.1008083
13. Stedden CG, Menegas W, Zajac AL, Williams AM, Cheng S, Özkan E, Horne-Badovinac S (2019) Planar-polarized Semaphorin-5c and Plexin A promote the collective migration of epithelial cells in *Drosophila*. Curr Biol 29:908–920.e6. https://doi.org/10.1016/j.cub.2019.01.049
14. Campos FC, Dennis C, Alégot H, Fritsch C, Isabella A, Pouchin P, Bardot O, Horne-Badovinac S, Mirouse V (2020) Oriented basement membrane fibrils provide a memory for F-actin planar polarization via the Dystrophin-Dystroglycan complex during tissue elongation. Dev 147:dev186957. https://doi.org/10.1242/dev.186957
15. Sherrard KM, Cetera M, Horne-Badovinac S (2021) DAAM mediates the assembly of long-lived, treadmilling stress fibers in collectively migrating epithelial cells in *Drosophila*. elife 10:e72881. https://doi.org/10.7554/elife.72881
16. Williams AM, Donoughe S, Munro E, Horne-Badovinac S (2022) Fat2 polarizes the WAVE complex in trans to align cell protrusions for collective migration.eLife 11:e78343. https://doi.org/10.7554/eLife.78343
17. Zajac AL, Horne-Badovinac S (2022) Kinesin-directed secretion of basement membrane proteins to a subdomain of the basolateral surface in *Drosophila* epithelial cells. Curr Biol 32:735–748.e10. https://doi.org/10.1016/j.cub.2021.12.025
18. Prasad M, Jang AC-C, Montell DJ (2007) A protocol for culturing *Drosophila melanogaster* egg chambers for live imaging. Protoc Exch. https://doi.org/10.1038/nprot.2007.233
19. Prasad M, Montell DJ (2007) Cellular and molecular mechanisms of border cell migration analyzed using time-lapse live-cell imaging. Dev Cell 12:997–1005. https://doi.org/10.1016/j.devcel.2007.03.021
20. Cetera M, Lewellyn L, Horne-Badovinac S (2016) *Drosophila*, methods and protocols. Methods Mol Biol Clifton N J 1478:215–226. https://doi.org/10.1007/978-1-4939-6371-3_12
21. Dixit R, Ross JL (2010) Chapter 27 – Studying plus-end tracking at single molecule resolution using TIRF microscopy. Methods Cell Biol 95:543–554. https://doi.org/10.1016/s0091-679x(10)95027-9
22. Chanet S, Huynh J-R (2020) Collective cell sorting requires contractile cortical waves in germline cells. Curr Biol. https://doi.org/10.1016/j.cub.2020.08.045
23. Wilcockson SG, Ashe HL (2019) *Drosophila* ovarian germline stem cell Cytocensor projections dynamically receive and attenuate BMP Signaling. Dev Cell 50:296–312.e5. https://doi.org/10.1016/j.devcel.2019.05.020
24. Wilcockson SG, Ashe HL (2021) Live imaging of the *Drosophila* ovarian germline stem cell niche. Star Protoc 2:100371. https://doi.org/10.1016/j.xpro.2021.100371
25. Drummond-Barbosa D, Spradling AC (2001) Stem cells and their progeny respond to nutritional changes during *Drosophila* oogenesis. Dev Biol 231:265–278. https://doi.org/10.1006/dbio.2000.0135
26. Mazzalupo S, Cooley L (2006) Illuminating the role of caspases during *Drosophila* oogenesis. Cell Death Differ 13:1950–1959. https://doi.org/10.1038/sj.cdd.4401892
27. Pritchett TL, Tanner EA, McCall K (2009) Cracking open cell death in the *Drosophila* ovary. Apoptosis 14:969. https://doi.org/10.1007/s10495-009-0369-z

Chapter 15

Detection and Assessment of *Wolbachia pipientis* Infection

Lindsay B. M. Nevalainen and Irene L. G. Newton

Abstract

Wolbachia pipientis is a widespread vertically transmitted intracellular bacterium naturally present in the model organism *Drosophila melanogaster*. As *Wolbachia* is present in a large number of *Drosophila* lines, it is critical for researchers to be able to identify which of their stocks maintain this infection to avoid any potential confounding variables. Here, we describe methods for detecting the bacterium and assessing the infection, including polymerase chain reaction (PCR) of DNA, multi-locus sequence typing (MLST) to identify strains, western blotting for protein detection, and immunohistochemistry and fluorescence in situ hybridization (FISH) of *Drosophila* ovaries to visually detect *Wolbachia* by fluorescence microscopy.

Key words *Drosophila*, *Wolbachia*, Endosymbiont, Microbiome, Fluorescence microscopy, Immunolabeling, FISH, PCR, MLST, Western blotting

1 Introduction

Drosophila melanogaster can be host to two bacterial endosymbionts: *Spiroplasma poulsonii* MSRO (*melanogaster* sex ratio organism) and *Wolbachia pipientis* *w*Mel. In this chapter, we focus on the detection and assessment of *Wolbachia* infection, as it is commonly found in widely used *Drosophila* stocks [1]. *Wolbachia* are alphaproteobacterial symbionts that live obligately intracellularly within *Drosophila* tissues. They are vertically transmitted, and a large number are present in the ovaries of an infected *Drosophila* female. *Wolbachia* colonize the germline stem cell niches, forming microcolonies in the germarium of each ovariole, localizing to the developing oocyte in each egg chamber, and can also be found in the hemolymph and brain, among other somatic tissues [2–9]. *Wolbachia pipientis*, as a species, is broadly infamous for the varying effects it can cause in the host, such as sperm-egg incompatibility (termed cytoplasmic incompatibility), parthenogenesis, pathogen blocking, and feminization [10–17]. In *Drosophila, Wolbachia* strain *w*Mel causes cytoplasmic incompatibility and can influence the viability of crosses as infected males have reduced fecundity

Michelle S. Giedt and Tina L. Tootle (eds.), *Drosophila Oogenesis: Methods and Protocols*,
Methods in Molecular Biology, vol. 2626, https://doi.org/10.1007/978-1-0716-2970-3_15,

when crossed to uninfected females. However, this defect is most pronounced in young male *Drosophila* (24 h old). *Wolbachia* infection is not typically listed on the entries for fly stocks at the national and international *Drosophila* Stock Centers or in FlyBase. It is therefore important that researchers be equipped with the tools to detect and assess the presence of *Wolbachia* in stocks prior to use. To avoid any potential confounding variables, infected stocks may be cleared of *w*Mel infection with antibiotics (e.g., 50 μg/mL tetracycline in their food for three generations) and repopulated with microbiome gut microbes [18]. Below, we detail the standard methods for detecting the symbiont and assessing the infection, including polymerase chain reaction (PCR) of DNA, multi-locus sequence typing (MLST) to identify strains, western blotting for protein detection, and immunohistochemistry and fluorescence in situ hybridization (FISH) of *Drosophila* ovaries to visually detect *Wolbachia* by fluorescence microscopy.

2 Materials

2.1 Detection of Wolbachia by PCR or MLST

1. DNA extraction kit.
2. Handheld drill.
3. Single-use sterile pestles (e.g., Axygen pestles from ThermoFisher).
4. DNA polymerase and buffers.
5. Thermocycler.
6. *Wolbachia* MLST primers (Table 1).
7. Agarose.
8. Ethidium bromide or sybersafe for visualizing DNA in the agarose gel.
9. Tris-borate (45 mM) and ethylenediaminetetraacetic acid (EDTA, 1 mM) (TBE) buffer (pH 8.0).
10. Agarose gel electrophoresis equipment.
11. Sanger sequencing facility.
12. MLST database to identify the strain (https://pubmlst.org/organisms/wolbachia-spp).

2.2 Detection of Wolbachia by Western Blotting

2.2.1 Western Sample Preparation

1. Lysis buffer: 10 mM Tris–HCl (pH 7.4), 150 mM NaCl, 1% Triton X-100, filter-sterilized.
2. 0.5 M EDTA and protease inhibitor cocktail (100×).
3. Handheld cordless drill.
4. 1.5 mL microcentrifuge tubes.

Table 1
MLST gene loci and primer sequences as abridged from Baldo et al. [20]

Gene	wMel code	Primer	Amplicon size (bp)
gatB	WD_0146	GAKTTAAAYCGYGCAGGBGTT	369
		TGGYAAYTCRGGYAAAGATGA	
coxA	WD_0301	TTGGRGCRATYAACTTTATAG	402
		CTAAAGACTTTKACRCCAGT	
hcpA	WD_0484	GAAATARCAGTTGCTGCAAA	444
		GAAAGTYRAGCAAGYTCTG	
ftsZ	WD_0723	ATYATGGARCATATAAARGATAG	435
		TCRAGYAATGGATTRGATAT	
fbpA	WD_1238	GCTGCTCCRCTTGGYWTGAT	429
		CCRCCAGARAAAAYYACTATTC	
Wsp	WD_1063	GTCCAATARSTGATGARGAAAC	546
		CYGCACCAAYAGYRCTRTAAA	

5. Sterile, single-use pestles that snugly fit the 1.5 mL microcentrifuge tubes.
6. 2× Laemmli loading buffer.
7. β-Mercaptoethanol.
8. Heating block for 1.5 mL centrifuge tubes.
9. Centrifuge.

2.2.2 Running the SDS-PAGE Gel for Western Blotting

1. Sodium dodecyl sulfate-polyacrylamide gel electrophoresis (SDS-PAGE) gel running apparatus.
2. 4–20% gradient SDS-PAGE gel, or 10% SDS-PAGE gel.
3. 10× SDS running buffer:30 g of Tris base, 144 g of glycine, and 10 g of SDS in 1000 mL of dH_2O.
4. Gel loading pipette tips.
5. Molecular weight protein ladder.
6. Tris-glycine-methanol transfer buffer: 3 g Tris base, 14.4 g glycine, 150 mL methanol, then add dH_2O to make a total of 1 L.

2.2.3 Membrane Preparation

1. Small plasticware storage tubs that will fit your membrane.
2. Methanol.
3. Polyvinylidene difluoride (PVDF) membrane.

2.2.4 Assembling and Assessing the Western Blot

1. Blocking reagent such as 3% bovine serum albumin (BSA) in Tris-buffered saline (TBS).
2. Primary and secondary antibodies (primary: anti-wsp (BEI resources NR-51684) at 1:10,000 in blocking reagent, secondary: goat anti-mouse HRP at 1:5000 in blocking reagent).
3. Sodium azide.
4. Flat-nosed forceps for handling the membrane.
5. Plastic wrap.
6. Kimwipes.
7. TBS with 0.1% Tween (TBST): 1 L 1× TBS buffer and 1 mL Tween-20.
8. 2% sodium azide solution.
9. Chemiluminescent substrate.
10. A way to visualize the membrane: imaging system or film, exposure cassette, and developing systems.
11. Stripping solution: 100 mM glycine, 0.1% ND-40, 1% SDS, pH 2.
12. Nutating mixer.

2.3 Visualization of Wolbachia Infection in Drosophila Ovaries by Immunohistochemistry

Prepare all solutions in sterile, distilled water. Filter sterilize as indicated.

2.3.1 Dissection of Ovaries

1. 3–5 day old female *Drosophila*.
2. Grape agar plates: 3% agar, 25% grape juice, and 0.3% sucrose – boil by heating in microwave and pour into Petri dishes.
3. Yeast paste: an approximately 1:1 mixture of commercial dry baker's yeast granules and water mixed to a thickness approaching that of peanut butter. For a thicker mixture: 10 g yeast granules/10 mL water. For a thinner mixture: 10 g yeast granules/20 mL water.
4. Mating cage: container with a mesh top to allow air in but plastic walls to keep the *Drosophila* within and a bottom to which a grape agar plate can be added.
5. Male and female *Drosophila* of the same genetic background.
6. Cold 1× phosphate-buffered saline (PBS).
7. Dissection dish.
8. *Drosophila* CO_2 pad or another method of anesthetization.
9. Two tungsten needles or two fine insect pins.
10. Glass Pasteur pipettes and bulbs.
11. 1.5 mL sterile microcentrifuge tubes.

12. Dissecting microscope.
13. Nutating mixer.
14. Centrifuge.

2.3.2 Fixation and Application of Antibodies

1. Devitt's solution: make using a fresh or under 10 days old ampule of electron microscopy grade 16% paraformaldehyde by diluting to 4% in 1× PBS. Add electron microscopy grade NP-40 added to a final concentration of 0.5%. Use the solution in 100 μL aliquots in combination with 600 μL heptane per sample (*see* **Note 1**).
2. 1× PBS.
3. Benchtop centrifuge, for example, "picofuge" or "mini centrifuge" (*see* **Note 2**).
4. Nutating mixer.
5. PBST: 0.2% Tween-20 added to 1× PBS.
6. PBST/BSA: 0.2% Tween-20 and 0.5% bovine serum albumin into 1× PBS.
7. Slides and coverslips.
8. Light-tight boxes or enough aluminum foil to obscure samples from light.
9. Primary and secondary antibodies (e.g., anti-wsp from BEI resources NR-51684, anti-mouse HRP conjugated antibody).
10. Glass Pasteur pipettes and bulbs.
11. Dissecting microscope.
12. Kimwipes.
13. Tungsten/dissection needles.
14. Mounting medium with 4′,6-Diamidine-2′-phenylindole dihydrochloride (DAPI) (*see* **Note 3**).

2.4 Visualization of Wolbachia Infection in Drosophila Ovaries by FISH

1. Fixative: 1.5 mL 16% paraformaldehyde, electron microscopy grade (freshly opened or within 10 days of opening ampule) and 4 mL diethylpyrocarbonate (DEPC)-treated PBS (DEPC-PBS), for a final concentration of 4% paraformaldehyde.
2. 50× Denhardts solution:1 g Ficoll 400, 1 g polyvinylpyrrolidone, 1 g BSA in 50 mL of DEPC-H_2O, filter sterilized and frozen in aliquots at −20 °C.
3. Hybridization solution: 50% formamide, 5× saline sodium citrate (SSC), 250 mg/mL salmon sperm DNA, 0.5× Denhardt's solution, 20 mM Tris–HCl, 0.1% SDS.
4. Wash: PBST (DEPC-PBS and 0.1% Tween-20).
5. Wash 1: 1× saline sodium citrate (SSC), 0.1% SDS, 20 mM Tris–HCl.

6. Wash 2: 0.5× SSC, 0.1% SDS, 20 mM Tris–HCl.
7. Proteinase K (25 mg/mL).
8. All materials in Subheading 2.4, **item 1**, DEPC-treated where applicable.
9. Mounting medium with DAPI (*see* **Note 3**).
10. RNase removal reagent.

2.5 Imaging Wolbachia Within Ovary Tissue

1. A fluorescence light microscope, with 60× or 100× objectives, and with the ability to detect wavelengths emitted by the FISH probes and/or the secondary antibody used in immunohistochemistry (IHC).
2. Lens paper.
3. Immersion oil.

3 Methods

3.1 Detection of Wolbachia by PCR or MLST

One can easily detect *Wolbachia* by using broad-spectrum primers and running out amplicons on an agarose gel. If one wants to identify the specific strain of *Wolbachia* present in an insect, the MLST primers can be used to type the bacterial infection.

3.1.1 DNA Extraction for PCR

We use the Qiagen DNeasy extraction kit but users should follow the manufacturer's instructions for whichever DNA extraction kit is used.

1. In a 1.5 mL microcentrifuge tube containing the *Drosophila* sample of interest (*see* **Note 4**), add 200 μL of buffer ATL and 20 μL of Proteinase K.
2. Using a handheld cordless drill and a single-use sterile pestle per sample, thoroughly homogenize the samples (*see* **Note 5**); the pestle fits where the drill bit would normally go. Follow the DNA extraction kit manufacturer's protocol to purify DNA. Elute the DNA in 100 μL final volume.
3. DNA can be extracted in advance and stored at −20 °C. Avoid numerous freeze-thaw cycles as this will decrease DNA quality and make it more difficult to amplify the template.

3.1.2 PCR

1. Assemble a 20 μL PCR reaction based on the polymerase manufacturer's instructions.
   (a) Per 20 μL mixture, add each deoxynucleoside triphosphate at a concentration of 0.2 mM, 1.5 mM $MgCl_2$, each primer at a concentration of 1 μM, 0.5 U Taq DNA polymerase, 1× PCR buffer, 2 μL of the DNA sample and PCR-grade water up to 25 μL. Mix well by pipetting up and down.

2. Perform the PCR using parameters indicated by the polymerase manufacturer, the melting temperature indicated in Tables 1 and 35 amplification cycles. For Taq polymerase, reaction cycles should begin at 94 °C for 2 min; followed by 35 cycles of 94 °C for 30s, the optimal annealing temperature for the primers for 45 s, 72 °C for 1.5 min; and a final elongation step at 70 °C for 10 min.
3. Run the PCR on an agarose gel—depending on the gel, you can load 15–25 μL. Detect the PCR amplification using the preferred detection method (e.g., ethidium bromide or SYBR-Safe) (*see* **Notes 6** and 7).

3.1.3 Determination of *Wolbachia* Strain by MLST

1. Extract DNA from the sample(s) as described in Subheading 3.1.1. Choose primers from Table 1 and run PCR reactions as described in Subheading 3.1.2.
2. Once an amplicon of the correct size is obtained, Sanger sequence it (and other MLST amplicons) using the same primers used for amplification.
3. Use the *Wolbachia* MLST database to identify the strain (https://pubmlst.org/organisms/wolbachia-spp).

3.2 Detection of *Wolbachia* by Western Blotting

Wolbachia infects many *Drosophila* tissues and can be detected in whole flies (*see* **Note 8**).

3.2.1 Western Sample Preparation

1. In a 1.5 mL microcentrifuge tube, add a minimal amount of Lysis Buffer, with 1× final HALT protease inhibitor and 1 mM EDTA, to submerge the samples (e.g., if a pool of three whole *Drosophila* is considered in one sample, then 200 μL is sufficient to fully submerge the sample).
2. Homogenize the samples using a sterile, single-use pestle attached to a handheld power drill at maximum speed (*see* **Note 5**).
3. Pellet debris and nuclei at 9000 g for 10 min in a centrifuge chilled to 4 °C.
4. Transfer 20 μL of the supernatant into a fresh 1.5 mL tube, avoiding the debris. Discard the rest of the sample.
5. Add 20 μL 2× Laemmli sample buffer and β-mercaptoethanol (BME) at 5% final concentration (v/v).
6. Heat the samples for 5 min at 95 °C.
7. Gently spin down the samples using a tabletop centrifuge to consolidate the condensate on the lids of the tubes.

3.2.2 Running the SDS-PAGE Gel for Western Blotting

1. Set up the apparatus to run the western samples on an SDS-PAGE gel.
   (a) Prepare 1 L of 1× SDS running buffer prepared from the 10× stock.
   (b) Assemble the gel cassette, ensuring that the shorter side of the gel faces inward and that the bottom of the gel is in firm contact with the lower gasket.
   (c) Once the assembled cassette has been placed into the running chamber, fill the inside of the cassette with 1× SDS running buffer to just past the tops of the wells (*see* **Note 9**).
   (d) Once the inner chamber has been filled, then fill the rest of the chamber to the appropriate line for the number of gels being run.
2. Load the samples and the molecular weight marker using loading tips (*see* **Note 10**).
3. Run the SDS-PAGE gel at 120 V until the loading buffer dye is near the bottom, ~1 cm from the edge, or in line with the gasket. This typically takes between 30 and 40 min.

3.2.3 Preparing the Membrane

1. While the gel is running, prepare the tris-glycine-methanol transfer buffer and other blotting materials (e.g., nitrocellulose membrane, plasticware tubs, blot cassette sponges, and filter paper).
2. Add methanol to the first smaller tub, and transfer buffer to both a second small tub and a larger tub.
3. Soak the PVDF membrane in methanol.
4. Once the PVDF membrane has fully absorbed the methanol, use flat-nosed forceps to transfer it to the second small plasticware tub that contains the transfer buffer (*see* **Note 11**).

3.2.4 Assembling and Assessing the Western Blot

1. When the gel has run far enough, disassemble the SDS-PAGE gel running apparatus, dispose of SDS running buffer, and prepare the transfer.
2. On a paper towel, use a razor blade to pry the gel plates apart at the corner.
3. Trim off the well tabs using the same razor blade, and then gently slide the gel off the plate into one of the plasticware tubs containing transfer buffer.
4. Ensure that the gel does not come into contact with the PVDF membrane until you assemble the "blot sandwich" transfer cassette as shown in Fig. 1 (*see* **Note 12**) within the larger plastic tub. Place the blot sandwich into the transfer apparatus.

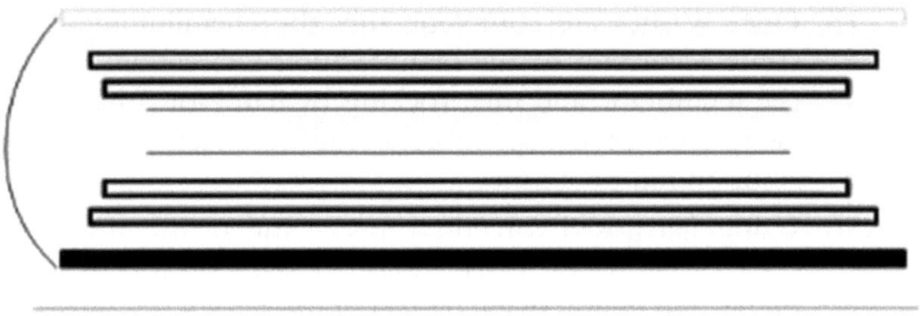

Fig. 1 Schematic of the correct assembly of the "blot sandwich" cassette appropriate for the BioRad mini gel system

5. Run the transfer at low voltage (30 V) for at least 3 h or overnight, either packed in a bucket with ice or in a 4 °C cold room (*see* **Note 13**).
6. When the transfer is finished, remove the "sandwich" cassette and dispose of the buffer. Orient the cassette so the gel is sitting on top of the PVDF membrane. Use a sharpie to denote the corners of the gel (and the gel number, if running multiple) onto the membrane. Dispose of the gel.
7. Using an appropriately sized (small) plasticware container, submerge the membrane in a blocking buffer (*see* **Note 14**) and incubate on a nutating mixer for at least 20 min.
8. Pour off the block and apply the primary antibody (anti-wsp at 1:10,000 in blocking reagent) for 1–2 h at room temperature or overnight/over the weekend at 4 °C on a nutating mixer (*see* **Note 15**).
9. Pour the primary antibody solution back into its container, as it is reusable when sodium azide is added at 0.02% w/v.
10. Wash the membrane with 1× TBST, 3× for 10 min each while nutating.
11. Incubate the blot in the appropriate secondary antibody at 1: 5000 in blocking buffer for 1–2 h at room temperature or overnight/over weekend at 4 °C while nutating (*see* **Note 15**).
12. Pour off and dispose of the secondary antibody solution.
13. Wash the membrane with 1× TBST, 3× for 10 min each while nutating.
14. While the washes are being performed, set up the blotting and developing materials.
    (a) Place 2 stacks of four kimwipes each onto a flat surface.
    (b) Using a razor blade, cut out a piece of plastic wrap approximately 2× the size of the membrane and place it on one side of the stacks of kimwipes.

15. Once washes are complete, mix the chemiluminescent substrate according to the manufacturer's directions. Pour onto the membrane and mix, rocking by hand for 3 min, ensuring every part of the membrane fully contacts the developing solution multiple times.
16. Using flat-nosed forceps, transfer the membrane onto one stack of kimwipes. Place the other stack on top of the membrane, and gently swipe on top of the stack 4× in one direction with the side of your hand. Move the membrane to the other side of the stack, cover, and repeat (*see* **Note 16**).
17. Transfer the membrane to the piece of saran wrap and fold it over securely. Trim off any excess with a razor blade, and tape the membrane face up into the exposure cassette. It is now ready to expose and develop in a darkroom with film. If using an image processor, no cassette or film is needed. Once developed, be sure to draw out the ladder markings and edges of the blot onto the film to ensure consistent interpretation.
18. If desired, the membrane can be stripped of antibodies by immersion in stripping solution for 1 h at room temperature, re-blocked, and then re-probed using a different primary and secondary antibody (*see* **Note 17**).

3.3 Visualization of Wolbachia Infection in Drosophila Ovaries by Immunohistochemistry

3.3.1 Dissection of Ovaries

1. Add a ~1 cm diameter dollop of yeast paste to the center of a grape agar plate.
2. Assemble a mating cage. Add females and males of the chosen *Drosophila* stock to a mating cage and cover it with a prepared grape agar plate. Allow them to mate for 3–4 days prior to dissection. This promotes ovary growth and development by stimulating egg production. Replace the grape agar plate with yeast paste daily.
3. Immediately preceding dissection, ensure all solutions are made and chilled where appropriate (*see* **Note 18**).
4. Pipette cold 1× PBS buffer into several wells of a dissection plate placed on ice (*see* **Note 19**).
5. Knockout flies using a CO_2 pad and separate females from males (*see* **Note 20**).
6. Using forceps, transfer 1 female into one of the wells of the cold dissection plate containing chilled PBS. Using the forceps and/or needles, maneuver the fly onto its back and grasp firmly around the thorax with the forceps to hold it in place.
7. Using either tungsten needles, dissection pins, or fine forceps, hold the fly on its back at the bottom of the dish and swipe two to four times with light pressure from the anterior to the posterior center abdomen to split the exoskeleton and free

the ovaries. Then using the tool of choice, gently push or lift the ovaries out of the abdomen into the PBS in the dissection well (*see* **Note 21**).

8. Transfer dissected ovaries into a separate well to isolate from the carcass and debris.
9. Repeat the dissection until five to ten sets of ovaries per genotype of interest are obtained.
10. Using a glass Pasteur pipette and bulb, gently transfer all dissected ovaries from the dissection dish into a 1.5 mL microcentrifuge tube.

3.3.2 Fixation and Application of Antibodies

1. Using a small tabletop centrifuge, spin the ovaries (short spin – max 100 g) into the bottom of the 1.5 mL microcentrifuge tube for 5–10 s.
2. Replace the PBS solution with 600 μL of heptane and 100 μL of premade Devitt solution.
3. Mix by hand vigorously by agitation for ~15–30 s. This will create a white, viscous emulsion with no phase line visible.
4. Nutate the samples for 10 min at room temperature.
5. Using a small tabletop centrifuge, gently spin the ovaries into the bottom of the 1.5 mL microcentrifuge tube for 5–10 s. Alternatively, allow the ovaries to settle to the bottom of the tube on their own. Remove the fixative/heptane solution.
6. Wash the samples in 500 μL of 1× PBST, three times for 10 min each on a nutator at room temperature.
7. Again, gently centrifuge the samples and replace the PBST wash with 500 μL PBST/BSA, then nutate for 15 min at room temperature.
8. Add the primary antibody (anti-wsp, detecting *Wolbachia* surface protein) diluted in PBST/BSA (500-750 μL volume), and incubate the samples at 4 °C overnight while nutating.
9. The following day, gently spin down the samples, remove the PBST/BSA, and replace it with PBST; this is a "quick" wash.
10. Again, gently centrifuge the samples, remove the PBST, and wash with fresh PBST three times for 15 min each on a rocker/nutator at room temperature.
11. Add the appropriate secondary antibody and/or additional stain (e.g., actistain), and incubate the samples rocking/nutating in the dark at room temperature for 2 h or rocking/nutating in the dark overnight at 4 °C (*see* **Note 22**).
12. Gently centrifuge the samples, remove the PBST/BSA, and wash with one "quick" and three 15 min washes with 500 μL PBST at room temperature in the dark while rocking/nutating.

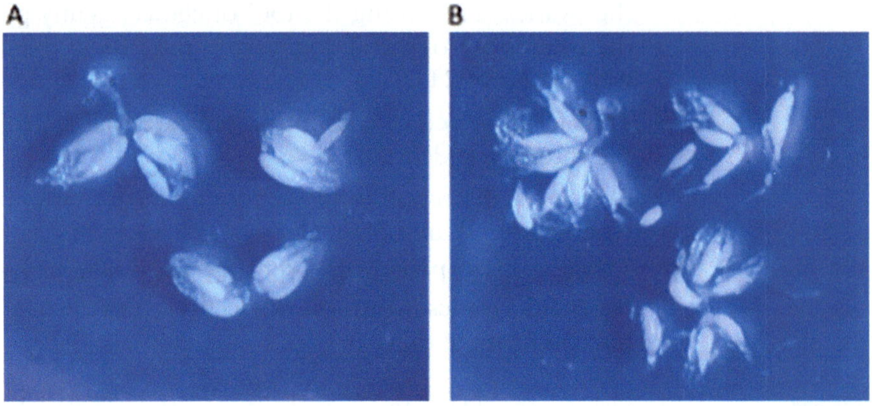

Fig. 2 Ovaries after processing for either immunohistochemistry or FISH. (**a**) Stained ovaries on a slide prior to being fanned out with tungsten needles or dissection tools. (**b**) Ovaries fanned out during the mounting process for ideal image capture conditions

13. Gently centrifuge the samples, remove the PBST, and add 20 μL of mounting solution. Incubate the samples in the dark at room temperature for 20–30 min to allow the DAPI to penetrate the cells before either mounting on slides or storing them in the dark at 4 °C until mounted.
14. To mount ovaries onto a slide for visualization, first transfer 5–10 ovaries onto a slide using a glass Pasteur pipette. Using a dissection scope, remove as much liquid mounting solution as possible with a pipette (*see* **Note 23**).
15. Once the ovaries are no longer floating and as much liquid has been wicked away as possible without completely drying the samples, add 20 μL of the preferred mounting medium with DAPI to the slide and ovaries.
16. Using the dissection scope, tease apart the ovaries with tungsten needles to fan out single ovarioles (Fig. 2).
17. Place a coverslip onto your sample once the ovarioles are sufficiently separated, and let the slide harden at 4 °C overnight (*see* **Note 24**). Ovaries prepared in this manner are best viewed within a few weeks or up to 1 month while being stored in a flat slide storage folder away from light at 4 °C.

3.4 *Visualization of* Wolbachia *Infection in* Drosophila *Ovaries by FISH*

All materials in this section are diethylpyrocarbonate (DEPC)-treated, and all washes are on a nutator. Hybridization and wash solutions should be warmed to 42 °C where indicated. All surfaces should be treated with RNase removal reagents.

1. Dissect ovaries as in Subheading 3.3.1, using DEPC-treated, filter-sterilized, chilled PBS, and ensure that all equipment is treated to remove RNases.

2. Add fixative to your sample for 20 min at room temperature, nutating.
3. Wash 3× 5 min each in a 50/50 mix of methanol and DEPC-PBS.
4. Wash 3× 5 min each in 100% methanol, then store overnight at −20 °C (*see* **Note 25**).
5. Gradually introduce fixative to the samples by:
   (a) Rehydrate the ovaries in a 1 mL total 70/30 mix of methanol and fixative for 5 min.
   (b) Remove this mixture and replace it with a 1 mL total 70/30 mix of fixative and methanol for 5 min.
   (c) Remove this mixture and replace it with 1 mL of fixative for 20 min.
6. Wash on a nutator three times for 5 min each in DEPC-PBST.
7. Treat the samples with proteinase K (1 μL of 25 mg/mL stock in 500 μL PBST) for 5 min at 37 °C.
8. Stop the proteinase K reaction with two quick rinses in PBST and a 5 min wash on a nutator in PBST at room temperature.
9. Replace solution with fresh fixative and incubate on a nutator for 25 min at room temperature.
10. Wash four times for 5 min each wash in PBST on a nutator at room temperature.
11. Gradually introduce the hybridization solution by:
    (a) Washing for 5 min in a 1/1 ratio of PBST and hybridization solution at room temperature.
    (b) Remove this mixture and wash for 5 min in the hybridization solution at room temperature.
12. Prewarm the hybridization solution to 42 °C. Prehybridize the sample by incubating it for 30 min at 42 °C in a new hybridization solution.
13. Mix 100 ng of the appropriate probe into 500 μL of pre-warmed hybridization solution per sample. Hybridize with either the eubacterial probe (AlexaFluor – EUB338) or the *Wolbachia*-specific probe.
14. Remove the prehybridization solution and replace it with the hybridization solution.
15. Nutate for 3 h at 42 °C in the dark (*see* **Note 26**).
16. Prewarm the wash solutions (#1 and #2).
17. Wash two times for 15 min each at 42 °C in Wash #1.
18. Wash two times for 15 min each at 42 °C in Wash #2.

19. Remove wash and add mounting media with DAPI, fan out and mount the ovaries as in Subheading 3.3.2 (Fig. 2) for observation.

3.5 Imaging and Analysis of *Wolbachia* Within Ovary Tissue

Ovary samples prepared for either immunohistochemistry or FISH can be visualized using a fluorescence microscope at 60× magnification. *Wolbachia* requires high magnification for visualization (60–100×). The instructions below are for an inverted microscope but the sample can be observed on an upright microscope; simply reverse the directions.

1. Turn on the microscope and place the slide onto the stage. Center the sample over the condenser.
2. Using a 10× objective (or other low-power objective), use the coarse focus to bring the objectives up to the top and slowly bring them down, bringing the sample into focus.
3. Once the sample is in focus at 10×, change the objective to 40× air. This time only use the fine focus. Move the stage to find the germarium, as this is the part of the ovary heavily colonized by *Wolbachia* and easiest to identify the infection.
4. Now move the 40× objective out of the way and add the appropriate immersion oil to the 60× objective. Rotate the 60× objective into position.
5. Switch on the fluorescence to the appropriate wavelength for the fluorophore (if using Alexa488, this would be green, for example). *Wolbachia* will appear as rod-shaped bacterial cells within the tissue (Fig. 3).
6. When finished, clean the objective with lens paper to remove any oil.

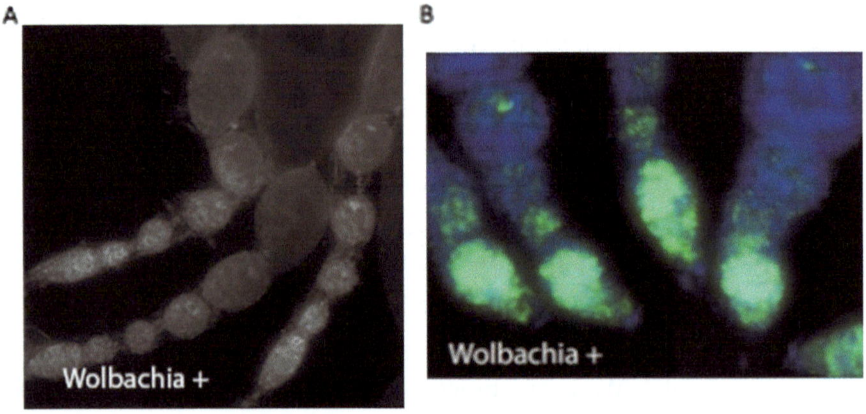

Fig. 3 Examples of immunohistochemistry (**a**) and FISH (**b**) of *Wolbachia*-positive *Drosophila melanogaster* ovarioles. (**a**) used with permission from MaryAnn Martin. (**b**). (Adapted from Sheehan et al. [19])

4 Notes

1. Prepare this solution fresh on the day of dissection.
2. If a small centrifuge is not available, the ovaries will eventually settle to the bottom of the tube.
3. Use ProLong Gold for taking micrographs, as it hardens over time, ensuring your sample will not move during imaging. Use SlowFade for storage at 4 °C before mounting.
4. This protocol can be used for samples containing a single individual *Drosophila* or three pooled individuals.
5. Some leg, wing, etc., remnants are acceptable.
6. The PCR amplicon of *wsp* is ~500 bp in length.
7. If PCR is not successful, then dilute the DNA two- or threefold and repeat the PCR.
8. Often, researchers will use whole flies to avoid unnecessary dissection upstream of western blotting. One to three flies can be used without overloading a western blot.
9. This should be performed first to check that the cassette does not leak as this can adversely affect the running of the current. If the running buffer is leaking out, then pour out the excess into a beaker, remove and readjust the gel, and try again until no running buffer leaks out.
10. Prior to this step, it can be useful to first gently flush out each well with a running buffer to eliminate bubbles or other debris present.
11. The membrane is hydrophobic, so use flat-nosed forceps to gently push the membrane into the liquid until it begins to soak it up.
12. If there is an odd number of gels, make a "blank" blot sandwich to insert during the transfer process, consisting of two pieces of filter paper sandwiched by two sponges.
13. Once the chamber is filled with transfer buffer, gently jostle the cassette holder to remove any bubbles prior to starting the transfer.
14. If desired, a Ponceau stain can be performed prior to blocking to confirm the protein sample is not degraded and that transfer to the PVDF membrane was successful.
15. If performing the overnight or over weekend 4 °C incubation, either use a plasticware container with an airtight lid or use plastic wrap and a rubber band to cover the container to prevent evaporation of the antibody solution.
16. Ensure that the membrane does not completely dry during this process as this will lead to the excessive background.

17. Membranes can be kept in their saran wrap in a sealed plastic bag at 4 °C for a month for potential re-probing to confirm results.
18. In particular, the paraformaldehyde solution must be made fresh the day of dissection.
19. This step can be done on top of a cold block or the dissection plate itself can be chilled prior to starting the protocol.
20. This dissection step should be completed in under 30 min.
21. Be sure to avoid also transferring the digestive tract.
22. It is important to note that samples will be light-sensitive from this step forward, so it is critical to sequester them away from light using either a light-tight box or by thoroughly wrapping them in aluminum foil.
23. First pipette off as much of the liquid as possible with a p10 micropipette until the ovaries are no longer floating. Then use a kimwipe to wick away more liquid by gently immersing an edge or corner into the remaining pool of liquid. Use caution to not wick up the ovaries or dry the sample entirely, so keep the kimwipe in contact with the liquid only briefly.
24. For best results, ensure that the surface the slide folder is sitting on is as level as possible to prevent the spread of ovaries from regrouping.
25. This step is to further fix the samples and serves as a convenient stop point when dissecting a large number of ovaries.
26. As this step still requires mixing while also necessitating sequestration from light, we have had success both using aluminum foil to cover a shaking heat block as well as covering the heat block entirely with an opaque container.

References

1. Clark ME et al (2005) Widespread prevalence of Wolbachia in laboratory stocks and the implications for Drosophila research. Genetics 170(4):1667–1675
2. Cho KO, Kim GW, Lee OK (2011) Wolbachia bacteria reside in host Golgi-related vesicles whose position is regulated by polarity proteins. PLoS One 6(7):e22703
3. Dobson SL et al (1999) Wolbachia infections are distributed throughout insect somatic and germ line tissues. Insect Biochem Mol Biol 29(2):153–160
4. Ferree PM et al (2005) Wolbachia utilizes host microtubules and Dynein for anterior localization in the Drosophila oocyte. PLoS Pathog 1(2):e14
5. Frydman HM et al (2006) Somatic stem cell niche tropism in Wolbachia. Nature 441(7092):509–512
6. Genty LM et al (2014) Wolbachia infect ovaries in the course of their maturation: last minute passengers and priority travellers? PLoS One 9(4):e94577
7. Pietri JE, DeBruhl H, Sullivan W (2016) The rich somatic life of Wolbachia. Microbiology 5(6):923–936
8. White PM et al (2017) Mechanisms of horizontal cell-to-cell transfer of Wolbachia spp. in Drosophila melanogaster. Appl Environ Microbiol 83(7):e03425–e03416
9. Xi Z et al (2008) Genome-wide analysis of the interaction between the endosymbiotic

bacterium Wolbachia and its Drosophila host. BMC Genomics 9:1

10. Caragata EP, Dutra HL, Moreira LA (2016) Inhibition of Zika virus by Wolbachia in Aedes aegypti. Microb Cell 3(7):293–295
11. Kaur R et al (2021) Living in the endosymbiotic world of Wolbachia: a centennial review. Cell Host Microbe 29(6):879–893
12. Mercot H, Charlat S (2004) Wolbachia infections in Drosophila melanogaster and D. simulans: polymorphism and levels of cytoplasmic incompatibility. Genetica 120(1–3): 51–59
13. Moreira LA et al (2009) A Wolbachia symbiont in Aedes aegypti limits infection with dengue, Chikungunya, and Plasmodium. Cell 139(7): 1268–1278
14. Rainey SM et al (2016) Wolbachia blocks viral genome replication early in infection without a transcriptional response by the endosymbiont or host small RNA pathways. PLoS Pathog 12(4):e1005536
15. Richardson KM et al (2016) Tropical Drosophila pandora carry Wolbachia infections causing cytoplasmic incompatibility or male killing. Evolution 70(8):1791–1802
16. Werren JH, Baldo L, Clark ME (2008) Wolbachia: master manipulators of invertebrate biology. Nat Rev Microbiol 6(10):741–751
17. Wong ZS, Brownlie JC, Johnson KN (2015) Oxidative stress correlates with Wolbachia-mediated antiviral protection in Wolbachia-Drosophila associations. Appl Environ Microbiol 81(9):3001–3005
18. Bryant KN, Newton ILG (2020) The intracellular symbiont Wolbachia pipientis enhances recombination in a dose-dependent manner. Insects 11(5):284
19. Sheehan KB et al (2016) Identification and characterization of a candidate Wolbachia pipientis type IV effector that interacts with the actin cytoskeleton. MBio 7(4):e00622–e00616
20. Baldo L, Werren JH (2007) Revisiting Wolbachia supergroup typing based on WSP: spurious lineages and discordance with MLST. Curr Microbiol 55(1):81–87

Chapter 16

Measuring Transposable Element Activity in Adult *Drosophila* Ovaries

Anikó Szabó, Péter Borkúti, Zoltán Kovács, Ildikó Kristó, Csilla Abonyi, and Péter Vilmos

Abstract

Transposons are genetic elements that use various mechanisms of transposition to move around the genome, thus posing a risk to genomic integrity. Repression of transposable elements (TEs) involves the complex PIWI pathway and several proteins associated with heterochromatinization. All players of TE repression are indispensable for proper reproductive fitness, as loss-of-function mutations in these genes result primarily in sterility and impaired reproductive development. When investigating the function of novel genes with similar phenotypes, elevated transposon expression in reproductive tissues can be a marker for involvement in the aforementioned processes. Here, we present a protocol for investigating TE levels in adult *Drosophila* ovaries, from dissection to data analysis.

Key words *Drosophila*, Oogenesis, Transposon, RT-qPCR, RNA isolation

1 Introduction

Maintenance of genomic integrity in the germline is of utmost importance in every organism as these cells are responsible for producing the next generation. Transposable elements (TEs) are repetitive DNA sequences capable of moving around in the genome, thus posing a risk to genomic integrity. Uncontrolled TE insertion in the germline can result in severe mutations leading to sterility, impaired development of the reproductive tissue, and decreased fitness [1–6]. However, several defensive mechanisms have evolved in the germline to restrict transposon mobility and maintain genetic stability. Members of the primary and secondary PIWI pathways [7] together with several other proteins responsible for heterochromatin establishment and maintenance, such as HP1a [8], Su(var)3-9 [9], eggless [10] histone methyltransferases, and linker histone H1 [11], are indispensable for TE silencing. Partial or complete loss of the activity of these proteins results in the

Michelle S. Giedt and Tina L. Tootle (eds.), *Drosophila Oogenesis: Methods and Protocols*,
Methods in Molecular Biology, vol. 2626, https://doi.org/10.1007/978-1-0716-2970-3_16,

elevated generation of transposon transcripts, leading to the appearance of the aforementioned phenotypes. When studying novel genes involved in oogenesis, whose mutations are characterized by sterility, impaired development of the reproductive tissue and decreased fitness, measuring TE transcript levels in the germline can be a reliable and relatively simple tool to assess whether the PIWI pathway and/or heterochromatinization are affected in the mutants. Transposable elements in *Drosophila* can be characterized by their expressional pattern, as some TEs are expressed predominantly in the germline (germline-specific TEs) or in the gonadal soma (somatic TEs), while others are transcribed in both tissue types (intermediate TEs) [12, 13]. It is important to note that TEs behave differently when crucial silencing factors are depleted or mutated [8, 11].

Reverse transcription-quantitative real-time PCR (RT-qPCR) is a widely used method to measure transcript levels in a wide variety of RNA samples [14]. To quantify the expression of specific gene products, total RNA is isolated from the tissue or cultured cell of interest, reverse-transcribed to cDNA, and expression is quantified by measuring the cDNA amount. Reverse transcription and quantification can be done separately in two steps—as described in this article – or combined in a single step. During the quantitative PCR step, double-stranded DNA-binding fluorescent dye or fluorescently labeled probes are added to the reaction mixture, along with target-specific primers. As amplification proceeds, the intensity of the fluorescent signal emitted by the dye or probes increases in proportion to the quantity of the targeted product. For each reaction, a Ct (threshold cycle) value is assigned which corresponds to the number of the PCR cycle when fluorescence intensity exceeds the detection threshold [15]. A commonly used method for RT-qPCR data analysis is comparative Ct ($2^{-\Delta\Delta Ct}$) [16, 17] where relative fold changes (rFC) are calculated from Ct value differences between the target and a reference gene, and then compared across treated and untreated or wild-type and mutant samples.

Here, we present a comprehensive protocol for the measurement of transposable element transcripts in adult *Drosophila* ovaries, from dissection to data analysis, and a collection of TE-specific primer sequences.

2 Materials

2.1 Dissection

1. Standard cornmeal, yeast, and sucrose *Drosophila* medium.
2. Dry yeast.
3. CO_2
4. Fly pad.

5. A pair of sharp forceps.
6. Glass or plastic Petri dish (*see* **Note 1**).
7. Schneider's *Drosophila* medium (*see* **Note 2**).
8. Stereomicroscope.
9. 1.5 ml centrifuge tubes.

2.2 RNA Isolation

1. Gloves.
2. RNase-free 1.5 mL centrifuge tubes.
3. RNase-free homogenizer pestle.
4. Refrigerated benchtop centrifuge.
5. RNase-free water (*see* **Note 3**).
6. TRIzol reagent (Invitrogen) or equivalent for guanidinium thiocyanate-phenol–chloroform extraction (*see* **Note 4**).
7. Chloroform.
8. Isopropanol (*see* **Note 5**).
9. 75% ethanol (*see* **Note 6**).
10. Spectrophotometer or fluorimeter for assessing RNA concentration.
11. Optional: nuclease-free glycogen (*see* **Note 7**).

2.3 DNase Treatment and Reverse Transcription

1. DNase treatment kit including RNase-free DNase I, 10× DNase buffer, and 50 mM EDTA.
2. Nuclease-free water.
3. RevertAid First Strand cDNA Synthesis Kit (Thermo Fisher Scientific) (*see* **Note 8**).
4. PCR tubes.
5. PCR machine.

2.4 RT-qPCR

1. SensiFAST™ SYBR® Hi-ROX Kit (*see* **Note 9**).
2. Forward and reverse primers (*see* Table 1).
3. Nuclease-free PCR-grade water.
4. QIAGEN Rotor-Gene-Q qPCR machine (*see* **Note 10**).

3 Methods

For an overview of the complete procedure, see Fig. 1.

3.1 Dissection

1. Collect and place ~5 virgin female flies and 5 males on fresh standard fly food media to allow the flies to mate as mating promotes oogenesis [24].

Table 1
List of *Drosophila* TE-specific primer sequences, plus a primer pair for the detection of the *rp49* internal control transcript

Primer name	Sequence 5'-3'	Type	References
rp49 F	CCGCTTCAAGGGACAGTATCTG	Reference gene	[18–21]
rp49 R	ATCTCGCCGCAGTAAACGC		
297 F	AAAGGGCGCTCATACAAATG	Germline	[22]
297 R	TGTGCACATAAAATGGTTCG		
accord F	ACAATCCACCAACAGCAACA		[20, 22]
accord R	AAAAGCCAAAATGTCGGTTG		
Burdock F	CGGTAAAATCGCTTCATGGT		[19]
Burdock R	ACGTTGCATTTCCCTGTTTC		
doc F	GGGTGACTATAACGCCAAGC		[22]
doc R	GCAAAATCGATCAGGTCTGG		
Het-A F	CGCGCGGAACCCATCTTCAGA		[19–21, 23]
Het-A R	CGCCGCAGTCGTTTGGTGAGT		
HMS-Beagle F	AATGCCCTTGTCGGACACGA		[23]
HMS-Beagle R	TGATGAAACACATTACCAGAACCTTGA		
I-element F	GACCAAATAAAAATAATACGACTT		[20]
I-element R	AACTAATTGCTGGCTTGTTATG		
412 F	CACCGGTTTGGTCGAAAG	Intermediate	[18, 20]
412 R	GGACATGCCTGGTATTTTGG		
idefix F	AACAAAATCGTGGCAGGAAG		[22]
idefix R	TCCATTTTTCGCGTTTACTG		
mdg1 F	AAACTCCAACTCCCAATCCG		[21]
mdg1 R	AGTGGTCCCTCGCAGTCGTT		
blood F	AACAATAGAAAGAAGCCACCGAAC		[21]
blood R	AGTCATGGACTATTGAGGGTGTTG		
gtwin F	TTCGCACAAGCGATGATAAG	Soma	[20, 22]
gtwin R	GATTGTTGTACGGCGACCTT		
gypsy F	GTTCATACCCTTGGTAGTAGC		[18–20]
gypsy R	CAACTTACGCATATGTGAGT		
tabor F	ACGTTGTTCACGACATTAGCCG		[23]
tabor R	GGGTTGGTTCGGATCTGACG		

The 14 TEs represent all classes of transposons active in the *Drosophila* ovarium

2. Feed flies at 25 °C for 2 days with dry yeast to fatten up the ovaries.
3. Fill a glass Petri dish of 100 mm diameter with room-temperature Schneider's *Drosophila* Medium.
4. Put the 2-day-old females to sleep using CO_2.
5. Pick up a female fly using forceps, submerge her into Schneider's *Drosophila* medium, and remove ovaries using a pair of sharp tweezers. Repeat to collect ~5 pairs of ovaries.

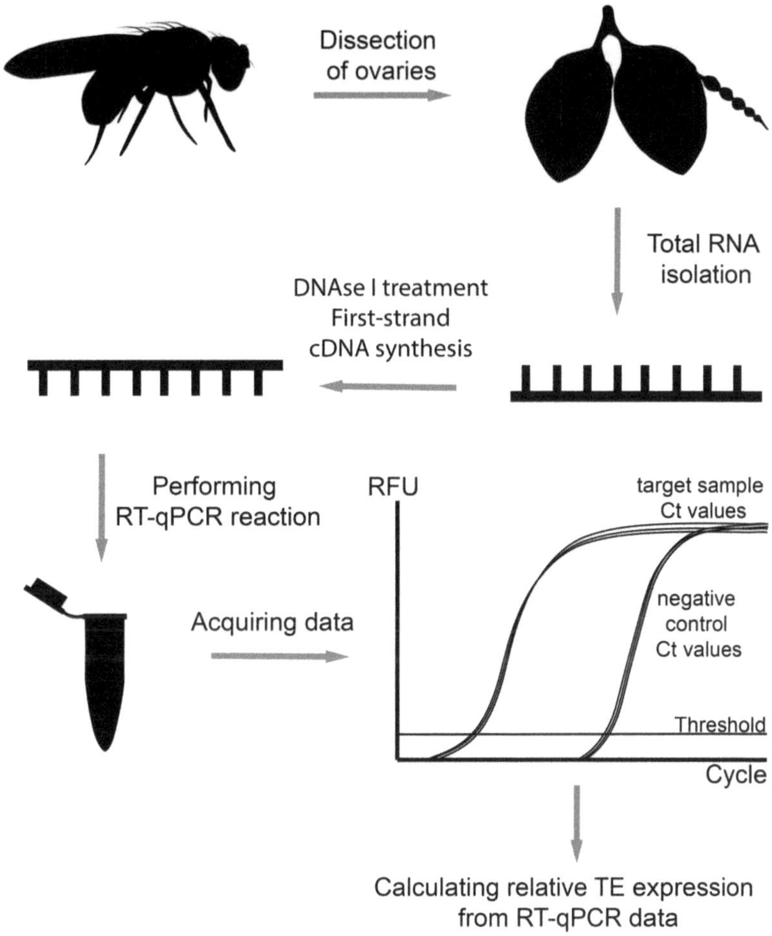

Fig. 1 The sequential stages of the workflow of measuring transposable element activity in adult *Drosophila* ovaries. RT-qPCR Quantitative real-time reverse transcription PCR, RFU relative fluorescence unit, Ct threshold cycle, TE transposable element

6. Transfer ~5 pairs of ovaries with the help of the forceps to a 1.5 mL centrifuge tube containing 300–500 μL of Schneider's medium.
7. Proceed immediately to RNA isolation (*see* **Note 11**).

3.2 RNA Isolation

1. Remove Schneider's *Drosophila* medium from the collected ovaries.
2. Homogenize ovaries in 250 μL TRIzol reagent (*see* **Note 12**).
3. Add 70 μL nuclease-free water and mix gently by inverting the tubes.
4. Incubate at room temperature for 5 min.

5. Add 70 μL chloroform to the sample and vortex vigorously. Incubate the sample at room temperature for 5 min.
6. Centrifuge sample at 18,000 g for 15 min at 4 °C.
7. After centrifugation, the liquid separates into the bottom phenol–chloroform phase (pink), white interphase, and the upper aqueous phase (colorless) which contains total RNA (*see* **Note 13**).
8. Carefully transfer the colorless top phase to a clean 1.5 mL centrifuge tube and add isopropanol equivalent to one sample volume. If necessary, add 10 μg of nuclease-free glycogen (*see* **Note 7**).
9. Incubate at room temperature for 10 min.
10. Centrifuge at 18,000 g for 10 min at 4 °C to pellet the RNA.
11. Discard the supernatant and wash the RNA pellet with 500 μL 75% ethanol (EtOH).
12. Centrifuge at 6800 g for 5 min at 4 °C and discard the EtOH.
13. Repeat **steps 10** and **11** to thoroughly wash the RNA pellet.
14. Remove EtOH and air-dry RNA pellets for 10–15 min. Do not over-dry the pellets to avoid difficulties during solubilization.
15. Dissolve RNA in 25–50 μL nuclease-free water and measure RNA concentration (*see* **Note 14**).
16. Proceed to DNase I treatment to remove contaminating DNA. Alternatively, RNA samples can be stored at −80 °C.

3.3 DNase I Treatment

1. Treat the RNA sample with 1 U of DNase I per 1 μg RNA at 37 °C for 30 min to remove contaminating DNA.
2. Add 1 μL of RNase-free EDTA to every 10 μL of the sample and incubate at 65 °C for 10 min to inactivate DNase I (*see* **Notes 15** and **16**).
3. The effectiveness of DNase I treatment can be assessed by performing standard PCR reactions using 30–50 ng of DNase I-treated RNA (*see* **Note 17**). If the treatment was successful, no amplification should take place.
4. Proceed to reverse transcription reaction or store DNase I-treated RNA samples at −80 °C.

3.4 Two-Step RT-qPCR

1. Prepare reverse transcription reaction in PCR tubes (*see* **Note 8**):
   - 1 μg total RNA.
   - 1 μL random hexamer primer.
   - 4 μL 5× reaction buffer.
   - 1 μL RiboLock RNase Inhibitor (20 U/μL).

- 2 μL 10 mM dNTP Mix.
- 1 μL RevertAid M-MuLV RT (200 U/μL).
- PCR-grade water to 20 μL final volume.

2. Perform cDNA synthesis in a PCR machine using the following program (*see* **Note 8**):
   - 25 °C 5 min (primer annealing).
   - 42 °C 60 min (reverse transcription).
   - 70 °C 5 min (termination of the reaction by inactivating the reverse transcriptase).
3. Prepare qPCR reactions by combining the following (*see* **Note 9**):
   - cDNA (*see* **Note 18**).
   - 0.8 μL forward and reverse primers (200 nM final concentration).
   - 10 μL 2× SensiFAST™ SYBR® Hi-ROX Kit.
   - Nuclease-free PCR-grade water to 20 μL final volume.
4. Perform reactions in duplicates or triplicates (*see* **Note 19**).
5. Perform a no-template control reaction with each primer pair by substituting cDNA with the same volume of nuclease-free PCR-grade water (*see* **Note 20**).
6. Perform the qPCR experiment according to the following program (*see* **Note 9**):
   - 95 °C 2 min (initial denaturation)
     - 40 cycles.
   - 95 °C 5 s (denaturation).
   - 60 °C 30 s (annealing, extension, and reading fluorescence).
7. Acquire Melt curve (60–95 °C) (*see* **Note 21** and Fig. 2). *See* **Note 22** for additional notes on setting the threshold and amplification plots.

3.5 Data Analysis

When using the $2^{-\Delta\Delta Ct}$ method, the Ct values of the target are normalized to the Ct values of a reference gene (*see* **Note 23**) by subtracting the Ct value of the reference gene from the Ct value of the target (ΔCt). Next, the ΔCt value of the control sample (to which all other samples are compared) is subtracted from the ΔCt value of the target sample (ΔΔCt). From this value, fold change is calculated by the 2 to the power of negative ΔΔCt ($2^{-\Delta\Delta Ct}$) formula [16, 17]. Table 2 summarizes the steps in data acquisition and analysis with example numbers. For a representative measurement of TE expression in *nos*-GAL4 > HP1a RNAi ovaries, *see* Figs. 2 and 3.

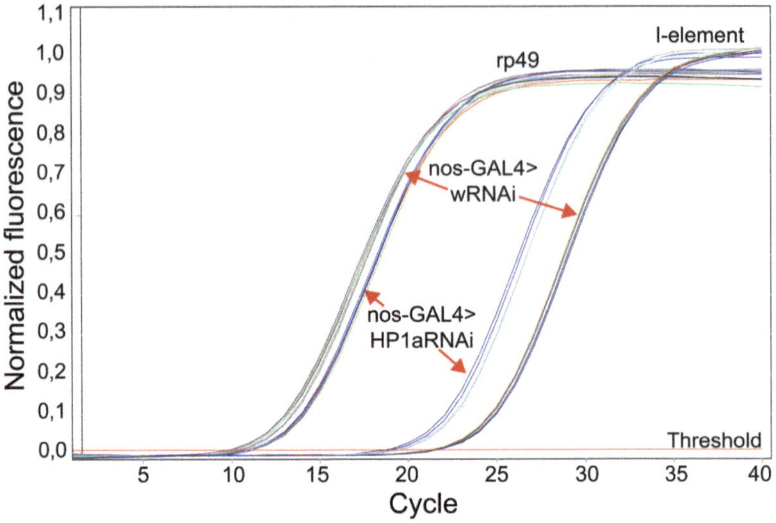

Fig. 2 Representative image of RT-qPCR measurement curves (linear scale). Curves have been obtained by analyzing the RT-qPCR data with the Rotor-Gene Q Series Software (Qiagen). Each colored line is a different measurement from *nos*-GAL4 > HP1aRNAi or *nos*-GAL4 > wRNAi samples, as indicated by the red arrows. The curves represent data obtained with primers specific to the *rp49* internal control gene and the I-element TE

4 Notes

A general guideline for RNA isolation is to only use tools and solutions which are RNase-free. Wearing gloves and careful handling of samples are a must for a successful experiment.

1. Glass tools are preferred over plastic ware because unfixed tissues tend to adhere to plastic surfaces.
2. Schneider's *Drosophila* medium can be substituted with 1× phosphate-buffered saline, Grace's medium, or Ringer solution. However, we found that the use of Schneider's medium [23] is the least stressful for dissected tissues, as it maintains ovarian cells in their natural environment.
3. Nuclease-free water can be obtained from several distributors. Alternatively, RNase-free water can be prepared by treating ultrapure water with diethylpyrocarbonate (DEPC) (1000-fold dilution). After treatment, DEPC must be inactivated by autoclave sterilization.
4. Alternatives such as TRI Reagent (Sigma-Aldrich) can be used for RNA isolation. In addition, several commercial kits are available for this purpose. It is important to note that kits typically have a set capacity for the maximum amount of RNA

Table 2
Calculating fold change from Ct values. In this example, we show the results of Ct calculations of two replicate experiments from three control samples and three samples of interest (SOI)

	Housekeeping gene measurements			Target TE measurements			Calculating fold change					
	Ct of R1	Ct of R2	**Av. Ct value**	Ct of R1	Ct of R2	**Av. Ct value**	ΔCt	Control samples av. ΔCt	ΔΔCt	FC $2^{\wedge-(\Delta\Delta Ct)}$	**Av. FC value**	SD
Control 1	10.22	10.13	**10.18**	28.42	28.35	**28.39**	18.21	18.33	−0.13	1.09	**1**	0.081
Control 2	9.55	9.39	**9.47**	27.85	27.95	**27.9**	18.43		0.1	0.94		
Control 3	9.88	9.85	**9.87**	28.14	28.33	**28.24**	18.37		0.03	0.98		
SOI 1	10.38	10.94	**10.66**	25.95	26.21	**26.08**	15.42		−2.91	7.53	**5.45**	1.804
SOI 2	10.48	10.84	**10.66**	26.85	26.91	**26.88**	16.22		−2.11	4.33		
SOI 2	10.61	10.57	**10.59**	26.74	26.77	**26.76**	16.17		−2.17	4.5		

Av. average, *R1* first replicate, *R2* second replicate, *FC* fold change, *SD* standard deviation of the FC values of biological replicates

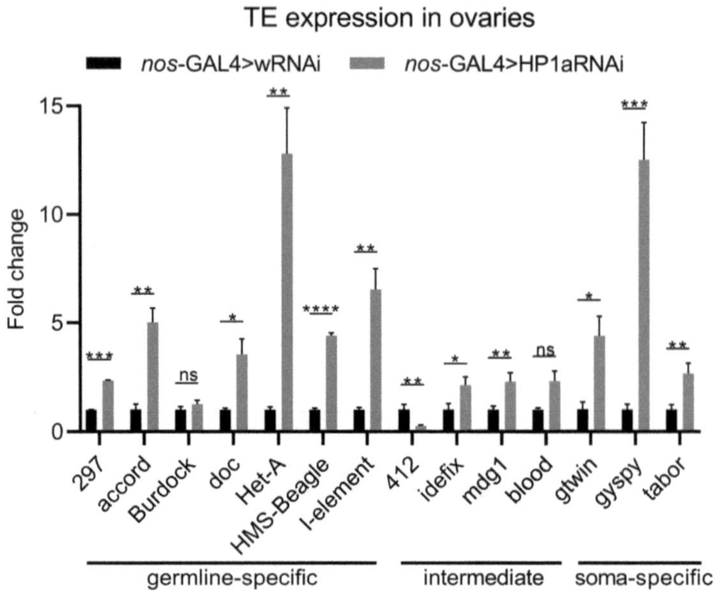

Fig. 3 TE expression compared in ovaries from *nos*-GAL4 > HP1aRNAi and *nos*-GAL4 > wRNAi animals. Data reveal that the silencing of the general factor responsible for heterochromatinization, HP1a, differently regulates the activity of the various TEs, as shown earlier also in [24]. In *nos*-GAL4 > HP1a RNAi animals, *HP1a* knockdown was induced by the GAL4 activator expressed under the regulation of the *nanos* (*nos*) promoter. As a control, we used *white* gene silencing (wRNAi) in the same fashion. Three independent samples have been prepared from both genotypes, according to the methods described in this paper. TE expressions were measured in duplicates on a Rotor-Gene Q qPCR machine (Qiagen). For data analysis and representation, Rotor-Gene Q Series Software and GraphPad Prism 9 [27] were used. For statistics, two-tailed, unpaired t-test was used. Error bars represent SD. *P* values: *<0.05; **<0.01; ***<0.001; ****<0.0001. *Drosophila* lines used in the experiments were obtained from the Bloomington Drosophila Stock Center, numbers 4937 (*nos*-GAL4), 36792 (HP1aRNAi), and 33623 (wRNAi)

isolated which should be considered for the experimental setup. The protocol for RNA isolation will be reagent specific.

5. For maximum efficiency, use fresh isopropanol (not older than 2 years from the date it was manufactured) for RNA precipitation, as old isopropanol can lower RNA yield.
6. Ethanol should be diluted with nuclease-free water.
7. When working with low sample volume, consider adding glycogen to your samples to aid precipitation and visualization of the RNA pellet.
8. Any cDNA synthesis kit can be used, but the manufacturer's instructions should be followed. The reaction protocol and the RT-PCR program will be specific to the kit.

9. Alternative qPCR reagent mixtures can be used as well, including reagents for One-Step qPCR applications. The qPCR reaction protocol and the qPCR program will be reagent specific.
10. Any other qPCR machine compatible with your qPCR reagent of choice can be used.
11. To preserve the integrity of the samples and thereby avoid RNA degradation, we recommend storing dissected ovaries on ice for no longer than 20 min before RNA extraction.
12. The volume of the TRIzol reagent can be scaled depending on the quantity of tissue used for RNA isolation, as recommended by the manufacturer. The added water and chloroform scale with the volume of TRIzol.
13. Inverted layers might appear if too much tissue sample is used or excess salt is present. In this case, add more TRIzol or nuclease-free water to the sample (~1/5 of total volume), vortex vigorously, and spin down again.
14. Using the method presented here, approximately 7.5–20 μg total RNA can be isolated, depending on the size and number of ovaries dissected. The exact volume of water, in which the RNA is dissolved, depends on the pellet size. Small pellets are dissolved in 25 μL, larger pellets can be dissolved in 50 μL water, yielding RNA concentration of 0.3–0.4 μg/μL. RNA purity can be assessed with a spectrophotometer. Pure RNA typically has OD 260/280 and OD 260/230 values very close to 2.0. Contaminating guanidinium thiocyanate from TRIzol is indicated by lower OD 260/230 ratio. In this case, another precipitation must be performed to fully remove contamination.
15. The addition of EDTA is important as RNA hydrolyses in the presence of divalent cations during heating.
16. As an alternative, DNase I can be removed by phenol/chloroform extraction.
17. For this purpose, a polymerase of choice and any primer pair that works on genomic DNA can be used. Amplification should not take place if no contaminating DNA is present. The PCR conditions greatly depend on the polymerase used. We recommend using the *rp49*-specific primer pairs (Table 1) at 200 nM final concentration, 60 °C annealing temperature, and Taq polymerase using the conditions described in the manufacturer's instructions. A positive control PCR reaction using genomic DNA should also be performed.
18. For the experiment shown in this paper, we have diluted our cDNA samples to 50 μL total volume with nuclease-free water and used 1.5 μL per qPCR reaction. The purpose of this step was to avoid pipetting very small amounts of samples.

19. It is important to replicate reactions at least two times to reduce variability in results and increase confidence level.
20. No template control (NTC) reactions are prepared to make sure that the reagents used in the experiment are free of contaminating DNA that could influence the results.
21. Melt curve analysis at the end of the run is used to assess whether the PCR reaction produced a single amplicon. When multiple peaks are plotted for a single PCR reaction, it typically means that the reaction produced more than one amplicon. In this case, running the reactions on an agarose gel is recommended to see if there are indeed multiple products, as amplicons with a high GC content can produce multiple peaks in the melt curve. This problem should not arise if using the primer pairs presented here. Follow the real-time instrument manufacturer's instructions for melting curve analysis. For an in-depth review of melt curve analysis, see [25, 26].
22. The threshold for detection is either set automatically or manually after the qPCR. The threshold is set above the baseline fluorescence in the exponential phase of all reactions. After the qPCR run has been completed, amplification plots are presented (*see* Fig. 3). If little or no amplification takes place, consider checking your calculations for the reaction mixture to rule out mistakes such as missing or limiting amounts of reagents. Low quantity or quality of the RNA samples can also result in no amplification. Amplification plots for technical replicates should be very similar with a difference of <0.5 between Ct values. Great variation between replicates suggests pipetting errors.
23. The reference gene used here is *rp49*, but other housekeeping genes can be used as well if they are not affected by your experimental setup.

Acknowledgments

This work was supported by the Hungarian National Research, Development and Innovation Office—NKFIH grants NKFIH-871-3/2020 (Péter Vilmos) and PD127968 (Ildikó Kristó).

References

1. McClintock B (1950) The origin and behavior of mutable loci in maize. Proc Natl Acad Sci U S A 36:344–355
2. Engels WR, Preston CR (1979) Hybrid dysgenesis in *Drosophila melanogaster*: the biology of female and male sterility. Genetics 92:161–174
3. Spradling AC, Rubin GM (1981) *Drosophila* genome organization: conserved and dynamic aspects. Annu Rev Genet 15:219–264

4. Hedges DJ, Deininger PL (2007) Inviting instability: transposable elements, double-strand breaks, and the maintenance of genome integrity. Mutat Res - Fundam Mol Mech Mutagen 616:46–59
5. Colnaghi R, Carpenter G, Volker M et al (2011) The consequences of structural genomic alterations in humans: genomic disorders, genomic instability and cancer. Semin Cell Dev Biol. Elsevier Ltd 22:875–885
6. Klein SJ, O'Neill RJ (2018) Transposable elements: genome innovation, chromosome diversity, and centromere conflict. Chromosom Res 26:5–23
7. Fejes Tóth K, Pezic D, Stuwe E et al (2016) The piRNA pathway guards the germline genome against transposable elements HHS public access. Adv Exp Med Biol 886:51–77
8. Teo RYW, Anand A, Sridhar V et al (2018) Heterochromatin protein 1a functions for piRNA biogenesis predominantly from pericentric and telomeric regions in *Drosophila*. Nat Commun 9:1–12
9. Lu X, Wontakal SN, Kavi H et al (2013) Drosophila H1 regulates the genetic activity of heterochromatin by recruitment of Su(var) 3-9. Science 80-) 340:78–81
10. Sienski G, Batki J, Senti KA et al (2015) Silencio/CG9754 connects the piwi-piRNA complex to the cellular heterochromatin machinery. Genes Dev 29:2258–2271
11. Iwasaki YW, Murano K, Ishizu H et al (2016) Piwi modulates chromatin accessibility by regulating multiple factors including histone H1 to repress transposons. Mol Cell 63:408–419
12. Czech B, Preall JB, McGinn J et al (2013) A transcriptome-wide RNAi screen in the *Drosophila* ovary reveals factors of the germline piRNA pathway. Mol Cell 50:749–761
13. Malone CD, Brennecke J, Dus M et al (2009) Specialized piRNA pathways act in germline and somatic tissues of the Drosophila ovary. Cell 137:522
14. Wong ML, Medrano JF (2005) One-step versus two-step real- time PCR. BioTechniques 39:75–85
15. Kim DW (2001) Real time quantitative PCR. Exp Mol Med 33:101–109
16. Livak KJ, Schmittgen TD (2001) Analysis of relative gene expression data using real-time quantitative PCR and the 2-ΔΔCT method. Methods 25:402–408
17. Schmittgen TD, Livak KJ (2008) Analyzing real-time PCR data by the comparative CT method. Nat Protoc 3:1101–1108
18. Klein JD, Qu C, Yang X et al (2016) c-Fos repression by Piwi regulates *Drosophila* ovarian germline formation and tissue morphogenesis. PLoS Genet 12:e1006281
19. Hur JK, Luo Y, Moon S et al (2016) Splicing-independent loading of TREX on nascent RNA is required for efficient expression of dual-strand piRNA clusters in *Drosophila*. Genes Dev 30:840–855
20. Qi H, Watanabe T, Ku HY et al (2011) The Yb body, a major site for Piwi-associated RNA biogenesis and a gateway for Piwi expression and transport to the nucleus in somatic cells. J Biol Chem 286:3789–3797
21. Sienski G, Dönertas D, Brennecke J (2012) Transcriptional silencing of transposons by Piwi and maelstrom and its impact on chromatin state and gene expression. Cell 151:964–980
22. Wang SH, Elgin SCR (2011) *Drosophila* Piwi functions downstream of piRNA production mediating a chromatin-based transposon silencing mechanism in female germ line. Proc Natl Acad Sci U S A 108:21164–21169
23. Akkouche A, Mugat B, Barckmann B et al (2017) Piwi is required during *Drosophila* embryogenesis to license dual-Strand piRNA clusters for transposon repression in adult ovaries. Mol Cell 66:411–419.e4
24. Avilés-Pagán EE, Orr-Weaver TL (2018) Activating embryonic development in *Drosophila*. Semin Cell Dev Biol 84:100
25. Downey N qPCR melt curve analysis: explaining multiple peaks | IDT, https://eu.idtdna.com/pages/education/decoded/article/interpreting-melt-curves-an-indicator-not-a-diagnosis
26. Lo YMD, Chiu RWK, Chan KCA et al (2006) Real-time PCR and melting curve analysis real-time polymerase chain reaction and melting curve analysis. Methods Mol Biol 336:19–32
27. GraphPad Prism 9.1.2 for Windows, https://www.graphpad.com/

Chapter 17

Preparation of *Drosophila* Ovarioles for Single-Cell RNA Sequencing

Nathaniel Meyer, Jobelle Peralta, and Todd Nystul

Abstract

The production of eggs in the *Drosophila* ovary requires complex interactions between multiple cell types that coexist within the same solid tissue. This cellular heterogeneity makes the ovary a rich subject of study, but also makes it challenging to identify transcriptional differences between individual cell types using methods such as bulk RNA sequencing. The development of single-cell RNA sequencing (scRNA-seq) techniques addresses this limitation by providing an avenue to profile genetic and functional heterogeneity at a cellular resolution. Here, we describe the isolation and preparation of the *Drosophila* ovary for scRNA-seq. This protocol emphasizes a short preparation time, high cell viability, prevention of RNA-degradation, and reduction of technical variation to achieve highly reproducible single-cell profiles.

Key words Single-cell sequencing, Drosophila, Ovary

1 Introduction

Over the past decade, there has been a revolution in the technologies that enable single-cell RNA sequencing (scRNA-seq) of whole tissues and organs [1–5]. These exciting developments make it possible to obtain transcriptional profiles of multiple different cell types in a heterogeneous tissue or organ, including rare cell types, such as stem and progenitor cells. At the same time, new bioinformatic methods are being developed to analyze the treasure trove of data produced by scRNA-seq experiments. Standard clustering algorithms help to identify transcriptionally distinct subsets of cells that correspond to distinct cell types [6, 7], and other tools build on the output from these algorithms to generate additional predictions, such as the lineage relationships, regulon activity, or cell signaling relationships in specific cell types of interest within the

Nathaniel Meyer and Jobelle Peralta contributed equally with all other contributors.

Michelle S. Giedt and Tina L. Tootle (eds.), *Drosophila Oogenesis: Methods and Protocols*,
Methods in Molecular Biology, vol. 2626, https://doi.org/10.1007/978-1-0716-2970-3_17,

tissue [8–13]. Taken together, these new technologies are providing an unprecedented level of detail about the cellular diversity that exists in complex tissues, as well as the ways in which these cellular populations change over time and in different conditions.

The key innovation that has made scRNA-seq possible is the development of methods to generate a unique cDNA library from each cell in the tissue. The cDNA libraries are generated using primers with a different "barcoding" sequence (or set of sequences) for each cell. This allows the cDNA libraries to be pooled together for sequencing and then sorted back out afterward by grouping sequences with common barcodes. To generate these cDNA libraries, the tissue is dissociated into a single-cell suspension, and then, each cell is separated into a well or oil droplet, where the reactions occur to generate the unique libraries [1–3]. Alternatively, rather than creating cDNA libraries for each cell separately, the cells can be repeatedly pooled in different combinations, so that each cell has a high probability of acquiring a unique set of barcodes [4]. Yet, another alternative is to perform RNA-seq on nuclei isolated from fresh or frozen tissue rather than on whole cells [14]. There are advantages and drawbacks to each approach, so the choice of which one to use depends on several factors, such as the number of cells and depth of sequencing that is desired, the size of the cells, and cost. This protocol was developed to prepare cells from freshly isolated *Drosophila* ovaries to be submitted for single-cell sequencing using the 10× Genomics droplet-based method, which is the method that was used in several recently published single-cell atlases of the *Drosophila* ovary [15–18]. However, it is likely that this protocol could be adapted for use with other strategies for scRNA-seq of whole cells with very few modifications (*see* **Note 1**).

2 Materials

Prepare all solutions using molecular grade water (free of RNases, DNases, and proteases). Eliminate RNase contamination from the working area using RNase decontamination sprays and wipes. Using freshly prepared reagents is ideal, however, aliquoted sterile fetal bovine serum (FBS), and solubilized insulin may be prepared beforehand and stored at −20 °C. If using frozen reagents, allow at least 1 h for reagents to thaw on ice prior to use. Preparation of all required buffers and solutions with thawed reagents will take approximately 1 h. All reagents should be kept on ice or at 4 °C until needed. Throughout this protocol, only use low binding microcentrifuge tubes and sterile, low binding filter tips.

2.1 Drosophila Culture

1. CO_2 fly pad and anesthesia apparatus.
2. Fly incubators.
3. Dissecting microscope.
4. Fly vials or bottles containing fly food.
5. Wet yeast paste: add active dry yeast and de-ionized water into a desired container. Mix well until a thick paste consistency is reached. Store at 4 °C.

2.2 Preparation of Buffers

1. Insulin in Acidified Water (1 mg/mL), for 15 mL: Add approximately 10 mL of molecular grade water to a 15 mL conical tube. Add 15 mg of lyophilized insulin powder from bovine pancreas. Mix and adjust pH with HCl to pH ~2.5. Allow solution to fully solubilize by gently turning the conical tube for 2–3 min at room temperature. Solubilize longer if needed. Add molecular grade, nuclease free water to a final volume of 15 mL. Aliquot and store at −20 °C (*see* **Note 2**).
2. S2 Buffer (10% FBS, 0.2 mg/mL insulin in Schneider's medium), for 7.2 mL: 5.28 mL Schneider's S2 *Drosophila* medium, 720 μL sterile, heat-inactivated fetal bovine serum (FBS) and 1.2 mL insulin (1 mg/mL in acidified water). Filter sterilize using a 0.22 μM filter syringe. Keep on ice (*see* **Note 3**).
3. Digestion Buffer (4 mg/mL Elastase, 2.5 mg/mL Collagenase (Type 1) in Cell Dissociation Buffer), for 1 mL: 4 mg of Elastase, 2.5 mg Collagenase (Type 1), and 1 mL Cell Dissociation Buffer (*see* **Note 4**). Allow to fully solubilize with gentle mixing. Keep on ice or at 4 °C until use (*see* **Note 5**).
4. Resuspension Buffer (0.04% Bovine serum albumin [BSA] in PBS), for 1 mL: 1 mL 1× PBS and 8 μL Ultra-pure BSA (50 mg/mL). Keep on ice or at 4 °C until use (*see* **Note 6**).

2.3 Ovary Dissection

1. Dissection kit with forceps (*see* **Note 7**).
2. Dissecting microscope.
3. P20, P200, P1000 pipettors.
4. Pipetman and serological pipettes.
5. Clear, 9-well dissection dish.
6. Flat ice pack (*see* **Note 8**).
7. Sterile, low binding, filter pipette tips.
8. Syringe filter, 0.22 μM, PVDF.
9. 1.5 mL low binding microcentrifuge tubes.
10. RNase purifying agents.
11. Insulin in Acidified Water (1 mg/mL).
12. S2 Buffer.

2.4 Tissue Digestion and Single-Cell Isolation

1. Pipetman and serological pipettes.
2. Benchtop, refrigerated centrifuge.
3. Scalpels (*see* **Note 9**).
4. Sterile, low binding, filter pipette tips.
5. 1.5 mL low binding microcentrifuge tubes.
6. Cell Dissociation Buffer, enzyme-free (*see* **Note 4**).
7. 1× Dulbecco's Phosphate Buffered Saline.
8. 30 μM filter.
9. Cell Filter, Cup Fitting 50 μM filter.
10. Digestion Buffer.
11. Resuspension Buffer.

2.5 Cell Counting

1. Hemocytometer.
2. 0.4% trypan blue solution.

3 Methods

3.1 Preparation of Flies

1. Female flies should be maintained on wet yeast in a bottle or vial for at least 5 days prior to dissections with sufficient males. We recommend a ratio of 2 females to 1 male. Feed flies with fresh wet yeast 3–4 h prior to dissection (*see* **Notes 10** and **11**).
2. To ensure rapid and efficient dissection (*see* **Note 12**), sort flies of the desired phenotype the night before dissections. Do not separate females from males prior to dissection.

3.2 Ovary Dissection and Enrichment of the Early (Non-Vitellogenic) Stages

1. Before beginning tissue digestion, precool a centrifuge to 4 °C to keep cells chilled during the spin step (*see* **Note 13**) and wipe down all work areas with RNase purifying agents.
2. Place a 9-well dissecting dish on a flat ice pack and add ~750 μL of cold S2 Buffer to each well. Keeping the samples cool facilitates cell survival during this process.
3. Isolate 60 pairs of ovaries using standard procedures [19] in the 9-well dissecting dish on a flat ice pack under a dissecting microscope (*see* **Note 14**). Do not separate ovarioles. Complete all dissections, including microdissections within 60 min (*see* **Note 15**).
4. Transfer isolated ovary pairs using forceps into a clean well in the 9-well dissecting dish filled with S2 Buffer (*see* **Note 16**).
5. Further dissect each ovary under a dissecting microscope to enrich for younger, non-vitellogenic stages. These younger stages can be identified by their transparent appearance. Using a scalpel, cut along the border of transparent

non-vitellogenic stages and the opaque vitellogenic stages to enrich for earlier stages of oogenesis (Fig. 1) (*see* **Note 17**). Transfer isolated tissues into a clean well filled with S2 Buffer.

6. Using a P200 filter tip pre-wet with S2 buffer, transfer the tissues into a 1.5 mL low binding microcentrifuge tube filled with 200 μL of S2 Buffer and keep on ice (*see* **Note 18**).
7. Once all 60 ovaries are dissected and enriched for non-vitellogenic stages, proceed immediately to Tissue Digestion.

3.3 Tissue Digestion and Single-Cell Isolation

1. Wash isolated ovaries with 750 μL of cold, sterile, Cell Dissociation Buffer. Briefly centrifuge tubes (~5 s) in a room temperature benchtop microcentrifuge to pool the tissue.
2. Using a P1000 pipette, carefully remove Cell Dissociation Buffer without disturbing the tissue pellet.
3. Add 1 mL of Digestion Buffer to each sample.
4. Using a P200 set to 175 μL, continuously pipette the sample up and down for 20 min without introducing air bubbles to facilitate dissociation (*see* **Notes 19** and **20**).
5. Pre-wet a P200 filter tip with Digestion Buffer then filter samples through a 50 μM filter into a new low binding microcentrifuge tube. Continue to digest the samples by continuously pipetting up and down for another 10 min without introducing air bubbles.
6. Pre-wet a P200 filter tip with Digestion Buffer then filter samples through a 30 μM filter into a new low binding microcentrifuge tube.
7. Quench the digestion by adding 500 μL of S2 Buffer and pellet cells via centrifugation (5 min, 4 °C, 3500 g). Using a P1000 pipette, remove the supernatant carefully without disturbing the cell pellet. Gently resuspend and wash cells once with 1 mL of S2 Buffer and pellet cells via centrifugation (5 min, 4 °C, 3500 g). Remove the supernatant carefully without disturbing the cell pellet.
8. Fully resuspend cells in 200 μL of Resuspension Buffer. Mix well using a P200 to prevent clustering of cells.

3.4 Cell Counting

1. Use a hemocytometer to count dissociated cells (Fig. 2) (*see* **Notes 21** and **22**).
   (a) Add 10 μL of Trypan Blue to 10 μL of cells from Subheading 3.3, **step 8**. This gives a dilution factor of 2. Trypan Blue is added to calculate cell viability.
   (b) Count 4 squares on the hemocytometer. Calculate the cell concentration:

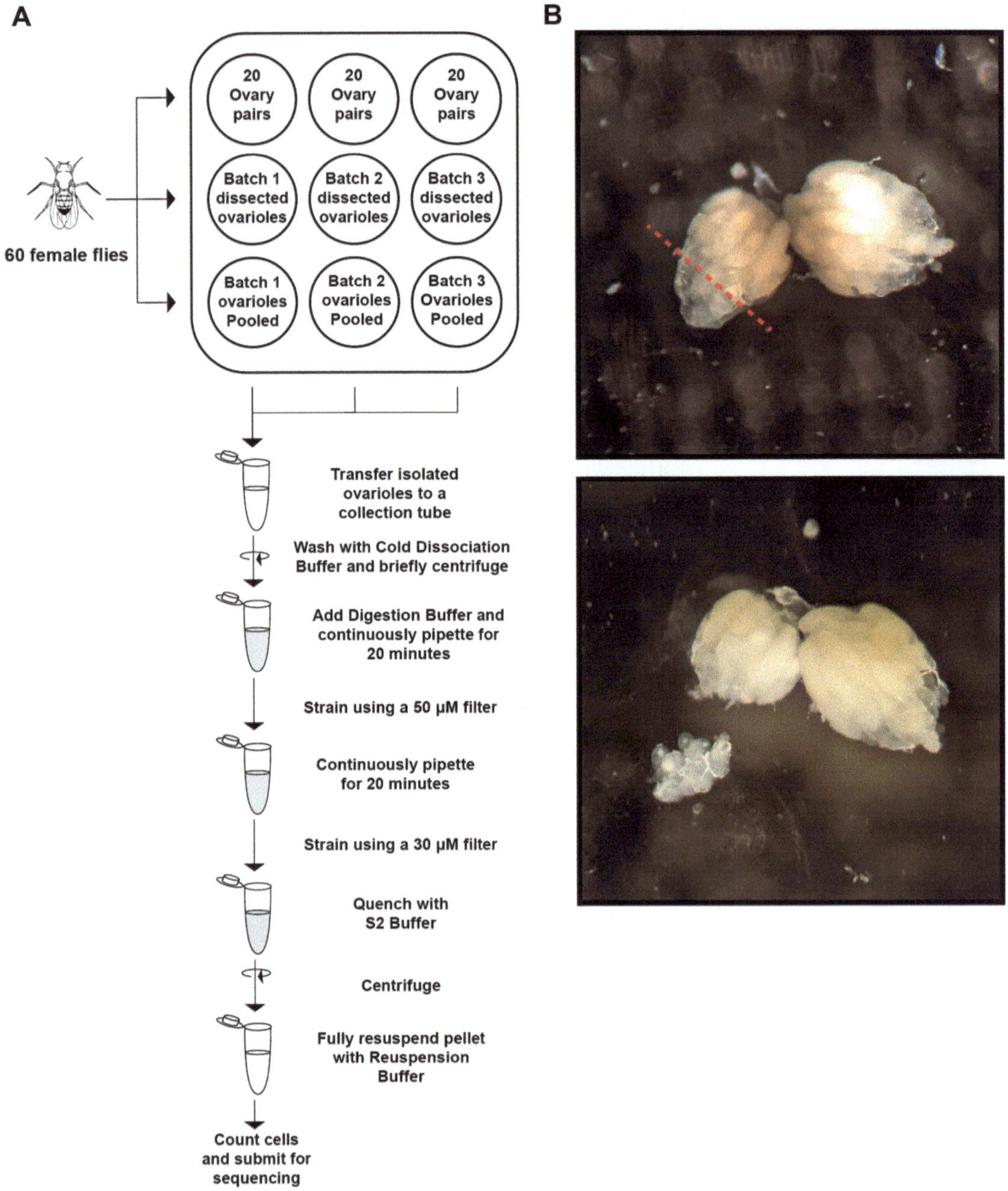

Fig. 1 Ovary dissection and enrichment for the early (non-vitellogenic) stages. (**a**) Schematic overview of the preparation of ovarioles for scRNA-seq. In brief, ovaries are isolated from female flies, microdissected to enrich for cell types of interest and are enzymatically dissociated to a single cell suspension ready for library preparation. (**b**) Microdissection strategy for the isolation of non-vitellogenic stages of *Drosophila* ovaries. Under a dissecting microscope, the non-vitellogenic stages of the ovarioles within the ovary can be identified by their transparent morphology. Cutting along the border of the transparent non-vitellogenic stages and opaque vitellogenic stages enriches for the region of the earlier stages of oogenesis. After cutting the tissue, the non-vitellogenic stages should remain attached as long as the ovarioles were not separated. This makes it easier to put the tissue into a tube

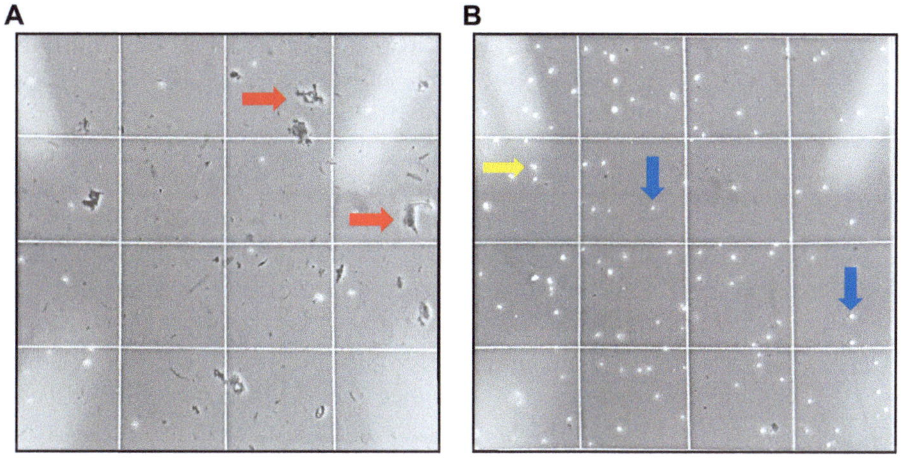

Fig. 2 Counting cells in a single cell suspension. (**a**, **b**) Representative single cell isolates from *Drosophila* ovarian tissue, enriched for the early stages of oogenesis. It is important to assess the quality of the single cell isolate prior to submission for sequencing. In a poor isolation (**a**), debris was present (red arrows) and there were not enough cells to meet the requirements for submission. In a high-quality isolation (**b**), less debris was observed, the overall number of cells that are fully dissociated (blue arrows) was high, and the number of incompletely dissociated cells (i.e. not in doublets or higher order clusters, yellow arrows) or dead cells was low

(i) $$\frac{\text{Unlabeled Cells Counted}}{\text{Squares Counted (4)}} \times \text{Dilution Factor (2)} \times 10{,}000 = \text{Live Cells per mL}$$

(ii) $$\frac{\text{Labeled Cells Counted}}{\text{Squares Counted (4)}} \times \text{Dilution Factor (2)} \times 10{,}000 = \text{Dead Cells per mL}$$

(c) Calculate the total cell number:

(i) $$\frac{\text{Live Cells per mL}}{1000\ \mu\text{L/mL}} \times \text{Volume Remaining (e.g.190}\ \mu\text{L)} = \text{Total Live Cells}$$

2. Dilute cells with Resuspension Buffer to a final concentration required by the single-cell genomics platform you are using (*see* **Note 23**).
3. The sample is now ready to be submitted for single-cell sequencing using a method, such as the 10× Genomics platform.

4 Notes

1. The germ cells and many of the follicle cell types in late-stage follicles are too big to be captured by the oil droplets in the 10× Genomics method, so these stages are removed in this protocol (see Subheading 3.2, **step 5**). However, other methods, such as those in which cells are sorted into individual wells of a multi-well plate, may be able to capture the larger cell types in the ovary. When using these methods, it may be desirable to skip **step 5** of Subheading 3.2 and retain the later stages.

2. Insulin has low solubility at neutral pH. Solubilize using hydrochloric acid in water at pH 2–3. Stock solutions can be prepared, aliquoted, and stored at −20 °C in single-use aliquots. Avoid freeze–thaw cycles.
3. The S2 Buffer is used both during the dissection as well as collection of each tissue sample. We found that approximately 7 mL of S2 Buffer per sample is sufficient for this protocol, including thoroughly washing and cleaning each tissue sample with minimal waste.
4. To minimize cell death due to exposure to enzymes, we recommend enzyme-free and chelating agent-free cell dissociation buffers, such as Gibco Cell Dissociation Buffer (Gibco, Cat #13151014).
5. Approximately 1 mL of Digestion Buffer per sample is required. It is important that the elastase and collagenase are fresh. Using expired enzymes in this step will reduce yield of single cells as well as decrease viability.
6. A minimum of 200 μL of Resuspension Buffer per sample is required. The addition of BSA helps prevent cells from clumping together or sticking to any tips or tubes. Make no less than 1 mL of Resuspension Buffer at once.
7. Dissecting kits in our lab include several tools, depending on the individual user preferences. These include small paint brushes, fine tip forceps (such as Dumont #5 Fine Forceps), standard forceps (Straight 13 cm), fine needle syringes, and fine tip tungsten needles. This protocol specifically mentions the usage of forceps but can be adapted for use with other dissection tools.
8. Flat ice packs/bricks from shipments were recycled for this purpose and used here to keep the samples cold during dissection.
9. We do not recommend using scissors for this protocol. We found scissors to be difficult to position carefully, leading to poor enrichment of the earlier stages in the final isolate. In comparison, a scalpel can be positioned much more precisely. We recommend scalpels with curved blades for use within the curved 9-well dissecting trays, such as scalpels with blade #10.
10. Maintaining flies with wet yeast causes the ovary to be healthy and plump up. Maintaining females with males in the same vial ensures a high level of fertility and ovary health [20].
11. Feeding with wet yeast for at least 5 days prior to dissection ensures the tissue is healthy. In addition, feeding the flies with fresh wet yeast 3–4 h prior to dissection helps enlarge the ovary, which makes the ovary isolation and microdissection steps easier.

12. In some cases, it may be time consuming to identify the flies of the correct genotype. In these cases, it may be beneficial to isolate a large number of flies with the required genotype before starting the dissection process. Maintain these flies with males and supplement with wet yeast (*see* **Note 10**).
13. The cells are kept at room temperature during the digestion step, because the enzymes are inefficient at cold temperatures, but the cells should be chilled whenever possible at other steps to minimize cell death.
14. The number of ovary pairs dissected can be changed. In cases where flies with the genotype of interest was relatively rare, we have used as little as 15 ovary pairs and found that we were still able to recover the target number of cells (20,000) for loading on the 10× Genomics platform. However, we recommend aiming for approximately 60 pairs of ovaries, if possible, to ensure that enough cells remain at the end of the protocol.
15. We recommend completing the tissue dissection and beginning digestion within 60 min. We found no noticeable difference in the amount of cell death during shorter dissection times. However, exceeding this time yielded noticeably larger amounts of dead cells in the final cell count. If cells from multiple different populations of flies (e.g. different genotypes or experimental conditions) will be submitted at the same time, we recommend organizing a "dissection party" in which each population of flies is assigned to a different person and the ovary dissections occur in parallel. This will help ensure that the process is completed within this timeframe and at the same time for all samples.
16. Moving the tissue through serial washes of S2 Buffer during the dissection helps clean the tissue. We find it easier to complete the dissection in 2–3 sets of 20–30 ovary pairs (Fig. 1). Preparing 2–3 sets of dissection wells with S2 Buffer also prevented the dissection wells from accumulating debris from the dissections.
17. The microdissection of the ovary to isolate the early stages can be difficult to perform in a quick, reproducible manner. The goal is to be able to isolate 60 pairs of ovaries, microdissect away the later stages, and pool the tissue together in a collection tube within 60 min. We recommend practicing these dissections prior to performing the full experiment.
18. Throughout this protocol, we recommend pre-wetting all pipette tips with the solution to be pipetted. The isolated tissue can easily stick to the sides of the pipette tip. Pre-wetting the pipette tips minimizes tissue loss.
19. Set a P200 to ~175 μL to prevent material contacting the filter while pipetting the tissue. During this process, the solution will become turbid, this is normal and expected.

20. If digesting more than one sample, place tubes on a nutator when not pipetting. We recommend that one person work on no more than two samples, pipetting each sample for 1–2 min before switching samples. If a single person processes more than two samples, the tissue may not be fully digested within the recommend time.
21. When preparing single-cell suspensions from 60 ovary pairs, we commonly achieved a yield of approximately 2–3 × 10^6 live cells and a viability ≥85%.
22. Use a hemocytometer and count manually rather than using an automated cell counter. In our experience, automated cell counters are not calibrated for the size of *Drosophila* ovarian cells, and thus give inaccurate cell counts. We also found that assessing the single-cell suspension manually allowed us to determine if the single-cell prep was performed well (see Fig. 2).
23. We recommend preparing excess sample for submission and keeping the single-cell isolation on ice until the sample is submitted. For example, our core facility requests that we provide 20 μL of cell suspension at a final concentration of 1000 cells per μL (totaling 20,000 cells). We try to submit 30,000 cells in 30 μL of Resuspension Buffer, so that there are extra cells in case any issues arise during processing.

Acknowledgements

This protocol is based on a protocol that was originally developed in our lab by Dr. Katja Rust, with input from Dr. Maija Slaidina. We are grateful to them for their contributions.

References

1. Zheng GXY, Terry JM, Belgrader P et al (2017) Massively parallel digital transcriptional profiling of single cells. Nat Commun 8:14049
2. Macosko EZ, Basu A, Satija R et al (2015) Highly parallel genome-wide expression profiling of individual cells using Nanoliter droplets. Cell 161:1202–1214
3. Klein AM, Mazutis L, Akartuna I et al (2015) Droplet barcoding for single-cell transcriptomics applied to embryonic stem cells. Cell 161:1187–1201
4. Rosenberg AB, Roco CM, Muscat RA et al (2018) Single-cell profiling of the developing mouse brain and spinal cord with split-pool barcoding. Science 360:176–182
5. Hwang B, Lee JH, Bang D (2018) Single-cell RNA sequencing technologies and bioinformatics pipelines. Exp Mol Med 50:1–14
6. Satija R, Farrell JA, Gennert D et al (2015) Spatial reconstruction of single-cell gene expression data. Nat Biotechnol 33:495–502
7. Kiselev VY, Andrews TS, Hemberg M (2019) Challenges in unsupervised clustering of single-cell RNA-seq data. Nat Rev Genet 20: 273–282
8. Qiu X, Mao Q, Tang Y et al (2017) Reversed graph embedding resolves complex single-cell trajectories. Nat Methods 14:979–982
9. Aibar S, González-Blas CB, Moerman T et al (2017) SCENIC: single-cell regulatory

network inference and clustering. Nat Methods 14:1083–1086

10. La Manno G, Soldatov R, Hochgerner H et al (2017) RNA velocity in single cells. bioRxiv:206052
11. Efremova M, Vento-Tormo M, Teichmann SA, Vento-Tormo R (2020) CellPhoneDB: inferring cell-cell communication from combined expression of multi-subunit ligand-receptor complexes. Nat Protoc 15:1484–1506
12. Browaeys R, Saelens W, Saeys Y (2020) NicheNet: modeling intercellular communication by linking ligands to target genes. Nat Methods 17:159–162
13. Jin S, Guerrero-Juarez CF, Zhang L et al (2021) Inference and analysis of cell-cell communication using CellChat. Nat Commun 12:1088
14. McLaughlin CN, Brbić M, Xie Q et al (2021) Single-cell transcriptomes of developing and adult olfactory receptor neurons in Drosophila. elife 10. https://doi.org/10.7554/eLife.63856
15. Li H, Janssens J, De Waegeneer M et al (2021) Fly Cell Atlas: a single-cell transcriptomic atlas of the adult fruit fly. bioRxiv:2021.07.04.451050
16. Rust K, Byrnes LE, Yu KS et al (2020) A single-cell atlas and lineage analysis of the adult Drosophila ovary. Nat Commun 11:5628
17. Jevitt A, Chatterjee D, Xie G et al (2020) A single-cell atlas of adult Drosophila ovary identifies transcriptional programs and somatic cell lineage regulating oogenesis. PLoS Biol 18:e3000538
18. Slaidina M, Gupta S, Banisch TU, Lehmann R (2021) A single-cell atlas reveals unanticipated cell type complexity in Drosophila ovaries. Genome Res 31:1938–1951
19. Wong LC, Schedl P (2006) Dissection of Drosophila ovaries. J Vis Exp 1:52
20. Yapici N, Kim Y-J, Ribeiro C, Dickson BJ (2008) A receptor that mediates the post-mating switch in Drosophila reproductive behaviour. Nature 451:33–37

Chapter 18

Chromatin Immunoprecipitation Experiments from *Drosophila* Ovaries

Maria Sokolova and Maria K. Vartiainen

Abstract

Chromatin is composed of DNA and its associated proteins, and has an essential role in all cellular processes, including those taking place during *Drosophila* oogenesis. In order to understand the molecular basis of chromatin-based processes, such as transcription, it is essential to be able to study how and when different proteins, such as transcription factors, histones and RNA polymerases, interact with chromatin. One of the most popular methods to study this is chromatin immunoprecipitation followed by next generation sequencing (ChIP-seq). Here, we describe a ChIP-seq protocol that has been optimized for *Drosophila* ovaries, focusing on sample preparation through preliminary data processing.

Key words Chromatin, Chromatin immunoprecipitation, Next-generation sequencing (NGS), ChIP-seq, *Drosophila* oogenesis, Ovaries

1 Introduction

Chromatin consists of the genetic material, DNA, and its associated proteins, such as histones and different factors that control chromatin compaction, maintain genomic integrity and transcribe the information contained within the genomes. In order to understand any chromatin-based process, it is essential to know, when and where specific proteins or their post-translational modifications bind the DNA. These types of studies can reveal, for example, how transcription factors and the RNA polymerase machinery control the expression of specific genes.

Chromatin immunoprecipitation followed by next-generation sequencing (NGS) (ChIP-seq) is one of the most widespread methods to study binding of different proteins to DNA [1]. In ChIP-seq, the chromatin is first crosslinked to preserve the interactions. Following chromatin fragmentation, DNA–protein complexes are immunoprecipitated with specific antibodies. Next, the captured DNA is purified, and a sequencing library prepared for next-

Michelle S. Giedt and Tina L. Tootle (eds.), *Drosophila Oogenesis: Methods and Protocols*,
Methods in Molecular Biology, vol. 2626, https://doi.org/10.1007/978-1-0716-2970-3_18,

generation sequencing (NGS). Finally, sequencing data are analyzed with different tools to acquire chromatin state annotation, transcription activation and processing. ChIP-seq can identify, for example, the binding sites for transcription factors, RNA polymerases, and chromatin remodeling and modifying complexes, as well as the position of histone modifications and histone variants on the genome. Finding a specific antibody that still recognizes its epitope in the crosslinked sample is one of the main limiting factors for what can be studied with this technique.

The occupancy of different proteins on the DNA is very cell type-specific [2]. Therefore, assays based on cultured cells, limited to only one cell type, are very useful to reveal basic principles of transcriptional regulation by decreasing the variation within the sample [3]. When using multicellular organisms or tissues as a starting material, the ChIP-seq signal might be diluted for some cell types and over-represented for others, and this must be considered when interpreting the results. Cell sorting-based technologies for ChIP sequencing have been developed for *Drosophila* embryos [4] but have not been used for ovaries most probably due to the complicated organization of this organ. Different ChIP-seq protocols have been applied for *Drosophila* embryonic cells [5] and embryos [6, 7] to study transcription factors and histone modifications [8, 9]. Adaptation of these protocols to ovaries requires some changes for sample preparation and data interpretation.

A *Drosophila* female has two ovaries, each of which is composed of 16–20 ovarioles. The ovariole is a string of egg chambers classified into 14 stages. Each egg chamber consists of 16 germ-line cells; one of them is the oocyte and the rest are nurse cells. The egg chamber is further covered by a layer of somatic follicle cells [10]. This big variation of cells' types and different developmental stages will result in unbalanced coverage signal in ChIP-seq that could be evaluated and clarified with some single-cell technologies [4, 11]. Another rather specific phenomenon relates to follicle cell development, which proceeds from mitotic cycle (stages 1–6), through endocycle (stages 7–10A), to amplification (stages 10B–13) [12]. The endocycle leads to under-replication of some genomic regions [13] and amplification results in multiple copies of chorion genes clusters [13, 14]. For this reason, input control (Fig. 1) is particularly important for any ChIP-seq experiments performed with *Drosophila* ovaries [15].

Here, we describe a ChIP-seq protocol for *Drosophila* ovaries, which we have optimized specifically for this sample type from a previously described method for cultured mammalian cells [16]. We concentrate on the sample preparation and immunoprecipitation part of the protocol. The subsequent steps, such as preparation of the sequencing library, can be done with any available DNA preparation kit for NGS, and several method papers and pipelines exist for the data analysis part [17–21]. The data

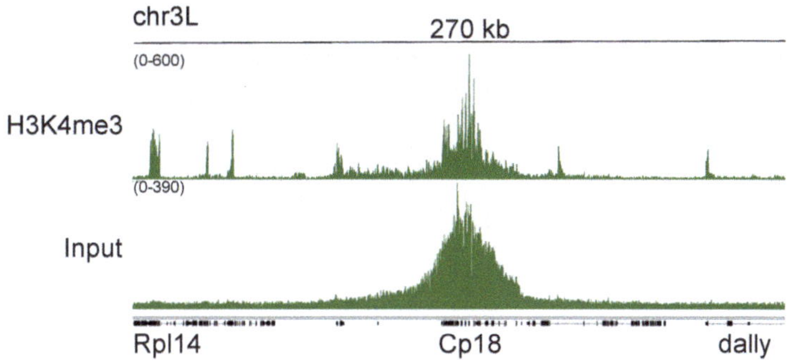

Fig. 1 Example of input and histone H3K4me3 coverage signal. ChIP-seq coverage of histone H3K4me3 and input is shown for 66D8–66E1 region of the fly genome, which is amplified during stages 10B–13 of follicle cell development [22]. The high peak in the input sample shows an amplified area of 66D8–66E1 region containing genes involved in eggshell formation. For proper evaluation of H3K4me3 profile, the signal must be normalized on the input

processing and visualization are shortly discussed at the end. The current protocol has been successfully applied for different histone modifications, RNA polymerase II carboxy-terminal domain (CTD) phosphorylation, and actin in *Drosophila* ovaries [22].

2 Materials

2.1 Sample Preparation

1. 1x Phosphate-buffered saline with 1% bovine serum albumin (PBS with 1% BSA): 10 mM Na_2HPO_4, 2 mM KH_2PO_4 pH 7.4, 137 mM NaCl, 2.7 mM KCl, 1% BSA.
2. Complete Protease Inhibitor tablets EDTA-free.
3. 36.5% formaldehyde.
4. 2.5 M Glycine prepared from powder in deionized water and kept at 4 °C.
5. 1.5 mL tubes.
6. Liquid nitrogen.
7. Fine forceps (Dumont #5).
8. Depression glass spot plates for dissection.
9. Adjustable-volume pipettes.
10. Pipet tips for adjustable-volume pipettes.
11. Rotator.
12. Ice.
13. –70–80 °C freezer.

2.2 Sonication Optimization

1. Seven fixed and frozen samples with 50 ovary pairs in each tube (see Subheading 3.1).
2. Sonication and IP buffer: 10 mM Tris–HCl pH 8.0, 140 mM NaCl, 1 mM EDTA, 1% Triton X100, 0.1% SDS, 0.1% Sodium deoxycholate, protease inhibitors.
3. Complete Protease Inhibitor tablets EDTA-free
4. RNase A.
5. Proteinase K.
6. 2 mL tubes with Phase Lock Gel.
7. 25:24:1 phenol/chloroform/isoamyl alcohol saturated with 10 mM Tris–HCl, pH 8.0.
8. 5 mg/mL glycogen.
9. 5 M NaCl.
10. 3 M NaAc, pH 5.2.
11. Absolute ethanol.
12. Ice cold 70%EtOH.
13. Tris–EDTA (TE) buffer: 10 mM Tris–HCl, pH 8.0, 1 mM EDTA.
14. Agarose gel electrophoresis equipment.
15. 1% agarose gel.
16. 100 bp DNA Ladder.
17. 1 kb DNA ladder.
18. SYBR Gold or any orange DNA dye.
19. Gel loading dye, 6× Orange.
20. Blue light transilluminator (wavelength 460–480 nm).
21. Adjustable-volume pipettes.
22. Pipet tips for adjustable-volume pipettes.
23. Rotator.
24. Diagenode Bioruptor.
25. Sonication tubes for Diagenode Bioruptor.
26. DNA LoBind tubes.
27. Ice.
28. Dry Block Heater for 1.5 mL tubes.
29. 1.5 mL tubes.
30. Vortex.
31. Centrifuge (*see* **Note 1**).
32. −20 °C freezer.
33. Cold room.

2.3 Chromatin Immunoprecipitation

1. Fixed and frozen samples with 50 ovary pairs in each tube.
2. Complete Protease Inhibitor tablets EDTA-free.
3. Primary antibodies (*see* **Note 2**).
4. Normal IgG (*see* **Note 3**).
5. Pellet pestles for tissues grinding.
6. Adjustable-volume pipettes.
7. Pipet tips for adjustable-volume pipettes.
8. Diagenode Bioruptor.
9. Sonication tubes for Diagenode Bioruptor.
10. Centrifuge (*see* **Note 1**).
11. −70–80 °C freezer.
12. −20 °C freezer.
13. 15 mL tubes.
14. Cold room.
15. Ice.
16. Dry Block Heater for 1.5 mL tubes.
17. 1.5 mL tubes.
18. DNA LoBind tubes.
19. Wheel rotator.
20. Protein A Sepharose.
21. RNase A.
22. Proteinase K.
23. 2 mL tubes with Phase Lock Gel.
24. 25:24:1 Phenol/chloroform/isoamyl alcohol saturated with 10 mM Tris–HCl, pH 8.0.
25. 5 mg/mL glycogen.
26. Sonication and IP buffer: 10 mM Tris–HCl pH 8.0, 500 mM NaCl, 1 mM EDTA, 1% Triton X100, 0.1% SDS, 0.1% sodium deoxycholate, protease inhibitors.
27. IP buffer with BSA: 10 mM Tris–HCl pH 8.0, 500 mM NaCl, 1 mM EDTA, 1% Triton X100, 0.1% SDS, 0.1% sodium deoxycholate, 1 mg/mL BSA.
28. Wash buffer 500: 10 mM Tris–HCl pH 8.0, 500 mM NaCl, 1 mM EDTA, 1% Triton X100, 0.1% SDS, 0.1% sodium deoxycholate
29. LiCl buffer: 10 mM Tris–HCl pH 8.0, 250 mM LiCl, 1 mM EDTA, 0.5% IGEPAL CA-630, 0.5% sodium deoxycholate.
30. TE buffer: 10 mM Tris–HCl, pH 8.0, 1 mM EDTA.
31. 5 M NaCl.

32. 3 M NaAc (pH 5.2).
33. Absolute ethanol.
34. Extraction buffer: 1% SDS in TE buffer, prepared fresh.
35. Ice cold 70% EtOH.
36. Agarose gel electrophoresis equipment.
37. 2% agarose gel.
38. 100 bp DNA Ladder.
39. SYBR Gold or any orange DNA dye.
40. Gel loading dye, 6× Orange.
41. Blue light transilluminator (wavelength 460–480 nm).

2.4 ChIP-seq Library Preparation

1. ChIP-seq library Prep kit for Illumina.
2. Nuclease-free water.
3. DNA LoBind tubes.
4. SYBR Gold or any orange DNA dye.
5. Agarose gel electrophoresis equipment
6. 2% agarose gel.
7. 100 bp DNA Ladder.
8. Gel Extraction Kit.
9. Surgical Razor Blade.
10. Gel loading dye, 6× Orange.
11. Blue light transilluminator (wavelength 460–480 nm).
12. TapeStation or Bioanalyzer.

2.5 ChIP-seq Data Processing

1. ChIP-seq data in FASTQ format.
2. FastQC [23].
3. BWA MEM [24] or Bowtie2 [25].
4. Integrative *Genomics* Viewer (*IGV*) [26].

3 Methods

3.1 Sample Preparation

This method is used to prepare samples for both sonication optimization and chromatin immunoprecipitation.

1. Add protease inhibitors to PBS with 1% BSA.
2. Using forceps dissect ovaries in PBS with 1% BSA and protease inhibitors at room temperature [27]. The amount of ovaries depends on the number of IPs. We recommend 50 pairs of ovaries per antibody (*see* **Note 4**). For sonication optimization in Subheading 3.2, prepare seven tubes of ovaries (*see* **Note 5**).

3. Transfer the ovaries with 400 μl PBS with 1% BSA and protease inhibitors into a 1.5 mL tube using pipette with 1 mL tip.
4. Add 11 μl of 36.5% formaldehyde to the tube to obtain 1% formaldehyde and mix gently with 200 μL pipetting or inverting the tube ten times. Crosslink ovaries for 15 min at room temperature using rotator with speed around 20 rpm.
5. Add 400 μl of 250 mM ice-cold glycine to stop the crosslinking reaction and mix gently by pipetting.
6. Incubate for 5 min on ice. Ovaries will settle to the bottom during incubation time.
7. Carefully remove the supernatant with pipet, leaving ovaries untouched.
8. Wash ovaries with 500 μl of cold PBS with 1% BSA and protease inhibitors three times on ice. Each time add cold PBS with 1% BSA and protease inhibitors, mix gently by pipetting, wait for a few minutes until ovaries settle to the bottom and then remove the supernatant leaving ovaries untouched.
9. Carefully remove all PBS leaving ovaries untouched.
10. Snap-freeze the ovaries in liquid nitrogen and store the samples at −70–80 °C. Samples can be stored for several months.

3.2 Sonication Optimization

1. Take seven tubes (*see* **Note 5**) with fixed ovaries from the freezer and immediately homogenize the ovary pellet on ice using pestle by rotating for at least 10 s. Monitor visually that ovaries are well ground and extend time for homogenizing if needed (*see* **Note 6**).
2. Add protease inhibitors in sonication buffer.
3. Resuspend the homogenate in 300 μl of sonication buffer and leave for lysis on rotator at 4 °C for 0.5–1 h.
4. Transfer six lysates into separate sonication tubes. The remaining tube with lysate should be left on ice without sonication as control for unsheared DNA.
5. For sonication with Diagenode Bioruptor, we recommend following parameters: power = HIGH, ON = 30 s, OFF = 30 s (*see* **Note 7**). If needed, follow Chromatin shearing Guide manual for Diagenode Bioruptor.
6. Perform 3, 6, 9, 12, 15 and 18 cycles of sonication. Sonicate samples with 3 cycles, take out one tube with lysate, mark it as *3 cycles* and replace with an identical tube with 300 μl of water. Sonicate again for next 3 cycles and take another tube with lysate, mark it as *6 cycles* and replace with an identical tube with 300 μl of water. Sonicate again for next 3 cycles and take another tube with lysate, mark it as *9 cycles* and replace with an identical tube with 300 μl of water. Continue until the last tube with lysate will be sonicated with 18 cycles.

7. Add 20 μl of 5 M NaCl (to obtain final concentration of 0.3 M) and 10 μg RNaseA to each tube and mix with either a pipet or vortex.
8. Incubate 1 h at 37 °C for RNA digestion.
9. Change temperature to 95 °C and incubate 15 min to reverse crosslinking.
10. Cool down and add 20 μg Proteinase K to each tube. Mix well and incubate at 56 °C for 1 h for protein digestion.
11. Add 300 μl of phenol/chloroform/isoamylalcohol into phase-lock tubes and add samples, vortex for 10 s and spin for 10 min at a minimum of 10,000 g.
12. Collect the upper phase, transfer it to 1.5 mL DNA LoBind tubes and add
    - 8.9 μl Glycogen
    - 75 μl 3 M NaAc, pH 5.2
    - 550 μl absolute EtOH.
13. Mix well with pipetting and leave for at least 4 h at −20 °C or overnight (*see* **Note 8**).
14. Centrifuge a minimum of 10,000 g for 20 min at 4 °C.
15. Discard supernatant and wash pellet with cold EtOH 70%.
16. Centrifuge a minimum of 10,000 g for 10 min at 4 °C.
17. Discard supernatant, dry pellet for about 5 min (*see* **Note 9**) and dissolve in 100 μl TE buffer.
18. Take 20 μl aliquot of each sample and load on a 1% agarose gel with both 100 bp and 1 kb DNA ladders to check the sonication efficiency.
19. Run the gel until DNA loading dye line has travelled approximately 80% of the gel and DNA ladder bands are well separated.
20. Stain the gel with SYBR gold or use any loading buffer with orange DNA dye.
21. Check the gel using blue light transilluminator.
22. From the gel picture, choose the number of sonication cycles where you obtain 150–300 bp average size of sheared DNA. Usually, 15 cycles is just enough (*see* **Note 10**).

3.3 Chromatin Immunoprecipitation

1. Take the required number of tubes from the freezer (*see* **Note 4**), homogenize the ovary pellets and prepare for sonication in the same way as for sonication optimization (Subheading 3.2, **steps 1–3**).
2. Sonicate using the optimized conditions (see Subheading 3.2) for obtaining average size of 150–300 bp for the shared DNA.

3. After sonication, centrifuge lysates at a minimum of 10,000 g for 5 min at 4 °C to remove debris. At this point, the supernatant can be transferred into the new tubes and frozen with liquid nitrogen and stored at −70–80 °C.
4. Carefully pool the lysates into one 15 mL tube and take 50 μl of lysate for input sample (*see* **Note 11**). Store input sample at −20 °C until reverse crosslinking (**step 17** below).
5. Add cold sonication buffer with protease inhibitors to get total amount of lysate in mL equal to the number of IP samples. For instance, for 5 IP samples you would need to have 5 mL of total lysate. Mix well by pipetting.
6. Split lysate into number of aliquots equal to the number of IP samples (min 2: normal IgG and antibody against protein of interest). The number of IP samples should be equal to the number of tubes taken at **step 1**. Each aliquot should contain 1 mL of lysate.
7. Add 5–10 μg of antibodies (*see* **Note 2**) into IP samples and incubate overnight on a rotating wheel with less than 20 rpm speed in cold room.
8. Prepare the Protein A slurry on the same day as the IP samples (*see* **Note 12**). Take desired amount of Protein A sepharose and prepare according to the manufacturer's protocol. We usually use 50 μl of Protein A slurry per IP sample. Wash the beads three times with IP buffer (the same as sonication buffer).
9. Preblock Protein A beads with IP buffer with BSA and incubate overnight on a rotating wheel in cold room together with IP samples.
10. The next day centrifuge the Protein A slurry for 5 min at 500 g at 4 °C. Remove supernatant and add 50 μl of cold IP buffer per IP sample. For instance, for 5 IP samples you would need to add 250 μl of total IP buffer. Mix well by pipetting.
11. For immunoprecipitation, add 50 μl of preblocked protein A sepharose into every sample.
12. Incubate for 4 h on a rotating wheel under 20 rpm speed at 4 °C.
13. Centrifuge for 5 min at 500 g at 4 °C.
14. Remove supernatant carefully, avoiding touching the beads.
15. Wash with the following buffers:
    (a) 1 mL of IP buffer three times
    (b) 1 mL of Wash buffer 500 two times
    (c) 1 mL LiCl buffer one time
    (d) 1 mL TE two times

Each time, wash for 5 min on a rotating wheel under 20 rpm speed in cold room. Centrifuge after each wash for 5 min at 500 g at 4 °C and remove supernatant carefully (*see* **Note 13**).

Important: Do not vortex sepharose beads with bound complex at any step before adding extraction buffer (see below).

16. Add 150 μL of extraction buffer to the IP sample beads and vortex.
17. Take input from freezer (**step 4**), thaw on ice and add 100 μL of extraction buffer to it. Vortex or mix well.
18. Add 10 μL of 5 M NaCl (to obtain the final concentration 0.3 M) and 10 μg RNaseA to each tube. Mix well.
19. Incubate 1 h at 37 °C for RNA digestion.
20. Add 20 μg Proteinase K to each tube, mix and incubate at 65 °C overnight to reverse crosslinking (*see* **Note 14**).
21. The next day centrifuge samples for 5 min at 500 g.
22. Transfer the supernatants into phase-lock tubes. Do not transfer the beads.
23. Add 300 μL phenol/chloroform/isoamylalcohol into phase-lock tubes with samples, vortex for 10 s and spin for 10 min at a minimum of 10,000 g.
24. Collect the upper phase, transfer it to 1.5 mL DNA LoBind tubes and add
    - 8.9 μL Glycogen
    - 75 μL 3 M NaAc, pH 5.2
    - 550 μL absolute EtOH.
25. Mix well by pipetting and leave for at least 4 h at −20 °C. We recommend leaving it overnight (*see* **Note 8**).
26. Centrifuge at a minimum of 10,000 g for 20 min at 4 °C.
27. Discard supernatant and wash pellet with cold 70% EtOH.
28. Centrifuge at a minimum of 10,000 g for 10 min at 4 °C.
29. Air-dry pellet for about 5 min and dissolve in 100 μL TE buffer.
30. Take a 20 μL aliquot of input and load on a 2% agarose gel (*see* **Note 15**) to confirm the sonication efficiency (Subheading 3.2, **steps 17–20** for details). If the size of the DNA is between 150 and 300 bp, continue with the library preparation. The amount of DNA from IP samples is very low and cannot be visualized with agarose gel.

3.4 ChIP-seq Library Preparation

1. Take 50 μL of IP sample and perform ChIP-seq library preparation using any available kit (*see* **Note 16**). We recommend starting with half of the IP sample (*see* **Note 17**) and 10% of input for library preparation. Input sample can be diluted in TE buffer up to 50 μL.

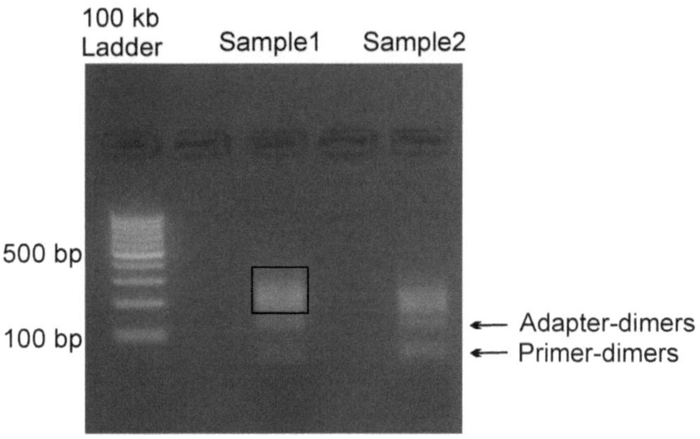

Fig. 2 ChIP-seq library after PCR amplification. A 2% agarose gel with two samples shows the result of ChIP-seq library amplification. The region recommended for excising the library is marked with a black box. Arrows indicate adapter-dimers and primer-dimers, which should be avoided when excising the library

2. Follow the manufacturer's protocol for PCR Enrichment of Adaptor-ligated DNA (*see* **Note 18**).
3. After cleanup of Adaptor-ligated DNA, take half of the sample volume and perform PCR amplification according to the ChIP-seq library preparation kit manual (*see* **Note 19**).
4. Load the whole PCR amplification sample on 2% agarose gel (*see* **Note 15** and **Note 20**).
5. Stain gel with SYBR gold or use any loading buffer with orange DNA dye.
6. With a razor blade, excise the DNA band corresponding to the library (the size is counted from sonication band + adapters with primers) using blue light transilluminators (Fig. 2). Do not cut either bands corresponding to primers (about 5–80 bp) or adaptor dimers (size is about 100–150 bp) (Fig. 2). If DNA band corresponding to the library is very faint, repeat the amplification step with 1–2 extra cycles (*see* **Note 19**).
7. Purify DNA with Gel extraction kit in 20 μL of extraction buffer.
8. Perform quality check of the ChIP-library with TapeStation (Fig. 3) or Bioanalyzer, pool together different samples, if it is required, and sequence (*see* **Note 21**).

3.5 ChIP-seq Data Processing

1. Once you get the ChIP-seq data in FASTQ format (it can be any of following format .fastq, .fq, .fastq.gz, .fq.gz), perform initial quality assessment with FastQC [24]. Continue with read trimming and filtering if necessary.

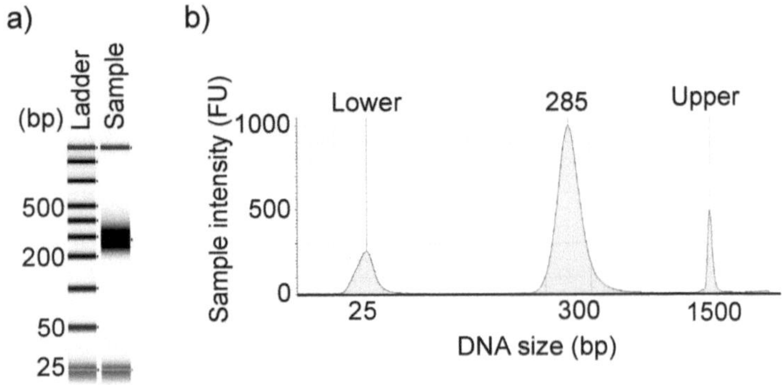

Fig. 3 TapeStation electrophoregram of ChIP-seq library. Example sample of final library with optimal size range and purity is shown as a gel image on left (**a**) and as intensity profile on the right (**b**)

2. Align good quality reads to *Drosophila* genome with any available software. We recommend using either BWA MEM [25] or Bowtie2 [26]. Both software produce .bam files with aligned reads. Aligned reads can be visualized with any genome browser. We recommend using Integrative *Genomics* Viewer (*IGV*) [27].
3. Check read distribution with genome browser (Fig. 4). Coverage of reads corresponding to the binding profile of protein of interest should be sufficient and well above background, IgG and input.
4. After evaluating read distribution, you can continue with data analysis. Several software can be used for enrichment analysis and peak calling [20, 28, 29].

4 Notes

1. The centrifuge must be capable of reaching up to 10,000 g and refrigerated.
2. We recommend choosing antibodies that have already been utilized for ChIP-seq in *Drosophila*. If you are going to use a custom-made antibody, verify their specificity with western blotting and immunostaining using either knockdown or knockout samples as control.
3. The use of normal IgG is necessary as a negative control along with input. The choice of normal IgG depends on source of the primary antibody used. If primary antibody is raised in rabbit, then normal rabbit IgG should be used as the negative control.
4. We recommend 50 pairs of ovaries per antibody (= IP sample). It is, therefore, convenient to fix sets of 50 pairs of ovaries into separate tubes and use one tube for one IP sample.

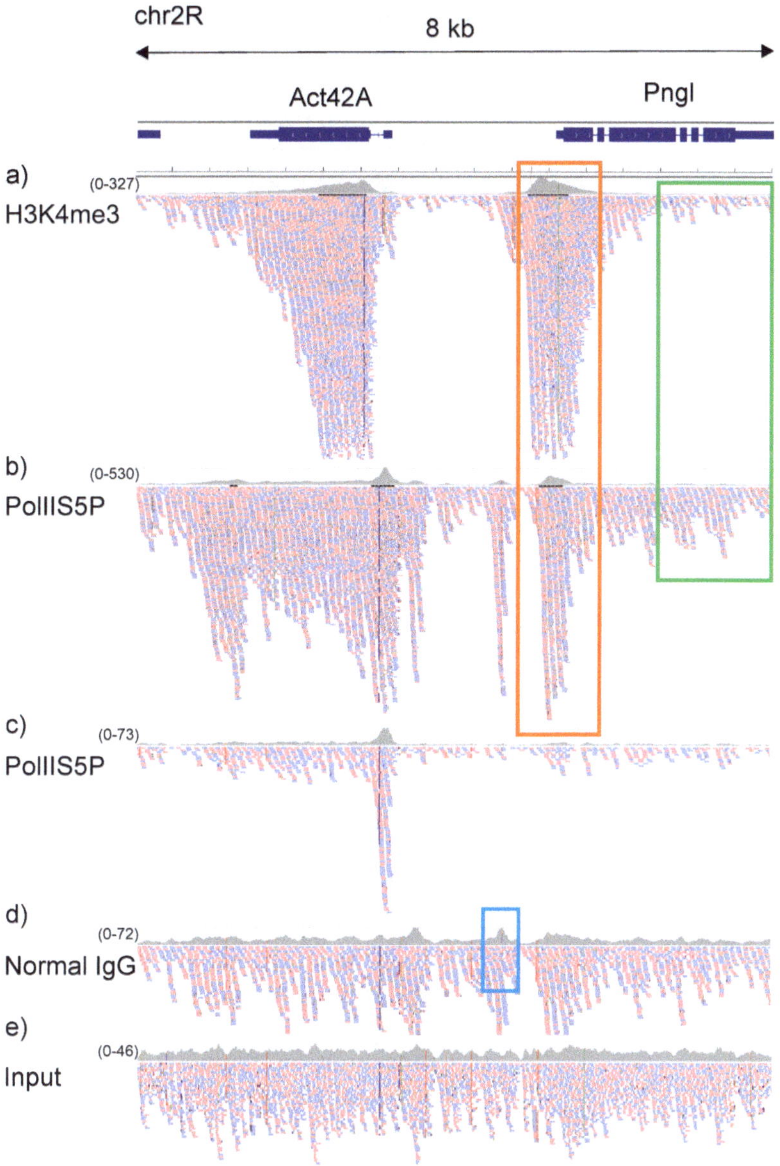

Fig. 4 Representative genome browser tracks of read coverage profiles. Coverage from .bam files for *Drosophila* genome region 42A7–42A8 containing *Act42* and *Pngl* genes visualized with IGV genome browser. Top two tracks with histone H3K4me3 (**a**) and PolII2S5P (**b**) have good coverage (with maximum level of 327 and 530 accordingly) with visible peaks at the gene start sites (orange box) above background level (green box) and can be processed further for enrichment and peak calling analysis. Next track with PolIIS5P (**c**) has a very poor coverage, resulting in complete absence of the peak at the promoter region of *Pngl* gene, and should not be processed for further data analysis. Normal IgG (**d**) and input (**d**) represent typical background coverage level. Unlike input, normal IgG can show weekly visible peaks (light blue box) as a result of non-specific binding during immunoprecipitation

5. Seven tubes with 50 ovary pairs each are needed for sonication optimization. It is important to have the same amount of material in each tube, as it can affect DNA shearing efficiency. Sonication parameters are always optimized for certain amount of chromatin in the lysate. Different genotypes or strains of the flies should not affect DNA shearing efficiency. However, if a fly strain has severe defects in oogenesis, and ovaries are much smaller compared to others, you need to optimize sonication parameters and/or number of ovaries used per IP for that fly strain separately.
6. For the best result, we recommend grinding the ovaries, while the pellet is still completely frozen, as this breaks the pellet more efficiently.
7. For every sonicator, you need to optimize the parameters (mainly number of cycles) in order to obtain 150–300 bp average size of DNA. If you use a different sonicator than Diagenode Bioruptor, follow the manufacturer's recommendations of power and cycle settings for chromatin shearing. The volume of the lysate can also be different, and you need to adjust the sonication protocol accordingly.
8. For precipitation of the sheared DNA, we recommend incubation for at least 4 h at −20 °C. If the timing for the rest of protocol is very tight, you can incubate the sample overnight.
9. Do not overdry the DNA pellet, as it might not dissolve.
10. The optimized sonication cycle is usually between 12 and 18 for Diagenode Bioruptor. If the DNA is sheared to 150–300 bp already at 10 cycles, it is most probably due to weak crosslinking efficiency. If DNA is not sheared to 150–300 bp with 18 cycles, it is most probably due to very strong crosslinking. In these cases, check the formaldehyde concentration, temperature and time for crosslinking. We recommend 15 min room temperature incubation with 1% formaldehyde for crosslinking.
11. Pooling together lysates helps to equalize the amount of chromatin between IP samples. You can pool together lysates only if they are coming from flies with the same genetic background. If you have ovaries coming from different strains of flies, pool them separately and take input sample for each strain.
12. Protein A will work with most of antibodies. However, check the recommendation for each antibody and general guidance for the use of Protein A or Protein G. You can also use a mixture of Protein A/G as well as Dynabeads.
13. If you are using Dynabeads, replace centrifugation step with magnetic stand to concentrate the beads.

14. The quick protocol used for optimization (Subheading 3.2 **steps 8** and **9**) is not recommended here, as it can result in uneven reversing of the crosslinks, and result in unspecific read distribution in sequencing data.
15. 2% agarose gel is required for confirmation of the sonication efficiency, estimation of DNA library size and its separation from adapter dimers (Fig. 2). For sonication optimization, 1% agarose gel is better for visualization of unsheared and differently sheared DNA on the same gel.
16. The amount of ChIP DNA is very low. Many sequencing facilities do not accept orders for ChIP-seq library preparation, or do not guarantee any results due to the very low amount of starting DNA. Therefore, we routinely prepare the sequencing library ourselves. Once the library is prepared, you can evaluate the size and concentration with TapeStation (Fig. 3) or Bioanalyzer prior to sequencing.
17. We recommend proceeding to the library preparation with half of each sample and leaving the other half for a backup in the freezer, if you need to repeat the procedure.
18. Follow the recommendation from the sequencing facility and DNA library preparation kit for using barcodes for samples.
19. We recommend using half of adaptor–ligated DNA for amplification and leaving the other half in the freezer, if you need to repeat the procedure. For instance, if the number of PCR cycles is too low for a sufficient DNA amount (you must see a clear band after amplification on the gel, Fig. 2), you can repeat the amplification with other half using more cycles. Usually, 15 cycles is enough. Never use more than 18 amplification cycles, as it can lead to PCR artifacts, such as increased level of duplicate reads.
20. For amplified library size selection, we recommend using agarose gel instead of PCR clean up column kit or magnetic beads (Fig. 2). Excising the library band from gel results in less adapter dimer contamination.
21. We recommend using a sequencing depth of 10–20 million of either single-end or paired-end reads for all samples including input. The 50–100 bp length of the reads is sufficient for the data analysis.

Acknowledgments

Work in the Vartiainen lab was supported by the Academy of Finland, Sigrid Juselius, Jane and Aatos Erkko as well as Finnish Cancer foundations.

References

1. Johnson DS, Mortazavi A, Myers RM, Wold B (2007) Genome-wide mapping of in vivo protein-DNA interactions. Science 316(5830):1497–1502. https://doi.org/10.1126/science.1141319
2. Kharchenko PV, Alekseyenko AA, Schwartz YB, Minoda A, Riddle NC, Ernst J, Sabo PJ, Larschan E, Gorchakov AA, Gu T, Linder-Basso D, Plachetka A, Shanower G, Tolstorukov MY, Luquette LJ, Xi R, Jung YL, Park RW, Bishop EP, Canfield TK, Sandstrom R, Thurman RE, MacAlpine DM, Stamatoyannopoulos JA, Kellis M, Elgin SC, Kuroda MI, Pirrotta V, Karpen GH, Park PJ (2011) Comprehensive analysis of the chromatin landscape in Drosophila melanogaster. Nature 471(7339):480–485. https://doi.org/10.1038/nature09725
3. Mendenhall EM, Bernstein BE (2008) Chromatin state maps: new technologies, new insights. Curr Opin Genet Dev 18(2):109–115. https://doi.org/10.1016/j.gde.2008.01.010
4. Bonn S, Zinzen RP, Perez-Gonzalez A, Riddell A, Gavin AC, Furlong EE (2012) Cell type-specific chromatin immunoprecipitation from multicellular complex samples using BiTS-ChIP. Nat Protoc 7(5):978–994. https://doi.org/10.1038/nprot.2012.049
5. Guertin MJ, Lis JT (2010) Chromatin landscape dictates HSF binding to target DNA elements. PLoS Genet 6(9):e1001114. https://doi.org/10.1371/journal.pgen.1001114
6. Ghavi-Helm Y, Zhao B, Furlong EE (2016) Chromatin immunoprecipitation for analyzing transcription factor binding and histone modifications in Drosophila. Methods Mol Biol 1478:263–277. https://doi.org/10.1007/978-1-4939-6371-3_16
7. Loubiere V, Delest A, Schuettengruber B, Martinez AM, Cavalli G (2017) Chromatin immunoprecipitation experiments from whole Drosophila embryos or larval imaginal discs. Bio Protoc 7(11):e2327. https://doi.org/10.21769/BioProtoc.2327
8. Slattery M, Ma L, Spokony RF, Arthur RK, Kheradpour P, Kundaje A, Negre N, Crofts A, Ptashkin R, Zieba J, Ostapenko A, Suchy S, Victorsen A, Jameel N, Grundstad AJ, Gao W, Moran JR, Rehm EJ, Grossman RL, Kellis M, White KP (2014) Diverse patterns of genomic targeting by transcriptional regulators in Drosophila melanogaster. Genome Res 24(7):1224–1235. https://doi.org/10.1101/gr.168807.113
9. mod EC, Roy S, Ernst J, Kharchenko PV, Kheradpour P, Negre N, Eaton ML, Landolin JM, Bristow CA, Ma L, Lin MF, Washietl S, Arshinoff BI, Ay F, Meyer PE, Robine N, Washington NL, Di Stefano L, Berezikov E, Brown CD, Candeias R, Carlson JW, Carr A, Jungreis I, Marbach D, Sealfon R, Tolstorukov MY, Will S, Alekseyenko AA, Artieri C, Booth BW, Brooks AN, Dai Q, Davis CA, Duff MO, Feng X, Gorchakov AA, Gu T, Henikoff JG, Kapranov P, Li R, HK MA, Malone J, Minoda A, Nordman J, Okamura K, Perry M, Powell SK, Riddle NC, Sakai A, Samsonova A, Sandler JE, Schwartz YB, Sher N, Spokony R, Sturgill D, van Baren M, Wan KH, Yang L, Yu C, Feingold E, Good P, Guyer M, Lowdon R, Ahmad K, Andrews J, Berger B, Brenner SE, Brent MR, Cherbas L, Elgin SC, Gingeras TR, Grossman R, Hoskins RA, Kaufman TC, Kent W, Kuroda MI, Orr-Weaver T, Perrimon N, Pirrotta V, Posakony JW, Ren B, Russell S, Cherbas P, Graveley BR, Lewis S, Micklem G, Oliver B, Park PJ, Celniker SE, Henikoff S, Karpen GH, Lai EC, DM MA, Stein LD, White KP, Kellis M (2010) Identification of functional elements and regulatory circuits by Drosophila modENCODE. Science 330(6012):1787–1797. https://doi.org/10.1126/science.1198374
10. Spradling A (1993) Developmental genetics of oogenesis. The Development of Drosophila Melanogaster
11. Rust K, Byrnes LE, Yu KS, Park JS, Sneddon JB, Tward AD, Nystul TG (2020) A single-cell atlas and lineage analysis of the adult Drosophila ovary. Nat Commun 11(1):5628. https://doi.org/10.1038/s41467-020-19361-0
12. Jia D, Huang YC, Deng WM (2015) Analysis of cell cycle switches in Drosophila oogenesis. Methods Mol Biol 1328:207–216. https://doi.org/10.1007/978-1-4939-2851-4_15
13. Royzman I, Orr-Weaver TL (1998) S phase and differential DNA replication during Drosophila oogenesis. Genes Cells 3(12):767–776. https://doi.org/10.1046/j.1365-2443.1998.00232.x
14. Orr-Weaver TL (1991) Drosophila chorion genes: cracking the eggshell's secrets. BioEssays 13(3):97–105. https://doi.org/10.1002/bies.950130302
15. Park PJ (2009) ChIP-seq: advantages and challenges of a maturing technology. Nat Rev Genet 10(10):669–680. https://doi.org/10.1038/nrg2641

16. Sokolova M, Turunen M, Mortusewicz O, Kivioja T, Herr P, Vaharautio A, Bjorklund M, Taipale M, Helleday T, Taipale J (2017) Genome-wide screen of cell-cycle regulators in normal and tumor cells identifies a differential response to nucleosome depletion. Cell Cycle 16(2):189–199. https://doi.org/10.1080/15384101.2016.1261765
17. Celniker SE, Dillon LA, Gerstein MB, Gunsalus KC, Henikoff S, Karpen GH, Kellis M, Lai EC, Lieb JD, MacAlpine DM, Micklem G, Piano F, Snyder M, Stein L, White KP, Waterston RH, mod EC (2009) Unlocking the secrets of the genome. Nature 459(7249): 927–930. https://doi.org/10.1038/459927a
18. ENCODE. https://www.encodeproject.org/
19. Bardet AF, He Q, Zeitlinger J, Stark A (2011) A computational pipeline for comparative ChIP-seq analyses. Nat Protoc 7(1):45–61. https://doi.org/10.1038/nprot.2011.420
20. Lerdrup M, Johansen JV, Agrawal-Singh S, Hansen K (2016) An interactive environment for agile analysis and visualization of ChIP-sequencing data. Nat Struct Mol Biol 23(4): 349–357. https://doi.org/10.1038/nsmb.3180
21. Afgan E, Baker D, Batut B, van den Beek M, Bouvier D, Cech M, Chilton J, Clements D, Coraor N, Gruning BA, Guerler A, Hillman-Jackson J, Hiltemann S, Jalili V, Rasche H, Soranzo N, Goecks J, Taylor J, Nekrutenko A, Blankenberg D (2018) The Galaxy platform for accessible, reproducible and collaborative biomedical analyses: 2018 update. Nucleic Acids Res 46(W1):W537–W544. https://doi.org/10.1093/nar/gky379
22. Sokolova M, Moore HM, Prajapati B, Dopie J, Merilainen L, Honkanen M, Matos RC, Poukkula M, Hietakangas V, Vartiainen MK (2018) Nuclear actin is required for transcription during Drosophila oogenesis. iScience 9: 63–70. https://doi.org/10.1016/j.isci.2018.10.010
23. Andrews S FastQC https://www.bioinformatics.babraham.ac.uk/projects/fastqc/
24. Burrows-Wheeler Alignment Tool http://bio-bwa.sourceforge.net/bwa.shtml#13
25. Langmead B, Salzberg SL (2012) Fast gapped-read alignment with Bowtie 2. Nat Methods 9(4):357–359. https://doi.org/10.1038/nmeth.1923
26. Robinson JT, Thorvaldsdottir H, Winckler W, Guttman M, Lander ES, Getz G, Mesirov JP (2011) Integrative genomics viewer. Nat Biotechnol 29(1):24–26. https://doi.org/10.1038/nbt.1754
27. Thompson L, Randolph K, Norvell A (2015) Basic techniques in Drosophila ovary preparation. Methods Mol Biol 1328:21–28. https://doi.org/10.1007/978-1-4939-2851-4_2
28. Ramirez F, Ryan DP, Gruning B, Bhardwaj V, Kilpert F, Richter AS, Heyne S, Dundar F, Manke T (2016) deepTools2: a next generation web server for deep-sequencing data analysis. Nucleic Acids Res 44(W1):W160–W165. https://doi.org/10.1093/nar/gkw257
29. Le Rolle V, Hernandez AI, Richard PY, Donal E, Carrault G (2008) Model-based analysis of myocardial strain data acquired by tissue Doppler imaging. Artif Intell Med 44(3): 201–219. https://doi.org/10.1016/j.artmed.2008.06.001

Chapter 19

Detection of Actin in Nuclear Protein Fraction Isolated from Adult *Drosophila* Ovary

Ildikó Kristó, Péter Borkúti, Zoltán Kovács, Anikó Szabó, Szilárd Szikora, and Péter Vilmos

Abstract

Much evidence supports the presence of cytoskeletal elements in the nucleus; however, the exact functions of these proteins in the nucleus are still uncertain. Of the cytoskeletal proteins, the activity and biological significance of nuclear actin has been the most extensively researched. It is now clear that actin performs essential tasks both in the cytoplasm and the nucleus, and that the dynamic balance between the large cytoplasmic and the significantly smaller nuclear actin pools is maintained by robust transport mechanisms. Therefore, the compartment-specific manipulation or investigation of actin has been an enormous challenge. Here, we present a protocol for the detection of actin in isolated nuclear protein fractions from *Drosophila* ovaries.

Key words *Drosophila*, Ovary, Nucleus, Actin, Western blot, Subcellular fractionation

1 Introduction

Of all the cytoskeletal proteins, actin is clearly the oldest and most intensively studied element. The assembly and organization of the dynamic network of actin cytoskeleton in the cytoplasm are key to essential cellular functions, such as cell movements, changes in cell shape, cell division, intracellular transport, formation of stable intercellular bridges, and the establishment of cell polarity [1, 2]. Similar to the situation in the cytoplasm, the recently discovered functions of actin in the nucleus are indispensable for the cell. Actin is present in almost all of the major nuclear complexes and it participates in important nuclear processes such as DNA repair, chromatin remodeling, mRNA and rRNA transcription, and RNA processing and export [3–6]. Recently, it has become clear that a dynamic balance between the cytoplasmic and nuclear actin pools sets actin levels in the two compartments. This fine-tuned balance is achieved through tightly regulated nuclear import

Michelle S. Giedt and Tina L. Tootle (eds.), *Drosophila Oogenesis: Methods and Protocols*,
Methods in Molecular Biology, vol. 2626, https://doi.org/10.1007/978-1-0716-2970-3_19,

and export mechanisms, and is the key to the proper functioning of actin in both compartments [7, 8].

The *Drosophila* ovary is a highly organized, complex organ. The individual egg chambers of the fly ovary are composed of 15 interconnected polyploid nurse cells and one oocyte; the nurse cells deposit a large amount of RNA and protein into the oocyte in late oogenesis. Surrounding each egg chamber is an epithelial monolayer of follicle cells. *Drosophila* oogenesis is characterized by a large variety of cytoskeletal and nuclear processes, and by extensive interaction between the two compartments, which make the *Drosophila* ovary an excellent model system to investigate subcellular molecular mechanisms and components. The relatively large cellular sizes in the fly ovary, especially in the case of the polytene nurse cells and the oocyte, together with the genetic tools available in *Drosophila*, contributed significantly to the research on actin cytoskeletal dynamics [9–12], as well as to the investigation of actin's nuclear localization and activities [13–16].

However, the compartment-specific investigation of actin with biochemical methods, such as analyzing its post-translational modifications, determining its interacting partners, or quantitatively measuring levels, are hindered by the abundant amount of actin in the cytoplasm and its robust and dynamic nuclear transport. Therefore, the faithful separation of cytoplasmic and nuclear actin pools and functions is challenging. Here, we present a comprehensive and powerful method for the isolation of clean nuclear protein fractions from *Drosophila* ovaries and the simultaneous detection of endogenous and epitope-tagged actin.

2 Materials

Prepare all solutions using distilled water and analytical grade reagents. Store all solutions at room temperature (RT) unless otherwise specified.

2.1 Dissection

1. Standard cornmeal, yeast, and sucrose Drosophila medium.
2. Dry yeast.
3. CO_2.
4. Fly pad.
5. A pair of sharp forceps (*see* **Note 1**).
6. Glass or plastic Petri dish (*see* **Note 2**).
7. 1× PBS (Phosphate-Buffered Saline): 137 mM NaCl, 2.7 mM KCl, 10 mM Na_2HPO_4, 1.8 mM KH_2PO_4.
8. Schneider's Drosophila Medium.
9. Stereomicroscope.
10. 1.5 mL centrifuge tubes.

2.2 Isolation of Cytoplasmic and Nuclear Protein Fractions

1. Gloves.
2. Pipette and tips.
3. Sterile polypropylene pestles.
4. 1.5 mL centrifuge tubes.
5. Refrigerated benchtop centrifuge.
6. 1× PBS.
7. Cytoplasmic extraction buffer (store at 4 °C): 15 mM HEPES-KOH pH 7.6, 10 mM KCl, 5 mM $MgCl_2$, 0.1 mM EDTA pH 8.0, 10% Glycerol, 1× Protease Inhibitor Cocktail.
8. Wash buffer 1 (store at 4 °C): 15 mM HEPES-KOH pH 7.6, 10 mM KCl, 5 mM $MgCl_2$, 0.1 mM EDTA pH 8.0, 10% Glycerol, 150 mM Sucrose, 1× Protease Inhibitor Cocktail.
9. Wash buffer 2 (store at 4 °C): 15 mM HEPES-KOH pH 7.6, 10 mM KCl, 5 mM $MgCl_2$, 0.1 mM EDTA pH 8.0, 10% Glycerol, 250 mM Sucrose, 1× Protease Inhibitor Cocktail.
10. Nuclear extraction buffer (store at 4 °C): 350 mM Sucrose, 15 mM HEPES-KOH pH 7.6, 385 mM KCl, 5 mM $MgCl_2$, 0.1 mM EDTA pH 8.0, 0.05% Tween 20, 10% Glycerol, 1× Protease Inhibitor Cocktail.

2.3 Measuring Protein Concentration with Bradford Assay

1. Pipette and tips.
2. 1.5 mL centrifuge tubes.
3. Bradford reagent.
4. 70% ethanol.
5. Bovine Serum Albumin (BSA).
6. Spectrophotometer instrument.

2.4 Western Blotting

1. Pipette and tips.
2. Protein gel electrophoresis and blot system.
3. PVDF transfer membrane of 0.2 or 0.45 μm pore size.
4. Cellulose blotting paper, and sponge.
5. Razor blade, and forceps with blunt tips.
6. 50 mL Conical flask or Falcon tube, and 15 mL Falcon tube.
7. 10% Sodium dodecyl sulfate (SDS)-polyacrylamide resolving gel (6.6 mL, for one pair of 10 × 10 cm glass plates with 1-mm-thick spacer).
8. Stacking gel (3 mL, for one pair of 10 × 10 cm glass plates with 1 mm thick spacer)
9. Butanol or isopropanol.
10. Protein molecular weight marker.

11. 2× SDS protein loading buffer: 120 mM Tris–HCl pH 6.8, 12.8% Glycerol, 4% sodium dodecyl sulfate (SDS), 200 mM dithiothreitol (DTT), 1 mg bromophenol blue (for 10 mL).
12. Power supply.
13. Glass or plastic dish (*see* **Note 3**).
14. 1× Running buffer, pH 8.3–8.5: 25 mM Tris base, 192 mM glycine, 0.1% SDS.
15. 1× Transfer buffer (does not require pH adjustment): 48 mM Tris base, 38.6 mM glycine, 0.03% SDS, 20% methanol.
16. TTBS, pH 7.5: 25 mM Tris base, 150 mM NaCl, 0.1% Tween 20.
17. Blocking solution: 5% low-fat milk powder dissolved in 1× TTBS.
18. Anti-Actin antibody (*see* **Note 4**).
19. Anti-Flag antibody to detect FLAG-tagged actin.
20. Anti-Tubulin antibody to control cytoplasmic contamination in the nuclear fraction.
21. Anti-Histone H3 (H3) or anti-Heterochromatin Protein 1 (HP1) antibody to detect nuclear contamination in the cytoplasmic fraction.
22. Horseradish peroxidase (HRP) conjugated secondary antibodies.
23. Chemiluminescent reagent.
24. Chemiluminescence imaging system.

3 Methods

Carry out all procedures at room temperature unless otherwise specified. For an overview of the complete procedure, see Fig. 1.

3.1 Isolation of Ovaries

1. Collect and place ~50 virgin female flies and 5 males (*see* **Note 5**) on fresh standard fly food media with dry yeast to allow the flies to mate as mating promotes oogenesis [17].
2. Incubate the flies at 25 °C for 5 days. On the fourth day feed them with fresh standard fly food media with dry yeast to fatten up the ovaries.

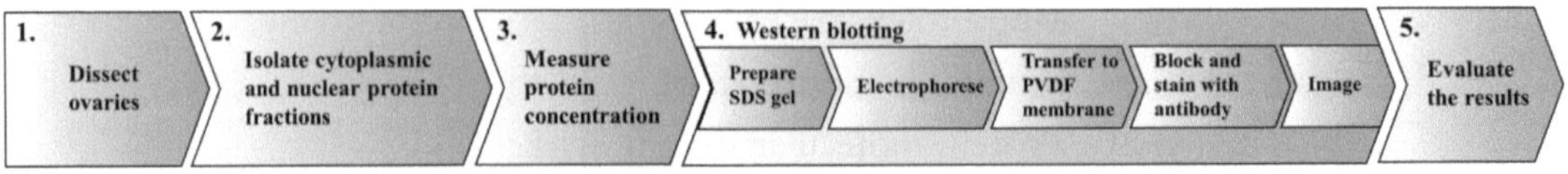

Fig. 1 Major steps of the workflow to detect Actin in the nuclear protein fraction of *Drosophila* ovaries

3. When ready to dissect, fill a glass petri dish of 100 mm diameter with room-temperature 1× PBS (*see* **Note 6**).
4. Put the 5-day-old flies to sleep using CO_2.
5. Pick up a female fly using forceps, submerge her into the 1× PBS, and remove ovaries using a pair of sharp tweezers. Repeat to collect 50 pairs of ovaries.
6. Using forceps, transfer 50 pairs of ovaries to a 1.5 mL centrifuge tube on ice containing 500 μL of Schneider's Drosophila Medium (*see* **Note 7**).
7. Proceed immediately to the isolation of cytoplasmic and nuclear protein fractions (*see* **Note 8**).

3.2 Isolation of Cytoplasmic and Nuclear Protein Fractions from the Ovaries

Solutions are pre-chilled before use and all steps in this protocol (*see* **Note 9**) must be carried out on ice. In order to minimize protein degradation, avoid unnecessary pipetting during isolation. All centrifugations must be performed at 4 °C during the isolation.

1. Wash the ovaries once with ice cold 1× PBS.
2. Homogenize gently the ovaries in 200 μL of cytoplasmic extraction buffer using a sterile polypropylene pestle for 1 min (*see* **Note 10**).
3. Centrifuge the sample at 600 × *g* for 1 min to pellet any residual tissue.
4. Transfer the supernatant which contains the cytoplasm and the nuclei into a new 1.5 mL centrifuge tube (*see* **Note 11**), centrifuge at 5000 × *g* for 8 min to collect the nuclei.
5. Keeping the tube with the nuclear pellet, transfer the supernatant into a new tube and clear it with an additional centrifugation at 12,000 × *g* for 10 min. Transfer the clean supernatant, which is the cytoplasmic protein fraction, into a new tube. Keep it on ice until protein concentration measurements, then store at −80 °C.
6. Resuspend the nuclear pellet of **step 4** in 150 μL of ice cold wash buffer 1, and centrifuge the sample at 5000 × *g* for 8 min. Repeat the wash step twice more.
7. Resuspend final pellet of **step 6** in 150 μL of ice cold wash buffer 2 and centrifuge the sample at 5000 × *g* for 8 min. Repeat the wash step twice more.
8. Resuspend the final pellet of **step 7** containing the nuclei in 50 μL nuclear extraction buffer, keep on ice for 5 min while vortexing it every minute for a few seconds.
9. Centrifuge at 12,000 × *g* for 10 min and transfer the supernatant into a new 1.5 mL centrifuge tube; this is the nuclear fraction. Keep it on ice during the protein concentration measurement, then store at −80 °C.

3.3 Measuring Protein Concentration with Bradford Assay

1. Prepare BSA standards with sterile distilled/ultrapure water for calibration with concentrations of 200, 400, 600, 800, 1000 and 1200 μg/ml (*see* **Note 12**).
2. Prepare 5× and 10× dilutions from the cytoplasmic and nuclear samples with sterile distilled/ultrapure water, as undiluted protein samples are usually too concentrated for the measurement.
3. Mix the Bradford reagent and diluted protein samples, as recommended by the manufacturer.
4. With a spectrophotometer measure the OD595 values of the standards and the diluted cytoplasmic and nuclear samples (*see* **Note 13**).
5. Create the calibration curve from the standards values and calculate the samples protein concentrations.

3.4 Western Blotting

Detection of endogenous and Flag epitope tagged actin in the nuclear fraction is carried out in Western blot analysis by using anti-Actin and anti-Flag antibodies (Fig. 2a). Cross-contamination of the isolated cytoplasmic and nuclear protein fractions can be checked by using cytoplasm (anti-Tubulin) and nucleus specific (anti-H3 or anti-HP1) antibodies (Fig. 2b).

3.4.1 Preparation of the SDS-Polyacrylamide Gel

1. Set up clean glass plates for pouring the SDS-polyacrylamide gel (*see* **Note 14**).
2. Prepare the resolving gel mixture in a 50 mL conical flask. After adding TEMED, which catalyzes polymerization, immediately mix it thoroughly and load it between the glass plates with a pipette (*see* **Note 15**). Make sure to leave about 0.5 cm of space between the resolving gel and the bottom of the comb for the stacking gel.
3. Layer 300 μL of butanol or isopropanol on the top of the gel to eliminate bubbles, ensure the gel surface is smooth and horizontal, and prevent contact with oxygen, which inhibits acrylamide polymerization.
4. After 20 min of polymerization (*see* **Note 15**), wash out the butanol or isopropanol with ultrapure water.
5. Remove water with a piece of filter paper.
6. Prepare the stacking gel mixture in a 15 mL Falcon tube. Load the stacking gel on the top of the resolving gel and immediately insert the comb between the plates without introducing air bubbles. Allow the gel to polymerize for 15 min.

3.4.2 Electrophoresis

To assemble the protein electrophoresis equipment with the gel, follow the manufacturer's instructions.

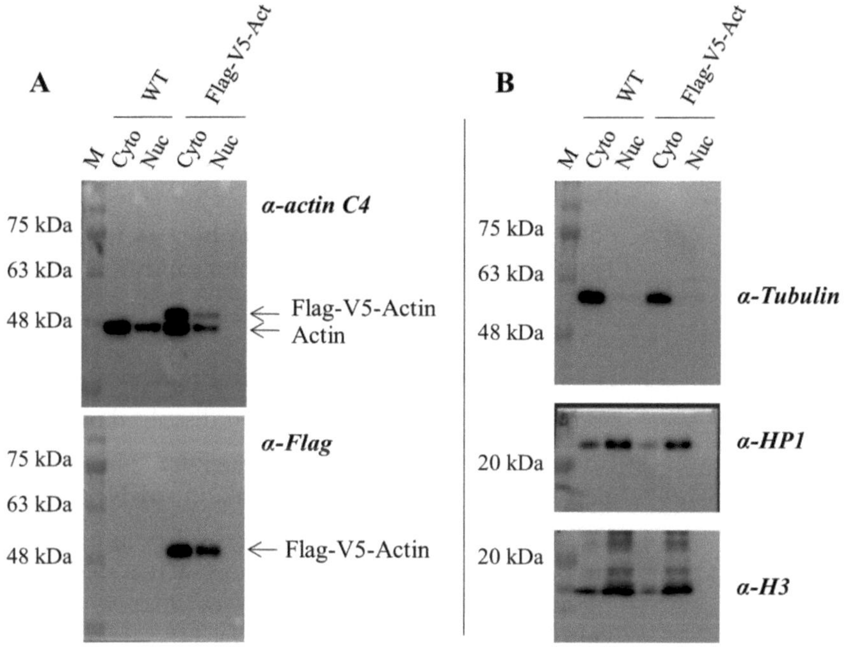

Fig. 2 Detection of endogenous Actin and Flag-V5-Actin in the nuclear protein fraction of *Drosophila* ovaries. (**a**) Detection of endogenous Actin and Flag-V5-Actin in the isolated cytoplasmic (Cyto) and nuclear (Nuc) protein fractions with anti-Actin C4 antibody (upper part). Detection of Flag-V5-Actin in the isolated cytoplasmic and nuclear protein fractions with the anti-Flag antibody (lower part). Note that the anti-Flag antibody detects Flag-V5-Actin in the nucleus much stronger than the C4 antibody (compare cytoplasmic and nuclear signals). (**b**) Purity control of the isolated fractions with anti-Tubulin (cytoplasm) and anti-HP1 or -H3 (nucleus) antibodies. Note that the cytoplasmic fraction has nuclear contamination but the nuclear fraction is devoid of any contamination as revealed by the Tubulin staining. WT wild type, M Protein molecular weight marker, Cyto cytoplasmic fraction, Nuc nuclear fraction

1. Fill up the tank of the electrophoresis instrument with 1× Running buffer, then remove the comb carefully from the gel.
2. Mix 3 μg of cytoplasmic and 30 μg of nuclear protein samples with 2× SDS protein loading buffer, boil them for 5 min and centrifuge at 10,000 × *g* for 2 min at RT. In case if the samples become viscous, incubate them on 50 °C for 15 s using a block thermostat.
3. Load your samples into the wells along with protein molecular weight marker.
4. Load 1× SDS protein loading buffer (see **Note 16**) in the remaining empty wells.
5. Electrophorese at 120 V, 60 mA for 60 min. (*see* **Note 17**).

3.4.3 Wet Transfer to PVDF Membrane

1. Following electrophoresis, pry the gel plates open with the use of a spatula. Rinse the gel with water and transfer carefully to a plastic or glass dish with 1× transfer buffer.

2. Cut a piece of PVDF membrane of the same size as the gel, and immerse it first in methanol, then in water, and finally in transfer buffer in a plastic or glass dish, for 3 min each at room temperature (RT) with shaking (*see* **Note 18**).
3. Assemble the blot sandwich, as recommended by the manufacturer.
4. Fill up the tank of the electrophoresis instrument with 1× Transfer buffer, and insert the blot sandwich.
5. Transfer the proteins at 300 mA for 1 h at 4 °C (*see* **Note 19**).

3.4.4 Immunoblotting

1. Following transfer, rinse the membrane in water.
2. Prepare 5% milk powder solution dissolved in 1× TTBS.
3. Place the membrane into the blocking solution in a plastic or glass dish for 1 h at RT with shaking (*see* **Note 20**).
4. If different secondary antibodies will be used, the membrane can be cut into pieces after blocking with a sharp blade, according to the size of the proteins of interest (Actin—42 kDa, Tubulin—50 kDa, HP1—22 kDa, H3—15 kDa).
5. Dilute primary antibody in blocking solution. Primary antibody working dilutions depend on the antibody used (*see* **Note 21**).
6. Incubate the membranes with the primary antibody in a plastic or glass dish for 2 h at RT (*see* **Note 22**) with shaking.
7. Wash the membrane twice with 1× TTBS, for 10 min each, at RT with gentle shaking.
8. Dilute HRP-conjugated secondary antibody in blocking solution. Working dilutions can vary depending on the antibody used (*see* **Note 23**).
9. Incubate the membranes with the secondary antibody in a plastic or glass dish for 1 h at RT with shaking.
10. Wash the membrane three times with 1× TTBS for 30 min each, at RT with gentle shaking.
11. Follow the manufacturer's instructions to prepare the chemiluminescent solution and pipette it onto the membrane. For a 10 × 10 cm PVDF membrane, 1 mL volume of solution is sufficient.
12. Remove the chemiluminescent solution from the membrane with the help of filter paper and lay the membrane on a piece of plastic wrap or glass without covering it.
13. Place the membrane together with the plastic wrap or glass in a chemiluminescence imaging system to detect the signal (*see* **Note 24**). Actin signal is visible at 42 kDa (*see* **Note 25**).

4 Notes

1. We recommend to use 110 mm long tweezers with a 0.05×0.01 mm tip.
2. Glass tools are preferred over plastic ware because unfixed tissues tend to adhere to plastic surfaces.
3. Dish size should fit to the size of the PVDF membrane to enable the submergence and shake of it in the working solutions.
4. We tested multiple commercially available anti-Actin antibodies on our protein samples. The results presented in Fig. 2 were obtained with the protein G purified version (Merck-Millipore, MAB1501R) of the widely used C4 anti-Actin monoclonal antibody (Merck-Millipore, MAB1501), because we experienced reduced background with the purified antibody. The rat (Babraham Tech, BT-GB-237P) and mouse (Cedarlane, C4, CLT9001) monoclonal antibodies detected actin weakly in our assays. Interestingly, the anti-Actin N-terminal polyclonal antibody (Merck, A2103 rabbit polyclonal) recognized the endogenous protein only, and did not detect N-terminally tagged Actin, independent of the tag used. We obtained no signal with the 2G2 mouse monoclonal anti-Actin antibody (Merck, MABT826).
5. We used females of wild type (*white*^{1118}, Bloomington Drosophila Stock Centre, #3605) and Flag-V5-Actin expressing flies (unpublished stock created in our laboratory) in the experiments presented here.
6. 1× PBS can be substituted with Schneider's Drosophila Medium, Grace's medium, or Ringer solution.
7. We found that use of Schneider's medium is the least stressful for dissected tissues, as it maintains ovarian cells in their natural environment [18].
8. To preserve the integrity of the samples and thereby avoid protein degradation, dissected ovaries should be kept on ice for no longer than 20 min before protein fraction isolation. To achieve this, we recommend that two to three colleagues carry out the dissection of the different genotypes in parallel.
9. The method described in [19] has been optimized to reduce cytoplasmic contamination in the nuclear fraction isolated from ovaries.
10. Gentle homogenization is pivotal in order to avoid shearing of the nuclei.
11. Total sample can be collected at this point, if desired.

12. As most spectrophotometers usually require few microliters of the sample for the measurement, we recommend preparing the dilutions of BSA standards and the cytoplasmic and nuclear protein fractions in a volume of 5–10 μL.
13. Bradford reagent can easily stain the glass surface of the instrument, which can affect the measurements. Therefore, the sample holder must be thoroughly cleaned after every measurement with 70% ethanol and a piece of filter paper.
14. Manufacturers provide glass plates for making gels of different thickness. We use 1 mm thick acrylamide gels. Glass plates should be thoroughly cleaned with detergent and alcohol before use.
15. It is important to avoid bubbles when mixing the gel and to pour the gel before polymerization starts. Depending on the temperature, the polymerization lasts about 20–30 min, and can be confirmed by checking the left-over gel mixture in the conical flask. We use gels freshly made up to 1 day prior to electrophoresis, but in case if the gel must be prepared several days ahead, store it at 4 °C in plastic wrap or plastic bag. Keep the comb inserted and remove it only before use.
16. 1× SDS protein loading buffer is diluted from the 2× SDS protein loading buffer with water.
17. Running conditions can vary depending on the percentage of acrylamide, thickness, and number of gels used. We typically use the same conditions for running the stacking and resolving gels.
18. Always handle membranes carefully and with clean forceps. Before you use them, clean all tools for incubations with methanol. Use enough reagents to prevent areas of the membrane from drying out during incubations and washes, because it can cause weak specific signal and increased background. Gently agitate during every incubation.
19. Transfer can be performed at 200 mA for 2 h or 35 mA overnight at 4 °C.
20. Blocking can be carried out overnight at 4 °C.
21. We used the following primary antibodies and dilutions: protein G purified anti-Actin C4 (Merck-Millipore, MAB1501R Mouse monoclonal, 1:3000), anti-beta Tubulin (DSHB, E7, 1:200), anti-H3 (Sigma-Aldrich, ab1791, 1:5000), anti-HP1 (DSHB, C1A9, 1:50), and anti-Flag (Sigma-Aldrich, F1804, 1:5000) antibody in blocking solution. Primary antibodies can be diluted also in 5% BSA dissolved in 1× TTBS.
22. Primary antibody can be incubated on the membrane overnight at 4 °C.

23. We used the following secondary antibodies and dilutions: anti-mouse horseradish peroxidase (HRP) conjugated (Jackson ImmunoResearch, 115-035-003) in 1:5000, and anti-rabbit HRP conjugated (Jackson ImmunoResearch, 111-035-144) in 1:10,000 dilution in blocking solution. Other secondary antibodies can be used as well. Do not exceed incubation time, as it can cause background.
24. It is also possible to develop radiographic films with developer and fixer solutions in a dark room.
25. Detecting nuclear actin in western blot experiments does not reveal any information about the polymerization status of nuclear actin because of the denaturing conditions.

Acknowledgments

This work was supported by NKFIH (National Research, Development and Innovation Office) through the National Laboratory for Biotechnology program, grant NKFIH-871-3/2020 (Péter Vilmos), and PD127968 (Ildikó Kristó).

References

1. Svitkina T (2018) The actin cytoskeleton and actin-based motility. Cold Spring Harb Perspect Biol 10:a018267
2. Dominguez R, Holmes KC (2011) Actin structure and function. Annu Rev Biophys 40: 169–186
3. Kristó I, Bajusz I, Bajusz C et al (2016) Actin, actin-binding proteins, and actin-related proteins in the nucleus. Histochem Cell Biol 145: 373–388
4. Percipalle P, Vartiainen M (2019) Cytoskeletal proteins in the cell nucleus: a special nuclear actin perspective. Mol Biol Cell 30:1781–1785
5. Hurst V, Shimada K, Gasser SM (2019) Nuclear actin and actin-binding proteins in DNA repair. Trends Cell Biol 29:462–476
6. Ulferts S, Prajapati B, Grosse R et al (2021) Emerging properties and functions of actin and actin filaments inside the nucleus. Cold Spring Harb Perspect Biol 13:1–16
7. Dopie J, Skarp KP, Rajakylä EK et al (2012) Active maintenance of nuclear actin by importin 9 supports transcription. Proc Natl Acad Sci U S A 109:E544
8. Bohnsack MT, Stüven T, Kuhn C et al (2006) A selective block of nuclear actin export stabilizes the giant nuclei of *Xenopus* oocytes. Nat Cell Biol 8:257–263
9. Jankovics F, Sinka R, Lukácsovich T et al (2002) MOESIN crosslinks actin and cell membrane in *Drosophila* oocytes and is required for OSKAR anchoring. Curr Biol 12: 2060–2065
10. Hudson AM, Cooley L (2002) Understanding the function of actin-binding proteins through genetic analysis of *Drosophila* oogenesis. Annu Rev Genet 36:455–488
11. Groen CM, Tootle TL (2015) Visualization of actin cytoskeletal dynamics in fixed and live *Drosophila* egg chambers. Methods Mol Biol 1328:113–124
12. Krishnan RK, Baskar R, Anna B et al (2021) Recapitulating Actin Module organization in the *Drosophila* oocyte reveals new roles for Bristle-Actin-Modulating proteins. Int J Mol Sci 22:4006
13. Kelpsch DJ, Groen CM, Fagan TN et al (2016) Fascin regulates nuclear actin during *Drosophila* oogenesis. Mol Biol Cell 27:2965–2979
14. Baarlink C, Plessner M, Sherrard A et al (2017) A transient pool of nuclear F-actin at mitotic exit controls chromatin organization. Nat Cell Biol 19:1389–1399
15. Sokolova M, Moore HM, Prajapati B et al (2018) Nuclear actin is required for transcription during *Drosophila* oogenesis. iScience 9: 63–70

16. Bajusz C, Kristó I, Abonyi C et al (2021) The nuclear activity of the actin-binding Moesin protein is necessary for gene expression in *Drosophila*. FEBS J 288:4812–4832
17. Avilés-Pagán EE, Orr-Weaver TL (2018) Activating embryonic development in *Drosophila*. Semin Cell Dev Biol 84:100–110
18. Akkouche A, Mugat B, Barckmann B et al (2017) Piwi is required during *Drosophila* embryogenesis to license dual-strand piRNA clusters for transposon repression in adult ovaries. Mol Cell 66:411–419.e4
19. Lo PL, Bonaccorso R, Onorati MC (2015) Nuclear and cytoplasmic soluble proteins extraction from a small quantity of Drosophila's whole larvae and tissues. Int J Mol Sci 16: 12360–12367

Chapter 20

STAMP: Spatio-Temporal Association Mapping of Proteins

Yuanbing Zhang, Bo Zhang, and Ji-Long Liu

Abstract

We describe a tool, Spatio-Temporal Association Mapping of Proteins (STAMP), for identifying protein interactomes via proximity labeling. For a proof-of-principle study, we use cytidine 5′-triphosphate synthase (CTPS) as an example. CTPS, a metabolic enzyme, forms filamentous structures termed cytoophidia in various tissues. We apply STAMP to a variety of developmental stages and tissues in *Drosophila* including adult ovaries. Using a cell-specific GAL4 driver, we verify that TurboID can biotinylate the bait protein CTPS, making possible the identification of protein–protein interactions (PPIs) in individual cells. Using the wild-type and mutant CTPS as bait proteins, STAMP results in two distinct sets of proximate proteomes. Our results suggest that STAMP is a feasible tool to catch in vivo PPIs in situ at a defined spatiotemporal resolution.

Key words *Drosophila*, Ovary, STAMP, TurboID, Proximity labeling

1 Introduction

Dynamic protein–protein interactions (PPIs) function in almost all cellular biological processes. However, the study of such interactions is difficult due to the number of transient and weak interactions among proteins, and therefore, our understanding of the protein interactions is incomplete. Traditional biochemical methods have long been used to elucidate these interactions, but they only capture in vitro snapshots. However, a recently developed novel method, termed proximate labeling, combines proximity-based biotinylation with mass spectrometry (MS) to identify in vivo and dynamic protein interactomes.

The proximity labeling system relies on the identification of biotinylated proteins. Biotin is a natural and non-toxic coenzyme that has a high affinity for avidin and its kind, which facilitates the routine detection or purification of biotinylated proteins. This makes it possible to study weak/transient protein interactions in vivo and in vitro. BioID, the first biotin-based proximity labeling system, was established by the laboratory of Brian Burke in 2012

Michelle S. Giedt and Tina L. Tootle (eds.), *Drosophila Oogenesis: Methods and Protocols*,
Methods in Molecular Biology, vol. 2626, https://doi.org/10.1007/978-1-0716-2970-3_20,

Table 1
Proximity labeling systems were developed from BioID- and APEX-tagged approaches

	Enzyme	Source	Substrates	Labeling activity	Size (kDa)
BirA*	Biotin ligase	*E. coli*	Biotin	Several hours	35
TurboID	Biotin ligase	*E. coli*	Biotin	Within 10 min	35
miniTurbo	Biotin ligase	*E. coli*	Biotin	Within 10 min	28
APEX	Ascorbate peroxidase	Pea (synthetic)	Biotin + H_2O_2	Several minutes	27

*BirA means a mutant biotin ligase BirA (Arg118Gly), and it is also termed BioID.

[1]. Using the *E. coli* biotin ligase BirA, they mutated a single amino acid (R11G) to generate BioID, which is also termed as BirA*. This mutation disrupts its association with biotin and promotes biotin binding to exposed lysines on a neighboring protein.

Another efficient proximity labeling system is derived from engineered ascorbate peroxidase (APEX), which biotinylates nearby proteins with exogenous biotin–phenol in the presence of hydrogen peroxide [2]. The main benefit of APEX is its speed, as proximity labeling occurs within several minutes, rather than many hours like BioID. However, it should be noted that the working reagents of APEX can be toxic to cells and organisms, which limits its use in vivo.

Recently, the Ting Lab utilized yeast display-based directed evolution of wild-type BirA and developed two promiscuous biotin ligase mutants, 35 kD TurboID and 28 kD miniTurbo, which dramatically increase the speed and efficiency of proximity labeling [3] (Table 1). TurboID and miniTurbo enable sufficient biotin-based proximity labeling in just 10 min, and have been applied in profiling protein interactomes in *S. pombe*, *N. benthamiana*, *Arabidopsis* and *Drosophila* [4–7].

cytidine 5′-triphosphate synthase (CTPS) forms filamentous structures termed cytoophidia across organisms [8–12], and these filaments have been extensively explored in *Drosophila* ovaries. Growing egg chambers in each ovariole can be divided into 14 different developmental stages. Cytoophidia are present in all major cell types in the egg chamber, including follicle cells, nurse cells and the oocyte. Normally, each follicle cell contains only one cytoophidium [9]. However, the length of cytoophidia in follicle cells varies greatly from egg chambers at different stages, which correlates with Myc expression in *Drosophila* follicle cells [13]. Both macro-cytoophidia and micro-cytoophidia can be observed in nurse cells and in oocytes, and exhibit various lengths and shapes [9, 14].

Here, we describe a tool, Spatio-Temporal Association Mapping of Proteins (STAMP), for identifying protein interactions in vivo and in situ. We describe the step-by-step STAMP method, TurboID-mediated highly effective proximate labeling in

Drosophila ovaries, including plasmid construction and transgenic fly generation, fly hybridization and feeding, verification the efficiency of proximity labeling in vivo, mass spectrometry and data analysis.

2 Materials

2.1 Experimental Instruments

1. PCR machine.
2. Rotary mixer.
3. Western blot imaging system.
4. Centrifuge.
5. NanoDrop.
6. Dry-bath.
7. Motor-driven tissue grinder.
8. Thermostatic incubator.
9. Bacterial incubator shaker.
10. Artificial climate chest.
11. Dissecting microscope.
12. Fluorescent microscope.
13. Semi-dry electrophoretic transfer machine.
14. Eppendorf thermoMixer.
15. Magnetic separation rack.
16. Vacuum concentrators.
17. Mass spectrometer.
18. R software (version 4.0.0).

2.2 RNA Isolation and Reverse-Transcription (RT) Kits and Reagents

1. RNA isolation kit.
2. RT reagent kit.
3. Chloroform.
4. Absolute ethanol.

2.3 Plasmid Preparation and Cloning

1. pUASt attB plasmid (*see* **Note 1**).
2. EcoRI.
3. XbaI.
4. CutSmart buffer (*see* **Note 2**).
5. LB (Luria–Bertani) broth.
6. LB agar Ampicillin (100 μg/mL) plates.
7. Super-fidelity PCR kit.
8. PCR cleanup kit.
9. Seamless cloning kit.

10. DH5α competent cell.
11. Plasmid extraction Kit.
12. EndoFree Plasmid Maxi Kit.

2.4 Immunostaining

1. Tweezers.
2. 16% Paraformaldehyde.
3. 1× PBS (phosphate-buffered saline) buffer.
4. Bovine serum albumin (BSA).
5. Triton X-100.
6. Glass coverslip (18 × 18 mm).
7. Standard microscope slides (75 × 25 mm, 1 mm thick).
8. Vectashield mounting medium.
9. Clear nail polish.
10. Hoechst 33342.
11. Anti-V5 antibody.
12. Alexa fluor 488-labeled secondary antibody.
13. Confocal microscope.

2.5 Western Blotting

1. RIPA buffer (radioimmunoprecipitation assay lysis buffer).
2. 1.5 mL microcentrifuge tubes.
3. Protease inhibitor cocktail.
4. Protein loading buffer.
5. 12% SDS–PAGE gel.
6. MOPS (3-(N-morpholino) propane sulfonic acid) running buffer powder.
7. PVDF (polyvinylidene fluoride) membranes.
8. 1× TBST (tris-buffered saline +0.1% Tween-20): 12.1 g Tris, 87.7 g NaCl, 10 mL Tween-20 in 1 L ddH_2O, pH 7.
9. Enhanced chemiluminescence.
10. Streptavidin–Cy3.
11. Streptavidin–HRP.
12. HRP-linked anti-mouse antibody.
13. Western blot imaging system.

2.6 Mass Spectrometry

1. Grace's insect medium.
2. 1.5 mL microcentrifuge tubes (ultra-low retention).
3. Urea powder.
4. 1 M DTT (dithiothreitol).
5. 1 M Iodoacetamide.

6. Streptavidin magnetic beads.
7. Buffer 1: 50 mM Tris 8.0, 8 M urea, 200 mM NaCl, 0.2% SDS.
8. Buffer 2: 50 mM Tris 8.0, 200 mM NaCl, 8 M urea.
9. Buffer 3: 50 mM Tris 8.0, 0.5 mM EDTA, 1 mM DTT.
10. Buffer 4: 100 mM ammonium carboxylate.
11. Trypsin.
12. 10% Formic acid.
13. Ziptip C18.
14. Acetonitrile.
15. 0.1% Trifluoroacetic acid.
16. 1.5 mL microcentrifuge tube heat block or incubator.
17. Magnetic separation rack.
18. Vacuum concentrator.
19. Access to mass spectrometry facility.
20. R 3.6.1.

2.7 Fly Strains and Fly Culture

1. w^{1118}.
2. attp2 fly line.
3. *da*-GAL4 (daughterless-GAL4) (*see* **Note 3**).
4. Fly incubator with light and humidity control.
5. Biotin.
6. Standard *Drosophila* cornmeal food in fly vials and bottles. Food recipe according to Bloomington *Drosophila* Stock Center.
7. 100 μM biotin-containing cornmeal food: dissolve biotin in 50 mL ddH_2O to final concentration of 100 mM in one 50 mL tube. Heat standard cornmeal food until melted and let it cool, then add 100 mM biotin into mixture to a final concentration of 100 μM and aliquot into clean bottles.

3 Methods

3.1 Plasmid Construction and Transgenic Fly Generation

We used this method to design two fusion protein constructs, pUASt $CTPS^{WT}$–TurboID-V5 and pUASt $CTPS^{H355A}$–TurboID-V5, for transgenic fly generation to identify proteins that interact with cytophidium (Fig. 1). The H355A mutation of CTPS destroys tetramer interactions in a dominant-negative manner [15–17] and disturbs the formation of endogenous filaments. Thus, this construct serves as a negative control for cytophidia protein interaction partners. Depending on the protein of interest for the interaction

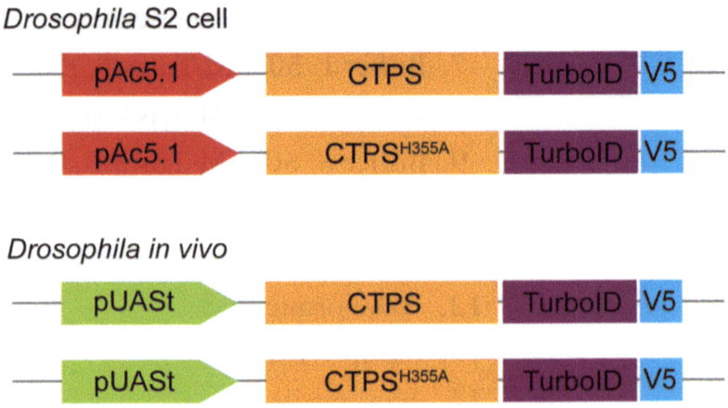

Fig. 1 Diagram of the expression cassettes used for proximity labeling. UASt sequence mediates the expression of the fusion protein (CTPS–turboID or CTPS–turboID) containing V5 tag in *Drosophila*

studies, the linker length, epitope tags, restriction enzyme selection, and localization sequences may need to be altered (*see* **Note 4**).

3.1.1 cDNA Generation

1. Collect several w^{1118} larvae or adults and isolate total RNA using RNA isolation kit according to the manufacturer's recommendations, and elute in 50 μL of nuclease-free water.
2. Generate the cDNA by performing RT reaction according to kit instructions (*see* **Note 5**).
3. PCR amplify the cDNA encoding the protein of interest with primers containing the desired linker sequence at the 3′ end. For example, for CTPS^{WT,} the primer contained GGGSG linker sequence.
4. If desired, perform site directed mutagenesis by PCR with specific primers to generate a mutant protein to impair or block protein interactions. For example, we generated CTPS^{H355A} to block cytophidium formation.
5. Commercially synthesize *Drosophila* codon-optimized TurboID sequences (*see* **Note 6**) containing the last 20 nt of the cDNA of interest gene (e.g., CTPS) as overlapped sequences at the 5′ end, and V5 sequence plus 20 nt sequence from the multiple cloning site of pUASt vector as overlapped sequences in at the 3′ end, which are required for subsequent seamless cloning (*see* **Note** 7). This yields the long sequence "-(last 20nt of cDNA)–TurboID-V5-(20nt from pUASt vector)-".

3.1.2 Plasmid Preparation and Cloning

1. Linearize the pUASt vectors by digesting with the restriction enzymes EcoRI and XbaI.
2. Clean up digested plasmids according to PCR cleaning kit manufacturer's instructions.

3. Insert the amplified cDNA and synthesized TurboID-V5 sequences into the digested pUASt vector by seamless cloning according to kit manufacture's instructions.
4. Transform the plasmid into DH5α competent cells by heat shock and plate the transformed cells on the appropriate LB plates containing 50ug/mL ampicillin. Incubate plates overnight at 37 °C.
5. Use individual colonies to inoculate 5 mL LB-amp and culture at 37 °C for 15 h.
6. Follow plasmid extraction kit manufacturer's instructions to purify the plasmids (*see* **Note 8**). Sequence the plasmids to verify cDNA presence and correct tagging.
7. Generate transgenic fly lines by mixing each plasmid with phiC31 integrase RNA, respectively, and injecting it into the embryonic fly germline (attP2) to obtain PhiC31 integrase-mediated site-specific integration. Alternatively, the plasmids can be sent to various companies that perform injection services.
8. Keep the embryos at room temperature or at 25 °C until adults hatch.
9. Screen F1 adult progeny for deep red eyes. Cross each deep red eyed fly to balancer flies to establish a stable stock, for example, +/+; TM6B/Sb.

3.2 Fly Hybridization and Feeding

1. Raise Turbo-ID and *da*-GAL4 flies in a 25 °C incubator, with 50% humidity under a 12 h/12 h light/dark cycle.
2. To drive the expression of the Turbo-ID fusion proteins, cross about 10 *da*-GAL4 virgin females with 10 males bearing UASt–TurboID in bottles containing either standard or 100 μM biotin-containing food.
3. Transfer the flies to fresh bottles every day. Retain the empty bottles, allowing the embryos to develop.
4. Collect F1 *da* >TurboID 14-day-old females and dissect ovaries in Grace's insect medium for further experiments (approximate 5 pairs for immunofluorescence and 20 pairs for western blot).

3.3 Verifying the Efficiency of Proximity Labeling In Vivo

To demonstrate TurboID proximate labeling feasibility of CTPS in *Drosophila*, immunofluorescence and western-blotting were used. Streptavidin–Cy3 and streptavidin–HRP signals reveal biotinylated proteins by TurboID, while V5 shows expression level of CTPS–TurboID fusion proteins (Fig. 2).

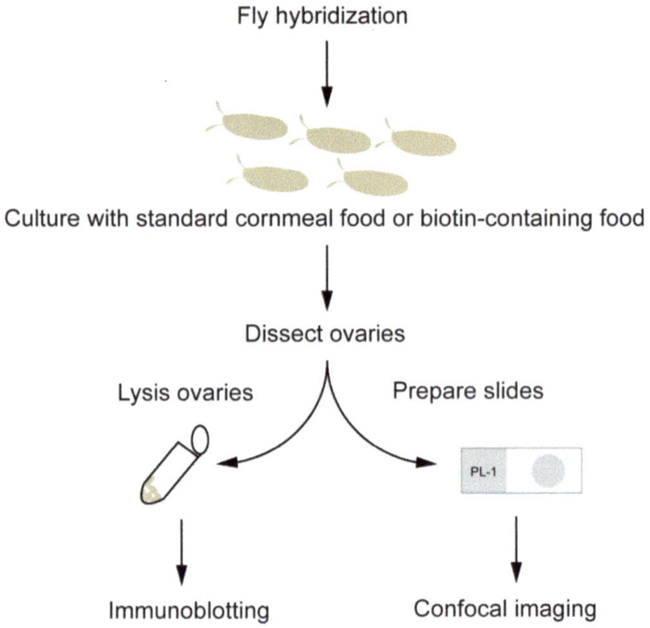

Fig. 2 Verification of the efficiency of proximity labeling in vivo. Transgenic flies were crossed with GAL4 driver flies to express TurboID ubiquitously or tissue specifically. Their recombinants were raised on either standard cornmeal food or 100 μM biotin-containing food accordingly. After biotinylation, ovaries are dissected and processed for verifying the efficiency of proximity labeling in two different ways: harvest protein for immunoblotting or prepare slides for confocal imaging

3.3.1 Verification of Ligase Expression and Labeling Efficiency by Immunofluorescence

1. Fix *da* >TurboID ovaries with paraformaldehyde 4% (w/v) in PBS for 20 min.
2. Wash three times with PBS.
3. Block with 5% BSA and 0.2% Triton X-100 in PBS for 15 min.
4. Incubate with diluted Anti-V5 antibody in 0.2% Triton X-100 in PBS and incubate overnight at room temperature.
5. Wash three times in PBS.
6. Incubate ovaries with diluted Alexa Fluor 488-labeled secondary antibody (*see* **Note 9**), Streptavidin–Cy3 and Hoechst 33342 (*see* **Note 10**) in 0.2% Triton X-100 in PBS for about 4 h at room temperature.
7. Transfer ovaries to coverslips and pipette excess liquid followed by adding ~20 μL mounting medium and sealing slide with clear nail polish.
8. Image by confocal microscopy.

3.3.2 Verification of Ligase Expression and Labeling Efficiency by Twice Western Blot

1. Lyse 10 pairs of *da* >TurboID ovaries with 100 μL RIPA buffer containing 1× protease inhibitor cocktail, and grind with a tissue grinder. Incubate on ice for 1 h.
2. Collect supernatants by centrifuging (10,000 g, 4 °C) for 15 min.
3. Boil with 1× protein loading buffer at 95 °C for 10 min.
4. Run samples on two SDS–PAGE gels (120 V, ~2 h).
5. Transfer proteins to PVDF membranes (25 V, 20 min).
6. Block the membranes with 5% BSA in 1 × TBST for 1 h.
7. To detect biotinylated proteins, incubate one PVDF membrane with Streptavidin–HRP diluted in 1 × TBST for 1 h (*see* **Note 11**).
8. Wash with TBST three times for 5 min each wash.
9. Visualize streptavidin–HRP signal via enhanced chemiluminescence and a Western Blot imaging system (*see* **Note 12**).
10. Incubate the other blocked membrane with anti-V5 antibody diluted in 1 × TBST at 4 °C overnight.
11. Wash with 1 × TBST three times for 5 min each wash.
12. Incubate membranes with HRP-linked anti-mouse antibody diluted in 1 × TBST for 1 h (*see* **Note 13**).
13. Analyze expression as described in **step 9** (*see* **Note 14**).

3.4 Mass Spectrometry and Data Analysis

After verifying expression of the TurboID fusion proteins and their labeling efficiency, please refer to the illustration of the proximity labeling strategy to identify the protein interactome (Fig. 3). TurboID was fused in-frame with wild-type and mutant CTPS. TurboID can use supplied biotin to biotinylate CTPS neighboring proteins. Biotinylated proteins are captured by streptavidin-based purification. Subsequently, enriched proteins are digested with trypsin followed by analyzing the peptides by MS.

1. Prepare 14-day-old females expressing the Turbo-ID fusion protein by incubating on 100 μM biotin-containing food (*see* **Note 15**).
2. Dissect approximately 60 pairs ovaries in Grace's insect medium and fix them in 4% paraformaldehyde (*see* **Note 16**).
3. Transfer ovaries into 1 mL RIPA buffer containing 1× protease inhibitor cocktail in a 1.5 mL microcentrifuge tube, homogenize with a tissue grinder, and lyse on ice for 1 h.
4. Centrifuge at 10,000 g, 4 °C for 10 min.
5. Transfer 800 μL supernatants into new 1.5 mL microtubes (ultra-low retention).

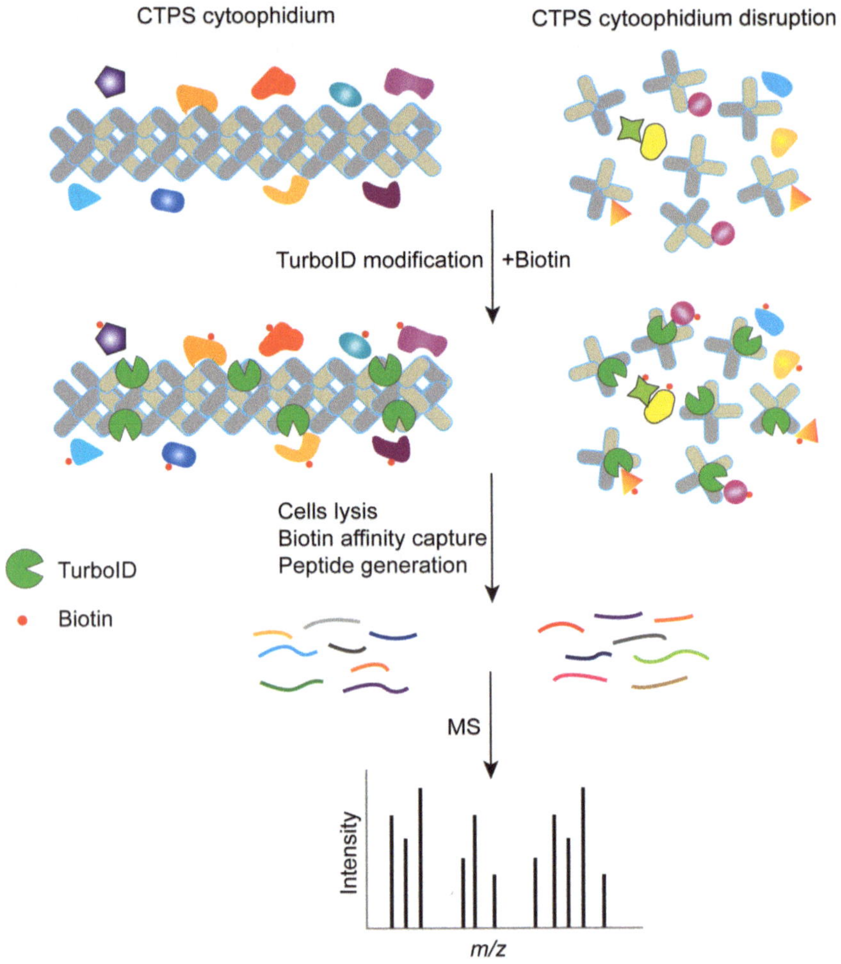

Fig. 3 Design of STAMP to map cytoophidia proximate proteomes in vivo. da-GAL4 drives fusion protein CTPSWT–TurboID-V5 or CTPS^{H355A}–TurboID-V5 expression in all tissues. TurboID uses biotin to biotinylate CTPS neighboring proteins. Ovaries are lysed and biotinylated proteins are enriched using streptavidin beads. Subsequently, small peptides are generated by trypsin digestion and peptides are analyzed by mass spectrometry. Note that just a finite number of TurboID are shown in CTPS cytoophidium and disrupted cytoophidium [7]

6. Add urea (powder: ~0.384 g) and DTT (stock: 1 M) to a final concentration of 8 M and 10 mM, respectively.
7. Incubate samples at 56 °C for 1 h.
8. Add iodoacetamide (stock: 1 M) to a final concentration of 25 mM and incubate samples in the dark for 45 min at room temperature to covalent modify reduced cystine with carbamidomethyl group.
9. Add DTT (stock: 1 M) to a final concentration of 25 mM and incubate for 30 min at room temperature to terminate the modification.

10. Wash 50 μL Streptavidin magnetic beads with 500 μL PBS three times in a new 1.5 mL microcentrifuge tube. After the wash, pellet beads using a magnetic separation rack and remove the PBS.
11. Resuspend the beads with the lysate.
12. Incubate on a rotator at 4 °C overnight.
13. Wash the beads twice with Buffer 1 on a rotator for 5 min each at room temperature.
14. Wash the beads once with Buffer 2 on a rotator for 5 min at room temperature.
15. Wash the beads twice with Buffer 3 for 5 min each on a rotator at room temperature.
16. Wash the beads three times with Buffer 4 for 5 min each on a rotator at room temperature.
17. Remove the last wash and resuspend the beads in 100 μL Buffer 4 containing 4 μg Trypsin and incubate on 37 °C shaker overnight to digest proteins and generate peptides.
18. Transfer the supernatant into new 1.5 mL tubes (ultra-low retention) and add formic acid to final concentration of 1%, which reduces the pH and deactivates the enzyme.
19. Wash Ziptip with 100 μL 100% Acetonitrile (ACN) gently.
20. Equilibrate Ziptip with 100 μL 0.1% Trifluoroacetic acid (TFA) twice.
21. Load peptides into Ziptip and pipette up and down slowly 4–5 times for full adsorption.
22. Wash peptides on the Ziptip three times with 100 μL 0.1% TFA for desalinization.
23. Elute peptides into new 1.5 mL microcentrifuge tubes (ultra-low retention) by adding 50 μL 70% ACN–0.1% TFA carefully and collecting the flow through. Repeat three times.
24. Fast dry the sample with a vacuum concentrator.
25. Send enriched peptides for analysis by mass spectrometry (Orbitrap Fusion).
26. Analyze and visualize data R 3.6.1 (Fig. 4). For analysis details, *see* refs. [7, 18] (*see* **Note 17**).

4 Notes

1. UASt and UASp both work well in somatic cells and germline cells from ovaries. We have tested the expression efficiency of UASt and chose it as an example in this chapter [7].

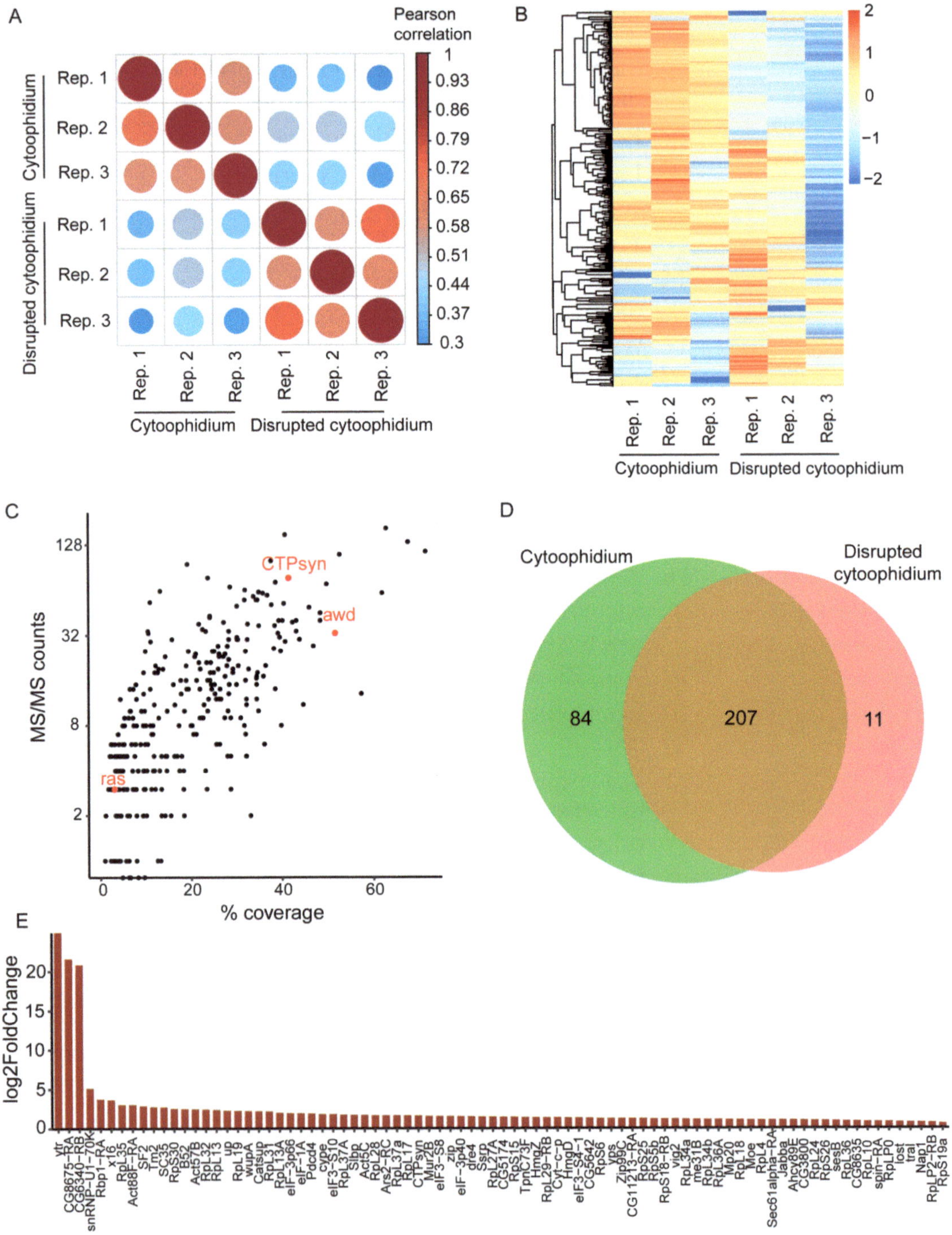

Fig. 4 Data analysis and visualization of cytoophidia proximate proteomes. (**a**) Pearson correlation coefficients between replicate (Rep.) mass spectrometry for cytoophidium and disrupted cytoophidium groups. (**b**) Hierarchical clustering of the proteome of CTPS cytoophidium and disrupted cytoophidium in three replicates. (**c**) Scatter plots by MS/MS count up the *y*-axis and percentage sequence coverage (amino acids) on the *x*-axis. Points corresponding to previously established CTPS interacting proteins in STRING database are labeled in red. (**d**) Venn diagram showing overlap and unique enriched proteins adjacent to CTPS cytoophidium and disrupted cytoophidium. (**e**) Bar plot list of 84 enriched neighboring proteins of CTPS cytoophidium compared to disrupted cytoophidium [7]

2. The selection of digestion buffer and optimal temperature depends upon the enzyme properties, and can be verified by product websites.
3. The *da*-GAL4 driver provides ubiquitous expression of TurboID-fusion proteins at a lower level than *actin*-GAL4. High levels of expression can sequester endogenous biotin and cause toxicity and, therefore, should be avoided.
4. Different tagging strategies may influence target protein folding and localization. Here, we identified that TurboID has little effects on CTPS cytoophidium conformation [7] and small V5 tag makes it easy to visualize the expression of fusion protein.
5. To perform gene expression analysis accurately, it is necessary to detect cDNA in samples without contaminating genomic DNA (gDNA). One means of removing gDNA is to use a RT Kit with reagent to degrade gDNA, such as PrimeScript RT reagent kit with gDNA eraser.
6. The TurboID sequences or plasmids can be obtained from Addgene. Codon optimization improves gene expression based on an organism's codon bias. A good tool for codon optimization is ExpOptimizer, https://www.novoprolabs.com/tools/codon-optimization.
7. Seamless cloning technology enables directional insertion of any DNA fragments into any linearized vector with high efficiency. Linearized vectors via enzyme digestion at the cloning site are required. A small sequence (~20 bp) overlapped with each end of the cloning site is added onto the DNA fragments through PCR. The fragments and the linearized vectors, with overlapped sequences (~20 bp) on both 5′- and 3′-ends, respectively, are mixed and incubated with recombinase.
8. Removal of endotoxin is required to yield plasmid DNA suitable for cell transfection and transgene injection into the fly embryo. Here, we recommend QIAGEN endofree plasmid mega kit, which harvests up to 10 mg endotoxin-free advanced transfection-grade plasmid.
9. Here, Alexa Fluor 488-labeled secondary antibody is specific for the mouse monoclonal Anti-V5 antibody. The secondary antibody should be against the host species of the chosen primary antibody.
10. Hoechst 33342 and Hoechst 33258 are membrane permeable, which gives them a huge advantage for staining live cells. Here, Hoechst 33342 can be used to stain fixed cells by substituting Hoechst 33342 for DAPI in general operation.
11. We find that it is best to incubate PVDF membranes with Streptavidin–HRP diluted in TBST, as commercial antibody dilution buffers can result in signal variability and/or high background.

12. Streptavidin–HRP blotting signal in flies raised with exogenous biotin are stronger than the control group, but there is always a band indicating the location of biotinylated CTPS–TurboID-V5.
13. The mouse monoclonal Anti-V5 antibody can react specifically with the V5-tagged recombinant fusion proteins. Therefore, we choose HRP-linked anti-mouse antibody as secondary antibody for Western blotting.
14. Anti-V5 antibody was used to detect the fused expression of CTPS–mCherry and CTPS–TurboID, which was expressed ubiquitously via da-GAL4 driver.
15. While higher concentrations of biotin can be used, we find raising flies on 100 μM biotin-containing food avoids exhausting endogenous biotin and causing toxicity.
16. This is an optional step. Formaldehyde-fixation is mandatory for proteomics samples prepared from diseased animals. Here, there may be some developmental or physiological defects in flies carrying $CTPS^{H355A}$ mutation and we performed this in our assay.
17. The data were log2 transformed and centered. Enriched proteins of CTPS cytoophidium with fold change >1.5 and $P < 0.05$ were defined as upregulated proteins and those with fold change <0.67 and $P < 0.05$ were defined as downregulated proteins for further study.

Acknowledgments

We thank the Molecular Imaging Core Facility (MICF), Multi-Omics Core Facility (MOCF) and the Molecular and Cell Biology Core Facility (MCBCF) at the School of Life Science and Technology, ShanghaiTech University for providing technical support. The research in the Liu Laboratory was supported by the Ministry of Science and Technology of China (grant no. 2021YFA0804700), the National Natural Science Foundation of China (grant no. 31771490), the Shanghai Science and Technology Commission (20JC1410500), and the UK Medical Research Council (grant nos. MC_UU_12021/3 and MC_U137788471) and ShanghaiTech University. This research was supported by Shanghai Frontiers Science Center for Biomacromolecules and Precision Medicine at ShanghaiTech University.

References

1. Roux KJ, Kim DI, Raida M, Burke B (2012) A promiscuous biotin ligase fusion protein identifies proximal and interacting proteins in mammalian cells. J Cell Biol 196(6):801–810. https://doi.org/10.1083/jcb.201112098
2. Rhee HW, Zou P, Udeshi ND, Martell JD, Mootha VK, Carr SA, Ting AY (2013) Proteomic mapping of mitochondria in living cells via spatially restricted enzymatic tagging. Science (New York, NY) 339(6125):1328–1331. https://doi.org/10.1126/science.1230593
3. Branon TC, Bosch JA, Sanchez AD, Udeshi ND, Svinkina T, Carr SA, Feldman JL, Perrimon N, Ting AY (2018) Efficient proximity labeling in living cells and organisms with TurboID. Nat Biotechnol 36(9):880–887. https://doi.org/10.1038/nbt.4201
4. Larochelle M, Bergeron D, Arcand B, Bachand F (2019) Proximity-dependent biotinylation mediated by TurboID to identify protein-protein interaction networks in yeast. J Cell Sci 132(11):jcs232249. https://doi.org/10.1242/jcs.232249
5. Zhang Y, Song G, Lal NK, Nagalakshmi U, Li Y, Zheng W, Huang PJ, Branon TC, Ting AY, Walley JW, Dinesh-Kumar SP (2019) TurboID-based proximity labeling reveals that UBR7 is a regulator of N NLR immune receptor-mediated immunity. Nat Commun 10(1):3252. https://doi.org/10.1038/s41467-019-11202-z
6. Mair A, Xu S-L, Branon TC, Ting AY, Bergmann DC (2019) Proximity labeling of protein complexes and cell-type-specific organellar proteomes in Arabidopsis enabled by TurboID. eLife 8:e47864. https://doi.org/10.7554/eLife.47864
7. Zhang B, Zhang Y, Liu JL (2021) Highly effective proximate labeling in Drosophila. G3 (Bethesda, Md) 11(5):jkab077. https://doi.org/10.1093/g3journal/jkab077
8. Ingerson-Mahar M, Briegel A, Werner JN, Jensen GJ, Gitai Z (2010) The metabolic enzyme CTP synthase forms cytoskeletal filaments. Nat Cell Biol 12(8):739–746. https://doi.org/10.1038/ncb2087
9. Liu J-L (2010) Intracellular compartmentation of CTP synthase in Drosophila. J Genet Genomics 37(5):281–296. https://doi.org/10.1016/S1673-8527(09)60046-1
10. Noree C, Sato BK, Broyer RM, Wilhelm JE (2010) Identification of novel filament-forming proteins in Saccharomyces cerevisiae and Drosophila melanogaster. J Cell Biol 190(4):541–551. https://doi.org/10.1083/jcb.201003001
11. Liu J-L (2016) The Cytoophidium and its kind: Filamentation and compartmentation of metabolic enzymes. Annu Rev Cell Dev Biol 32:349–372. https://doi.org/10.1146/annurev-cellbio-111315-124907
12. Zhou S, Xiang H, Liu JL (2020) CTP synthase forms cytoophidia in archaea. J Genet Genomics 47(4):213–223. https://doi.org/10.1016/j.jgg.2020.03.004
13. Aughey GN, Grice SJ, Liu JL (2016) The interplay between Myc and CTP synthase in drosophila. PLoS Genet 12(2):e1005867. https://doi.org/10.1371/journal.pgen.1005867
14. Zhang Y, Liu J, Liu JL (2020) The atlas of cytoophidia in Drosophila larvae. J Genet Genomics 47(6):321–331. https://doi.org/10.1016/j.jgg.2020.06.004
15. Sun Z, Liu J-L (2019) Forming cytoophidia prolongs the half-life of CTP synthase. Cell Discov 5(1):32. https://doi.org/10.1038/s41421-019-0098-6
16. Zhou X, Guo CJ, Hu HH, Zhong J, Sun Q, Liu D, Zhou S, Chang CC, Liu JL (2019) Drosophila CTP synthase can form distinct substrate- and product-bound filaments. J Genet Genomics 46(11):537–545. https://doi.org/10.1016/j.jgg.2019.11.006
17. Liu J, Zhang Y, Zhou Y, Wang Q-Q, Ding K, Zhao S, Lu P, Liu J-L (2022) Cytoophidia coupling adipose architecture and metabolism. Cell Mol Life Sci 79(10) 534. https://doi.org/10.1007/s00018-022-04567-w
18. Zhang Z, Sun W, Shi T, Lu P, Zhuang M, Liu JL (2020) Capturing RNA-protein interaction via CRUIS. Nucleic Acids Res 48(9):e52. https://doi.org/10.1093/nar/gkaa143

Chapter 21

Using *Drosophila* Oogenesis in the Classroom to Increase Student Participation in Biomedical Research

Jayson A. Cortes and Christina I. Swanson

Abstract

Students that participate in undergraduate research benefit in multiple ways, including improved learning outcomes, increased enthusiasm for science, technology, engineering, and mathematics (STEM) fields, and increased likelihood of continuation into a STEM career. These benefits are even more pronounced for students that are traditionally under-represented in STEM, although these students often face barriers to participation in traditional apprenticeship-style research experiences. Course-based undergraduate research experiences (CUREs) are a promising and increasingly popular approach to increase undergraduate participation in research in a way that is inclusive of all students. Here, we describe how *Drosophila* oogenesis can be used as the basis for CUREs in a wide variety of courses. We provide an overview of our own oogenesis-based CURE, as well as suggestions for how this CURE could be adapted to accommodate a variety of schedules, course sizes, and institution types. Our goal is to simplify the process for CURE implementation in the hopes that a greater number of instructors choose to implement a CURE in their own courses.

Key words Course-based undergraduate research experience, CURE, Undergraduate research, STEM education, *Drosophila* oogenesis, Inclusive pedagogy

1 Introduction

1.1 Introduction to Course-Based Undergraduate Research Experiences (CUREs)

An expansive body of research shows that undergraduate students benefit greatly from participation in undergraduate research experiences (UREs). Research experiences can increase student understanding of core concepts and improve course outcomes [1, 2], increase their understanding of the scientific process [3–6], increase their enthusiasm for science in general [4, 5], and increase their continuation into a science, technology, engineering, and mathematics (STEM) career [6–8]. These advantages are often particularly pronounced for students that belong to groups that are traditionally under-represented in STEM; multiple studies have shown that participation in high impact practices, including undergraduate research, is important for the retention of students who identify as Black, Indigenous, or people of color (BIPOC) in STEM

Michelle S. Giedt and Tina L. Tootle (eds.), *Drosophila Oogenesis: Methods and Protocols*,
Methods in Molecular Biology, vol. 2626, https://doi.org/10.1007/978-1-0716-2970-3_21,

programs, and increases the likelihood that these students will go on to pursue a career in STEM [9–12]. Given these benefits, colleges and universities should strive to provide research opportunities for as many undergraduate students as possible.

Apprenticeship-style undergraduate research experiences (UREs) are the classic model of undergraduate research. In traditional UREs, students receive one-on-one training from a mentor—typically a faculty member, or perhaps a postdoctoral fellow or graduate student—as they work on an independent research project [13]. UREs can be enriching experiences that allow students to develop close mentorship relationships and feel that they are making important contributions to their lab's research goals, both of which help students develop their STEM identity [14–16]. However, UREs are very labor-intensive for the research mentor, and this can be especially challenging at primarily undergraduate institutions (PUIs), where faculty bear heavy teaching loads and may have limited financial resources to conduct research [17–20]. As a result, UREs are often restricted in number, and in many cases, the students that are excluded from these opportunities are the very students that might benefit most. Female students, first generation students, and BIPOC students can face significant barriers to participation in UREs, such as a lack of role models, a sense of alienation from campus culture, and financial and/or cultural constraints that limit extracurricular involvement [11, 21–23]. Therefore, there is a need for alternative models for undergraduate research that are more inclusive and can accommodate a greater number of students.

Course-based research experiences (CUREs) are intended in part to address the limitations of traditional UREs. CUREs are research experiences that occur within classrooms and teaching labs [24–26]. They are distinguished from UREs in that students participate within the context of a course that they are taking, working collaboratively with other students enrolled in the course. CUREs are distinguished from other types of teaching lab activities, even those that qualify as inquiry-based, in that the students work on previously unanswered scientific questions and collect data that are novel not just to themselves, but to the scientific community at large. Students also directly participate in multiple aspects of the scientific process throughout the CURE. This could include reading scientific literature relevant to their research question, contributing to experimental design, collecting and analyzing data, interpreting results, and communicating their findings. As a result, students participating in a CURE gain much of the same knowledge and experience as students participating in UREs, and enjoy many of the same benefits [2, 7, 27–29]. Because CUREs have few barriers to participation—students must simply enroll in the course—they have great potential to broaden student access to authentic research experiences. In fact, CUREs are seen as an important tool in the ongoing effort to make STEM education more inclusive and to improve outcomes for a diverse array of STEM students [21, 30, 31].

While the benefits for students participating in CUREs are evident, there are many advantages for instructors as well, particularly those at PUIs that must balance heavy teaching loads with research productivity. CUREs present an opportunity to optimize time and collect valuable research data while teaching: faculty can use the CURE to make progress on an existing research program, to collect preliminary data for potential future projects, or to expand their scholarship into pedagogical research [32, 33]. CUREs can also be valuable mechanisms for recruiting students into independent research labs [32, 33]. When a CURE is used to replace a "cookbook-style" lab, the overall course is improved and enriched by the addition of critical thinking, data analysis, and data presentation exercises [33, 34]. And finally, many instructors report that the CURE makes the course more engaging and enjoyable for both faculty and students [33, 34].

Many faculty are hesitant to integrate CUREs in their own courses, because they lack the time to design and/or implement an appropriate project [33, 35, 36]. However, this hesitancy can be overcome as more and more educators share their own ideas for easily adopted, widely applicable CUREs, reducing the need for individuals to invent their own. *Drosophila* oogenesis provides an excellent basis for CUREs in a wide variety of classroom settings. The process of oogenesis involves steps that are relevant for multiple courses, including introductory biology (reproduction), genetics (meiosis), cell biology (stem cells), and physiology (nutrition, aging). Universities with a large number of students, or limited resources, can use simple methods to assess oogenesis, such as embryonic hatch assays. In smaller, upper-level courses where the instructor prefers to teach more technical skills, ovaries can be dissected, fixed, stained, and examined using microscopy. This chapter provides a guideline for the design and implementation of an oogenesis-based CURE, with suggestions for adaptations to a variety of courses.

1.2 Design and Implementation of an Oogenesis-Based CURE

1.2.1 Experimental Questions and General Approaches

The CURE described here is an exploratory project, seeking preliminary answers to the question "is gene *X* required for *Drosophila* oogenesis?" (for suggested variations *see* **Note 1**). This question was addressed using loss-of-function experiments. Loss-of-function experiments are easy and affordable to conduct in a variety of research settings due to the wide range of mutant alleles and RNAi lines available from various stock centers. Two different assays were chosen to assess oogenesis: embryonic hatch assays, and dissection and DAPI staining of adult ovaries. Both assays can be easily implemented in an undergraduate lab, require only simple reagents and resources, and allow students to learn different skills. Hatch assays are quantitative and provide opportunities for students to carry out

Table 1
Overview of CURE structure

Timeline	Instructor tasks	Student tasks
One week before Lab Week 1	Cross parents to generate offspring of desired genotype	None
Lab Week 1	CURE Introduction Remove parents from vials	CURE Introduction—background reading and preparatory assignment
Lab Week 2	Prepare lab reagents, monitor crosses if needed	Set up embryo collection chambers Set up females for ovary dissections
Lab Week 3	Maintain embryo collection chambers, prepare reagents and set up lab	Set up and complete hatch assays Dissect, stain, and examine ovaries
Lab Week 4	Distribute data	Data analysis, and prepare for presentations
Lab Week 5	None	Presentations

statistical evaluation of their data, while examination of stained ovaries allows students to practice microdissection and fluorescent microscopy. Furthermore, because the outcomes of CURE experiments are often unknown to either the instructor or the students, the combination of assays is more likely to provide interesting data than either experiment alone.

1.2.2 Introduction of the CURE and Preparatory Assignments

The CURE, which occupied five weekly lab sessions, began with a brief presentation and an accompanying assignment intended to introduce the CURE and to begin to involve students in the scientific process (*see* Table 1 for overview of CURE timeline). The presentation included a general explanation of the goal of the CURE: to conduct research that builds on previous findings and generates novel data that will make a valuable contribution to the scientific community. The presentation also reviewed the basic background of the project, including an overview of *Drosophila* oogenesis [37, 38], how it is related to topics covered in class, and the name of the specific gene of interest. However, the presentation did not include the specific hypothesis or experimental plan, so that students had an opportunity to independently practice these aspects of the experimental process in the accompanying assignment.

Students were then expected to read the relevant scientific literature about the gene under investigation. This typically included a combination of published papers and posters from students that participated in the CURE in previous semesters. This process is similar to how students would begin to familiarize themselves with their project in an URE. However, while students in an

URE might be expected to locate sources themselves, the CURE students were provided with the most relevant sources for their project. The students read this background information and completed an assignment (due in lab the following week) that required them to practice components of the scientific process. They were asked to (1) propose a hypothesis about the role of gene "X" in oogenesis; (2) propose an experiment to test their hypothesis; and (3) predict possible experimental outcomes of that experiment.

1.2.3 CURE Experiments, Data Collection, and Data Analysis

During weeks 2 and 3 of the CURE, students worked in groups to conduct their CURE experiments (Table 1). During week 2 of the CURE, students collected female flies and prepared them for dissection, and they also set up embryo collection chambers. During week 3 of the CURE, they dissected, fixed, and stained ovaries, examined the stained ovaries on the microscope, and completed hatch assays using their embryo collection chambers. In week 4 of the CURE, students met with their groups to analyze their data, prepare figures to depict their results, and discuss conclusions (Table 1, and *see* Fig. 1 for an example of a student-generated data figure). Most of this discussion occurred amongst the members of the lab group, but the instructor also circulated to provide assistance as needed, and students were encouraged to share their results with members of other lab groups as well, since science is a collaborative process. Students could also use this time to begin to work on their presentation for the final week of the CURE.

A

Maternal genotype	Total # Eggs	# Eggs Unhatched	# Eggs Hatched	Hatch rate (%)
Control	20	3	17	85
Mutant	20	20	0	0

B

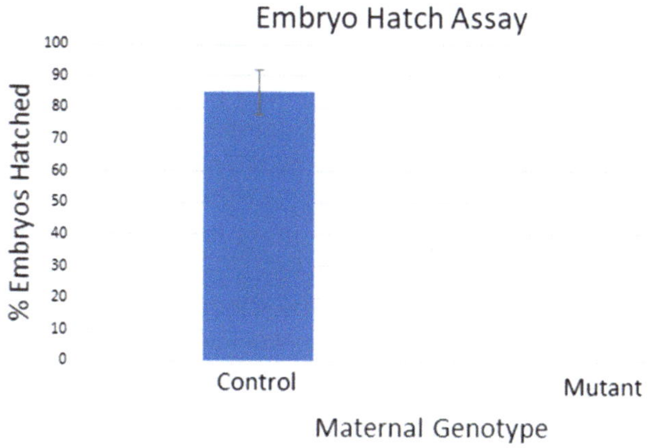

Fig. 1 Example of student-generated data. Hatch assays results are depicted in both a table (**a**) and a graph (**b**). Table includes data from just one group, while graph includes data from multiple groups that have been averaged. Error bars show standard deviation

1.2.4 CURE Presentations

During the final week of the CURE, each student group prepared a presentation to share the results of their experiments (Table 1). During the presentations, students summarized background knowledge, explained how that led to their hypothesis, explained their experimental methods, described their results, stated their conclusions, and proposed follow-up questions and experiments. As a result, the presentation addressed many aspects of the scientific process: connecting their research to previous findings, explaining their hypothesis and experimental approach, describing and interpreting their data, planning ahead for future iterations of the project, and participating in the process of scientific communication. While most presentations were brief (5–10 min), each student was required to present, ensuring that each student participated and, therefore, experienced the same learning gains.

1.2.5 Student Learning Outcomes

The CURE described here was intended to increase student engagement and understanding of the scientific process, as opposed to a focus on concept mastery. The success of the CURE was reflected in student responses on assignments and course evaluations.

Student understanding of the scientific process was assessed via assignments that accompanied the CURE. The first assignment, given during the week of the CURE Introduction, asked students to evaluate background information, use what they have learned to form a hypothesis, and suggest an experiment to test that hypothesis. These questions addressed key components of the scientific process. Virtually, all students formed appropriate hypotheses, typically stating that "gene *X* is required for normal oogenesis". Importantly, all students suggested loss-of-function experiments to test their hypothesis, indicating that they identified the most appropriate experimental approach for testing gene requirement during a specific developmental process. Most students described loss-of-function experiments to compare the fertility of mutant with wild-type female flies, but several students proposed more specific experiments, such as immunostaining to label cell types in the ovary, and predicted loss of specific structures (such as the oocyte) in the ovaries of mutant females. These responses suggest that students understood the scientific basis of the CURE and were able to successfully and independently complete essential steps of the scientific process, including hypothesis formation and experimental design.

Additional aspects of the scientific process were assessed during the presentations. Students worked in groups to analyze, interpret, and present their data. Almost all students reached conclusions that accurately represented their results, and some students even progressed to the step of discussing how their results agreed or disagreed with previously published data. Finally, although students in a one-semester CURE do not have the opportunity to repeat their experiments or carry out follow-up experiments, they were asked to include suggestions for future student projects at the end of their presentations to incorporate the iterative nature of scientific research into the CURE. Most students accurately identified questions for further investigation, in particular the need to better characterize the observed phenotype, and many provided useful suggestions for those follow-up experiments, such as immunostaining for specific cell types and structures in the ovary to better understand the defects that occurred to cause infertility. Overall, student responses indicated that they had gained experience with multiple steps of the scientific process through the course of the CURE.

Student evaluations suggested that students enjoyed participating in the CURE. Although the CURE was not evaluated independently of the course, multiple students mentioned it specifically in their course evaluation comments. For example, one student stated that "the gene research was really interesting", and another wrote that "I really enjoyed studying the uncharacterized gene!". These comments suggest that participation in the CURE increased student engagement and left them with a positive impression of the scientific process.

1.3 Summary of the CURE

The CURE described here is easy to implement and can be flexibly adapted to a variety of courses, taught at a variety of levels and at a range of institutions. Once the CURE has been adopted, the benefits for the students are numerous; our assessments indicate that participating students are highly engaged and gain valuable experience with the scientific process throughout the course of the CURE. The instructor that adopts the CURE benefits greatly as well. Courses are more enjoyable for both students and instructors when the students are engaged, and increased student engagement yields positive course evaluations for the instructor. Furthermore, the CURE can be used to advance the instructor's scholarly goals. Taken together, the collective benefits of CUREs make a strong argument for the integration of CUREs into as many courses as possible.

2 Materials

2.1 Fly Maintenance and Manipulation

1. Fly stocks. This CURE included w^{1118} (Bloomington Stock #3605), maternal triple driver–Gal4 (MTD–Gal4, Bloomington Stock #31777), UAS–GFP (Bloomington Stock #4775), and UAS–RNAi flies (*see* **Note 1**).
2. Incubator that can be maintained at 25 °C.
3. Fly culture vials and plugs.
4. Fly food (*see* **Note 2**).

5. FlyNap (or other anesthesia if preferred).
6. Small white paper plates (approximately 6 in in diameter).
7. Small paintbrushes (for moving flies; round head watercolor paintbrushes, 4/32″ (size 6) or similar are typically appropriate).
8. Stereomicroscopes/dissecting microscopes.

2.2 Ovary Dissection, Staining, and Microscopy

1. Classroom projector or individual screens to show students ovary dissection video [39].
2. Microdissection forceps (extra-fine point preferred).
3. Nine well glass spot plate or small petri dish (such as 30 mm × 15 mm).
4. 1× phosphate-buffered saline (PBS): prepare by diluting 10× PBS 1:10 in dH_2O.
5. Stereomicroscopes/dissecting microscopes.
6. Transfer pipets.
7. Small white paper plates (approximately 6 in diameter).
8. Microcentrifuge tubes.
9. 3.7% formaldehyde: prepare by diluting 37% formaldehyde 1: 10 in 1× PBS.
10. Nutating mixer.
11. 1× PBS-Tx: prepare by adding Triton X-100 to 1× PBS at a concentration of 0.1% v/v.
12. 1 μg/mL 4′,6-diamidino-2-phenylindole (DAPI): first prepare a 1 mg/mL stock by dissolving 10 mg of DAPI in 10 mL of dH_2O. Prepare the 1 μg/mL working solution by diluting the 1 mg/mL stock 1:1000 in dH_2O.
13. Nitrile gloves.
14. Microscope slides.
15. Coverslips (size 1.5, 22 mm × 22 mm, or other if required for your fluorescent microscope).
16. 200 μL micropipette and accompanying pipette tips.
17. 70% Glycerol: prepare by diluting glycerol 7:10 in dH_2O.
18. Clear nail polish.
19. Fluorescent microscope fitted with DAPI filter and equipped with digital camera.

2.3 Hatch Assays

1. Embryo collection chambers (for example, Genesee Scientific, Catalog #59-105, or other).
2. Petri dishes in a size complementary to your embryo collection chambers.
3. Grape agar: prepare using a pre-made mix according to directions, or according to Cold Spring Harbor Protocol [40].

4. Dry baker's yeast.
5. Yeast paste (any baker's yeast mixed with water, to reach peanut butter consistency).
6. Dissecting probes (straight tip, approximately 6 in).
7. Stereomicroscopes/dissecting microscopes.

3 Methods

3.1 Implementation of the CURE

3.1.1 Advance Preparation by Instructor

1. At least 6 weeks in advance of the CURE, order fly stocks from stock centers. This CURE utilized w^{1118} flies, the MTD–Gal4 maternal driver, UAS–GFP flies, and UAS–RNAi lines obtained from the Bloomington Stock Center (for additional options and suggestions, *see* **Note 1**).
2. Prepare fly food in vials according to product directions and store in a refrigerator to use as needed (*see* **Note 2**).
3. Four weeks prior to the beginning of the CURE, expand all fly cultures (w^{1118}, MTD–Gal4, UAS–GFP, and UAS–RNAi) by transferring adults to fresh vials, setting up multiple vials if possible (dependent on number of available adults; each vial should have 5–10 females and at least 5 males). Adults can be transferred to new vials every 2 days to increase reproductive rates.
4. Two weeks before CURE experiments are scheduled to start, set up parental crosses to generate desired phenotypes for CURE experiments (*see* Table 1). Cross MTD–Gal4 males with UAS–GFP virgin females to generate MTD > GFP control flies, and cross MTD–Gal4 males with UAS–RNAi virgin females to generate MTD > RNAi experimental flies (*see* **Note 3**). Each cross should include 5–10 virgin females and at least 5 males. Set up at least one of each cross for each lab group to ensure sufficient flies for CURE experiments. Store crosses in a 25 °C incubator to ensure progeny emerge prior to Lab Week 2 (*see* Table 1).
5. Prepare grape agar plates for embryo collection chambers by following product directions or as previously described [39]. Grape agar plates should be prepared in petri dishes that fit the embryo collection chambers. Grape agar plates can be stored in a refrigerator, in a covered container or sealed plastic bag, typically for up to 1 month.
6. Prepare lab solutions required for CURE experiments, including 1× PBS, 1× PBS-Tx, 1 mg/mL DAPI, 70% glycerol, and yeast paste. DAPI and yeast paste should be stored in the refrigerator until used, but all other solutions can be stored at room temperature.

3.1.2 Instructor Guidelines for Implementation of the CURE

The specific organization of any CURE is directly affected by course and laboratory schedule. This CURE took place in a course that met for lab for 3 h, once a week (*see* **Note 4** for suggestions for adapting the CURE to alternate schedules).

1. During Lab Week 1:
   (a) Instructor: remove parent flies from crosses before progeny enclose to ensure that only progeny flies (which have the desired genotypes) will be used during CURE experiments (*see* Table 1).
   (b) Instructor: present CURE background and assign students reading and Preparatory Assignment to be completed prior to following lab meeting.
2. During Lab Week 2:
   (a) Students: collect progeny, prepare female flies for dissection, and set up embryo collection chambers (*see* Table 1).
   (b) Instructor: maintain the embryo collection chambers in between Lab Weeks 2 and 3 by replacing the old grape agar plate with a new grape agar plate, topped with a dime-sized smear of yeast paste, every day.
3. During Lab Week 3:
   (a) Instructor: replace grape agar plates within 24 h of the lab to ensure that students have young embryos to set up hatch assays (*see* Table 1). If the lab meets in the morning, the instructor should place fresh grape agar plates in the embryo collection chambers late in the afternoon the day before. Alternatively, if the lab meets in the afternoon, the instructor should place fresh grape agar plates in the embryo collection chambers early that morning.
   (b) Students: dissect, fix, and stain ovaries (*see* **Note 5**); examine ovaries on the microscope; and set up hatch assays (*see* Table 1). To optimize time, students should start their dissections immediately, then use the incubation period, while ovaries are being fixed to set up their hatch assays. Hatch assays must be checked approximately 24 h later to determine hatch rates, a process that typically only takes about 5 min. Students should be allowed into the lab on their own time the following day to complete their hatch assays (*see* **Note 6** for suggestions for alternate schedules).
4. During Lab Week 4:
   (a) Students: analyze their data and prepare figures (*see* Table 1).
   (b) Instructor: facilitate the data analysis process by collecting and sharing hatch assay results with the entire class to increase sample size for statistical analyses (if desired). It may be necessary to download microscopy files and

convert them into .jpg files that are more likely to be compatible with student software. Finally, communicate details and expectations regarding the presentations for the following week's lab session.

5. During Lab Week 5, student groups present their results.

3.2 Lab Protocols and Assignments for Students

Subheading 3.2.1 provides student instructions for Lab Week 2, while Subheadings 3.2.2 and 3.2.3 provide student instructions for Lab Week 3.

3.2.1 Preparation of Females for Dissection and Embryo Collection Chambers

1. Retrieve three vials of flies from the instructor. There will be one vial of control flies (MTD > GFP), one vial of experimental flies (MTD > RNAi), and one vial of w^{1118}.
2. Prepare two vials of fly food by sprinkling extra dry baker's yeast at the bottom. Label one "MTD > GFP females" and the other "MTD > RNAi females". These vials will be used to store females for next week's dissections.
3. Prepare two embryo collection chambers by labeling one "MTD > GFP females" and the other "MTD > RNAi females". Prepare two grape agar plates by adding a dime-sized smear of yeast paste to the center of each.
4. Use FlyNap (or other anesthesia method) to anesthetize the MTD > GFP flies, place on a small white paper plate, and view under a stereomicroscope. Sort males from females (females do not need to be virgins) (*see* **Note 7**). Transfer 10 females to the prepared vial and 20 females to the prepared embryo collection chamber.
5. Use FlyNap (or other anesthesia method) to anesthetize the MTD > RNAi flies, place on a small white paper plate, and view under a stereomicroscope. Sort males from females (females do not need to be virgins) (*see* **Note 7**). Transfer 10 females to the prepared vial and 20 females to the prepared embryo collection chamber.
6. Use FlyNap (or other anesthesia method) to anesthetize the w^{1118} flies, place on a small white paper plate, and view under a stereomicroscope. Sort males from females. Add five males to each of the vials, and ten males to each of the embryo collection chambers.
7. Close vials, leave them on their sides until flies awaken, and then turn upright. Seal embryo collection chambers with grape agar plates. Store vials and chambers at 25 °C until next week.
8. The instructor will maintain the embryo collection chambers by replacing the grape agar plates with fresh yeast paste regularly until the next lab session.

3.2.2 Ovary Dissection, Fixation, and Staining

1. Watch the video of ovary dissection [39] (*see* **Note 8**).
2. Prepare a dissecting dish (nine well plate or petri dish) by using a transfer pipet to add 1× PBS (about 300 μL) to two wells (or dishes).
3. Anesthetize female flies using FlyNap (or other anesthesia method) and place them on small paper plates.
4. Observe the anesthetized flies through the stereomicroscope. Use forceps to pull abdomen away from thorax as shown in the video [39]. Transfer the abdomen to one of the wells of 1× PBS in the dissecting dish (or to one petri dish). Repeat for remaining flies.
5. Observe abdomens in the dissecting dish (or petri dish) through the stereomicroscope. Use forceps to squeeze or push ovaries out of abdomen as shown in video [39]. Use forceps to transfer ovaries to a second well (or second petri dish) of 1× PBS (to separate ovaries from debris). Repeat until all abdomens have been dissected.
6. Use a transfer pipet to transfer dissected ovaries to an empty microcentrifuge tube (*see* **Note 9**).
7. Use a transfer pipet to remove PBS, leaving ovaries behind in microcentrifuge tube.
8. Wearing gloves, add 300 μL of 4% formaldehyde to ovaries. Place the tube on nutating mixer for 20 min.
9. Allow ovaries to sink to the bottom of the tube. Wearing gloves, remove formaldehyde and place in an appropriate waste container, leaving ovaries behind in the microcentrifuge tube.
10. Add 300 μL of PBS-Tx to ovaries and place on nutating mixer for 5 min.
11. Allow the ovaries to sink to the bottom of the tube. Remove PBS-Tx, leaving ovaries behind in tube.
12. Wearing gloves, add 300 μL of 1 μg/mL DAPI to ovaries. Place on nutating mixer for 2 min.
13. Allow the ovaries to sink to the bottom of the tube. Wearing gloves, remove DAPI and place in an appropriate waste container, leaving ovaries behind.
14. Add 300 μL of PBS-Tx to ovaries and place on nutating mixer for 5 min.
15. Use a transfer pipet to transfer ovaries to glass slides, then use the 200 μL micropipette (set to 100 μL) to carefully remove as much of the liquid from the slide as possible while leaving the ovaries behind on the slide.

16. Use a transfer pipet to cover the ovaries with approximately 100 μL of 70% glycerol, then cover the ovaries with a coverslip.
17. Use the clear nail polish to seal just around the edges of the coverslip, so that the coverslip does not move during microscopy. Allow the nail polish to dry for at least 2 min.
18. View ovaries on fluorescent microscope and capture images of the staining.

3.2.3 Hatch Assays

1. Remove the grape plate with embryos from the embryo collection chamber, replacing it with a fresh grape plate to seal the chamber. Label the bottom of this plate with the embryos with a Sharpie to indicate the genotype of the embryos.
2. Label the bottom of a clean grape agar plate with your name and the appropriate genotype.
3. Use the dissecting probe to gently gather embryos from the first plate and transfer them to the clean plate by following these steps:
   (a) Hold the dissecting probe parallel to the surface of the plate with embryos, and gently run the probe along the surface to pick up the embryos—the embryos should be slightly sticky and should adhere to the probe with only gentle contact.
   (b) Transfer the embryos to the clean plate by gently running the probe (with attached embryos) horizontally over the surface of the clean plate, essentially wiping the embryos off onto the clean surface.
4. While looking through microscope, use the dissecting probe to gently push/roll the embryos along the plate to line up the embryos in rows, so they can be easily counted. When finished, use the Sharpie to write the number of embryos on the bottom of the plate (*see* **Note 10**).
5. Incubate plates at 25 °C overnight.
6. The following day, examine the plates under the stereomicroscope. Count eggs that have not hatched: they will look the same as they did yesterday, whereas hatched eggs will have left behind a flattened, shrunken eggshell.
7. Calculate the number of eggs hatched by subtracting unhatched eggs from total number of eggs.
8. Calculate hatch rate as follows: (number of eggs hatched/total number of eggs) × 100%.
9. Generate a data table as shown in Table 2.

Table 2
Embryonic hatch assay data

Maternal genotype	Total # eggs on plate	# eggs unhatched	# eggs hatched	Hatch rate (%) = #eggs hatched/ total #eggs) x 100%

4 Notes

1. This CURE utilized pre-existing MTD–Gal4 and UAS–RNAi flies to perform loss-of-function experiments in the female germline. The MTD–Gal4 driver was selected, because it activates robust gene expression within all germline cells, but multiple alternative maternal drivers exist and could be used instead. RNAi target genes were chosen by the instructor based on the instructor's research interests, but if desired, the target gene could be selected to link the CURE to course material instead (such as a meiosis gene in an Introductory Biology or Genetics course, a signaling pathway gene in a Developmental Biology or Cell Biology course, etc). The instructor should make sure that the selected RNAi construct can be expressed in the germline. The instructor could alternately choose to carry out Gal4–UAS gain-of-function experiments, which would require similar genetic crosses, or conduct classic loss-of-function experiments using genetic mutants, although in that case, the crossing scheme would need to be adjusted. Students could also use wild-type flies to investigate environmental effects on fertility, testing conditions such as maternal age, maternal diet, or maternal chemical exposure, and adjusting the CURE structure and/or timeline to include environmental treatments. Any of these experiments could be designed by the instructor, of course, but students could also be included in the experimental design process if desired. If students select the gene of interest, they could do so at the beginning of the semester. The instructor could immediately order the appropriate stocks, receive them 2–3 weeks later, and the CURE could be completed within the final 4–6 weeks of the semester. For experiments exploring the effects of the environment on oogenesis and fertility, students could be allowed to choose the specific chemicals or maternal conditions, and as long as the necessary reagents are easily accessible, the CURE could readily be completed within a single semester.

2. Fly food was purchased as a mix and prepared according to manufacturer instructions. Instructions for making fly food from scratch can be found on the Bloomington Stock Center website [41].

3. Crosses were set up to generate flies of the desired experimental genotype (MTD > RNAi) as well as an appropriate control (MTD > GFP). The specific crosses, as well as the appropriate controls, depend on the specific experiment (*see* **Note 1** for suggestions of alternative experiments).

4. This CURE took place over the course of five weekly, 3-h lab meetings, but that schedule could easily be adapted. For example, the CURE could be expanded to occupy an entire semester if desired. Options for expansion including allowing students to have more input in the experimental plan, expanding introductory assignments, asking students to do more of the preparatory lab work, stretching experiments out to take place over multiple weeks by carrying out the hatch assays and dissections in separate lab meetings, and allowing students more time to analyze their data, assemble figures, and prepare presentations at the end of the CURE. For those labs that meet twice a week for a shorter period of time, weekly work could easily be split into two shorter segments each week. Alternatively, an instructor could condense or simplify the CURE by doing more of the preparatory work themselves, or choosing to do only one of the two assays. Simplification of the CURE, in particular by removing the ovary dissection, might also be more feasible for first year students, large courses, or institutions without access to a fluorescent microscope. Such a modification would reduce instructor effort while still yielding valuable and interesting data.

5. Ovaries were stained with DAPI alone for simple morphological examination. However, the instructor could choose to incorporate immunostaining into the CURE to allow students to identify specific cell types. A wide range of antibodies are available to label specific cell types within the ovary, many of which are very affordably available from the Developmental Studies Hybridoma Bank at the University of Iowa [42]. Immunostaining typically cannot be completed within a single lab session, so inclusion of immunostaining would likely extend the length of the CURE by an additional lab meeting/ week.

6. If there is a concern about completing the work within a single lab period, the work could be split amongst lab groups to increase efficiency. For example, half of the groups could conduct the experiments with only the control flies, while the other half could conduct the experiments with only the experimental flies. Alternatively, lab activities can be spread out over 2 or

more weeks, or instructors could choose to complete only one of the two experiments during their CURE. Finally, hatch assays must be checked approximately 24 h later to determine hatch rates, a process that typically only takes about 5 min. If students cannot come into the lab on their own time to check the plates, the plates could be checked by the students during lecture time or could be checked by the instructor outside of class time.

7. Genotypes may vary depending on the experiment (*see* **Note 1** for more detail). Female flies do not need to be virgins for the experiments described here, because sibling males are expected to be wild type. However, if there is concern about females mating with sibling males, students can be instructed to use only virgin females to ensure that they only mate with true wild type males (such as the w^{1118} males used here) prior to ovary dissection and embryo collection.
8. The video referenced here is an excellent depiction of ovary dissection; instructors who have not performed ovary dissections before may benefit from watching the video and practicing the technique in advance.
9. When using plastic transfer pipets, we have not experienced significant issues with ovaries getting stuck inside the pipet. However, if this becomes a problem, we recommend pre-coating the inside of the pipet with either a 2% bovine serum albumin solution or a 10% normal goat serum solution before using the pipet to transfer the ovaries.
10. Students should attempt to include as many embryos as possible in their hatch assays to have a sample size large enough to detect subtle differences in hatch rates (30–50 embryos is a feasible goal). Student groups can share their results and conduct statistical analyses to increase the value of this experiment.

References

1. Fechheimer M, Webber K, Kleiber PB (2017) How well do undergraduate research programs promote engagement and success of students? CBE Life Sci Educ 10:156–163
2. Peteroy-Kelly MA, Marcello MR, Crispo E, Buraei Z, Strahs D, Isaacson M, Jaworski L, Lopatto D, Zuzga D (2017) Participation in a year-long CURE embedded into major core genetics and molecular biology laboratory courses results in gains in foundational biological concepts and experimental design skills by novice undergraduate researchers. J Microbiol Biol Educ 18(1):18.1.1
3. Seymour E, Hunter A, Laursen SL, Deantoni T (2004) Establishing the benefits of research experiences for undergraduates in the sciences: first findings from a three-year study. Sci Educ 88(4):493–534
4. Lopatto D (2007) Undergraduate research experiences support science career decisions and active learning. CBE Life Sci Educ 6: 297–306
5. Russell SH, Hancock MP, McCullough J (2007) Benefits of undergraduate research experiences. Science 316:548–549
6. Odera E, Lamm AJ, Odera LC, Duryea M, Davis J (2015) Understanding how research experiences foster undergraduate research skill development and influence STEM career choice. NACTA J 59(3):180–188

7. Rodenbusch SE, Hernandez PR, Simmons SL, Dolan EL (2016) Early engagement in course-based research increases graduation rates and completion of science, engineering, and mathematics degrees. CBE Life Sci Educ 15:1–10
8. Wilson AE, Pollock JL, Billick I, Domingo C, Fernandez-Figueroa EG, Nagy ES, Steury TD, Summers A (2018) Assessing science training programs: structured undergraduate research programs make a difference. Bioscience 68(7): 529–534
9. Villarejo M, Barlow AEL, Kogan D, Veazey BD, Sweeney JK (2008) Encouraging minority undergraduates to choose science careers: career paths survey results. CBE Life Sci Educ 7:394–409
10. Eagan MK, Hurtado S, Chang MJ, Garcia GA, Herrera FA, Garibay JC (2013) Making a difference in science education: the impact of undergraduate research programs. Am Educ Res J 50(4):683–713
11. Carpi A, Ronan DM, Falconer HM, Lents NH (2017) Cultivating minority scientists: undergraduate research increases self-efficacy and career ambitions for under-represented students in STEM. J Res Sci Teach 54(2): 169–194
12. Hernandez PR, Woodcock A, Estrada M, Schultz PW (2018) Undergraduate research experiences broaden diversity in the scientific workforce. Bioscience 68(3):204–211
13. Sadler TD, McKinney L (2010) Scientific research for undergraduate students: a review of the literature. J Coll Sci Teach 39(5):43–49
14. Hunter A, Laursen SL, Seymour E (2006) Becoming a scientist: the role of undergraduate research in students' cognitive, personal, and professional development. Sci Educ 91(1): 36–74
15. Thiry H, Laursen SL, Hunter A (2011) What experiences help students become scientists? A comparative study of research and other sources of personal and professional gains for STEM undergraduations. J High Educ 82(4): 357–388
16. Robnett RD, Nelson PA, Zurbriggen EL, Crosby FJ, Chemers MM (2018) Research mentoring and scientist identity: insights from undergraduates and their mentors. Int J STEM Educ 5:41
17. Eagan MK, Sharkness J, Hurtado S, Mosqueda CM, Chang MJ (2011) Engaging undergraduates in science research: not just about faculty willingness. Res High Educ 52:151–177
18. Wei CA, Woodin T (2011) Undergraduate research experiences in biology: alternatives to the apprenticeship model. CBE Life Sci Educ 10:123–131
19. Laursen S, Seymour E, Hunter A (2012) Learning, teaching, and scholarship: fundamental tensions of undergraduate research. Change 44(2):30–37
20. Awong-Taylor J, D'Costa A, Giles G, Leader T, Pursell D, Runck C, Mundie T (2016) Undergraduate research for all: addressing the elephant in the room. CUR Q 37(1):11–19
21. Bangera G, Brownell SE (2014) Course-based undergraduate research experiences can make scientific research more inclusive. CBE Life Sci Educ 14:602–606
22. Ahmad AS, Sabat I, Trump-Steele R, King E (2019) Evidence-based strategies for improving diversity and inclusion in undergraduate research labs. Front Psychol 10:1305
23. Pierszalowski S, Bouwma-Gearhart J, Marlow L (2021) A systematic review of barriers to accessing undergraduate research for STEM students: problematizing under-researched factors for students of color. Soc Sci 10(9):328
24. Auchincloss LC, Laursen SL, Branchaw JL, Eagan K, Graham M, Hanauer DI, Lawrie G, McLinn CM, Palaez N, Rowland S, Towns M, Trautmann NM, Varma-Nelson P, Weston TJ, Dolan EL (2014) Assessment of course-based undergraduate research experiences: a meeting report. CBE Life Sci Educ 13:29–40
25. Corwin LA, Graham MJ, Dolan EL (2015) Modeling course-based undergraduate research experiences: an agenda for future research and evaluation. CBE Life Sci Educ 14:1–13
26. Dolan EL (2016) Course-based undergraduate research experiences: current knowledge and future directions. National Research Council Commissioned Paper
27. Shapiro C, Moberg-Parker J, Toma S, Ayon C, Zimmerman H, Roth-Johnson EA, Hancock SP, Levis-Fitzgerald M, Sanders ER (2015) Comparing the impact of course-based and apprentice-based research experiences in a life science laboratory curriculum. J Microbiol Biol Educ 16(2):186–197
28. Mader CM, Beck CW, Grillo WH, Hollowell GP, Hennington BS, Staub NL, Delesalle VA, Lelo D, Merritt RB, Griffin GD, Bradford C, Mao J, Blumer LS, White SL (2017) Multi-institutional, multidisciplinary study of the impact of course-based research experiences. J Microbiol Biol Educ 18(2):1–11
29. Krim JS, Coté LE, Schwartz RS, Stone EM, Cleeves JJ, Barry KJ, Burgess W, Buxner SR, Gerton JM, Horvath L, Keller JM, Lee SC, Locke SM, Rebar BM (2019) Models and impacts of science research experiences: a review of the literature of CUREs, UREs, and TREs. CBE Life Sci Educ 18(ar65):1–14

30. Pierszalowski S, Vue R, Bouwma-Gearhart J (2018) Overcoming barriers in access to high quality education after matriculation: promoting strategies and tactics for engagement of under-represented groups in undergraduate research via institutional diversity action plans. J STEM Educ 19(1):49–56
31. Jayabalan M, Caballero ME, Cordero AD, White BM, Asalone KC, Moore MM, Irabor EG, Watkins SE, Walters-Conte KB, Taraboletii A, Martings MR, Chow BY, Saeed BA, Bracht KA, Bracht JR (2021) Unrealized potential from smaller institutions: four strategies for advancing STEM diversity. Cell 184: 5854–5850
32. Shortlidge EE, Bangera G, Brownell SE (2016) Faculty perspectives on developing and teaching course-based undergraduate research experiences. Bioscience 66(1):54–62
33. Shortlidge EE, Bangera G, Brownell SE (2017) Each to their own CURE: faculty who teach course-based undergraduate research experiences report why you too should teach a CURE. J Microbiol Biol Educ 18(2):1–11
34. Genné-Bacon EA, Wilks J, Bascom-Slack C (2020) Uncovering factors influencing instructors' decision process when considering implementation of a course-based research experience. CBE Life Sci Educ 19(ar13):1–19
35. Spell RM, Guinan JA, Miller KR, Beck CW (2014) Redefining authentic research experiences in introductory biology laboratories and barriers to their implementation. CBE Life Sci Educ 13:102–110
36. Wang JTH (2017) Course-based undergraduate research experiences in molecular biosciences – patterns, trends, and faculty support. FEMS Microbiol Lett 364(15):1–9
37. Bastock R, St Johnson D (2008) *Drosophila* oogenesis. Curr Biol 18(23):R1082–R1087
38. McLaughlin JM, Bratu DP (2015) *Drosophila melanogaster* oogenesis: an overview. Methods Mol Biol 1328:1–20
39. Parker DJ, Moran A, Mitra K (2017) Studying mitochondrial structure and function in *Drosophila* ovaries. J Vis Exp 119:54989
40. Cold Spring Harbor Protocols (2007) *Drosophila* fruit juice egg plates. Cold Spring Harb Protoc. http://cshprotocols.cshlp.org/content/2007/9/pdb.rec11113.full. Accessed 15 Mar 2022
41. BDSC Fly Food and Methods (2021) Bloomington *Drosophila* Stock Center. https://bdsc.indiana.edu/information/recipes/index.html. Accessed 17 Mar 2022
42. Lie-Jenson A, Haglund K (2016) Antibody staining in *Drosophila* germaria. Methods Mol Biol 1457:19–33

Chapter 22

Gaining Wings to FLY: Using *Drosophila* Oogenesis as an Entry Point for Citizen Scientists in Laboratory Research

Dara M. Ruiz-Whalen, Christopher P. Aichele, Ebony R. Dyson, Katherine C. Gallen, Jennifer V. Stark, Jasmine A. Saunders, Jacqueline C. Simonet, Erin M. Ventresca, Isabela M. Fuentes, Nyellis Marmol, Emly Moise, Benjamin C. Neubert, Devon J. Riggs, Ava M. Self, Jennifer I. Alexander, Ernest Boamah, Amanda J. Browne, Iliana Correa, Maya J. Foster, Nicole Harrington, Troy J. Holiday, Ryan A. Henry, Eric H. Lee, Sheila M. Longo, Laurel D. Lorenz, Esteban Martinez, Anna Nikonova, Maria Radu, Shannon C. Smith, Lindsay A. Steele, Todd I. Strochlic, Nicholas F. Archer, Y. James Aykit, Adam J. Bolotsky, Megan Boyle, Jennifer Criollo, Oren Eldor, Gabriela Cruz, Valerie N. Fortuona, Shreeya D. Gounder, Nyim Greenwood, Kayla W. Ji, Aminah Johnson, Sophie Lara, Brianna Montanez, Maxwell Saurman, Tanu Singh, Daniel R. Smith, Catherine A. Stapf, Tarang Tondapu, Christina Tsiobikas, Raymond Habas, and Alana M. O'Reilly

Abstract

Citizen science is a productive approach to include non-scientists in research efforts that impact particular issues or communities. In most cases, scientists at advanced career stages design high-quality, exciting projects that enable citizen contribution, a crowdsourcing process that drives discovery forward and engages communities. The challenges of having citizens design their own research with no or limited training and providing access to laboratory tools, reagents, and supplies have limited citizen science efforts. This leaves the incredible life experiences and immersion of citizens in communities that experience health disparities out of the research equation, thus hampering efforts to address community health needs with a full picture of the challenges that must be addressed. Here, we present a robust and reproducible approach that engages participants from Grade 5 through adult in research focused on defining how diet impacts disease signaling. We leverage the powerful genetics, cell biology, and biochemistry of *Drosophila* oogenesis

Michelle S. Giedt and Tina L. Tootle (eds.), *Drosophila Oogenesis: Methods and Protocols*,
Methods in Molecular Biology, vol. 2626, https://doi.org/10.1007/978-1-0716-2970-3_22,

to define how nutrients impact phenotypes associated with genetic mutants that are implicated in cancer and diabetes. Participants lead the project design and execution, flipping the top-down hierarchy of the prevailing scientific culture to co-create research projects and infuse the research with cultural and community relevance.

Key words Citizen science, Oogenesis, Genetic screen, Diet, Equity and inclusion

1 Introduction

Science, Technology, Engineering, and Math (STEM) fields are growing at leaps and bounds, with an expected 1 million new jobs by 2030. Geographic areas with the highest growth rates are scrambling for qualified employees already, with anticipated growth a concern in several key STEM fields [1]. Importantly, STEM fields have the lowest rates of unemployment and higher salaries than non-STEM fields [1]. The demand for STEM workers is referred to as a "crisis", with access to adequately trained professionals a challenge that will only get worse over time. In healthcare, 16% growth is expected by 2030, with 16 million new jobs to be filled [2]. Multiple explanations are given for the challenges within the STEM workforce, including loss of interest in STEM majors, switching majors during college to a less time intensive career path, lack of STEM preparation in pre-college education, and systemic bias [3–7]. Notably, diversity is severely lacking in all STEM fields [1, 4, 8], with urgent consequences in biomedicine [9]. Currently, only 12% of physicians and 19% of licensed registered nurses are from underrepresented minority (URM) backgrounds [10, 11]. URM faculty in life sciences research represents only 3.5% of the total, with only 39% women [4]. Recent data suggest that 69% of high school students are unprepared for STEM majors, in large part due to a lack of context and access to inquiry-based experimentation [12, 13]. These numbers have shown little improvement over decades, despite numerous programs to recruit and retain these populations of scientists in academic biomedicine. This lack of diversity has been shown to contribute to disparities in healthcare, including poor communication, misunderstanding of the impact of culture on care decisions, and bias [14]. In contrast, diversity in the workforce promotes interactions among workers from different backgrounds to establish critically important cultural competency, improves communication and trust between patients and providers, helps address language and cultural barriers, enables approaches to ensure adherence of patients to treatment, and provides better access to care for underserved communities [14, 15]. The ability to address social determinants of health results in improved health outcomes, strongly emphasizing the importance of enhancing diversity in the biomedical workforce [14]. Most importantly, enhanced diversity in the scientific research workforce is expected to broaden culturally relevant perspectives in the design of research questions and projects.

The simplicity of the solution is almost shocking. Students need opportunity, training, and support to succeed. Early research experiences for undergraduates yield impressive gains including:

- Improving interest, motivation, and commitment of undergraduates to pursue research careers [16–22].
- Leveling the playing field across gender, race, and ethnicity, with students performing at equivalent levels in STEM courses and programs after completion of research experiences [23–28].
- Promoting sustainable changes, with students maintaining interest in pursuing additional research experiences and careers in science [24, 25, 29].
- Higher gains for women and URM students, emphasizing the importance of these opportunities for addressing research disparities [24, 25].

Still, access to research experiences for URM students is too limited. URM, disadvantaged, and first-generation college students are more likely to be underprepared for college, resulting in higher need for remedial coursework, difficulty passing college-level courses, and feelings of not belonging [28, 30–32]. These challenges result in high drop-out rates in the first year and dramatic disparities in graduation rates [28]. In addition, lack of awareness of research as an option and feelings of intimidation or insecurity when approaching faculty to obtain competitive research positions are barriers for participation, even for students with intended STEM majors [28]. Finally, many research positions are unpaid and time intensive, precluding participation for students, whose financial situation requires paid work or who have caregiving obligations at home [31, 32].

Minoritized scientists enter biomedicine to give back to their communities as a general principle [33–36]. This drive to make a difference can be lost in the midst of research training, with mentors directing project choices based on interests of the lab as a whole, rather than the intended purpose of the minoritized student. Perhaps, more strikingly, disparities in funding for community-centered projects conducted by minoritized scientists prevent progress on necessary research and promote attrition of scientists from the field of biomedicine [33, 37]. Here, we describe a citizen science program that leverages *Drosophila* oogenesis as a means to engage, celebrate, and retain under-represented students in life sciences careers. Citizen science is an approach that provides iterative levels of participation for non-scientists to contribute to ongoing professional research. Our program is designed to integrate the cultural WHY of participants while providing them rigorous, interdisciplinary training that will advance their questions and projects in a fundable manner. The goal is to develop a committed and funded cadre of community-focused life science research professionals who will advance efforts to diversify biomedical research, address health disparities, and expand the breadth of discovery with new questions and approaches built from life experience.

1.1 Enter Citizen Science

Citizen science is founded on the principle that collaboration between professional scientists and members of the lay public can enhance and speed research projects that require input of data points on a large scale [38]. Initial citizen involvement is generally very accessible, such as sharing computer use or entering observations about the natural world into a phone app. Scientists gain valuable access to resources and data that would otherwise be prohibitively costly while engaging a new audience in a particular scientific problem. This Level 1 citizen science concept of crowdsourcing is a powerful way to inform the public in a way that involves them in the research process, allowing them to gain new understanding of both the science and how research works. A well-designed citizen science research project also provides opportunities for increased citizen involvement. Interactive databases and opportunities to discuss observations and data allow citizens from different communities to share experiences and ideas relevant to the research project. Opening channels of reciprocal exchange between scientists and participants advance citizen science through three subsequent levels (Fig. 1): Level 2 (Distributed Intelligence): citizens learn basic research skills necessary for accurate and useful data collection, Level 3 (Participatory Science): citizens and scientists work together to decide which research questions to pursue and

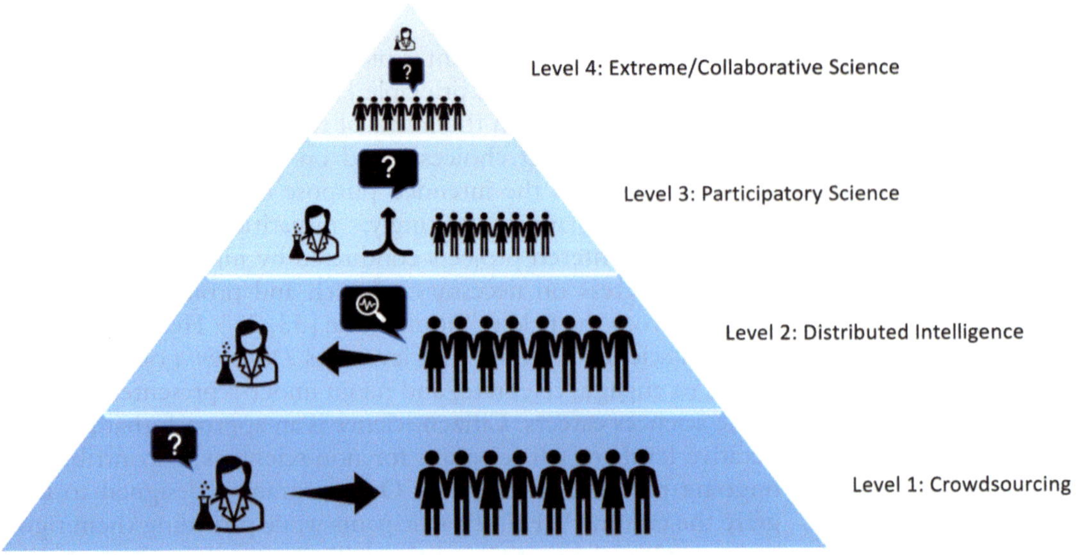

Fig. 1 Levels of citizen science in biomedical laboratory research. (Figure reprinted with permission from eCLOSE Institute)

what types of data should be collected, and Level 4 (Extreme or Collaborative Science): citizens design their own research questions and execute data collection plans independently or in collaboration with professional scientists.

1.1.1 Contributory and Collaborative Citizen Science for Biomedicine

Citizen science for biology emerged in 1999, with non-scientists providing computer access to large-scale data analysis programs [39]. Quickly, scientists realized the value of this approach as a method for mass data collection at low cost [39], with striking advances made in diverse fields such as conservation, ecology, and astronomy [39]. Relevant to our work, FoldIT, a citizen science protein modeling project, successfully pairs computer analysis with human logic to solve problems that are beyond the scope of standard computational analysis [40]. Eyewire is another gamified citizen science project, where players help map retinal neurons in three-dimensional space [41]. Stall Catchers [42] and Cell Slider-Trailblazer [43] use video or pathology samples to allow participants to identify defects in blood flow in a mouse model of Alzheimer's disease or pick out cancer cells from a tissue slice, respectively. These contributory programs mirror environmental citizen science projects, where participants provide data in a contributory fashion [44–47].

More recently, programs that use laboratory-based data collection have been developed, advancing participation from simple data contribution to collaboration, where analysis and interpretation are provided to some degree by participants. In the biomedical realm, collaborative Level 2 citizen science projects include SEA-PHAGES [48], the Wolbachia Project [49], BioEyes [50], and yEvo [51]. Collaborative citizen science is well documented to advance both the learning and the research [52]. Given the general skepticism among scientists regarding the value of public contributions to "real science" [53, 54], it is important to note that these projects result in robust numbers of peer-reviewed scientific publications in reputable journals, emphasizing the rigorous and reproducible nature of the data collection and analysis process [55].

Wonderful opportunities have been provided by these "contributory" citizen science projects, with creative, committed, professional scientists designing the scope of participation and overseeing project execution and publication of results. The process still relies on scientific hierarchy, however, with senior scientists designing the research questions and participants providing data [56]. This, together with significant ethical challenges surrounding health data privacy [57], makes it difficult to delve directly into the health inequity problems that generally bring URM students into science.

1.1.2 Co-created Citizen Science

Our goal was to flip the equation, with co-created research projects [52] initiated upon entry into the program. In co-created projects, citizens are actively involved in each step of the process, with their knowledge, life experience, culture, and skills valued on par with the scientific expertise of the professional scientist partners. The approach upends the view that scientific knowledge is untouchable, inaccessible, and in some ways superior to other knowledge. Instead, scientists with broad research questions engage the ideas, questions, experimentation and interpretation of participants, and jointly create new research projects based on novel and interesting discoveries. Our experience supports the idea that celebrating the "WHY" of participants to create new, community-focused scientific directions is a powerful approach to retain URM scientists in research fields. Notably, the approach leverages the scaling potential of citizen science, with opportunities for one thousand participants to contribute to research projects at the cost of less than 1/2 of a starting postdoctoral fellow salary. In parallel, the approach capitalizes on large-scale engagement of students who would never have pursued (or heard of!) research, providing potential for unprecedented gains in understanding research, inclusion in the scientific community, and enhanced preparation for STEM careers.

Our vision of an ideal co-created citizen science program (Fig. 2) is built on the following criteria:

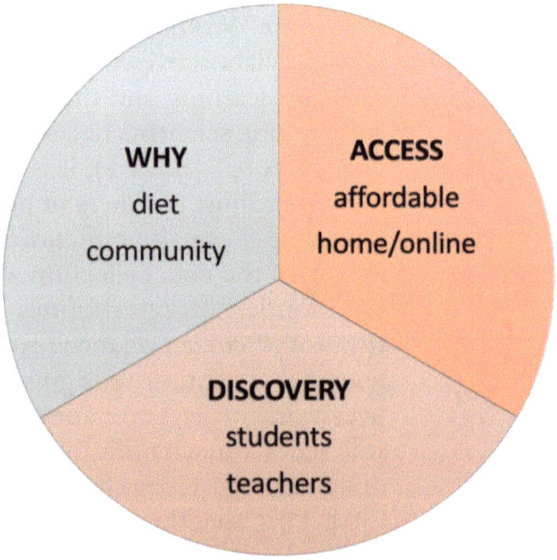

Fig. 2 Criteria of a co-created citizen science program. (Figure reprinted with permission from eCLOSE Institute)

1. *Celebrate the WHY*

A research question that is broad enough for any participant to contribute is an essential pillar of a co-created citizen science effort. The few examples of co-created citizen programs feature local issues that a passionate group of citizens bring to scientists, and the citizen–scientist team tackles the problem together [52]. Most of our participants are students in grades 5-undergraduate, with little or no access to advanced science education or lab equipment and a long history of marginalization from science [28–32]. As such, we developed a broad framework for entry into research focused on the idea that "everybody eats". A diet-focused program starts with familiar words and concepts, reducing the intimidation factor of "doing science" to more of a conversation that could take place in a kitchen or at a lunch table. Every participant has an idea of a food that is "healthy" or "unhealthy" and can easily grasp the overall scientific goal of understanding how diet impacts health. For inner city communities such as Philadelphia, blame abounds for the egregious health disparities experienced by citizens [58–61], the majority of whom hail from URM backgrounds. Obesity, poor food choices, food deserts, lack of access to food outside of school, financial challenges, and perceived preferences for junk food or fast food are often cited as drivers of the health disparities within the community. Conversely, little is known about the genetic drivers of these diseases in URM communities, or how diet intersects with disease risk and incidence. The topic of diet opens up conversation from the starting point of the students, enabling discussion of what actually (1) defines "healthy" and why, (2) how diet is linked to disease, (3) the impact of systemic racism and education deficits on health, (4) blame and shame regarding health challenges, and (5) how students can advocate for and improve their own health through an improved understanding of diet [62]. An additional benefit of celebrating the WHY from the student perspective is engagement of entire families and communities in the research effort. Students investigate cultural or familial dietary practices, particularly those that are used to promote health. Grandparents are a frequent source of advice in this area, providing opportunities for students to learn about their family or community history, what health challenges arise most frequently, and how diet might improve the outcomes for the newest generation.

What does the science gain? Science organization charts (Fig. 3a) are aggressively hierarchical, with the newest entrants at the very bottom of a long list of superiors [63]. The decades required to achieve a faculty position in an STEM field highlight the survival characteristics and drive embodied by most senior faculty [64–66]. All others on the organization chart accept the dominance order, yielding to those more senior and correcting their course accordingly. While this structure often results in high

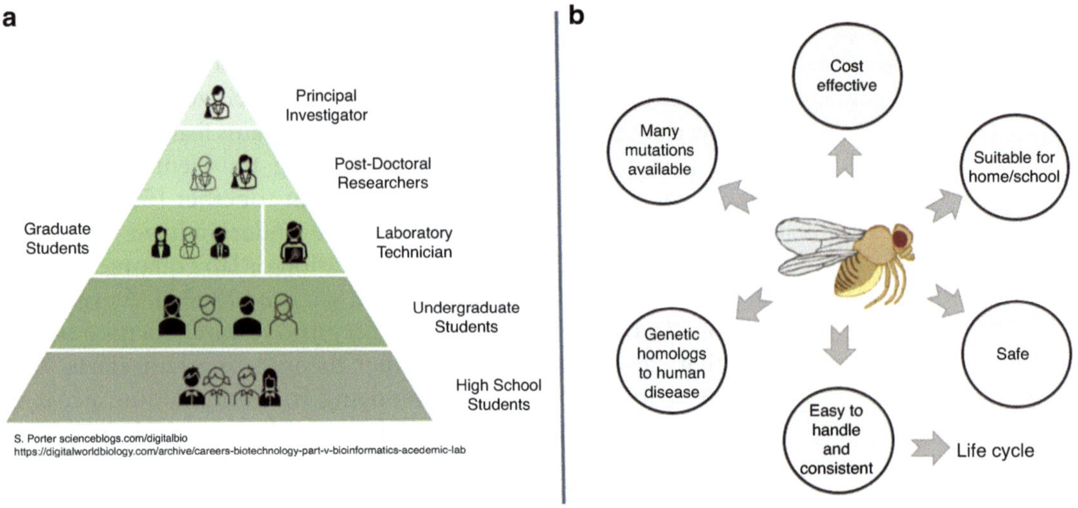

Fig. 3 (**a**) Historical hierarchy of laboratory researchers, (**b**) *Drosophila melanogaster* as a model for citizen science research. (Figures reprinted with permission from eCLOSE Institute)

productivity in individual labs, the research questions and solutions tend to be expensive and inaccessible to non-scientists, especially those from minoritized backgrounds, as well as creating access barriers that prevent inclusion of new ideas or people into the research effort [65, 66]. Building programs, where research-naive citizens leverage their own experience to address health challenges introduces new questions and interventions into the broader scientific focus. In our experience, this expands the research into new, very exciting and unexpected directions while simultaneously providing a direct connection to the community being impacted.

2. *Unlimited Access*

In addition to the limits on access imposed by hierarchy, research costs money and time [67]. Most inner city or rural schools operate on limited budgets, with pricey STEM programs not being an option. Professional scientists generally work in urban centers, where hospitals, universities, pharmaceutical companies, and biotech companies are located, geographically limiting physical access of many students to laboratory-based research [1, 68]. Citizen science projects are usually low cost, with a low barrier to entry even for those located in remote areas or in neighborhoods experiencing substantial financial challenges [69]. App-based citizen science projects are particularly accessible, with any phone or device capable of connecting to bona fide research. We tackled the access problem in three ways. First, we developed programming that costs as little as \$15–25/student, a small investment for immersion in original research relative to expensive, pre-determined experiments offered by for-profit companies. Second, we created comprehensive research kits in boxes that can be delivered by mail (Lab-in-a-Box).

Each research kit is tailored for particular educational and experimental programs and can be used in classrooms or at home. Finally, we adapted programming for a virtual learning environment, providing access to participants via a Wi-Fi-enabled device in a geographically unrestricted manner. Pairing of Lab-in-a-Box research kits that can be utilized anywhere with access via online communication platforms enables participants everywhere to contribute to research projects while learning professional lab techniques and conceptual skills to prepare them for STEM success.

3. *Original Research*

Middle and high school courses use dog-eared labs built on discoveries made up to a century ago, where they are graded on repeating experiments originally conducted by scientific geniuses. Students that get the "wrong" answer are penalized, an approach that is a stellar way to push students out of science. The limitless questions and infinite unknowns in biology make this seem silly. Why not ask students to think of their own questions and conquer new ground in scientific discovery, and grade them on their creativity, innovation, rigor and reproducibility, and significance instead [70]? Even better, extend the effort to address environmental impacts, engaging the students' knowledge and concern over social, genetic, diet, financial, and local environment to address challenges specific to their own communities and neighborhoods.

Our framework leverages *Drosophila* lines bearing mutations in genes that are known drivers of diseases in humans, providing a model system, that is, (1) inexpensive, (2) allowed in schools and homes, (3) non-hazardous, (4) robust and reproducible, (5) well-documented as a model for human disease, and (6) mutations are available in almost every gene, with extensive genetic and phenotypic overlap with human conditions (Fig. 3b). The most important feature of the *Drosophila* system is the ability to quantify specific stages in the life cycle. Starting with simple tasks, such as counting flies or pupal cases, increasingly advanced analysis of phenotypes ranging from fertility to atomic changes in signaling pathways can be evaluated to ascertain impacts of diet-derived compounds or environmental changes on wild-type or mutant flies (specific protocols are detailed below). A particularly robust assay is scoring of pupal cases on day 14 after transferring flies to new fly food (Fig. 4). Figure 4 represents variation in pupal counts over a 9-year period, with scoring done by students ranging from 5th to 12th grade and adults participating in citizen science outreach efforts. None of the participants had prior research experience but were able to engage and contribute to the research following a 90 min training session. Use of this simple assay presents an entry point into research that enables collection of publishable data while focusing on questions and hypotheses of high impact to a particular community of participants.

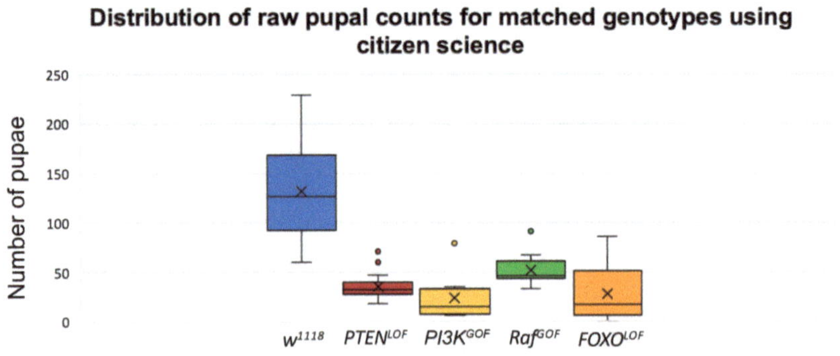

Fig. 4 Example pupal count data. Wild-type (w^{1118}, blue), $PTEN^{C076}/CyO$ (red), *GMR-PI3K92E* (yellow), heat-shock-inducible Raf (green), and $FOXO^{BG10108}$ (orange) pupal counts on day 14 after adding flies to control food vials. Ideally, the differences between pupal counts in "wild-type" controls and mutant lines will be clear. A relatively tight distribution (e.g. for PTEN, PI3K, and Raf) is ideal. Very low pupal counts or substantial variability (e.g. for FOXO) makes interpretation of data challenging for students entering the research field, especially given time constraints for conducting repeat experiments. Note that dots above the box plots indicate outlying data points: in most cases these are caused by errors in scoring or record keeping and provide important opportunities to have students re-visit their approach

1.2 Enter Oogenesis

The pupal counting described above is a reproducible and robust method for initial measurement of fertility, with fewer (or more) pupae indicating an impact of a treatment and mutant combination. Once an effect is observed, the next goal is to use more refined assays like counting eggs laid during a set amount of time to narrow down the period of impact. For example, if the numbers of eggs are approximately the same as the number of pupae that are observed, then any defects in pupal number are likely due to production of fewer eggs that all develop normally through larval and pupal stages. Alternatively, production of substantial eggs that fail to develop into pupae indicates developmental defects. Using a simple 24 h egg laying assay (*see* below), conditions that pinpoint novel effects of environmental change on egg development can be identified and investigated further.

Oogenesis itself is a beautifully orchestrated sequence of signaling events that control the myriad developmental events necessary to produce a fertilized egg (*cite this book!). The reliance of specific developmental events on individual signaling pathways makes analysis of oogenesis an exceptional approach to defining how environmental conditions and/or genetic mutations affect signal transduction in a precise manner. Direct examination of eggs can quickly indicate defects, for example: a primary focus of our work has been epidermal growth factor signaling, a critically important developmental signaling mechanism that is co-opted in the breast, lung, and colorectal cancers [71] that are of high interest to many of our student participants. Mutations in the epidermal growth factor receptor (EGFR) cause measurable changes in dorsal appendages [72], long, tubular extensions on eggs that enable

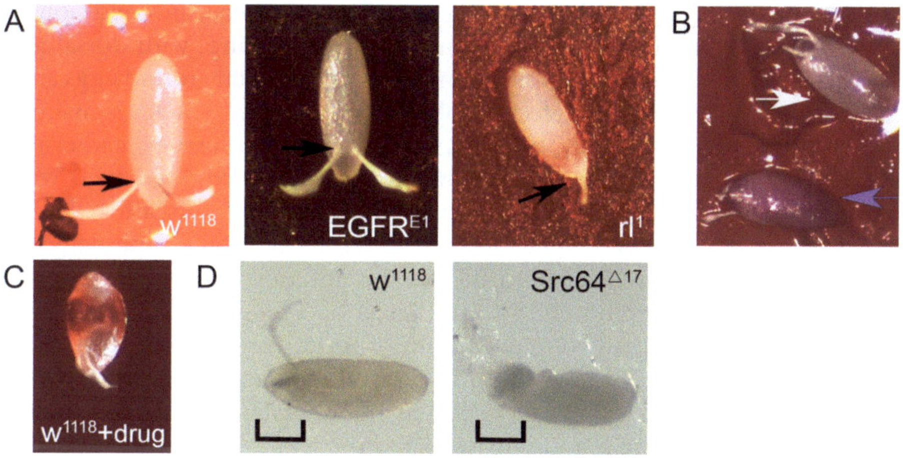

Fig. 5 Egg defects point to signaling pathways of interest. (**a**) Laid eggs of the indicated genotypes; black arrows point to dorsal appendages. WT (w^{1118}) egg (left). In WT (w^{1118}; left) eggs have a white, shiny appearance and presence of two long dorsal appendages on the anterior end of the egg. WT eggs are oblong, with clear distinctions between anterior and posterior poles. Eggs laid by flies bearing a constitutively activated *EGFR* ($EGFR^{E1}$, middle) exhibit dorsal appendages that are farther apart relative to WT. Loss of function mutation in EGFR effectors such as *rolled* (rl^{1}, right) result in fusion of dorsal appendages at the midline, often appearing as a single, thick dorsal appendage. (**b**) Absorption of color into the egg can indicate eggshell defects. WT flies treated with various edible compounds can yield variable eggshell defects, with some eggs appearing normal (white arrow) and some with a porous eggshell (purple arrow). (**c**) Round eggs are associated with defects in planar cell polarity (PCP). Compounds that cause WT eggs to appear round may be involved in PCP regulation. (**d**) Dumping defects, indicated in *Src64* mutants ($Src64^{\Delta 17}$, right), result in short eggs that often have excess cytoplasm attached. This contrasts with WT eggs (left), where germ cell cytoplasm transferred completely

respiration. Whereas gain of EGFR pathway function leads to dorsal appendages that are further apart than those observed on wild-type eggs, decreased EGFR activity reduces the space between dorsal appendages, often causing them to fuse at the midline (Fig. 5a) [73]. Identifying conditions that lead to fused dorsal appendages can thus be used as an initial approximation of defective EGFR pathway function, as well as elimination of conditions that have little or no effect.

Similarly, close examination of eggs using simple light microscopy can indicate defects in follicle cell function. Follicle cells are epithelial cells that surround each developing egg and its supporting germ cells (called nurse cells). During late stages of oogenesis, follicle cells secrete components of the hard eggshell that surrounds each egg [74–76]. Conditions that affect either secretion or production of these components can result in thin or spotty eggshells that fail to protect the egg from absorption of colored dyes. Thus, quantitation of eggs that absorb color is a simple assay to identify mutants or supplements that affect late follicle cell function (Fig. 5b).

Egg shape can indicate impacts of specific conditions on development as well. Normal eggs are oblong, with dorsal appendages extending from the anterior end. Perfectly round eggs are often the result of defective planar cell polarity (PCP) signaling [77] in follicle cells, due to a failure of egg elongation earlier in development. Figure 5c shows a very round egg laid by a wild-type fly treated with a dietary supplement that may affect PCP. Another revealing egg shape is referred to as a "dumping defect", where a pointed tail remains on the anterior end of laid eggs (Fig. 5d). Dumping defects are caused by a failure of nurse cells to transfer all their contents to the oocyte during stage 10b of oogenesis [78, 79], resulting in a developmental lag in resorption of used nurse cells prior to egg laying. Transfer of nurse cell contents to the oocyte is a highly regulated process, whereby actin-rich ring canals that link adjacent germ cells produce a sunburst array of actin filaments at stage 10b of oogenesis. These actin cables hold the nucleus in place while cortical actin contracts, squeezing the contents through the four largest ring canals into the oocyte [78, 79]. Dumping defects thus indicate potential impact on actin cytoskeletal regulators, enabling focused pursuit of more detailed mechanism. Finally, simple assays that enable visualization of the morphology of cells within the developing ovary and their nuclei (Fig. 6) can be leveraged to map changes that occur in many pathway-specific processes that occur during oogenesis at a more detailed level. For example, ovaries in which the nuclei are labeled with propidium iodide and actin are labeled with phalloidin, a toxin that binds to actin filaments, and help students determine effects on signaling pathways of interest (Fig. 6b).

1.3 Signaling Pathways in Oogenesis

1.3.1 EGFR Signaling

EGFR is critically important for most aspects of development, including oogenesis, and a major contributor to diseases of interest to student populations, including breast, lung, and colorectal cancer. The success of drugs targeting EGFR in cancer treatment [80] emphasizes the importance of identifying compounds that impact this pathway, a concept easily grasped by new scientists and seasoned professionals alike. During oogenesis, EGFR is an important regulator of the somatic cells that support oocyte development. Two easily identifiable processes that depend on EGFR are (1) oocyte polarity and (2) development of dorsal appendages and tubular structures that enable respiration in fertilized eggs and embryos (Fig. 5). The central role of EGFR in oogenesis presents many opportunities for analysis of the effects of environmental changes on the signaling pathway, with initial impacts on dorsal appendage formation as good indicators and more focused phenotypes ideal for follow-up analysis (*see* below).

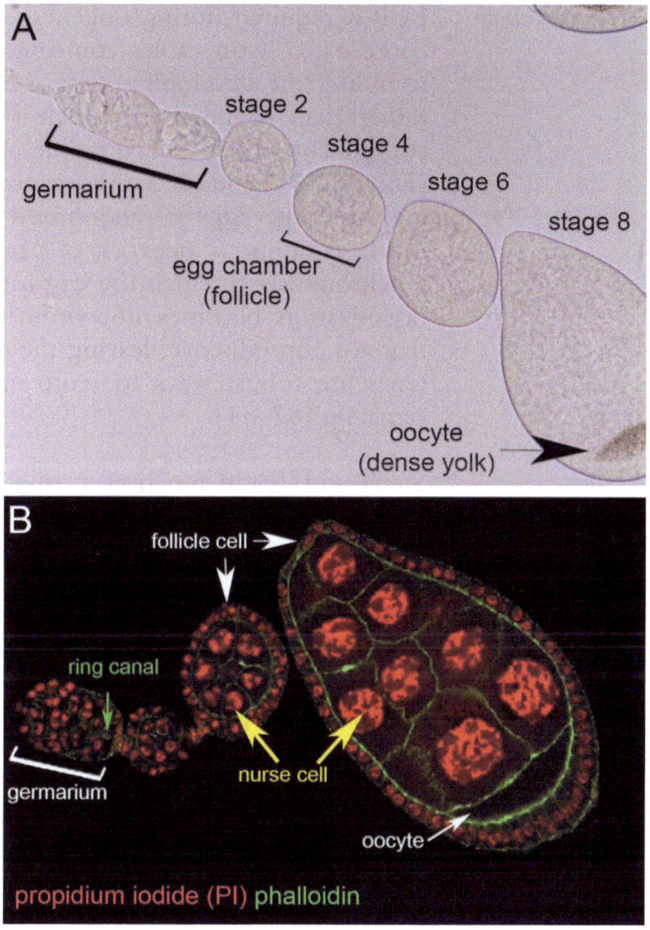

Fig. 6 Light microscopy and simple labels can reveal information about impacts on signaling. (**a**) Image of an ovariole taken with light microscopy. The germarium is located at the anterior-most end and houses two populations of stem cells. The germline stem cells (GSCs) produce germline cysts of 16-interconnected cells, one of which becomes the oocyte. Follicle stem cells (FSCs) are somatic cells that produce the follicular epithelium, or "follicle cells" that surround developing germline cysts throughout oogenesis. Follicles called "egg chambers" remain connected throughout development, enabling analysis of impacts of treatment or genetic conditions in a spatial and temporal manner. Oocytes accumulate yolk, which is denser than other cytoplasm under a light microscope. Egg chambers develop through 14 stages (up to stage 8 is shown) to form a mature egg. (**b**) Stains such as propidium iodide (red) can be used to visualize nuclei, enabling precise identification of follicle cells and germ cells in developing egg chambers. Use of the toxin phalloidin coupled to a fluorescent molecule (green) enables visualization of the actin cytoskeleton and interesting structures, such as polarized actin cables, ring canals, or the thickened actin that surrounds the oocyte

1.3.2 Planar Cell Polarity (PCP)

PCP is required during oogenesis for egg elongation [81], with coordinated actin cables coursing through the follicle cells surrounding the developing eggs constricting to create an oblong, rather than rounded egg shape [82]. Defects in PCP pathway components result in two visible phenotypes: (1) eggs are round (Fig. 5c) [81] and (2) actin organization in follicle cells surrounding late-stage eggs is randomized [76]. Rather than forming a nearly continuous network of parallel actin fibers that contract synchronously to cinch the egg into its final shape, cellular actin organization becomes uncoordinated, resulting in contractions that are unproductive, leaving the egg to round up. Both phenotypes are robust ways to score impacts of treatments on PCP signaling [82–84].

1.3.3 Src/Tec/Cytoskeleton Signaling

Src64 and Tec29 function at ring canals (Fig. 7) [85–89]. Ring canals are formed by incomplete cytokinesis during germ cell division early in oogenesis. Germ cells undergo 4 cycles of cell division, producing cysts of 16 interconnected cells stabilized by ring canals [90]. As cysts develop, ring canals grow accordingly, a process that depends on the activity of Src64 and Tec29 (Fig. 7). Under conditions where ring canals fail to grow, the increased pressure of squeezing cytoplasm through tiny ring canals often causes rupture of nurse cell membranes and aberrant fusion of nurse cells [79, 86]. In addition, cytoplasmic transfer is incomplete, with associated dumping defects [78, 79, 86]. Ring canal defects can be confirmed by (1) visualizing the actin cytoskeleton and collecting images to measure ring canal size (Fig. 7a–d), (2) scoring numbers of late-stage egg chambers with fused nurse cells and disrupted ring canals (Fig. 7e), (3) using activation-specific antibodies that target Src64, Tec29, or phosphotyrosine to evaluate changes in their ring canal localization status (Fig. 7g), or

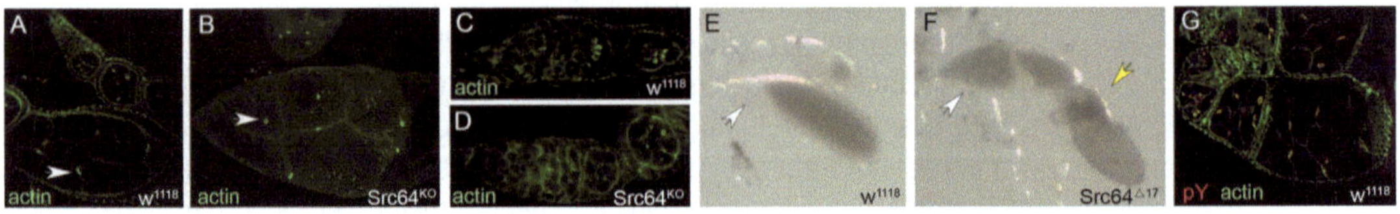

Fig. 7 Src64 controls the actin cytoskeleton in germ cells. (**a**) WT (w^{1118}) egg chambers exhibit large pores between germ cells called ring canals (white arrow) that are supported by a Src64-dependent actin–cytoskeleton (green). (**b**) Ring canals in *Src64* loss-of-function mutants ($Src64^{KO}$) are tiny, inhibiting dumping. (**c**) Ring canals in WT germline cysts form in the germarium as a result of incomplete cytokinesis followed by Src64-dependent assembly and maintenance of the actin cytoskeleton. (**d**) Ring canals are already small in the germarium in *Src64* mutants. (**e**) WT stage 14 eggs exhibit complete "dumping", with little or no excess nurse cell cytoplasm remaining (white arrow). (**f**) *Src64* mutants fail to dump the contents of the nurse cells into the oocyte (white arrow), resulting in short eggs with tails of lumpy cytoplasm attached (yellow arrow). (**g**) Ring canals are phosphorylated by Src64, an effect that can be visualized using anti-phosphotyrosine (red) antibodies

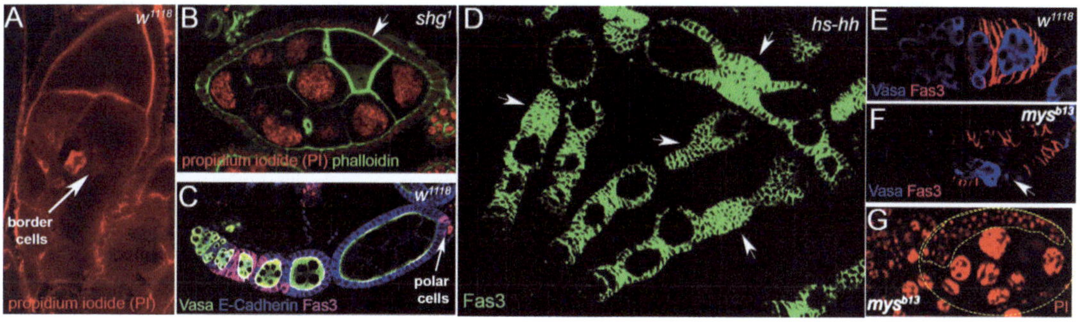

Fig. 8 Phenotypes characteristic of core signaling pathway are accessible using simple stains. (**a**) Stage 9 follicle stained for propidium iodide (red). Arrow points out the border cells. (**b**) E-Cadherin, encoded by the *shotgun* (*shg*) gene, is required for proper localization of oocytes to the posterior pole of each egg chamber. *shg* homozygous mutants (*shg*^{1}) exhibit noticeable oocyte localization defects (arrow). (**b**) E-Cadherin (blue) localizes to the lateral membranes of follicle cells, providing an excellent approach to assess epithelial cell polarity impacts. Germ cells are labeled with Vasa (green) and early follicle cells with Fas3 (pink). (**c**) Ectopic activation of Hh signaling leads to massive over proliferation of follicle cells (Fas3, green), which accumulate in stalks between egg chambers (arrows). (**d, e**) Integrins are required to establish polarity in developing follicle cells, with columnar epithelial cells forming in WT (*w*^{1118}, **d**) germarium that are grossly disrupted (arrow) in homozygous mutants for the β-integrin, β-PS (**e**), encoded by *myospheroid* (*mys*). Follicle cells (red) are labeled with Fas3 and germ cells (blue) with Vasa. (**f, g**) The polarity disruptions caused by integrin mutation early in development result in multilayering of follicle cells (yellow-boundary) and gaps in the epithelium

(4) directly visualizing actin cytoskeleton components or regulators using antibodies [85–89]. Viable homozygous mutants exist for both kinases, enabling enhancer–suppressor chemical screens to be conducted to identify treatments that impact Src/Tec signaling.

1.3.4 E-Cadherin

E-Cadherin is encoded by the gene *shotgun* (*shg*) in flies [91]. It mediates cell–cell interactions via homotypic adhesion between cells. *shg* mutants have myriad phenotypes during development [92], including reduced egg laying. Two easy to discern phenotypes provide opportunities for student analysis: (1) *shg* is required for migration of a sub-population of follicle cells called border cells [93–95]. During mid-oogenesis, border cells at the anterior end of each egg chamber delaminate from the follicular epithelium and crawl toward the oocyte in an E-Cadherin-dependent manner [96] (Fig. 8a). Upon arrival at the anterior end of the oocyte, the cells coalesce and develop into the micropyle, or sperm entry structure. *shg* mutant border cells fail to migrate, with stalled cells visible with actin or nuclear staining, or with specific antibodies against proteins expressed in border cells [93, 94]. In addition, *shg* mutant eggs are sterile, due to failure to produce a micropyle [97]. (2) *shg* is required earlier in oogenesis for oocyte localization and specification (Fig. 8b) [98, 99]. While E-Cadherin is expressed in both follicle cells and germ cells, the highest expression levels are in polar cells, a subset of follicle cells located at the poles of each egg chamber (Fig. 8c). Reducing E-Cadherin in polar cells leads to mislocalization of oocytes away from the posterior pole. Thus, mislocalized oocytes are a clear indication of defects in E-Cadherin function that can be easily observed by newly minted scientists.

1.3.5 Hedgehog

Hedgehog (Hh) signaling leads to surplus follicle cell formation (Fig. 8d) [100, 101]. Hh is expressed and released by apical cells at the tip of the ovary and received by follicle stem cells (FSCs), the mother cells of the follicular epithelium [101]. Within FSCs, Hh promotes proliferation and generation of new follicle cells to accompany developing oocytes throughout development [100–102]. Accumulation of follicle cells in long stalks (Fig. 8d) is an easy-to-identify Hh mutant phenotype using simple assays (e.g. Fas3, actin, propidium iodide or methylene blue staining, *see* protocols below) that indicate impact of a particular treatment on Hh signaling.

1.3.6 Discs Large/Lethal (2) Giant Larvae/Scribble

Septate junctions, the fly equivalent of mammalian tight junctions, are adhesion complexes that control cell–cell interactions in many developing tissues. Loss of tight junction components including *scribble*, *discs large*, or *lethal(2)giant larvae* in fly ovaries results in dramatic invasion phenotypes during early oogenesis [103], with mutant follicle cells swarming amidst germ cells and disrupting egg chambers. The dramatic phenotype and clear association with metastasis in cancer make the septate junction complex a useful tool for students to delve into interventions that suppress cell migration and invasion.

1.3.7 Integrins

Integrins are cell-adhesion receptors that control many physiological processes [92, 104]. Integrins mediate adhesion between cells and the extracellular matrix, with dynamic regulation controlling the cell's adhesive strength, enabling cell migration, establishment of cellular polarity, and signaling [105]. Analysis of genetic or environmental impacts on integrin signaling in oogenesis can be achieved using simple actin staining [106]. FSCs lacking expression of the β-integrin subunit, encoded by the *myospheroid* (*mys*) gene, exhibit aberrant location of FSCs and disrupted formation and polarity of the follicular epithelium and interfollicular stalks (Fig. 8e–g) [107]. These major phenotypic changes can be used for students to identify novel contributors to the mechanism of integrin regulation of oogenesis.

1.3.8 Notch

During oogenesis, Notch is critical for a process known as endoreplication in follicle cells. Cells essentially remain in S-phase without division for multiple rounds of replication, resulting in retention of multiple copies of each chromosome for use in large-scale production of specific proteins (Fig. 9a). In the absence of

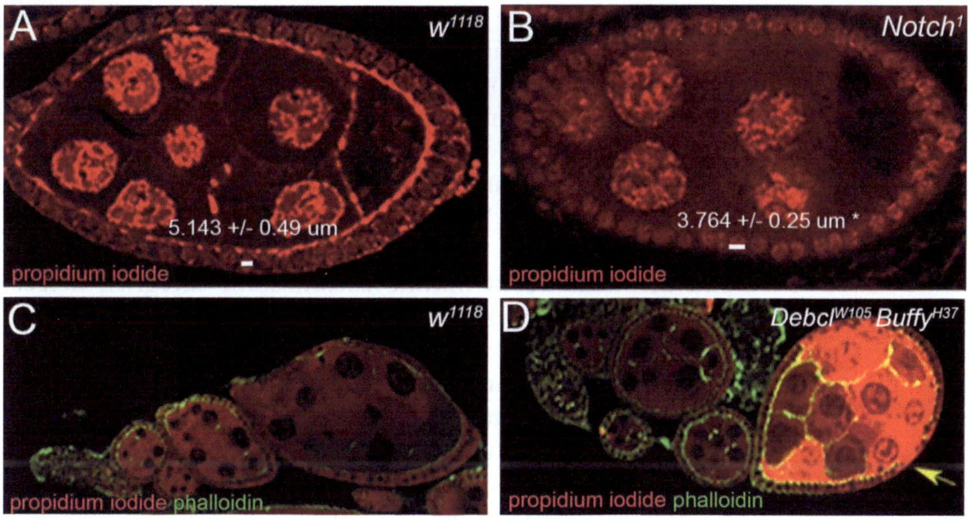

Fig. 9 Aberrant phenotypes can be scored by students for quantitative results on the impact of conditions on development. (**a**) WT (w^{1118}) follicle cells at stage 8 have an average nuclear diameter of ~5 μm, whereas (**b**) *Notch* mutant follicle cells have limited endoreplication resulting in smaller nuclei ~3.8 μm. Programs such as Image J or the confocal system used to collect images are straightforward approaches for measurement of parameters from fluorescent images. (**c**, **d**) Propidium iodide (red) enters dying cells in unfixed tissue, providing a robust method for measuring cell death in various conditions. After treatment with propidium iodide and washing, tissue can be fixed and stained with phalloidin (green) to visualize the actin cytoskeleton. Flies bearing mutations in *Debcl* and *Buffy* (**d**) have increased rates of cell death relative to WT (w^{1118}, **c**)

Notch, endoreplication is inefficient, resulting in follicle cells with smaller nuclei (Fig. 9b) [108–110]. As nuclei are relatively easy to label and image, measurement of the diameter of follicle cell nuclei (Fig. 9a, b) and identification of conditions that yield smaller size is a solid approach to uncover new Notch regulators.

1.3.9 Bcl-2 Family

The Bcl-2 family of cell death regulators provides an accessible inroad for students to consider how the balance between cell proliferation and cell death impact development and disease. In mammals, Bcl-2 is an inhibitor of apoptotic cell death [111]. Its loss, therefore, enables uncontrolled apoptosis and massive cell death. Similar effects are seen in flies, with Buffy (from Buffy the Vampire Slayer) acting as an anti-apoptosis regulator [112]. A second Bcl-2 family member, Death by Executioner (Debcl) is pro-apoptotic [113], with the balance between the two proteins determining cell survival versus cell death status. Using a fly mutant line that reduces expression of both *Debcl* and *Buffy*, our students observed dramatically increased apoptosis during oogenesis (Fig. 9c, d). Multiple nutrients can induce cell death during oogenesis, providing an exciting and virtually unexplored area of investigation to link diet to the control of cell survival and cell death.

1.4 New Genes and Phenotypes

1.4.1 Brca2

BReast CAcer Gene (BRCA) family members are key regulators of homologous recombination, a critical mechanism in control of DNA damage responses. The BRCA family represents one of the best studied mechanisms of inherited cancer risk [114, 115]. Notably, BRCA family members are responsible for significant incidence of breast and ovarian cancers in URM patients, as well as in well-studied Ashkenazi Jewish populations, where BRCA mutations were first described. *Brca2* in flies is a known regulator of meiosis, and maintenance of mitotic chromosome integrity, defects that affect eggshell development [116]. Our students' examination of homozygous *Brca2* viable mutants revealed heightened apoptosis in late-stage egg chambers and increased prevalence of mispackaged egg chambers containing too few germ cells surrounded by a normal-appearing follicular epithelium (Fig. 10a–c). As null mutations in *Brca2* have dramatic effects on development in early stages of oogenesis, these viable mutants may be revealing roles for *Brca2* later in development, with further work needed to understand its full function during oogenesis.

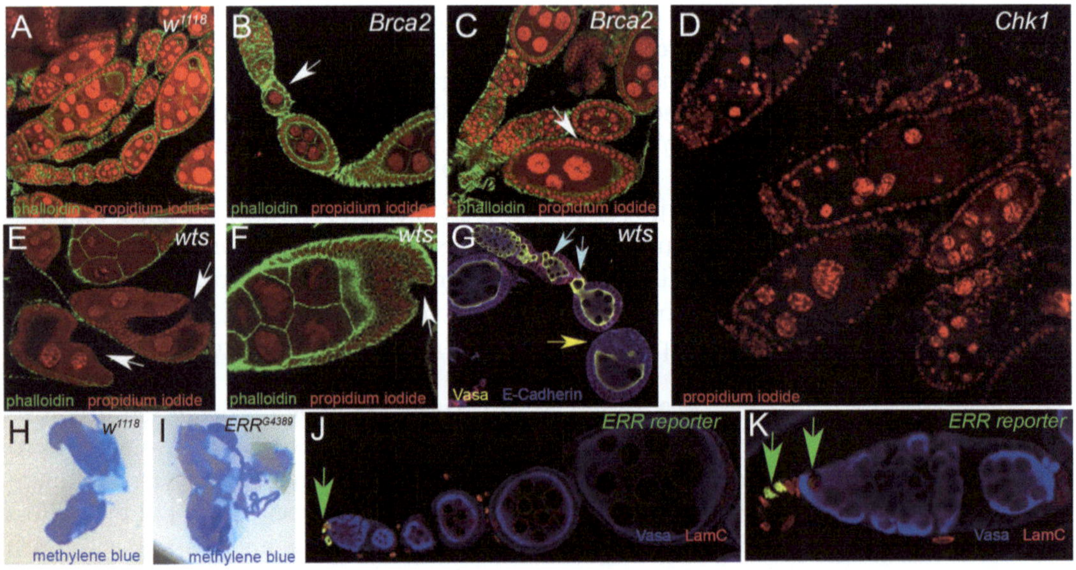

Fig. 10 Novel phenotypes revealed via citizen science. (**a**–**c**) *Brca2* mutants exhibit egg chambers with fewer than 16 germ cells (arrow in **b**, **c**) relative to WT (w^{1118}, **a**). Nuclei (PI, red), actin (phalloidin, green). (**d**) Cell cycle regulators such as Chk1 are required for normal development. Here, egg chamber organization, polyploidy, and nuclear structure are all disrupted in *Chk1* mutants, visualized by nuclear labeling with propidium iodide. (**e**, **f**) *warts* mutants (wts^{1}) exhibit a "cookie monster" phenotype with epithelial cells invading germ cells in a manner that appears like a bite (white arrows in **e**, **f**). (**g**) *wts* mutants also exhibit aberrant egg chamber formation (blue arrows) and excess production of invasive follicle cells (yellow arrow). (**h**, **i**) Live ovaries stained with methylene blue (blue) enable size comparisons between conditions. Here, WT (w^{1118}) ovaries (**h**) exhibit reduced size upon drug treatment but ERR homozygous mutants (**i**) are unaffected. (**j**, **k**) GFP-based activity reporters exist for many proteins, including ERR. Visualization of GFP (green, arrows) reveals activation in terminal filament and cap cells at the anterior-most end of the ovary. Rare activity was observed in germ cells or Follicle Stem Cells, suggesting functional roles for *ERR* during oogenesis. Nuclei are outlined with LamC (red) and germ cells (blue) are labeled with Vasa

1.4.2 Cell Cycle Regulators

Cancer, at its base, is a disease of uncontrolled proliferation. As might be expected, defects in cell cycle regulators affect germ cell division [117]. For example, mutations in the CDC25 homolog, *String*, result in germline cysts of only eight cells [118]. We have begun to investigate phenotypes for homozygous viable mutants in cell cycle genes, including *String*, *Cdk4/6*, *Cyclin D*, *p53*, *Rb*, and *Chk1*, finding both expected and unexpected phenotypes. For example, *Chk1* mutants, which regulate cell cycle progression by surveilling DNA damage, exhibit morphological defects, aberrant nuclear size, lack of stalks, and defects in interaction between follicle cells and germline cysts (Fig. 10d). Cyclin-dependent kinases (Cdks) are the most well-known example of cell cycle regulators to pre-college students. While well-understood mechanistically, the impact of specific cell cycle regulators on developmental events in oogenesis is not well studied. Examination of mechanisms that are well-studied in mammalian cells and are conserved in flies provides an underutilized method for screening new compounds or genetic interactions with potential impact in cancer.

1.4.3 Hippo/Warts/Yki

The Hippo/Warts/Yorkie signaling pathway (referred to as "the Hippo pathway") is a driver of disease, including cancer [119]. Emerging literature supports roles for the Hippo pathway in FSC regulation [120], where it functions downstream of Hh signaling to control proliferation. Hippo has also been shown to regulate polar cell fate specification and migration of border cells [121]. We found that viable mutations in *warts*, a kinase that acts in the Hippo pathway, had a striking "cookie monster" phenotype, where egg chambers appeared to have large bites taken out of them (Fig. 10e, f), as well as egg chambers with too few cells or massive overgrowth of follicle cells and invasion of germline cysts (Fig. 10g). These unexpected phenotypes may indicate new roles for Hippo signaling in regulation of follicle cell–germ cell interactions, altered mechanical or adhesion properties, or suggest Hippo-independent functions of Warts during egg chamber formation.

1.4.4 Estrogen-Related Receptor (ERR)

Estrogen-related receptors (ERRs) were initially identified due to their structural similarities to estrogen receptors (ERs) [122]. Unlike estrogen receptors, however, ERRs are a family of orphan nuclear receptors, with no known ligand. ERR is highly expressed in "triple negative" breast cancers (TNBC) which have lost dependency on hormones including estrogen and EGF, and ERR expression is associated with high rates of recurrence and poor prognosis [123, 124]. Thus, ERR is a prime candidate for targeting aggressive, intractable TNBC [125], a disease that particularly impacts young, Black women [126, 127]. Our students observed increased apoptosis in developing egg chambers of ERR loss-of-function mutants (data not shown), consistent with the idea that overexpression or gain-of-function might promote resistance to cell death. Notably, ERR homozygous mutants exhibited resistance to treatments that reduced the size and limited development of wild-type ovaries (Fig. 10h, i), suggesting high potential for using oogenesis as a screening approach for drugs that target ERR mutants. Using transgenic fly lines that report ERR activity, students visualized ERR activity in cap cells, escort cells, germ cells, and on rare occurrences in FSCs (Fig. 10j, k). These early results suggest that oogenesis may be an important model with the capacity to mimic aspects of ERR overexpression in cancer and screen for compounds that target cells overexpressing ERR.

1.5 Defining Mechanisms in Molecular Detail

Once a pathway of interest is identified, students can pursue mechanism using genetics, biochemistry, and imaging. We are particularly interested in identifying direct targets of particular nutrients. To accomplish this, individuals or groups of students screen mutants representing sequential steps in a signaling cascade for effects, a pseudo-epistasis-like approach that allows them to pinpoint which effectors in a given pathway yield phenotypic changes, and which do not. Mapping phenotypic outcomes enable hypotheses to be generated regarding the precise step in a signaling pathway that is targeted by the nutrient or intervention. Going one step further, isolation of ovaries and use of immunoblotting enable quantitation of impacts on signaling pathways. For example, antibodies targeting the EGFR effector ERK reveal patterns of pathway activity upon treatment of wild-type or mutant flies (Fig. 11a). The same ERK antibodies used for immunoblotting also work robustly for immunostaining (Fig. 11b), allowing teams of students to address key questions using multiple, complementary techniques.

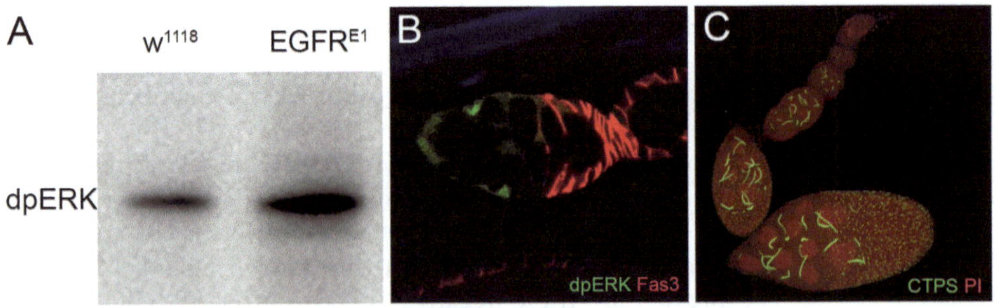

Fig. 11 Protein activity in specific pathways can be measured in a temporal–spatial manner during oogenesis. (**a**) Immunoblot analysis of EGFR pathway activity using antibodies against activated (doubly phosphorylated) ERK (dpERK). WT (w^{1118}) ovaries exhibit basal ERK activation that is increased upon expression of a constitutively activated form of EGFR ($EGFR^{E1}$). (**b**) Same anti-dpERK antibodies can be used on fixed tissue, with activity (green) observed in the germarium. Follicle cells are labeled (red, Fas3). (**c**) GFP-tagged proteins such as CTPS (green) reveal complex protein assemblies with impact on development. Nuclei are labeled (red, PI)

An additional complement to these approaches is use of fluorescently labeled proteins that reveal localization and organization of key regulators of specific cellular processes [128].

Observation and analysis of unexpected mechanisms are exciting and engaging. For example, CTP synthase (CTPS), a nucleotide biosynthesis enzyme which is essential for CTP production, has been found to form large filaments in organisms from bacteria to humans including in Drosophila (Fig. 11c) [129]. In humans, these filaments have been found in some cancer cells, but their function is poorly understood [130]. These filaments are present in both nurse cells and follicle cells in a regulated manner [131, 132], making this an important model system to explore filament function and regulation. There is also a GFP-tagged CTPS fly line which can be used with fluorescent imaging to view the CTPS filaments during oogenesis. Thus, the possibilities are essentially limitless given access to fairly standard lab equipment and/or a fluorescent microscope.

1.6 Moving the Research into the Home Laboratory

The COVID-19 pandemic closed the doors of research labs worldwide, creating a huge gap in scientific progress. A small positive outcome was the development of protocols and approaches to conduct research at home. We refer to the process as "kicking it old school", reviving protocols that depend on vital dyes that were developed nearly 150 years ago. A simple methylene blue stain reveals cellular outlines and nuclei in live ovaries (Fig. 12a). These can be photographed using magnifying pocket scopes or even cell phones outfitted with a magnifying lens (Fig. 12b). While the

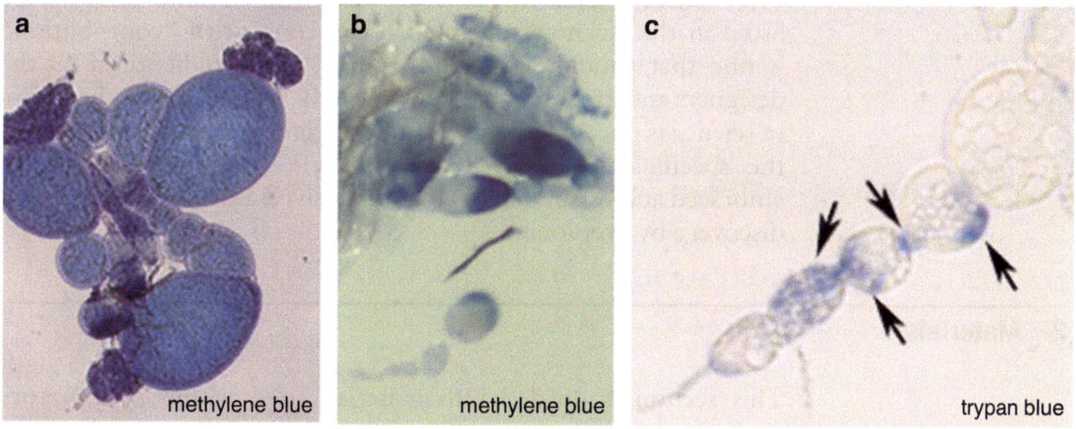

Fig. 12 Vital dyes enable cutting edge research to be conducted anywhere. (**a**) Ovaries stained with methylene blue exhibit faint nuclear outlines and dark blue/purple follicle cells. Egg chamber stages are easy to recognize in images taken on a professional light microscope outfitted with a camera. (**b**) Participants can perform experiments using methylene blue staining at home and collect images using a pocketscope or cell phone camera. Developmental stages are clearly visible, yolk is heavily stained, and germ cell architecture is straightforward to analyze, even at low magnification. (**c**) Trypan blue treatment of live ovaries reveals dying cells that can be identified (follicle cell versus germ cell) and scored

degree of detail is obviously less than can be observed on a high-end fluorescent microscope, students can observe changes in cell numbers, organization of cells, and fairly specific phenotypes, including dumping defects and aberrant replication. Drilling down further, trypan blue dye only enters dying cells (Fig. 12c), enabling scoring of apoptosis and other types of cell death. Initial studies at home can thus provide a mechanistic focus to be followed up when access to more technologically advanced equipment becomes available.

1.7 An Untapped Resource

Citizen science provides a way for non-scientists not only to learn about scientific techniques and concepts, but to take a huge step further to make a contribution to advance our understanding of biology. There are 16 million high school students in the United States and 550 million students worldwide. The great majority of these students have never heard of research. In fact, most students view science as boring, difficult, and a barrier to their pursuit of more interesting career paths. Just imagine if citizen science approaches could flip their viewpoint, providing immersive involvement in important research as the drivers of their first science experience. Not only would more students pursue scientific experiences, but the breadth and depth of the research questions throughout the scientific community would expand [62]. In our experience, participation enhances research and health literacy, with huge gains in problem solving, collaborative, and communication skills. Tremendous positive impact is reported by all participants, even those who do not pursue STEM careers. Participants who generated data from self-designed projects are listed as authors on this chapter, and participants in our global citizen science effort who collected data on eCLOSE-designed research projects are listed in the Acknowledgements. The concept of "contribution" is one that should be taken seriously, with credit given to the designers and executors of novel research. Our vision for the future of science is one, where the cultural and life experiences paired with the specific and highly focused "WHY" of each student are embraced and celebrated to create inclusive, safe spaces for scientific discovery by everyone.

2 Materials

This section is divided into materials used within a laboratory environment (indicated as "in-lab") and those that are safe to use in classrooms or homes ("in-home"). Protocols for both approaches are provided below.

2.1 Pupal Counting

1. 10× Phosphate buffered saline (PBS), pH 7.4. This can be purchased or prepared [133] for 1 L: 80 g NaCl, 2 g KCl,

14.4 g Na_2HPO_4, 2.4 g KH_2PO_4, and dH_2O up to 1 L. Dilute to a working concentration of 1×.

2. Fly food supplemented with test compounds (*see* **Notes 1** and **2**).
3. Grape juice plates: 50% purple grape juice, either 1% acetic acid (lab grade) for in-lab or 20% white vinegar for in-home, 3% agar, and 0.1% methylparaben in ethanol.
4. Fly morgue: add 70% ethanol (in-lab) or white vinegar (in-home) to a wide mouth bottle or flask. Insert a funnel into the open top.
5. Standard microwave or stirring hotplate.
6. Microwave/heat safe beaker (stir plate) or flask (microwave), 250 mL.
7. Magnetic stir bar, if using stir plate.
8. Digital balance with weigh boat.
9. Graduated cylinder.
10. 35 × 10 mm petri dishes.
11. Pipet/pipette filler.
12. Tips/micropipette.
13. 50 mL conical tubes with screw top caps.
14. Parafilm.
15. Magnifying lens.
16. Push pins.
17. Stereoscope (camera-optional).
18. Narrow *Drosophila* vials with Flugs.
19. Chemical fly anesthesia.
20. Wood applicator sticks for stirring.
21. Collection cages: 50 mL conical tubes are used as collection cages. With a push-pin, make eight holes in the pointed end of the tube to allow for air circulation.

2.2 Ovary Dissection

1. 1× PBS: dilute with dH_2O from 10× Stock.
2. Grace' Insect Medium, unsupplemented (in-lab).
3. Gloves and safety goggles.
4. 4% formaldehyde fixative solution, in PBS (in-lab): for 1 mL, add 40 μg of paraformaldehyde powder to 1 mL of 1× PBS. Heat to 95 °C. Mix frequently by vortexing until powder is dissolved. Let cool to room temperature. Alternatively, add 108 μL of 37% formaldehyde to 892 μL of 1× PBS.
5. PBS-T (in-lab): 1× PBS, 0.3% Triton.

6. Dissecting dish, silicone-based.
7. 1.5 mL microfuge, snap-cap tubes.
8. 2 extra-fine point forceps.
9. *Drosophila* anesthetic: chemical-based or CO_2.
10. Stereoscope (camera-optional).
11. Rocker or Nutator.
12. Fume hood (in-lab).
13. Personal protective equipment (PPE; in-lab): gloves, goggles, and lab coat.

2.3 Phalloidin and Propidium Iodide Staining (In-Lab)

1. PPE.
2. Propidium iodide: dilute 2 μL in 200 μL of 1× PBS.
3. RNAse A: add 6.4 μL of a 25 mg/mL stock into 200 μL 1× PBS.
4. Fluorescein–phalloidin.
5. 1× PBS.
6. PBS-T.
7. Tube rocker or nutator (optional).
8. Mounting medium (*see* **Note 3**).
9. Kimwipes.
10. Glass slides.
11. 22 × 22 #1.5 coverslips.
12. Slide box.
13. 100 mL glass bottle.
14. Nail polish.
15. Fluorescent or confocal microscope.

2.4 Immunostaining (In-Lab)

1. PPE.
2. 5% bovine serum albumin (BSA) in 1× PBS.
3. Anti-dpERK primary antibody (*see* **Note 4**).
4. Anti-rabbit secondary antibody coupled to a fluorophore.
5. PBS-T.
6. Tube rocker or nutator (optional).
7. Refrigerator with electrical outlet or cold room (optional).
8. Glass slides.
9. 22 × 22 #1.5 coverslips.
10. Slide box.
11. 100 mL glass bottle.
12. Nail polish.

13. Mounting medium (*see* **Note 3**).
14. Kimwipes.
15. Fluorescent or confocal microscope.

2.5 Trypan Blue and Methylene Blue Staining (In-Lab and In-Home)

1. 1% Trypan Blue Stain (purchased from manufacturer).
2. 0.2% Methylene Blue Stain (purchased from manufacturer).
3. Glycerol.
4. Glass slides.
5. 22 × 22 glass coverslips #1.5.
6. Stereomicroscope.
7. Nail polish.

3 Methods

3.1 Pupal Counting (In-Lab and In-Home)

A simple pupal counting assay is a first step in identifying genetic and environmental conditions that affect fertility. The underlying premise is that differences in the ability to develop to the pupal stage indicate a defect between conditions, enabling further analysis, as described below. Our focus is diet. Foods, dietary supplements, or purified chemicals found in food can be screened for effects on wild-type or mutant flies using the following approaches (*see* **Note 2**). The same approach can be applied to other environmental impacts, such as stress, UV irradiation, light–dark cycles, toxins, or other items of interest.

3.1.1 Prepare Fly Food Vials by Adding Test Foods/Beverages

1. Heat standard fly food in a microwave to melt. One 50 mL bottle will take about 1 min to melt, pause and stir every 15 s, and is enough to make 4 vials.
2. Prepare control vials and, if using, prepare liquid experimental vials (*see* **Note 5**):
   (a) Pipet 9 mL of melted fly food into a narrow plastic vial.
   (b) Let the vial cool slightly (*see* **Note 6**), and then add 1 mL of 1× PBS or your test compound and mix thoroughly with a wooden stir stick.
   (c) Cool vials enough to safely touch with the inner wrist.
   (d) Insert Flug halfway into the vial to allow for easy removal.
3. If using, prepare solid food experimental vials (*see* **Note 5**):
   (a) Weigh out 0.5 g of item to be tested (or a specific amount based on the experimental question, *see* **Note 7**).
   (b) Pipet 10 mL of melted fly food into a narrow plastic vial.

(c) Let the vial cool slightly (*see* **Note 6**) and then mix in 0.5 g of the test food with a wooden applicator stick.

(d) Insert Flug halfway into the vial to allow for easy removal.

4. Allow food vials to set:

(a) Wipe water droplets off the sides of the vials with a lint-free tissue, if necessary.

(b) Allow food to set overnight at room temperature with the Flugs inserted halfway into the vial. Food may be stored, premade, in the refrigerator for up to 10 days sealed with Flugs and wrapped in plastic.

3.1.2 Pupal Counting

1. Anesthetize flies either with CO_2 in a lab environment or with chemical fly anesthetic at home (Fig. 13b) (*see* **Note 8**).
2. Sort flies, selecting ten females and five males of the desired genotypes (*see* **Note 8**), and add to the cooled and set fly food vials.
3. Keep the vials on their sides with Flugs inserted until the flies wake up and then tip them upright into a rack for safekeeping.
4. Record this as "Day 0" in your lab notebook. In addition, record: (a) genotype, (b) treatments, (c) date of experiment, (d) dose of treatment compounds, and (e) any additional observations.

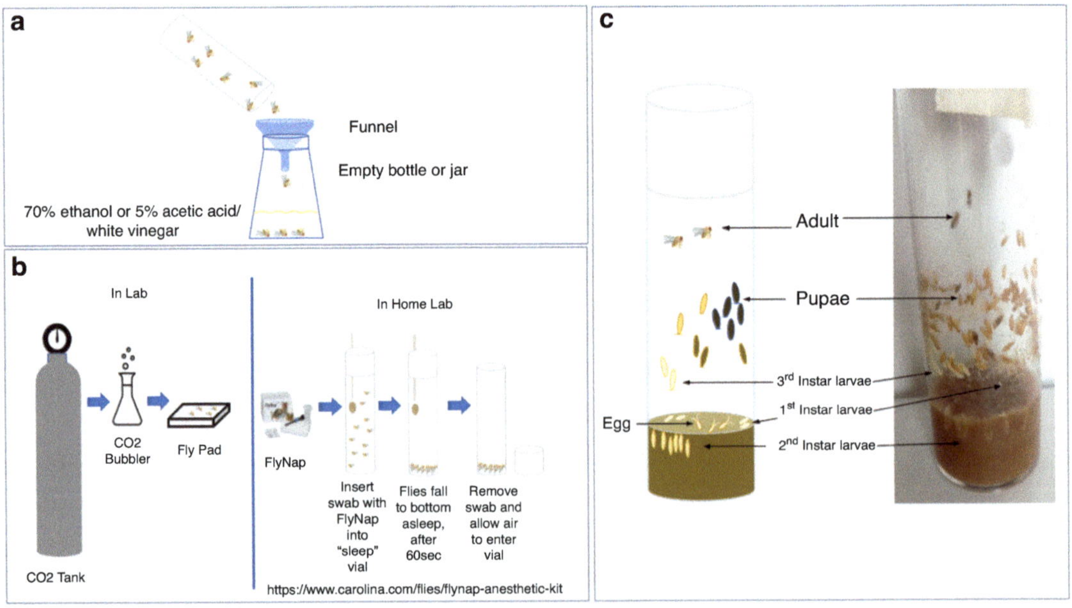

Fig. 13 (**a**) Morgue assembly. Dump flies no longer being used in the experiment into 5% white vinegar (in-home) or 70% ethanol (in-lab). (**b**) Modes of anesthetizing *Drosophila* in labs and at home. (**c**) Method 1—Pupal Counting. (Figures reprinted with permission from eCLOSE Institute)

Table 1
Score sheet for pupal counts

Pupal Score: EXAMPLE		
Day 14		
DATE		
GENE	Control: 1X PBS	Experimental: FOOD
(NAME)	(SCORE: total# of pupae)	(SCORE: total # of pupae)
Ex. *w1118*	158	62

5. After 7 days, remove adult flies from the food vials. The adult flies can be transferred for use in the 24 h egg laying assay (Subheading 3.2) or dumped into a morgue.
6. Record this in your notebook as "Day 7" with the date.
7. Score total numbers of pupae on day 14 (be sure to record the date in your lab notebook). Circle small groups of ~10–15 pupae with a permanent marker, then count the pupae inside the circles, placing a dot on each one, as it is counted (Fig. 13c).
8. Record total pupal counts in a lab notebook (refer to Table 1 for an example score sheet and *see* **Note 9**).
9. Analyze the data (*see* **Note 10**).

3.2 Egg Analysis

3.2.1 Prepare Grape Juice Agar Plates

Microwave Preparation of Grape Juice Agar Plates, Active Time 20 min (In-Lab or In-Home)

1. Add 30 mL distilled water, 50 mL grape juice, 20 mL 5% vinegar to 250 mL Pyrex flask (*see* **Note 11**).
2. Heat, in microwave, on high 30 s.
3. Using hot mitt, remove and place onto heat-safe surface.
4. Carefully, add 1 g agar to the center of flask-mouth, and swirl gently to incorporate the agar (*see* **Note 12**).
5. Repeat **step 4** twice to incorporate a total of 3 g of agar; the solution will be cloudy.
6. Microwave at 10 s intervals, swirling between each (using a hot mitt), until the solution is clear, indicating all the agar are dissolved.
7. Cool the heated solution for 90 s. In lab add 1 mL 10% methylparaben. In home, proceed directly to **step 8** (*see* **Note 13**).
8. Using a pipet filler, dispense warm agar solution into both lids and bottoms of 35 × 10 mm petri dishes. Stop dispensing when agar touches all sides of the dishes (*see* **Notes 13** and **14**).

Hotplate Version, Active Time 60 min

1. Carefully place a stir bar into a 250 mL Pyrex beaker.
2. Add 30 mL distilled water, 50 mL grape juice, 20 mL 5% vinegar (*see* **Note 11**). In-lab, in a fume hood, add 1% acetic

acid; this may NOT be done in a school or home-lab setting! Once the acetic acid is diluted in the grape juice solution, the protocol may be completed outside the hood.

3. Place the mixture on stirring hot plate and mix the solution. Set heat knob to medium-high and the stir knob to low-medium.
4. Once you see small bubbles at the bottom edge of the beaker, carefully, add 1 g agar to the center of the beaker and continue stirring to incorporate the agar.
5. Repeat **step 4** twice to incorporate a total of 3 g of agar and stir until the solution is clear.
6. Turn off heat, allow solution to cool for 90 s. Then, if in-lab, add 1 mL 10% methylparaben (*see* **Notes 13** and **14**). If in home, proceed to **step 7**.
7. Remove the beaker from the heat and place onto heat-resistant surface or lab bench. Cool for 3 min.
8. Using a pipet filler, dispense warm agar solution into both lids and bottoms of 35 × 10 mm petri dishes. Stop dispensing when agar touches all sides of the dishes (*see* **Note 13**).

3.2.2 24-h Egg Lay and Visualization of Egg Structures

Once conditions are identified that indicate a defect in development up to the pupal stage, the next step is to determine whether the difficulty arose before or after oogenesis was complete. To do this, the numbers of eggs laid by a group of female flies during a set time period is measured. As egg laying is a variable behavior, short timepoints tend to be unreliable, depending on the time of day and conditions of the environment at the time of the assay. Most labs use a 24 h timepoint, both for ease of data collection, because most females will deposit their eggs at some point during the day, increasing the reliability and reproducibility of the assay.

1. Use Method Subheading 3.1.1 to prepare vials and sort ten females and five males of flies. One genotype per vial. Alternatively, use the adult flies previously set up for pupal counting for use in the egg laying assay.
2. Transfer adult flies into the collection cage (Fig. 14a).
3. Gently place a grape juice plate on the open end of the collection cage, with the grape juice facing the inside of the cage (Fig. 14b).
4. Gently seal the grape juice plate to the collection cage by pulling a piece of parafilm around the interface (*see* **Note 15**).
5. Flip the collection cage such that the grape juice plate stands on the bench and the pointed end with holes is facing up.
6. Allow the flies to lay eggs for 24 h, preferably in a space, where the temperature will be consistent throughout the experiment (*see* **Note 16**).

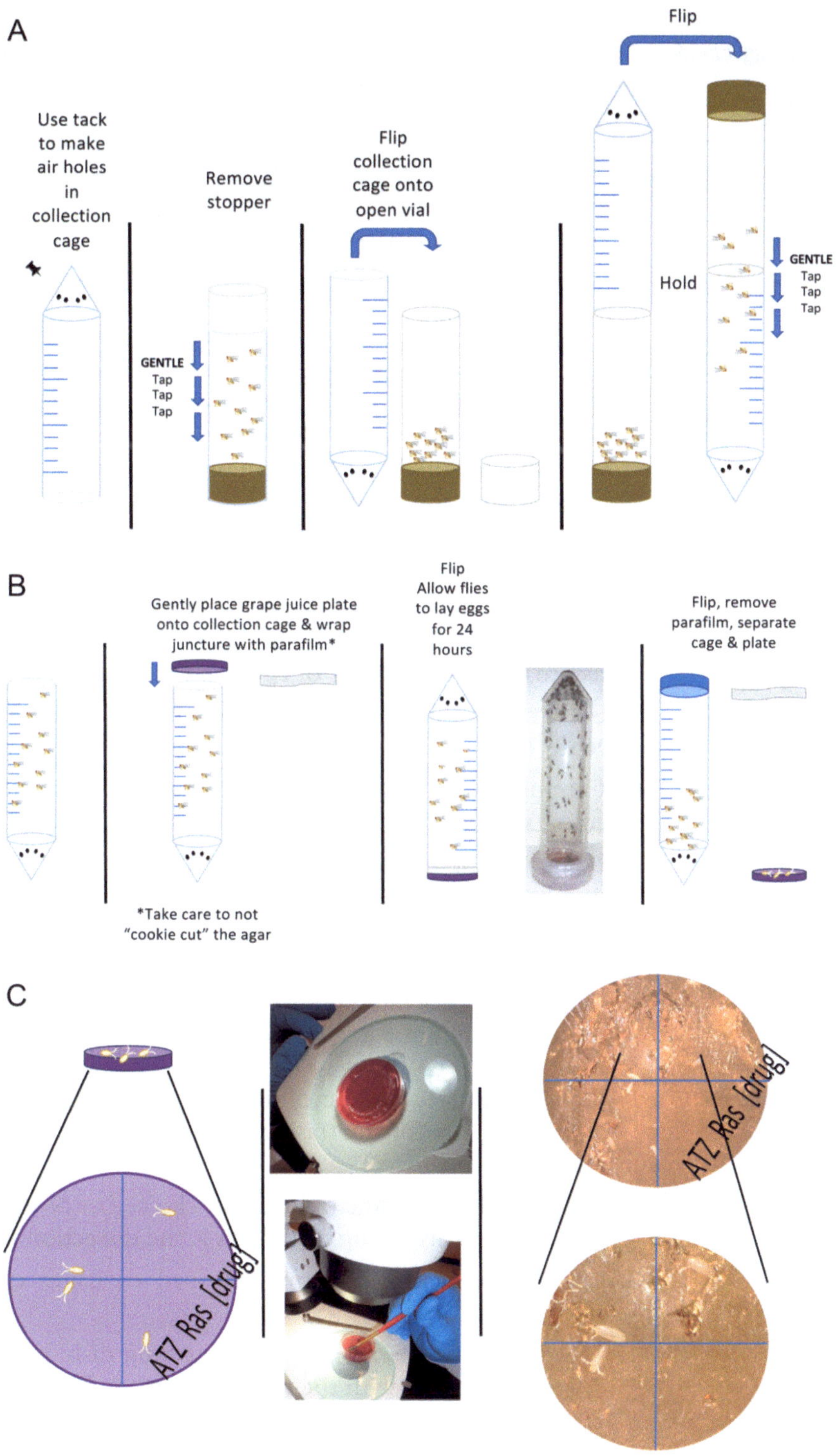

Fig. 14 Method 2—24-h egg lay. (**a**) Transfer flies to collection cage. (**b**) Assembly and disassembly of collection cage. (**c**) Visualization of eggs. (Figure reprinted with permission from eCLOSE Institute)

Table 2
Scoresheet for egg counts

Egg Score: EXAMPLE		
24 hours		
DATE		
GENE	Control: 1X PBS	Experimental: FOOD
(NAME)	(SCORE: total# of eggs)	(SCORE: total # of eggs)
Ex. *w1118*	97	84

7. At the 24 h timepoint, place the collection cage into a 50 mL conical tube rack with the grape juice plate at the top. Tap the flies to the bottom and quickly remove the grape juice plate. Continue tapping the flies down until they can be flipped into a morgue and discarded (Fig. 13a). If you need extra time, use the original screw-cap and re-close the cage.
8. With a permanent marker, draw four quadrants on the bottom of the petri dish of the grape juice plate. Count the eggs in each quadrant using either a magnifying glass or a microscope and record the total number of eggs on each plate (Fig. 14c).
9. Use a table (Table 2), record egg counts for each condition (*see* **Note 10**).
10. Carefully examine the eggs on the grape juice plate for eggshell color, size, and morphology.
11. If any phenotypes are observed, quantify them by counting the numbers of eggs that are "normal" versus "abnormal" for each condition above (*see* **Notes 11**, **17**, and **18**).
12. Image eggs on a stereomicroscope with an attached camera, or by carefully placing a mobile phone above the eyepieces to capture images in a lower cost, highly accessible manner.

3.3 Ovary Dissection and Fixation

3.3.1 Dissection

1. Pour a coin-size puddle of 1× PBS at the outer edge of the dissecting plate for the trash pile (pieces of fruit flies not being used in the protocol).
2. Make a 400 μL puddle of buffer of choice (*see* **Note 19**) in the center of the dissecting dish. This is the dissection puddle.
3. Anesthetize female flies on CO_2 pad or using chemical fly anesthetic.
4. Move individual females to the dissecting dish, just under the dissection puddle with the forceps. Remove heads (Fig. 15a).
5. Submerge the rest of the fly in the dissection buffer puddle. Place forceps at the anterior-most position with one hand. With the other hand, grasp the middle of the abdomen. Pull in opposite directions to make the ovaries pop out (Fig. 15b).

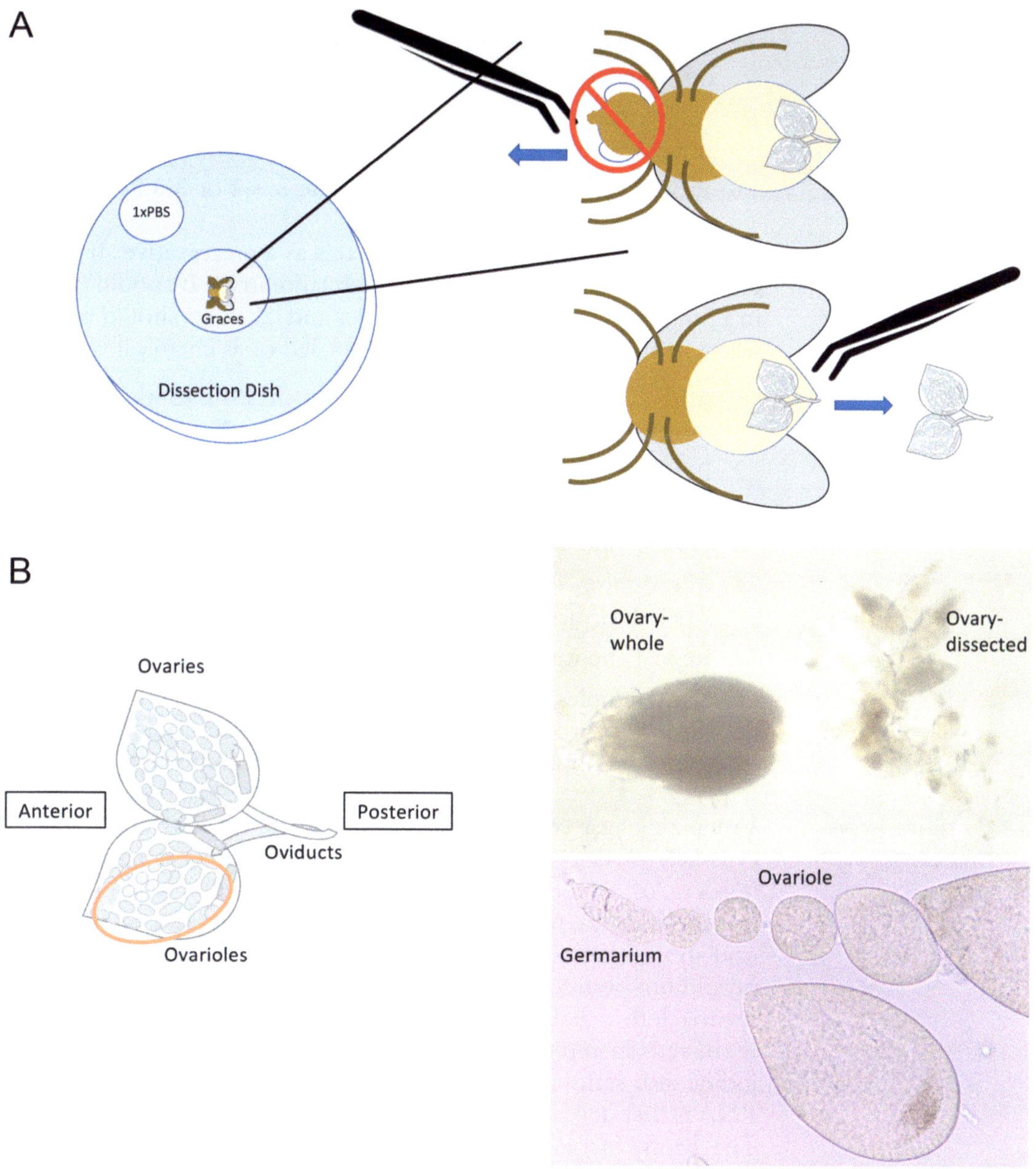

Fig. 15 Method 3—visualizing oogenesis. (**a**) Ovary dissection, (**b**) ovarian anatomy. (Figure reprinted with permission from eCLOSE Institute)

6. Leave the ovaries in the dissection buffer and discard the rest of the fly parts in the trash puddle.
7. Repeat for a total of five or more pairs of ovaries (*see* **Note 20**).
8. Once the ovaries are removed, cut the end off a 200 μL micropipette tip and use aspiration to collect the ovarioles very carefully. Some ovaries may be damaged or lost during the process, so it is best to start with more than needed for the final experiment.

9. Transfer the ovaries, in 30 μL volumes at a time, to a microfuge tube and gently pipet ovaries up and down using a cutoff yellow tip to separate individual ovarioles (*see* **Note 21**).
10. Allow the ovarioles settle to the bottom of the tube and then remove as much buffer as possible into a trash bottle.
11. Proceed immediately to Subheading 3.3.2 or 3.3.5.

3.3.2 Fixation (In-Lab)

Formaldehyde embalms tissue and acts as a preservative. It should NEVER be used outside of a research laboratory. It should be used in a fume hood, gloves, safety goggles and lab coats should be worn when handling, and it should be disposed of as chemical waste.

1. Add 200 μL of 4% formaldehyde in PBS to the dissected ovaries. Incubate for 10 min, rocking (*see* **Note 22**).
2. Remove fixative into a chemical waste container labeled specifically for formaldehyde in fume hood.
3. Wash tissue: add 400 μL of PBST and rock for 5 min.
4. Allow samples to settle for 5 min and remove liquid into a labeled formaldehyde waste container in a fume hood.
5. Repeat the washing process twice, for a total of three washes.
6. Store samples in the refrigerator in a tube rack or box or move directly on to Subheading 3.3.3 (propidium iodide and phalloidin staining) or Subheading 3.3.4 (Immunostaining).

3.3.3 Propidium Iodide and Phalloidin Staining (In-Lab)

Visualizing just two cellular structures, the nucleus and the actin cytoskeleton can reveal a tremendous amount of information about impacts of genetic mutations or environmental conditions on oogenesis (Fig. 6b). Propidium iodide intercalates into nucleic acid strands including DNA and, to a lesser degree, RNA. Adding propidium iodide to fixed, washed ovary samples is a robust approach to visualize nuclei, as propidium iodide is autofluorescent in the red–far-red spectrum when excited by green light. Propidium iodide also stains RNA, making messy looking samples unless the RNA is first degraded by treatment with an enzyme called RNAse. The protocol below aims to stain nuclei for clear imaging and analysis, and thus includes an important step of degrading RNA. Phalloidin is a toxin isolated from poisonous mushrooms that acts by binding directly to actin filaments and stabilizing them. Phalloidin fused to a fluorescent molecule thus provides an exceptional method to label actin filaments in fixed and washed cells. Since both propidium iodide and phalloidin are potent toxins, gloves must be worn for this protocol and caution must be taken to dispose of waste in a properly labeled waste container inside a fume hood. As such, this protocol requires access to a professional laboratory with the capability of handling hazardous waste.

1. Degrade the RNA in the fixed and washed ovary samples by adding 6.4 μL of a 25 mg/mL stock of RNAse A to 200 μL PBST in a microfuge tube and incubate for 10 min at room temperature (*see* **Note 23**).
2. During the incubation, make a solution of propidium iodide in a microfuge tube by adding 2.0 μL of propidium iodide to 200 μL of 1× PBS (*see* **Note 24**).
3. During the incubation, prepare the phalloidin solution (*see* **Note 24**). Pipet 2 μL fluorescein–phalloidin into a microfuge tube and place in the hood with the lid open to evaporate the methanol (*see* **Note 25**). Once dry, add 200 μL 1× PBS and vortex thoroughly.
4. To the sample, add 100 μL of propidium iodide solution and rock for 10 min (*see* **Note 22**).
5. Allow ovaries to settle to the bottom of the tube for 5 min. Remove propidium iodide solution from tube and place into waste bottle for propidium iodide and/or phalloidin in the fume hood (*see* **Note 26**).
6. Add 400 μL of 1× PBS and incubate for 5 min, rocking to wash (*see* **Note 22**).
7. Repeat the washing process twice (3 washes total).
8. Remove as much of the PBS as possible after the third wash.
9. Add 100 μL of phalloidin to each tube (*see* **Note 24**), and incubate for 10 min, rocking (*see* **Note 22**).
10. Allow ovaries to settle fully to the bottom of the tube for 5 min. Remove phalloidin and put into appropriate waste container in the fume hood (*see* **Note 26**).
11. Wash by adding 400 μL of PBS and rocking for 5 min (*see* **Note 27**).
12. Allow ovaries to settle to the bottom of the tube for 5 min and then remove all liquid.
13. Using a 200 μL pipet tip with the very end cut off (*see* **Note 28**), add 30 μL mounting medium to the tubes containing each sample.
14. Pipet ovaries in mounting medium on to a clean glass slide. Cover gently with a 22 × 22 #1.5 coverslip.
15. For examination of early stages of oogenesis, place folded kimwipe on the cover slip to absorb any excess liquid. Apply even pressure to the cover slip by placing an inverted slide box flat on top of the kimwipe, and then gently place an empty 100 mL bottle as a weight on top of the slide box. Allow the weight to "squash" the sample gently for 2–3 min.

16. Remove the weight, the slide box, and the kimwipe from the stack.
17. Paint around the edges of the cover slip with nail polish to seal.
18. Collect images of the samples on a fluorescent or confocal microscope.

3.3.4 Antibody Staining (In-Lab)

Our work has centered on EGFR signaling, with additional investigation of Src, Piwi, Hedgehog, ERR, Integrins, and CTPS, leveraging use of antibodies for analysis. The example presented here will focus on activation of the EGFR signaling pathway, with dually phosphorylated ERK (dpERK) as a read-out (Fig. 11b). ERK becomes dually phosphorylated after a four-step signaling cascade is activated by binding of EGF ligand to EGFR. Once dually phosphorylated, ERK translocates to the nucleus to activate target genes. Antibodies raised against dpERK are widely used in many systems to track its activity using both cell biology and biochemistry approaches. To successfully detect changes in signaling pathway activity, all of the steps of the following protocol are important (*see* Correa et al. for a fluorescent imaging protocol [134]).

1. Collect fixed, unstained ovary samples from Subheading 3.3.2 and set in rack.
2. Prepare the primary antibody master mix for two samples: 20 μL 5% BSA, 0.2 μL rabbit anti-dpERK antibody, 179.8 μL PBS-T (*see* **Note 29**).
3. Carefully remove all liquid from the samples and add 100 μL of the primary master mix to each sample.
4. Rock the tubes at room temperature for 2 h or in the refrigerator/cold room overnight.
5. Allow ovaries to settle to the bottom of the tubes and remove as much liquid as possible into a non-hazardous waste container.
6. Wash 3× in PBS-T by adding 400 μL of PBS-T to each sample tube, rock for 10 min at room temperature, allow ovaries to settle to the bottom of the tube, remove liquid, and repeat.
7. Create a secondary antibody master mix for two samples: 20 μL 5% BSA, 1 μL anti-rabbit-FITC antibody, 179.0 μL PBS-T (*see* **Note 30**).
8. Remove all liquid from samples and add 100 μL of the secondary antibody master mix to each tube.
9. Rock samples at room temperature for 2 h or in refrigerator/cold room overnight (*see* **Note 31**).
10. Wash 2× in PBS-T as in **step 6**.
11. During the washes, make a solution of propidium iodide in a microfuge tube by adding 2.0 μL of propidium iodide to 200 μL of 1× PBS (*see* **Note 24**).

12. Let samples settle to the bottom of the tubes and remove all liquid.
13. Add 100 μL of diluted stock of propidium iodide to each sample and rock at room temperature for 10 min.
14. Place tubes in a rack and make sure all the ovarioles settle to the bottom.
15. Remove the propidium iodide into a waste bottle designated for propidium iodide and phalloidin in the hood (*see* **Note 26**).
16. Add 400 μL of PBS to each tube and rock for 5 min (*see* **Note 22**).
17. Prepare, mount samples onto slides, and image as described in Subheading 3.3.3 **steps 12–18**.

3.3.5 Methylene Blue and Trypan Blue Staining (In-Lab or In-Home)

Access to professional labs for large-scale citizen science efforts is a major challenge. The immense benefit of engaging non-professional scientists in research drove us to reincarnate dye-based protocols for visualizing cells and nuclei that have faded somewhat into the past of developmental biology history. This protocol describes two blue dyes, methylene blue and Trypan blue.

Methylene blue can be added to freshly dissected, unfixed ovaries to label cell membranes and nuclei (Fig. 12a, b). Depending on the power of the microscope used for analysis, most of the morphological features that can be detected with propidium iodide/phalloidin staining can be visualized and imaged using this approach. A major difference is that methylene blue is used on live tissues, making it possible to determine impact without needing dangerous fixatives or toxic compounds for staining.

Trypan blue is a vital dye. This means that it is excluded from healthy live cells but can enter cells that are dead or dying (Fig. 12c). Treatment of freshly dissected ovaries with Trypan blue thus measures the quality of the ovaries dissected and can also be used to quantify the number of dying or dead cells in controls versus treatment conditions. Note that propidium iodide can be used as a vital dye in live samples to determine the numbers of dying cells as well (Fig. 9d).

The protocols for use of methylene blue or Trypan blue staining are the same:

1. Using freshly dissected, unfixed ovaries (Subheading 3.3.1), let the tissue settle to the bottom of the tube and remove as much liquid as possible.
2. Add 100 μL of methylene blue OR Trypan blue to the sample and incubate for 3 min at room temperature. Tubes may be gently rocked by hand every 30 s to provide maximal exposure of the tissue to the stain (*see* **Note 32**).

3. Remove stain and wash the samples in PBS by adding 400 μL 1× PBS to the tube and gently rocking the tube by hand for 5 min (*see* **Note 33**).
4. Let the ovaries settle for 5 min. Carefully remove the liquid to a waste container and repeat the wash with 400 μL of fresh 1× PBS at least twice (*see* **Note 34**).
5. Once the sample is adequately washed, remove as much liquid as possible.
6. Add 30 μL of glycerol to each tube as a mounting medium.
7. With a cutoff tip (*see* **Note 28**), pipet ovaries in mounting medium on to a clean glass slide. Cover gently with a 22 × 22 #1.5 coverslip.
8. Apply nail polish around the edges of the cover slip to seal it (*see* **Note 35**).
9. Use an inexpensive stereomicroscope or other school-grade microscope to evaluate phenotypes. Images can be collected using devices designed for microscope objectives or with a mobile phone camera.

4 Notes

1. Our food preparation follows the Janelia Research Campus recipe and can be purchased from Lab Express (Fly Food J) in 50 mL volume bottles. Fly food can be stored at room temperature for several weeks and longer if kept chilled (18 °C).
2. It is useful to make 2 vials of control (PBS) supplemented fly food and 2 vials of test compound supplemented fly food. This allows the use of wild-type flies +/− PBS or "test" as one important control and mutant +/− PBS or "test" for comparison. Prep time: 0–150 min. Active time: 30–45 min.
3. For mounting medium, Vectashield, Glycerol, or other preferred mediums can be used. Aqueous media (e.g. glycerol) with anti-fade chemicals to protect from fluorescence photobleaching is most appropriate for fluorescently labeled ovaries prepared in PBS.
4. Anti-dpERK antibodies from Cell Signaling Technology have been validated for use in fly ovaries for both immunostaining and immunoblotting.
5. Stay away from dairy-based and meat-based products. These will create an unpleasant, odiferous condition that will be noticed through the porous stopper.
6. After removal of bubbling fly food from the microwave, place vials on the table or bench to cool for ~1 min before adding PBS or test food or compounds. Be sure to add the PBS or test

compounds before the food begins to solidify to ensure thorough mixing.

7. Sample amounts can be optimized by starting with a high amount of test compound or food and then titrating down to determine the dose curve of impact. Observation of phenotypic changes in the linear range of the dose curve will establish ideal conditions for detailed analysis.
8. Limit time exposure to chemical anesthetic to no more than 60 s. Flies will still move slightly, but this will enable them to wake up and for you to stand vials upright in a shorter session.
9. Substantial differences in pupal counting may indicate an impact of the treatment on the flies. If differences are observed, the experiment should be repeated to ensure that the results are reproducible, with student t tests used to determine significance.
10. The standard method for reporting pupal counts is to record the number of pupae divided by the number of female flies ($N = 10$) used for the experiment.
11. Juices other than grape may be used (e.g. apple juice or even Gatorade if the student has nothing else), but purple grape makes the eggs more visible.
12. Adding agar to hot liquids all at once will cause clumping. Be sure to swirl the flask and add the agar very slowly to avoid this.
13. Grape juice plates can be stored in the refrigerator in an air-tight bag for 1 week without methylparaben added, and longer if methylparaben is added.
14. Alternate protocol: http://cshprotocols.cshlp.org/content/2015/6/pdb.rec086876.short [135].
15. Make sure not to press down or the edge of the 50 mL conical tube will cut the grape juice agar and release a plug of agar into the collection cage. If this happens, start over with a new collection cage and grape juice plate and work more gently.
16. Many mutant fly lines are temperature sensitive. Different results will be attained in summer versus winter, especially if delivering samples by mail and/or relying on home/school temperature control for experimentation. Keeping a strict record of temperature recordings and setting up experiments with matched controls is critical for data interpretation.
17. If the eggs absorb the purple color of the grape juice plates, there may be defects in the eggshell.
18. Defects such as short or missing dorsal appendages, round eggs, purple eggs (*see* **Note 17**), or fewer eggs indicate an effect of the mutation or treatment on fecundity.
19. Use 1× PBS in lieu of Grace's Insect Medium when working at school or in a home lab.

20. There is no limit on the total number of ovaries, but do not leave ovaries on the dish for more than 30 min without fixing as the tissue starts to degrade when outside of the fly for a long period of time.
21. Take up ovaries in small volumes (about 30 μL) to avoid sticking to the tip. It is better to collect them in small batches than try to collect all of them in one shot.
22. Rocking can be accomplished by placing tubes on a lab rocker or nutator, or by inverting by hand at 30 s intervals during the incubation period.
23. Without RNAse digestion, you will see a large amount of stain in the cytoplasm instead of a clean nuclear DNA stain.
24. Propidium iodide and phalloidin are toxic agents. Gloves, safety glasses, and lab coats must be worn when handling.
25. In a hood, it takes about 15 min for the methanol to fully evaporate. Make sure that the methanol used to dissolve the phalloidin is fully evaporated. Residual methanol will corrupt the actin cytoskeleton.
26. Refer to your institution's policies for hazardous waste disposal and proper labeling of waste containers.
27. Do NOT overwash after incubating with phalloidin since phalloidin–actin binding is reversible and the phalloidin can wash out.
28. Ovaries can shear inside a normal pipet tip. Cutting the small end off permits the ovaries to be taken up and pipetted out of the tip without damage.
29. For analysis of other proteins, replace the rabbit anti-dpERK antibody with the antibody of choice, following manufacturer's instructions for volume and concentration. This recipe is for two samples, but you can multiply it if you have more.
30. For examination of other proteins, be sure to make the secondary antibody match the primary antibody. For example: secondary anti-Rabbit antibodies only bind to antibodies made in rabbits. If your primary antibody is made in mouse, guinea pig, rat, chicken, sheep, or goat, use the secondary antibody coupled to a fluorophore that will bind to your primary antibody. Keep the master mixes wrapped in foil to protect from light.
31. If a rocker is not available, then rest the tubes horizontally on a shelf of the refrigerator to maximize surface area and incubate until the next session.
32. Be sure to keep track of which dye you are using for staining, as Trypan blue and Methylene Blue are both dark blue stains, but give vastly different information e.g. Trypan blue stains dying cells and methylene blue stains all cells.

33. PBS-T contains detergent that will burst the cells in unfixed ovaries. A very common mistake is to use PBS-T instead of PBS, which will not disrupt cell membranes. Be sure to always use PBS for any protocols involving live tissue samples.
34. Three washes is often sufficient to enable visualization of individual cells and key morphological features. If the samples still appear dark blue, repeat washes until the supernatant is mostly clear.
35. Unlike fixed ovaries where gravity pressure is applied using a small a weight, live ovaries cannot be squashed without ruining the samples.

Acknowledgments

We would like to thank the congregation of Lion Zion Baptist Church, the UCScience Center's BULB community, Ms. Dyson's ninth grade environmental science students at Abraham Lincoln High School, the eCLOSE Institute CitSci Public Outreach participants, and students and teachers in the School District of Philadelphia, Kingsway Regional High School, Cheltenham High School, Abington Junior High School, and Nazareth Academy High School for lively discussions and contributions. This work was funded by grants from NICHD (R01 HD065800 (AOR)), NCI (T32 CA009035, P30 CA06927), Howard Hughes Medical Institute (RH, AOR), Genetics Society of America (FCCC trainees), Howard Lockhart Seiple Trust (eCLOSE), Anna T. Jeanes Foundation of Temple University Health System (eCLOSE), and donations to ISP and/or eCLOSE Institute from the Kicking Cancer Foundation (ISP), Janssen Pharmaceutical Companies of Johnson & Johnson (eCLOSE), Giant/Eagle Supermarkets (eCLOSE), Coriell Institute (eCLOSE), Oriental Trading Company (eCLOSE), Michael and Judith Bolotsky Foundation (ISP and eCLOSE), Dr. Benjamin Neel and Dr. Phyllis Koton Neel (eCLOSE), (...funding sources to be added).

References

1. National Science Board NSF (2021) The STEM labor force of today: scientists, engineers and skilled technical workers. NSB-2021-2
2. The Bureau of Labor Statistics (2021) Occupational outlook handbook. U.S. Department of Labor. https://www.bls.gov/ooh/about/ooh-developer-info.htm
3. Herman A (2019) America's STEM crisis threatens our national security. Am Aff 3(1): 127–148
4. Lambert WM, Wells MT, Cipriano MF, Sneva JN, Morris JA, Golightly LM (2020) Career choices of underrepresented and female postdocs in the biomedical sciences. eLife 9: e48774. https://doi.org/10.7554/eLife.48774
5. Board NS (2020) Science and engineering indicators 2020: the state of U.S. science and engineering, vol NSB-2020-1. National Science Foundation, Alexandria

6. Rosenzweig EQ, Hecht CA, Priniski SJ, Canning EA, Asher MW, Tibbetts Y, Hyde JS, Harackiewicz JM (2021) Inside the STEM pipeline: changes in students' biomedical career plans across the college years. Sci Adv 7(18):eabe0985. https://doi.org/10.1126/sciadv.abe0985
7. McGee EO (2020) Interrogating structural racism in STEM higher education. Educ Res 49(9):633–644
8. Fry R, Kennedy B, Funk C (2021) STEM jobs see uneven progress in increasing gender, racial and ethnic diversity. Pew Research Center
9. LaVeist TA, Pierre G (2014) Integrating the 3Ds–social determinants, health disparities, and health-care workforce diversity. Public Health Rep 129 Suppl 2(Suppl 2):9–14. https://doi.org/10.1177/00333549141291S204
10. Castillo-Page L, Conrad SS, Caulfield M, Dill M, Eliason J, Jones K, Sims AM, Ciera Brown L (2019) Diversity in medicine: facts and figures 2019. Association of American Medical Colleges, Washington, D.C.
11. Smiley R, Lauer P, Bienemy C, Berg J, Shireman E, Reneau K, Alexander M (2018) The 2017 National Nursing Workforce Survey. J Nurs Regul 9(3):S1–S88
12. Shah Harsh R, Martinez Luis R (2016) Current approaches in implementing citizen science in the classroom. J Microbiol Biol Educ 17(1):17–22. https://doi.org/10.1128/jmbe.v17i1.1032
13. OECD (2019) PISA 2018 assessment and analytical framework. OECD publishing
14. Byington CL, Lee V (2015) Addressing disparities in academic medicine: moving forward. JAMA 314(11):1139–1141. https://doi.org/10.1001/jama.2015.10664
15. Nelson B (2014) The data on diversity. Commun ACM 57(11):86–95. https://doi.org/10.1145/2597886
16. Bangera G, Brownell SE (2014) Course-based undergraduate research experiences can make scientific research more inclusive. CBE Life Sci Educ 13(4):602–606. https://doi.org/10.1187/cbe.14-06-0099
17. Collins TW, Grineski SE, Shenberger J, Morales X, Morera OF, Echegoyen LE (2017) Undergraduate research participation is associated with improved student outcomes at a Hispanic-serving institution. J Coll Stud Dev 58(4):583–600. https://doi.org/10.1353/csd.2017.0044
18. Graham Mark J, Frederick J, Byars-Winston A, Hunter A-B, Handelsman J (2013) Increasing persistence of college students in STEM. Science 341(6153):1455–1456. https://doi.org/10.1126/science.1240487
19. Jones MT, Barlow AEL, Villarejo M (2010) Importance of undergraduate research for minority persistence and achievement in biology. J High Educ 81(1):82–115. https://doi.org/10.1080/00221546.2010.11778971
20. Linn Marcia C, Palmer E, Baranger A, Gerard E, Stone E (2015) Undergraduate research experiences: impacts and opportunities. Science 347(6222):1261757. https://doi.org/10.1126/science.1261757
21. Rodríguez Amaya L, Betancourt T, Collins KH, Hinojosa O, Corona C (2018) Undergraduate research experiences: mentoring, awareness, and perceptions—a case study at a Hispanic-serving institution. Int J STEM Educ 5(1):9. https://doi.org/10.1186/s40594-018-0105-8
22. Strayhorn TL (2010) Undergraduate research participation and STEM graduate degree aspirations among students of color. New Dir Inst Res 2010(148):85–93. https://doi.org/10.1002/ir.364
23. Carpi A, Ronan DM, Falconer HM, Lents NH (2017) Cultivating minority scientists: undergraduate research increases self-efficacy and career ambitions for underrepresented students in STEM. J Res Sci Teach 54(2):169–194. https://doi.org/10.1002/tea.21341
24. Chen J, Call GB, Beyer E, Bui C, Cespedes A, Chan A, Chan J, Chan S, Chhabra A, Dang P, Deravanesian A, Hermogeno B, Jen J, Kim E, Lee E, Lewis G, Marshall J, Regalia K, Shadpour F, Shemmassian A, Spivey K, Wells M, Wu J, Yamauchi Y, Yavari A, Abrams A, Abramson A, Amado L, Anderson J, Bashour K, Bibikova E, Bookatz A, Brewer S, Buu N, Calvillo S, Cao J, Chang A, Chang D, Chang Y, Chen Y, Choi J, Chou J, Datta S, Davarifar A, Desai P, Fabrikant J, Farnad S, Fu K, Garcia E, Garrone N, Gasparyan S, Gayda P, Goffstein C, Gonzalez C, Guirguis M, Hassid R, Hong A, Hong J, Hovestreydt L, Hu C, Jamshidian F, Kahen K, Kao L, Kelley M, Kho T, Kim S, Kim Y, Kirkpatrick B, Kohan E, Kwak R, Langenbacher A, Laxamana S, Lee C, Lee J, Lee SY, Lee TH, Lee T, Lezcano S, Lin H, Lin P, Luu J, Luu T, Marrs W, Marsh E, Min S, Minasian T, Misra A, Morimoto M, Moshfegh Y, Murray J, Nguyen C, Nguyen K, Nodado E 2nd, O'Donahue A, Onugha N, Orjiakor N, Padhiar B, Pavel-Dinu M, Pavlenko A, Paz E, Phaklides S, Pham L,

Poulose P, Powell R, Pusic A, Ramola D, Ribbens M, Rifai B, Rosselli D, Saakyan M, Saarikoski P, Segura M, Singh R, Singh V, Skinner E, Solomin D, Soneji K, Stageberg E, Stavchanskiy M, Tekchandani L, Thai L, Thiyanaratnam J, Tong M, Toor A, Tovar S, Trangsrud K, Tsang WY, Uemura M, Unkovic M, Vollmer E, Weiss E, Wood D, Wu S, Wu W, Xu Q, Yackle K, Yarosh W, Yee L, Yen G, Alkin G, Go S, Huff DM, Minye H, Paul E, Villarasa N, Milchanowski A, Banerjee U (2005) Discovery-based science education: functional genomic dissection in Drosophila by undergraduate researchers. PLoS Biol 3(2):e59. https://doi.org/10.1371/journal.pbio.0030059

25. Hernandez PR, Woodcock A, Estrada M, Schultz PW (2018) Undergraduate research experiences broaden diversity in the scientific workforce. Bioscience 68(3):204–211. https://doi.org/10.1093/biosci/bix163
26. Lopatto D (2004) Survey of undergraduate research experiences (SURE): first findings. Cell Biol Educ 3(4):270–277. https://doi.org/10.1187/cbe.04-07-0045
27. Lopatto D (2007) Undergraduate research experiences support science career decisions and active learning. CBE Life Sci Educ 6(4): 297–306. https://doi.org/10.1187/cbe.07-06-0039
28. Royster P (2015) Timing is everything: getting students Back on track to college readiness in high school. High Sch J 98(3): 208–225
29. Estrada M, Burnett M, Campbell AG, Campbell PB, Denetclaw WF, Gutiérrez CG, Hurtado S, John GH, Matsui J, McGee R, Okpodu CM, Robinson TJ, Summers MF, Werner-Washburne M, Zavala M (2016) Improving underrepresented minority student persistence in STEM. CBE Life Sci Educ 15(3):es5. https://doi.org/10.1187/cbe.16-01-0038
30. Hein V, Smerdon B, Sambolt MW (2013) Predictors of postsecondary success. Retrieved from https://www.cde.state.co.us/postsecondary/americaninstitutes
31. Kricorian K, Seu M, Lopez D, Ureta E, Equils O (2020) Factors influencing participation of underrepresented students in STEM fields: matched mentors and mindsets. Int J STEM Educ 7(1):16. https://doi.org/10.1186/s40594-020-00219-2
32. Rainey K, Dancy M, Mickelson R, Stearns E, Moller S (2018) Race and gender differences in how sense of belonging influences decisions to major in STEM. Int J STEM Educ 5(1):10. https://doi.org/10.1186/s40594-018-0115-6
33. Hoppe TA, Litovitz A, Willis KA, Meseroll RA, Perkins MJ, Hutchins BI, Davis AF, Lauer MS, Valantine HA, Anderson JM, Santangelo GM (2019) Topic choice contributes to the lower rate of NIH awards to African-American/black scientists. Sci Adv 5(10): eaaw7238. https://doi.org/10.1126/sciadv.aaw7238
34. Page-Reeves J, Cortez GL, Ortiz Y, Moffett M, Deerinwater K, Medin D (2019) Situating giving back for native Americans pursuing careers in STEM: you don't just take, you give something back. Intersect Crit Issues Educ 3(1):4. https://digitalrepository.unm.edu/intersections/vol3/iss1/4
35. Seymour E, Hunter A-B (2019) Talking about Leaving Revisited: persistence, relocation, and loss in undergraduate STEM education. Springer
36. Stanford N, Carlock S, Jia F (2021) The role of Community in Black Identity Development and Occupational Choice. Societies 11(3): 111. https://doi.org/10.3390/soc11030111
37. Taffe MA, Gilpin NW (2021) Racial inequity in grant funding from the US National Institutes of Health. eLife 10:e65697. https://doi.org/10.7554/eLife.65697
38. Bonney R, Cooper CB, Dickinson J, Kelling S, Phillips T, Rosenberg KV, Shirk J (2009) Citizen science: a developing tool for expanding science knowledge and scientific literacy. Bioscience 59(11):977–984
39. Hand E (2010) Citizen science: people power. Nature 466(7307):685–687
40. Cooper S, Khatib F, Treuille A, Barbero J, Lee J, Beenen M, Leaver-Fay A, Baker D, Popović Z, Players F (2010) Predicting protein structures with a multiplayer online game. Nature 466(7307):756–760. https://doi.org/10.1038/nature09304
41. Kim JS, Greene MJ, Zlateski A, Lee K, Richardson M, Turaga SC, Purcaro M, Balkam M, Robinson A, Behabadi BF, Campos M, Denk W, Seung HS (2014) Space-time wiring specificity supports direction selectivity in the retina. Nature 509(7500):331–336. https://doi.org/10.1038/nature13240
42. Falkenhain K, Ruiz-Uribe NE, Haft-Javaherian M, Ali M, Stall C, Michelucci PE, Schaffer CB, Bracko O (2020) A pilot study investigating the effects of voluntary exercise on capillary stalling and cerebral blood flow in the APP/PS1 mouse model of Alzheimer's

disease. PLoS One 15(8):e0235691–e0235691. https://doi.org/10.1371/journal.pone.0235691

43. Bracko O, Njiru BN, Swallow M, Ali M, Haft-Javaherian M, Schaffer CB (2020) Increasing cerebral blood flow improves cognition into late stages in Alzheimer's disease mice. J Cereb Blood Flow Metab 40(7):1441–1452. https://doi.org/10.1177/0271678x19873658
44. Bonney R, Phillips TB, Ballard HL, Enck JW (2016) Can citizen science enhance public understanding of science? Public Underst Sci 25(1):2–16
45. Miller-Rushing A, Primack R, Bonney R (2012) The history of public participation in ecological research. Front Ecol Environ 10(6):285–290
46. Shirk JL, Ballard HL, Wilderman CC, Phillips T, Wiggins A, Jordan R, McCallie E, Minarchek M, Lewenstein BV, Krasny ME (2012) Public participation in scientific research: a framework for deliberate design. Ecol Soc 17(2):29
47. Brossard D, Lewenstein B, Bonney R (2005) Scientific knowledge and attitude change: the impact of a citizen science project. Int J Sci Educ 27(9):1099–1121
48. Jordan TC, Burnett SH, Carson S, Caruso SM, Clase K, DeJong RJ, Dennehy JJ, Denver DR, Dunbar D, Elgin SC (2014) A broadly implementable research course in phage discovery and genomics for first-year undergraduate students. MBio 5(1):e01051–e01013
49. Lemon A, Bordenstein SR, Bordenstein SR (2020) Discover the microbes within! The Wolbachia project: citizen science and student-based discoveries for 15 years and counting. Genetics 216(2):263–268. https://doi.org/10.1534/genetics.120.303649
50. Shuda JR, Butler VG, Vary R, Farber SA (2016) Project BioEYES: accessible student-driven science for K-12 students and teachers. PLoS Biol 14(11):e2000520. https://doi.org/10.1371/journal.pbio.2000520
51. Taylor MB, Skophammer R, Warwick AR, Boyer JM, Students Y, Walson M, Large CRL, Hickey AS-M, Rowley PA, Dunham MJ (2021) yEvo: experimental evolution in high school classrooms selects for novel mutations and epistatic interactions that impact clotrimazole resistance in *S. cerevisiae*. bioRxiv:2021.2005.2002.442375. https://doi.org/10.1101/2021.05.02.442375
52. Gunnell JL, Golumbic YN, Hayes T, Cooper M (2021) Co-created citizen science: challenging cultures and practice in scientific research. J Sci Commun 20(5):1–17
53. Golumbic YN, Orr D, Baram-Tsabari A, Fishbain B (2017) Between vision and reality: a study of scientists' views on citizen science. Citiz Sci Theory Pract 2(1):6
54. Riesch H, Potter C (2013) Citizen science as seen by scientists: methodological, epistemological and ethical dimensions. Public Underst Sci 23(1):107–120. https://doi.org/10.1177/0963662513497324
55. Cohn JP (2008) Citizen science: can volunteers do real research? Bioscience 58(3):192–197. https://doi.org/10.1641/b580303
56. Cooper CB, Dickinson J, Phillips T, Bonney R (2007) Citizen science as a tool for conservation in residential ecosystems. Ecol Soc 12(2):11
57. Bowser A, Wiggins A, Shanley L, Preece J, Henderson S (2014) Sharing data while protecting privacy in citizen science. Interactions 21(1):70–73
58. Carethers JM, Sengupta R, Blakey R, Ribas A, Souza G (2020) Disparities in cancer prevention in the COVID-19 era. Cancer Prev Res 13(11):893. https://doi.org/10.1158/1940-6207.CAPR-20-0447
59. Minas TZ, Kiely M, Ajao A, Ambs S (2021) An overview of cancer health disparities: new approaches and insights and why they matter. Carcinogenesis 42(1):2–13. https://doi.org/10.1093/carcin/bgaa121
60. Zavala VA, Bracci PM, Carethers JM, Carvajal-Carmona L, Coggins NB, Cruz-Correa MR, Davis M, de Smith AJ, Dutil J, Figueiredo JC, Fox R, Graves KD, Gomez SL, Llera A, Neuhausen SL, Newman L, Nguyen T, Palmer JR, Palmer NR, Pérez-Stable EJ, Piawah S, Rodriquez EJ, Sanabria-Salas MC, Schmit SL, Serrano-Gomez SJ, Stern MC, Weitzel J, Yang JJ, Zabaleta J, Ziv E, Fejerman L (2021) Cancer health disparities in racial/ethnic minorities in the United States. Br J Cancer 124(2):315–332. https://doi.org/10.1038/s41416-020-01038-6
61. Livengood K, Diez Roux A, Mullin G, Acharya B, Bettigole C, Moore K, Rollins H, Washington R (2020) Community brief: cancer in Philadelphia. Drexel University Urban Health Collaborative
62. Jenkins LL (2011) Using citizen science beyond teaching science content: a strategy for making science relevant to students' lives. Cult Stud Sci Educ 6(2):501–508. https://doi.org/10.1007/s11422-010-9304-4

63. Greshake Tzovaras B, Rera M, Wintermute EH, Kloppenborg K, Ferry-Danini J, Aidelberg G, Aronoff R, Lindner A, Misevic D (2021) Empowering grassroots innovation to accelerate biomedical research. PLoS Biol 19(8):e3001349. https://doi.org/10.1371/journal.pbio.3001349
64. Fisher AJ, Mendoza-Denton R, Patt C, Young I, Eppig A, Garrell RL, Rees DC, Nelson TW, Richards MA (2019) Structure and belonging: pathways to success for underrepresented minority and women PhD students in STEM fields. PLoS One 14(1):e0209279–e0209279. https://doi.org/10.1371/journal.pone.0209279
65. Clark US, Hurd YL (2020) Addressing racism and disparities in the biomedical sciences. Nat Hum Behav 4(8):774–777. https://doi.org/10.1038/s41562-020-0917-7
66. Hayes CA (2021) Black scientists are not the door to diversity. ACS Chem Neurosci 12(13):2256–2260. https://doi.org/10.1021/acschemneuro.1c00375
67. Safdari R, Ehtesham H, Robiaty M, Ziaee N (2018) Barriers to participation in medical research from the perspective of researchers. J Educ Health Promot 7:22–22. https://doi.org/10.4103/jehp.jehp_26_17
68. Nov O, Arazy O, Anderson D (2014) Scientists@home: what drives the quantity and quality of online citizen science participation? PLoS One 9(4):e90375. https://doi.org/10.1371/journal.pone.0090375
69. Petersen C, Austin RR, Backonja U, Campos H, Chung AE, Hekler EB, Hsueh P-YS, Kim KK, Pho A, Salmi L, Solomonides A, Valdez RS (2019) Citizen science to further precision medicine: from vision to implementation. JAMIA Open 3(1):2–8. https://doi.org/10.1093/jamiaopen/ooz060
70. Means B, Wang H, Wei X, Lynch S, Peters V, Young V, Allen C (2017) Expanding STEM opportunities through inclusive STEM-focused high schools. Sci Educ 101(5):681–715. https://doi.org/10.1002/sce.21281
71. Sigismund S, Avanzato D, Lanzetti L (2018) Emerging functions of the EGFR in cancer. Mol Oncol 12(1):3–20. https://doi.org/10.1002/1878-0261.12155
72. Pyrowolakis G, Veikkolainen V, Yakoby N, Shvartsman SY (2017) Gene regulation during Drosophila eggshell patterning. Proc Natl Acad Sci 114(23):5808–5813
73. Wasserman JD, Freeman M (1998) An autoregulatory cascade of EGF receptor signaling patterns the Drosophila egg. Cell 95(3):355–364
74. Cavaliere V, Bernardi F, Romani P, Duchi S, Gargiulo G (2008) Building up the Drosophila eggshell: first of all the eggshell genes must be transcribed. Dev Dyn 237(8):2061–2072
75. Wu X, Tanwar PS, Raftery LA (2008) Drosophila follicle cells: morphogenesis in an eggshell. Semin Cell Dev Biol 19(3):271–282. https://doi.org/10.1016/j.semcdb.2008.01.004
76. Horne-Badovinac S, Bilder D (2005) Mass transit: epithelial morphogenesis in the drosophila egg chamber. Dev Dyn 232(3):559–574. https://doi.org/10.1002/dvdy.20286
77. Matis M, Axelrod JD (2013) Regulation of PCP by the fat signaling pathway. Genes Dev 27(20):2207–2220. https://doi.org/10.1101/gad.228098.113
78. Gutzeit HO (1986) The role of microfilaments in cytoplasmic streaming in Drosophila follicles. J Cell Sci 80(1):159–169
79. Cooley L, Verheyen E, Ayers K (1992) chickadee encodes a profilin required for intercellular cytoplasm transport during Drosophila oogenesis. Cell 69(1):173–184
80. Xu MJ, Johnson DE, Grandis JR (2017) EGFR-targeted therapies in the post-genomic era. Cancer Metastasis Rev 36(3):463–473. https://doi.org/10.1007/s10555-017-9687-8
81. Horne-Badovinac S, Hill J, Gerlach G 2nd, Menegas W, Bilder D (2012) A screen for round egg mutants in Drosophila identifies tricornered, furry, and misshapen as regulators of egg chamber elongation. G3 (Bethesda) 2(3):371–378. https://doi.org/10.1534/g3.111.001677
82. Horne-Badovinac S (2014) The Drosophila egg chamber—a new spin on how tissues elongate. Integr Comp Biol 54(4):667–676. https://doi.org/10.1093/icb/icu067
83. Duhart JC, Parsons TT, Raftery LA (2017) The repertoire of epithelial morphogenesis on display: progressive elaboration of Drosophila egg structure. Mech Dev 148:18–39. https://doi.org/10.1016/j.mod.2017.04.002
84. Barlan K, Cetera M, Horne-Badovinac S (2017) Fat2 and Lar define a basally localized planar signaling system controlling collective cell migration. Dev Cell 40(5):467–477.e465. https://doi.org/10.1016/j.devcel.2017.02.003
85. Cooley L (1998) Drosophila ring canal growth requires Src and Tec kinases. Cell 93(6):913–915

86. Dodson GS, Guarnieri DJ, Simon MA (1998) Src64 is required for ovarian ring canal morphogenesis during Drosophila oogenesis. Development 125(15):2883–2892
87. Guarnieri DJ, Dodson GS, Simon MA (1998) SRC64 regulates the localization of a Tec-family kinase required for Drosophila ring canal growth. Mol Cell 1(6):831–840
88. Lu N, Guarnieri DJ, Simon MA (2004) Localization of Tec29 to ring canals is mediated by Src64 and PtdIns(3,4,5)P3-dependent mechanisms. EMBO J 23(5): 1089–1100
89. O'Reilly AM, Ballew AC, Miyazawa B, Stocker H, Hafen E, Simon MA (2005) Src64 and CSK are required in the germline for cyst encasuplation during Drosophila oogenesis. submitted to Development
90. Hinnant TD, Merkle JA, Ables ET (2020) Coordinating proliferation, polarity, and cell fate in the drosophila female germline. Front Cell Dev Biol 8(19):19. https://doi.org/10.3389/fcell.2020.00019
91. Tepass U, Gruszynski-DeFeo E, Haag TA, Omatyar L, Török T, Hartenstein V (1996) Shotgun encodes Drosophila E-cadherin and is preferentially required during cell rearrangement in the neurectoderm and other morphogenetically active epithelia. Genes Dev 10(6):672–685
92. Bulgakova NA, Klapholz B, Brown NH (2012) Cell adhesion in Drosophila: versatility of cadherin and integrin complexes during development. Curr Opin Cell Biol 24(5): 702–712. https://doi.org/10.1016/j.ceb.2012.07.006
93. Niewiadomska P, Godt D, Tepass U (1999) DE-Cadherin is required for intercellular motility during Drosophila oogenesis. J Cell Biol 144(3):533–547. https://doi.org/10.1083/jcb.144.3.533
94. Cai D, Chen S-C, Prasad M, He L, Wang X, Choesmel-Cadamuro V, Sawyer JK, Danuser G, Montell DJ (2014) Mechanical feedback through E-cadherin promotes direction sensing during collective cell migration. Cell 157(5):1146–1159. https://doi.org/10.1016/j.cell.2014.03.045
95. Fulga TA, Rørth P (2002) Invasive cell migration is initiated by guided growth of long cellular extensions. Nat Cell Biol 4(9): 715–719. https://doi.org/10.1038/ncb848
96. Stuelten CH, Parent CA, Montell DJ (2018) Cell motility in cancer invasion and metastasis: insights from simple model organisms. Nat Rev Cancer 18(5):296–312. https://doi.org/10.1038/nrc.2018.15
97. Gupta T, Schüpbach T (2001) Two signals are better than one: border cell migration in Drosophila. Dev Cell 1(4):443–445. https://doi.org/10.1016/S1534-5807(01)00062-4
98. Godt D, Tepass U (1998) Drosophila oocyte localization is mediated by differential cadherin-based adhesion. Nature 395(6700): 387–391. https://doi.org/10.1038/26493
99. González-Reyes A, St Johnston D (1998) The Drosophila AP axis is polarised by the cadherin-mediated positioning of the oocyte. Development 125:3635–3644
100. Zhang Y, Kalderon D (2000) Regulation of cell proliferation and patterning in Drosophila oogenesis by Hedgehog signaling. Development 127(10):2165–2176
101. Forbes AJ, Lin H, Ingham PW, Spradling AC (1996) Hedgehog is required for the proliferation and specification of ovarian somatic cells prior to egg chamber formation in Drosophila. Development 122(4):1125–1135
102. Zhang Y, Kalderon D (2001) Hedgehog acts as a somatic stem cell factor in the Drosophila ovary. Nature 410(6828):599–604. https://doi.org/10.1038/35069099
103. Bilder D, Li M, Perrimon N (2000) Cooperative regulation of cell polarity and growth by Drosophila tumor suppressors. Science 289(5476):113–116. https://doi.org/10.1126/science.289.5476.113
104. Santa-Cruz Mateos C, Valencia-Expósito A, Palacios IM, Martín-Bermudo MD (2020) Integrins regulate epithelial cell shape by controlling the architecture and mechanical properties of basal actomyosin networks. PLoS Genet 16(6):e1008717
105. Cooper J, Giancotti FG (2019) Integrin signaling in cancer: mechanotransduction, stemness, epithelial plasticity, and therapeutic resistance. Cancer Cell 35(3):347–367
106. O'Reilly AM, Lee HH, Simon MA (2008) Integrins control the positioning and proliferation of follicle stem cells in the Drosophila ovary. J Cell Biol 182(4):801–815. https://doi.org/10.1083/jcb.200710141
107. Lovegrove HE, Bergstralh DT, St Johnston D (2019) The role of integrins in Drosophila egg chamber morphogenesis. Development 146(23):dev182774. https://doi.org/10.1242/dev.182774
108. Xu J, Gridley T (2012) Notch signaling during oogenesis in Drosophila melanogaster. Genet Res Int 2012:648207–648207. https://doi.org/10.1155/2012/648207
109. López-Schier H, Johnston DS (2001) Delta signaling from the germ line controls the

proliferation and differentiation of the somatic follicle cells during Drosophila oogenesis. Genes Dev 15(11):1393–1405

110. Deng W-M, Althauser C, Ruohola-Baker H (2001) Notch-Delta signaling induces a transition from mitotic cell cycle to endocycle in Drosophila follicle cells. Development 128(23):4737–4746

111. Kale J, Osterlund EJ, Andrews DW (2018) BCL-2 family proteins: changing partners in the dance towards death. Cell Death Differ 25(1):65–80. https://doi.org/10.1038/cdd.2017.186

112. Quinn L, Coombe M, Mills K, Daish T, Colussi P, Kumar S, Richardson H (2003) Buffy, a Drosophila Bcl-2 protein, has anti-apoptotic and cell cycle inhibitory functions. EMBO J 22(14):3568–3579. https://doi.org/10.1093/emboj/cdg355

113. Colussi PA, Quinn LM, Huang D, Coombe M, Read SH, Richardson H, Kumar S (2000) Debcl, a proapoptotic Bcl-2 homologue, is a component of the Drosophila melanogaster cell death machinery. J Cell Biol 148(4):703–714

114. Stoppa-Lyonnet D (2016) The biological effects and clinical implications of BRCA mutations: where do we go from here? Eur J Hum Genet 24(1):S3–S9. https://doi.org/10.1038/ejhg.2016.93

115. Krais JJ, Johnson N (2020) BRCA1 mutations in cancer: coordinating deficiencies in homologous recombination with tumorigenesis. Cancer Res 80(21):4601–4609. https://doi.org/10.1158/0008-5472.Can-20-1830

116. Klovstad M, Abdu U, Schüpbach T (2008) Drosophila brca2 is required for mitotic and meiotic DNA repair and efficient activation of the meiotic recombination checkpoint. PLoS Genet 4(2):e31. https://doi.org/10.1371/journal.pgen.0040031

117. Jia D, Huang Y-C, Deng W-M (2015) Analysis of cell cycle switches in Drosophila oogenesis. Methods Mol Biol 1328:207–216. https://doi.org/10.1007/978-1-4939-2851-4_15

118. Edgar BA, Datar SA (1996) Zygotic degradation of two maternal Cdc25 mRNAs terminates Drosophila's early cell cycle program. Genes Dev 10(15):1966–1977

119. Calses PC, Crawford JJ, Lill JR, Dey A (2019) Hippo pathway in cancer: aberrant regulation and therapeutic opportunities. Trends Cancer 5(5):297–307

120. Huang J, Kalderon D (2014) Coupling of Hedgehog and Hippo pathways promotes stem cell maintenance by stimulating proliferation. J Cell Biol 205(3):325–338. https://doi.org/10.1083/jcb.201309141

121. Lin TH, Yeh TH, Wang TW, Yu JY (2014) The Hippo pathway controls border cell migration through distinct mechanisms in outer border cells and polar cells of the Drosophila ovary. Genetics 198(3):1087–1099. https://doi.org/10.1534/genetics.114.167346

122. Hong H, Yang L, Stallcup MR (1999) Hormone-independent transcriptional activation and coactivator binding by novel orphan nuclear receptor ERR3. J Biol Chem 274(32):22618–22626

123. Berman AY, Manna S, Schwartz NS, Katz YE, Sun Y, Behrmann CA, Yu JJ, Plas DR, Alayev A, Holz MK (2017) ERRα regulates the growth of triple-negative breast cancer cells via S6K1-dependent mechanism. Signal Transduct Target Ther 2:17035. https://doi.org/10.1038/sigtrans.2017.35

124. Ma JH, Qi J, Lin SQ, Zhang CY, Liu FY, Xie WD, Li X (2019) STAT3 targets ERR-α to promote epithelial-mesenchymal transition, migration, and invasion in triple-negative breast cancer cells. Mol Cancer Res 17(11): 2184–2195. https://doi.org/10.1158/1541-7786.Mcr-18-1194

125. Du Y, Song L, Zhang L, Ling H, Zhang Y, Chen H, Qi H, Shi X, Li Q (2017) The discovery of novel, potent ERR-alpha inverse agonists for the treatment of triple negative breast cancer. Eur J Med Chem 136:457–467. https://doi.org/10.1016/j.ejmech.2017.04.050

126. Cho B, Han Y, Lian M, Colditz GA, Weber JD, Ma C, Liu Y (2021) Evaluation of racial/ethnic differences in treatment and mortality among women with triple-negative breast cancer. JAMA Oncol 7(7):1016–1023. https://doi.org/10.1001/jamaoncol.2021.1254

127. Dietze EC, Sistrunk C, Miranda-Carboni G, O'Regan R, Seewaldt VL (2015) Triple-negative breast cancer in African-American women: disparities versus biology. Nat Rev Cancer 15(4):248–254. https://doi.org/10.1038/nrc3896

128. Zhang Q, Huang H, Zhang L, Wu R, Chung C-I, Zhang S-Q, Torra J, Schepis A, Coughlin SR, Kornberg TB, Shu X (2018) Visualizing dynamics of cell signaling in vivo with a phase separation-based kinase reporter. Mol Cell 69(2):334–346.e334. https://doi.org/10.1016/j.molcel.2017.12.008

129. Aughey GN, Liu J-L (2015) Metabolic regulation via enzyme filamentation. Crit Rev

Biochem Mol Biol 51(4):282–293. https://doi.org/10.3109/10409238.2016.1172555

130. Chang CC, Jeng YM, Peng M, Keppeke GD, Sung LY, Liu JL (2017) CTP synthase forms the cytoophidium in human hepatocellular carcinoma. Exp Cell Res 361(2):292–299. https://doi.org/10.1016/j.yexcr.2017.10.030

131. Liu J-L (2010) Intracellular compartmentation of CTP synthase in Drosophila. J Genet Genomics 37(5):281–296

132. Strochlic TI, Stavrides KP, Thomas SV, Nicolas E, O'Reilly AM, Peterson JR (2014) Ack kinase regulates CTP synthase filaments during Drosophila oogenesis. EMBO Rep 15(11):1184–1191. https://doi.org/10.15252/embr.201438688

133. Phosphate-buffered saline (PBS) (2006) Cold Spring Harb Protoc 2006(1):pdb.rec8247. https://doi.org/10.1101/pdb.rec8247

134. Correa I, Wang M, Lee EH, Ruiz-Whalen DM, O'Reilly AM, Singh T (2021) Protocol for evaluating autophagy using LysoTracker staining in the epithelial follicle stem cells of the Drosophila ovary. STAR Protoc 2(2):100592. https://doi.org/10.1016/j.xpro.2021.100592

135. Grape Juice Agar Plates (2015) Cold Spring Harb Protoc 2015(6):pdb.rec086876. https://doi.org/10.1101/pdb.rec086876

Index

Michelle S. Giedt and Tina L. Tootle (eds.), *Drosophila Oogenesis: Methods and Protocols*,
Methods in Molecular Biology, vol. 2626, https://doi.org/10.1007/978-1-0716-2970-3,

O

P

R

S

T

W

MIX
Papier aus verantwortungsvollen Quellen
Paper from responsible sources
FSC® C105338

If you have any concerns about our products,
you can contact us on
ProductSafety@springernature.com

In case Publisher is established outside the EU,
the EU authorized representative is:
Springer Nature Customer Service Center GmbH
Europaplatz 3, 69115 Heidelberg, Germany

Printed by Libri Plureos GmbH
in Hamburg, Germany